Emiliy

Insights into the Social Ecology of ICT Impleme... Schools

Emily M. L. Wong

Insights into the Social Ecology of ICT Implementation in Schools

A Quantitative Approach

VDM Verlag Dr. Müller

Imprint

Bibliographic information by the German National Library: The German National Library lists this publication at the German National Bibliography; detailed bibliographic information is available on the Internet at http://dnb.d-nb.de.

Cover image: www.purestockx.com

Publisher:
VDM Verlag Dr. Müller Aktiengesellschaft & Co. KG
Dudweiler Landstr. 125 a, 66123 Saarbrücken, Germany
Phone +49 681 9100-698, Fax +49 681 9100-988, Email: info@vdm-verlag.de

Produced in USA and UK by:
Lightning Source Inc., La Vergne, Tennessee, USA
Lightning Source UK Ltd., Milton Keynes, UK

ISBN: 978-3-8364-3509-3

ABSTRACT

The purpose of the present study is to develop models to deepen our understanding of how ICT acts as a lever to bring about changes in student learning. Informed mainly by research in three major domains, namely the constructivist views of learning, organisational approach to educational change, and the perspective of looking at ICT use in an organisational context, ICT implementation is framed in an activity system characterised by the interplay between the ICT implementation strategies, pedagogical intervention, and organisational intervention in achieving changes in student learning.

With the questionnaire survey data collected from over 1000 teachers in 130 schools located in various districts of Hong Kong, the present study used structural equation modelling (SEM) techniques to construct a theoretical model depicting the social ecology of ICT implementation by mapping the relationships among ICT implementation strategies, the pedagogical and organisational factors, and changes in student learning. The study was successful in demonstrating that the collegial capacity of ICT implementation strategies mediated the effect of organisational interventions on changes in classroom practices. Specifically, the collegial capacity of school's ICT implementation strategies affected directly the perceived changes in learning from a teacher-centred approach to a student-centred approach, and indirectly through pedagogical intervention. To deepen our understanding on the SEM model developed, hierarchical linear modelling (HLM) was used to construct a 2-level model to examine the multilevel effects on changes in learning (the outcome variable). Results of the multilevel analysis showed that the regression results obtained from each individual school (in the level-1 model) were consistent with those in the SEM analysis. In addition, the multilevel analysis showed that 14% of the total variation found in changes in student learning was at the school level, which was largely attributable to school-level variables of socio-cultural in nature rather than to the structural characteristics of schools.

ACKNOWLEDGMENTS

Firstly, I would like to thank Dr. Sandy S C Li for his useful advice and guidance.

I am also very grateful to Dr. Kong Chit-Kwong for the exchange of professional experience in my research design. Special thanks should also be given to Mr. Simon Siu and Mr. Victor Lee for their valuable help.

Finally, I would like to dedicate this book to my husband, Y K Yee, and the almighty God. Without their love and support, this book would never have been.

TABLE OF CONTENTS

LIST OF TABLES

LIST OF FIGURES

Chapter 1: Introduction

Background

The ongoing, unprecedented growth of information and communication technology (ICT), coupled with the globalisation of the economy, has created a huge challenge for education. The common belief that ICT has the potential to support certain fundamental changes in learning has been guiding and shaping educational reforms (Panel On Education Technology, 1997). Many countries have already drawn up their master plans for the development of ICT in education. In Hong Kong, the commitment of promoting the application of ICT across the school curriculum was manifested through the introduction of "Information Technology for Learning in a New Era: Five-year Strategy – 1998/99 to 2002/03" in 1998 (Education and Manpower Bureau, 1998).

Despite this, significant concerns have been raised with regard to whether ICT is simply reinforcing old pedagogy (Lankshear, Snyder, & Green, 2000; Papert, 1997), or even distracting the students from studying the subject matter (Oppenheimer, 2003). As indicated from the findings of the report "Preliminary Study on Reviewing the Progress and Evaluating the Information Technology in Education Projects, December 2000 to August 2001" submitted to the Education and Manpower Bureau of HKSAR Government (Centre for Information Technology in School and Teacher Education, 2002), many schools regarded ICT in education as an effort to simply replace chalk and board by multimedia presentations/ animations. Nevertheless, schools that had history and background in curriculum innovation and strong leadership were able to describe a much wider range of plans for teaching and learning activities using ICT.

This suggests that pedagogical and organisational issues need to be addressed if ICT use is to go beyond fancy tools (Papert, 1997), or a kind of apolitical technology adoption lacking reflection on how ICT may change the social environment (Ferneding, 2003). Moreover, it also brings to our attention the debatable issue of evaluating the effort of ICT implementation. In this respect, some researchers have argued for a transformation of classroom practices by taking advantage of ICT rather than superficial translation of existing practices (Thomas, Carswell, Price, & Petre, 1998), and some others have further pointed out the need to understand how ICT may help to support the kinds of sustained and substantial inquiry and analysis that we all want our children to achieve (Honey, Culp, & Carrigg, 1999). Following this line of thought, some researchers have suggested that effective ICT use is embedded in a larger process of school change (Honey et al., 1999), and effort of ICT implementation needs to be examined as part of multifaceted school reform rather than in isolation (Bober, 2002).

It has been believed that an understanding of how educational change takes place is vital to successful ICT implementation (Hall, 1995). A review of the literature on educational change and school reform has suggested that although educators/policymakers shared the same basic vision of effecting changes in student learning, they had divergence in the way of implementing the changes, with the extremes of mere emphasis on instructional methods in 1970s (Cheng, 2002a), and of mere emphasis on system-wide restructuring without making connections between new governance structures and the teaching and learning process (Fullan, 1995a). It has been cautioned that sometimes education reforms might divert us from the very basic issues they were purportedly established to address (Cizek & Ramaswamy, 1999; Fullan, 1993, 1995b; Levin, 2001). While acknowledging the core of educational change was best manifested by changes in the daily teaching and learning activities (Hall, 1995), education policymakers have begun to view educational change in a wider institutional context (McLaughlin, 1991). In adopting a coherent approach, it has been believed that attention to second-order changes like school restructuring is essential to the first-order changes related to classroom practices (Leithwood, 1994). Because of this, the focus of educational change has been described as being shifted from the classroom level to the organisational level since 1980s (Cheng, 2002a).

As a result of the widespread interest in studying educational change in an institutional context, many researchers have adopted an organisational approach to study the issue. According to Fullan (1993), effective change started firstly in the organisational culture, which in turn spread and led to changes in structure. Educational change was construed as changes in instructional practices and in the culture of teaching toward greater collaborative relationships among students, teachers and other potential partners (Fullan, 1993). At the same time, increasing attention has also been given to conceptualising educational policy implementation as a form of individual and organisational learning (Fullan, 1995b; Hall & Carter, 1995; Senge et al., 2000). In addition, some researchers have also adopted a system approach to educational change. Education is often regarded as an open system, linked to other systems through feedback loops (Hall & Carter, 1995). In addition, the role of school leadership has been gaining increasing importance in the context of educational change (Leithwood, 1994). In 1980s, instructional leadership conceptualisation drawn from the effective school literature was dominant, with a leadership focusing on teachers' behaviours that related to activities directly affecting the growth of students (Leithwood & Duke, 1999). Since the 1990s, many researchers have adopted new perspectives in understanding school leadership other than an instructional leadership conceptualisation (Hallinger & Heck, 1998). Leadership has been conceptualised as an organisational quality that is embedded in a social and cultural context (Ogawa & Bossert, 1995). The dominant model has been transformational leadership that focuses on increasing the organisation's capacity to innovate (Hallinger & Heck, 1998), and adopting a commitment-building strategy for school restructuring (Leithwood, 1994).

How are the concepts of educational change related to the use of ICT in education? In a review of studies on ICT in education, it was found out that the focus has been shifted from studying particular technologies and their impact on student learning, to studying how technology is integrated into education settings (Honey, Culp, & Carrigg, 2000). While such progression indicates an increased awareness of viewing the use of ICT in context, conceptualisation of ICT implementation is still inadequate, in a sense that there is a lack of a holistic conceptualisation of ICT implementation that links the use of ICT to educational change. We suggest that answering these questions requires the re-conceptualisation of a range of concepts relevant to educational change: (1) effective use of ICT cannot be accomplished by technological fixes alone, instead it is embedded in the larger process of educational change; (2) effective use of ICT requires pedagogical intervention from teachers, which involves a conceptual change of teaching and learning from a teacher-centred

4

approach to a student-centred approach; (3) effective use of ICT requires organisational intervention, which involves a conceptual change of leadership from instructional leadership to transformational leadership, and a conceptual change of professionalisation from teacher training to human and social capital development of teachers.

If effective ICT use is embedded in a larger process of school change (Honey et al., 1999), then effort of ICT implementation needs to be examined as part of multifaceted school reform rather than in isolation (Bober, 2002). Arguably, studies seeking to demonstrate the impact of particular software on student learning do not help much in this respect. Despite frequent attempts to address educational changes and the roles of ICT in effecting changes in student learning, few have provided a framework that conceptually integrates ICT implementation with changes in learning within a school context of managing change.

Purpose of the Study

The purpose of the present study is to develop models that conceptually integrate ICT implementation with changes in student learning within a context of managing change in schools, with a view to understanding how ICT acts as a lever to bring about a paradigm shift in student learning in terms of teacher's perceptions as well as organisational features of schools. In building the models, we relied on teacher perceptions on the contextual factors and we also used teacher judgements of student learning as an outcome variable. While studies based on perceptions of teachers have its limitations with regard to the actual impact on teaching and learning, we believe perceived impacts are important to examine in school studies. First and foremost, the gist of the study is not on measuring improvements in traditional processes and knowledge as measured by attainment tests, but rather on discerning whether there are changes in teaching and learning towards a constructivist approach or student-centred approach as a result of ICT implementation. While many education systems recognise the constructivist approach in teaching and learning, few have formally adopted new modes of assessment with greater emphasis in measuring learning attitude and learning ability in terms of active construction of knowledge and collaborative work (Balanskat, Blamire, & Kefala, 2006; Valli, Cooper, & Frankes, 1997). To overcome the inherent difficulties in measuring changes in teaching and learning towards a student-centred approach as a result of ICT use, some researchers resort to gathering evidence from perceptions of teachers. Such opinion-based approach has been regarded as informative and was used by a number of Nordic studies in investigating perceived impacts of ICT on student learning such as team work, independent learning and critical thinking (Balanskat et al., 2006). In fact, the practice of eliciting perceptions of participants is quite common and has been reflected by a huge body of literature on leadership, school improvement and managing change (Owens, 2001; Smylie & Hart, 1999). According to Owens (2001), while perceptions themselves are not objective reflections of "reality", the point is that whatever people in the organisation perceive as their experience is the reality to be described.

Quantitative approach is adopted as the methodology of the study. It begins with adopting an ecological perspective to conceptualise ICT implementation as a process of interactions among a variety of contextual factors with a view to targeting changes in student learning in schools. The conceptualisation goes beyond the boundary of simply identifying factors affecting the use of ICT in schools and takes a further step to scrutinise the interplay of the contextual factors.

Following the above conceptualisation, the next step is to identify the key institutional factors affecting ICT implementation as well as changes in student learning. An organisational and sociocultural approach is adopted as past research tended to ignore both the social contexts and institutional culture in which teachers situated (Windschitl & Sahl, 2002), and therefore the term "social ecology of ICT implementation" is introduced. In adopting a situative perspective, ICT implementation practice is constituted in the interaction of individuals and their situation in the execution of particular tasks (Greeno, 1998; Spillane, Reiser, & Todd, 2002). The analysis is shifted from isolated individuals to their interactions in the social collective or activity system (Greeno, 1998). In the social ecology of ICT implementation, we have identified five sociocultural factors: school leadership, school climate, effectiveness of government ICT policy, collegial capacity of school's ICT implementation strategies, and changes in pedagogy. In addition, this study also adopts a constructivist perspective in evaluating the outcome variable changes in learning as suggested by the HKSAR official document "Learning to Learn: Life-long Learning and Whole-person Development"(Curriculum Development Council, 2001).

Developing an instrument to collect data is the next move, in which the factors comprising the social ecology of ICT implementation will be transformed into operational measurement in the form of questionnaire survey. Following this is data collection and then validation of the instrument for measuring the abstract concepts of school leadership, school climate, effectiveness of government ICT policy, collegial capacity of school's ICT implementation strategies, changes in pedagogy, and changes in learning. As the structural equation-based approach has been considered as a powerful method for addressing construct validity (Anderson & Gerbing, 1988; Bagozzi, Yi, & Phillips, 1991; Bollen, 1989), the present study relies mainly on the structural equation-based approach and its complementary methods for construct validation.

Afterwards, the researcher builds a model using the technique of structural equation modelling (SEM) to delineate the structural relationships among the constructs. In this way, the SEM model helps us to understand the social ecology of ICT implementation among teachers pooled from the entire sample, irrespective to which school the teachers belonged. However, if we regard school as a single unit and do a regression analysis for each individual school, we may obtain a very different set of regression coefficients as compared to those obtained from the regression over the data pooled from the entire sample. To deepen our understanding on the SEM model, the study also constructs a 2-level model with hierarchical linear modelling (HLM) to find out whether there are any multilevel effects on the outcome variable, which is to see (1) whether the regression results obtained from each individual school are consistent with those in the SEM analysis and (2) how much variation found in the outcome variable can be attributed to school-level variables. These school-level effects can significantly affect the results as it has been reported that inconsistencies found in principal leadership research results was due to the neglect of multilevel nature of schooling (Hallinger & Heck, 1998). In the process of selecting school-level variables, the researcher attends to structural characteristics of schools as well as socio-cultural factors shaping the social ecology of ICT implementation in schools. The purpose of the study is four-fold:

1. In a context of managing change in schools with ICT implementation, what are the factors affecting changes in student learning from a teacher-centred approach to a student-centred approach?

2. In an overall picture (the social ecology of ICT implementation), how do the above factors interact with one another to bring about the paradigm shift in student learning?

3. In the social ecology of ICT implementation within schools, how does ICT act as a lever to bring about the paradigm shift in student learning?

4. Among schools, what are the organisational features of schooling facilitating the paradigm shift in student learning?

Significance of the Study

The present study has theoretical as well as practical significance, in particular in the field of educational administration and ICT implementation. Theoretically, despite frequent attempts to address educational changes and the roles of ICT in effecting changes in student learning, few have acknowledged that effective ICT use needs to be embedded in a larger process of school change (Honey et al., 1999). The structural and hierarchical models developed in this study enable researchers to conceptualise ICT implementation as a process embedded in the school context of managing change, and as part of multifaceted school reform rather than in isolation (Bober, 2002). The present study entrenches a conceptual framework for understanding the complexity of managing change and to unfold the ecological dynamics of various contextual factors in ICT implementation. In the past, research tended to ignore both the social contexts and institutional culture in which teachers situated (Windschitl & Sahl, 2002). Viewing from a sociocultural perspective, the present study is able to measure the abstract concept of ICT implementation strategies by conceptualising it as the capacity of implementation strategies in enhancing collegial ICT exchange and in mobilising resources.

By employing SEM techniques, the present study is able to model the complex relationships in the social ecology of ICT implementation in schools given that SEM has the advantage of allowing the researcher to accommodate multiple interrelated dependence relationships in a single model that are not possible with other popular multivariate techniques (Hair, Anderson, Tatham, & Black, 1998). Nevertheless, the data set in the present study is in the form of teachers nested within schools, and structural analysis does not take into consideration the multilevel nature of such typical kind of educational data (Raudenbush & Bryk, 2002). It has been pointed out that some discrepancies found in principal leadership research results may be attributed to the limitation of methodological tools relating to unit of analysis problem and the multilevel nature of schooling (Hallinger & Heck, 1998). In employing HLM technique to build a 2-level model, the present study is able to deepen our understanding of the SEM model by finding out what school effects are behind the scene. To address the school-level effect, the HLM analysis has the advantage of providing a means of partitioning the outcome variable's variance into within-units (teacher level: variance among teachers within the same school) and between-units (school level: variance among schools); and within the analysis, a means of assigning variables to different levels (Heck & Thomas, 2000). In the 2-level model of the present study, "changes in student learning" was viewed as a function of both teacher-level (level-1) variables and school-level (level-2) variables. In other words, variation in "changes in student learning" was explained in terms of teacher-level variables (perception of teachers) as well as school-level variables (teacher aggregate of the same school and structural characteristics from other sources). Without pooling data across all schools, the level-1 model enables us to obtain an estimation of effects within teachers of the same school. Allowing the level-1 intercept and slope coefficients to vary from school to school, the level-2 analysis enables us to obtain an estimation of effect across schools (Kreft & De Leeuw, 1998; Raudenbush & Bryk, 2002).

Practically, results of the present study will have significant implications on understanding and evaluating ICT implementation effort in an education setting. As the outcome variable is proposed as perceived changes in student learning from a teacher-centred approach to a student-centred approach, the findings of this study will also have significant implications on understanding how ICT, in a context of managing change in schools, interacts with contextual factors to act as a lever to bring about a paradigm shift in learning. At the same time, the multi-level model will shed light on the role of organisational factors of schooling on bringing about the paradigm shift.

Organisation of the Thesis

To begin with, this chapter has given brief reasons for exploring the implementation of ICT in education from the perspective of managing change. The study attempts to bridge the knowledge gap by relating perceived changes in learning outcome to ICT implementation as well as other contextual factors. Built on the architecture of change management, the present research approaches the issue of ICT implementation in education in a more holistic way, in which ICT implementation in schools is examined as a total package designed to change the school into a learning organisation that targets at teaching effectiveness and enhanced learning outcome (Fullan, 1993, 1995b).

Chapter 2 will discuss the emerging perspectives on educational change, and the various ways to achieve changes in student learning outcome, in particular psychological constructivism with a social focus. It goes on to discuss the significance of adopting an organisational approach to study educational change, in particular the ideology of learning organisations, school leadership, organisational culture, as well as human and social capacity building. In addition, it examines the emerging trends of research in ICT in education, and the different approaches to analyse ICT implementation. Lastly, it concludes with the introduction of a theoretical framework that conceptually integrates ICT implementation with changes in learning within a context of managing change in schools, which is a result of activity between pedagogical intervention and organisational intervention to achieve changes in student learning. Implementation dynamics emerges from the elements interacting in the activity system, in which the collegial capacity of ICT implementation strategies is the central focus.

Chapter 3 will proceed to give a detailed account of the samples and procedures to collect data, the measures to be used, and the data analysis to be performed. It begins with introducing the research design, which is of quantitative approach using a cross-sectional questionnaire survey administered to teachers from schools located in various districts in Hong Kong. It goes on to introduce the research objective, which is exploratory in nature (as the research empirically tests alternative theoretical models that conceptualise and unfold the dynamics of ICT implementation in schools), and is also explanatory in nature (as the plausible model(s) seek to explain how ICT implementation strategies interact with organisational intervention and pedagogical intervention to enhance changes in student learning from a teacher-centred approach to a student-centred approach). This chapter will also give an account on how the items for the survey instrument are conceptually defined on a theoretical basis, and indicate the content domains for each of the six constructs. The six constructs are transformational leadership, climate for collaboration and experimentation, effectiveness of government ICT policy, collegial capacity of school's ICT implementation strategies, changes in teacher pedagogy, and changes in student learning.

Chapter 4 will address the issue of reliability and validity for the instrument, and its central role in quantitative studies that rely on questionnaire survey. The chapter will give a detailed account on how the structural equation-based approach and its complementary methods demonstrated the unidimensionality, reliability, convergent validity, and discriminant validity of the instrument. In particular, it will introduce how the constructs were refined in response mainly to the discriminant validity checks to ultimately attaining overall construct realiability and validity.

Chapter 5 is a chapter dedicated to unfolding the process of finding out a plausible structural model to depict the social ecology of ICT implementation in schools. Before proceeding to examine the alternative models, it gives reasons for choosing the kinds of strategies, input matrix, estimation method, and evaluation criteria for model building and comparison. Of the six proposed models, the chapter concludes with finding out a dichotomy categorisation pattern from the four plausible models - The better-fit group of models shares the same proposition of placing the collegial capacity of ICT implementation strategies as a central focus, which affirms the central role of the collegial capacity of ICT implementation strategies in the social ecology of ICT implementation.

Chapter 6 will address the methodological power of using the statistical technique of HLM to address the unit of analysis problem inherited in the nested data, and how the HLM technique may enrich our understanding of the social ecology of ICT implementation in the aspect of organisational features of schooling. The chapter gives the rationale in adopting an exploratory procedure and "step-up" strategy for model building, and in defining a small set of variables measuring background of teachers/structural characteristics of schools and those measuring the sociocultural factors relating to ICT implementation. Competing models are constructed, those with structural characteristics as variables, and those with socio-cultural factors as variables. The chapter concludes with finding out a plausible model, from which this research indicates that among schools, higher school mean for changes in student learning is related to socio-cultural factors shaping the social ecology of ICT implementation rather than the structural characteristics of schools. Specifically, the average changes in pedagogical practices and the average school climate are shown to have positive effect on the school mean for changes in student learning.

From the research findings of the present study, Chapter 7 will focus on gaining insights into understanding ICT implementation, with a view to helping educators and policymakers to unleash the power of organisational interventions and pedagogical interventions to advance changes in student learning. To begin with, the chapter will sum up the individual results from construct validation, SEM analysis, and HLM analysis to reach a number of conclusions, from which a 2-level-3-factor conceptual model was constructed to unfold the social ecology of ICT implementation as interactions of contextual variables in an activity system that comprised two levels and 3 factors. It goes on to discuss the theoretical, methodological, and practical implications drawn from the findings. Lastly, the chapter opens its way to other related issues that may lend themselves to future quantitative and qualitative studies, which will be helpful in unleashing irreversible changes in schools.

Chapter 2: Literature Review and Theoretical Framework

Introduction

The present study was informed mainly by research in three major domains, namely the constructivist views of learning, organisational approach to educational change, and the perspective of looking at ICT use in an organisational context. Heading in these directions, this chapter will lead us to understand the concept of ICT implementation, and why it is plausible to explore ICT implementation in an organisational context of managing change in schools. It begins with examining the emerging perspectives on educational change, and the various ways to achieve changes in student learning outcome, in particular psychological constructivism with a social focus. It goes on to discuss the significance of adopting an organisational approach to study educational change, in particular the ideology of learning organisations, school leadership, organisational culture, as well as human and social capacity building. In addition, the emerging trends of research in ICT in education, and the different approaches to analyse ICT implementation will be examined. Lastly, the chapter concludes with the introduction of a theoretical framework that conceptually integrates ICT implementation with changes in learning within a context of managing change in schools, with a view to understanding how ICT acts as a lever to bring about a paradigm shift in student learning in terms of teacher's perceptions as well as organisational features of schools.

An Overview of Educational Change

Entering into a new millennium, the discourse of expecting schooling and society to produce lifelong learners with continuous development of new knowledge and skills pervades the policy language and governmental pronouncements of many countries (A. Hargreaves, 1999). To this end, schools are increasingly expected to meet challenges arising from shifts in society and family that affect children. Although the future roles of schools are hard to tell under the fast changing world, the safest prediction is change (Senge et al., 2000). Compared to the change agenda of 1980s, recent trends in educational change have been far more complex and uncertain (Sleegers, Geijsel, & van den Berg, 2002). On the other hand, under global economic competition, education systems are receiving reduced support from governments (A. Hargreaves, 1999; Mok & Welch, 2002). Under such policy context, schools are increasingly governed by market ideologies and shaped by corporate discourse of efficiency and effectiveness. The discourse of education is said to be largely dominated by the ideas of economic rationalism and managerialism (Mok & Welch, 2002; Tse, 2002). While the millennial thinking of expecting schools to change runs through contemporary educational discourse, some educational writers, especially historians, have cast doubts on it (A. Hargreaves, 1999). It has been argued that the high expectations for radical and fundamental change in schooling are unrealistic, in a sense that they are prone to idealise the future but to overlook the continuities inherited from the past (MuCulloch, 1997). Significant concerns have also been raised with regard to whether the dominant ideas of economic rationalism and managerialism will worsen the issues in relation to equity, humanity, and educational outcomes (Chan, 2002; Mok & Welch, 2002; Tse, 2002).

A review of the recent history of educational reform indicates that great efforts have been made to undertake significant educational reform among many countries in the West and elsewhere (A. Hargreaves, 1999). Probing deeply, many educators believe that the driving forces behind have come from social and economic shifts rather than results of educational research (A. Hargreaves, 1999; Mok & Welch, 2002; Pipho, 2000). Conceptually, it has been proposed that the recent history of educational reform can be traced by six ideas, namely equity, educational excellence, accountability, choice, discipline, and professionalisation (Duke, 2004). Likewise, three tracks have been used to trace the strategic intentions of educational change: the shifting of significant authority, responsibility, and accountability to schools; an unrelenting focus on learning outcomes; and finally the creation of schools for knowledge society (Caldwell, 1999). Specifically, Fullan (1996) has illustrated the new context of education using eight emerging trends: a worldwide trend toward self-managing schools with collaborative work cultures, new forms of school-community governance intertwined with family and community, a trend to reduce dependence outside bureaucracy and regulation, new centralist roles taken by the state on curriculum and standards, reinventing teacher professionalism with increased standards of practice, massive expansion of ICT with new learning opportunities, new learning outcomes defined more in terms of teaching for understanding and performance, and issue of equity arising from multi-racial, gender and sexual politics (Fullan, 1996). In another analysis, it has also been suggested that educational reform is successfully initiated by five interacting factors, namely leadership theory, local politics and governance, state and national school politics, organisational theory and change theory (McAdams, 1997).

In exploring ICT implementation in an organisational context of managing change in schools, this literature review is going to focus on some selected issues mentioned above in relation to educational change, with some relevant issues being merged into new dimensions: (1) new learning outcomes defined more in terms of teaching for understanding and performance, with particular reference to constructivist teaching and learning; (2) self-managing schools with collaborative work cultures, and the creation of schools for knowledge society, with particular reference to the ideology of learning organisations and knowledge management; (3) leadership theory, organisational theory and change theory, with particular reference to transformational leadership; (4) teacher professionalisation, with particular reference to human and social capital development perspective; (5) massive expansion of ICT with new learning opportunities, with particular reference to the emerging roles of ICT in teaching and learning as well as the perspective of looking at ICT use in an organisational context.

Emerging Concept of Learning

As discussed in the previous section, educational change points to a new direction of conceptualising how organisations change and how learning takes place. A review of literature indicates that school restructuring in general calls for a greater emphasis on transformational leadership and learning organisation in order to improve teaching and learning (van den Berg & Sleegers, 1996). As the basic issue of educational change is to advance changes in learning (Fullan, 1995b), the typical direction of organisational change will naturally centre on the emerging concept of learning. In Hong Kong's school-based management consultation document "Transforming Schools into Dynamic and Accountable Professional Learning

Communities", it explicitly states that "We expect that the recommendations of this consultation document will contribute towards enhanced school effectiveness and ultimately improve student learning outcomes in Hong Kong. We are confident that, with community participation and, in particular, with the concerted efforts of the key stakeholders, our vision for quality education will be realised." (Education and Manpower Bureau, 2000).

What is the emerging concept of learning? Entering the new millennium, there is an increased tendency to view learning as a whole person phenomenon that encompasses behavioural, emotional, and cognitive aspects of learning (McCombs, 2000). As indicated by Fullan (1996), learning outcomes are defined more in terms of teaching for understanding and performance. Schools are urged to change from reproducing knowledge to building knowledge, the essence of which is on the development of understanding (Scardamalia & Bereiter, 1993). All these converge on a dominant theory of learning: constructivism.

Constructivism

At the instructional level, constructivist approach to teaching and learning is one of the emerging perspectives of educational change. Constructivism is more of a philosophy of learning that holds a few major guidelines as central to the process (Tiene & Ingram, 2001), and is in fact represented by a collection of theories of learning over the years (Newby, Stepich, Lehman, & Russell, 2000; Prawat & Peterson, 1999), with different founders and advocates, schools, foci, and disciplinary approaches (Richardson, 2003). Although constructivism is conceptualised differently by various groups of theories depending on their foci (Windschitl, 2002), literature review suggests that there are in general two differentiated forms of constructivism, namely social constructivism and psychological constructivism (Phillips, 2000; Richardson, 2003). While both approaches assume that knowledge is actively constructed in the human mind, they use different lenses to view the process of knowledge construction. The lens used by social constructivism focuses on power, economic, social and political forces, whereas that used by psychological constructivism focuses on the ways in which meaning is created within the individual mind and within a group process (Richardson, 2003). In acknowledging the social context in which knowledge is constructed, psychological constructivism is developed into one with a social focus, which has been regarded as an important contribution in teaching and learning (Richardson, 2003). The constructivism mentioned in the present study refers to the psychological constructivism with a social focus.

Roots of constructivism. Constructivist learning theory has its roots in the work of Piaget (1970) and Vygotsky (1978). The roots of constructivism lie primarily in developmental psychology, which studies human learning very differently from either behaviourism or cognitive psychology. Developmental psychologists are of the opinions that people develop and grow over time in interactions with their environments, and therefore they stress the importance of each individual's autonomy as a thinker and the importance of the social context of learning. Unlike behaviourism and cognitive science that stress the passive role of learners in the learning process, constructivism emphasises the importance of the learner's initiative (Tiene & Ingram, 2001).

Constructivist approach to knowledge construction. Same as developmental psychologists who attach high importance to interactions of the environment on the process of learning, constructivists propose that knowledge is constructed by individuals through their experiences of that object in a particular context (Honebein, Duffy, & Fishman, 1993). Grounded on constructivism, the notion of situated learning asserts further that the knowledge we construct includes information about the context of the experience. As a result, knowledge construction results from activity and is anchored in the context and culture (Brown, Collins, & Duguid, 1989). It is believed the context and authentic activities that are inherently a part of apprenticeship provide meaningful experiences for learners and help them transfer skills and knowledge to the world outside of schools (Honebein et al., 1993). Likewise, constructivist curriculum favours activities, in particular group activities designed in part to foster collaborative skills, which are essential within contemporary work environments. In addition, cooperative groups are thought to provide opportunities for students to form communities of inquiry that allow them to discuss and explain and hence are rich sources to learn with understanding (Cognition and Technology Group at Vanderbilt, 1993).

Constructivist Approach to Teaching and Learning

While constructivism as a learning theory has a history of a number of decades, constructivist teaching has only received attention in the last decade (Richardson, 2003). Constructivists assume that learners typically undertake substantial authentic tasks presented in a realistic context, and are required to apply various sorts of knowledge and skills to construct meaning out of learner-initiated inquiries driven at least in part by their own curiosity. Explicit attention is also given to the acquisition of higher-order thinking and problem-solving skills. Teaching in a constructivist fashion is therefore very different from the comparatively traditional view of instruction as a process involving the transmission of isolated facts (Panel On Education Technology, 1997). Richardson (2003) has defined constructivist pedagogy as "the creation of classroom environments, activities, and methods that are grounded in a constructivist theory of learning, with goals that focus on individual students developing deep understandings in the subject matter of interest and habits of mind that aid in future learning" (p. 1627). In promoting a child-friendly and activity-oriented approach to education, constructivism played a prominent role in the 1960s educational reform discourse (Prawat & Peterson, 1999). Since then, constructivism has grown to include a substantial body of research work relating to various teaching and learning issues, and also the proposition of new ways of framing the act of teaching, for example, as co-constructing knowledge with students (Windschitl, 2002).

Concerns over Constructivist Teaching and Learning

Despite this, concerns have been raised with respect to difficulties that may arise from translating a theory of learning into a theory or practice of constructivist pedagogy (Richardson, 2003). While a large part of the educational research community has in recent years converged on the constructivist model of learning, it will be misleading to suggest that it is unanimous in endorsing the principles and practice of constructivism without qualification. According to Richardson (2003), as roots of constructivism are western, liberal, and individualistic, it is not clear whether the less privileged cultures are interested in the strong individualistic approach as suggested in the constructivist teaching approach. Extending from this culture critique of constructivist pedagogy is the concern of imposing a dominant pedagogy on teachers who may have good reasons to operate differently (Richardson, 2003). Constructivist model of learning is also represented by cooperative learning groups, which has been criticised as being inherent with potential disadvantages such as the existence of differential benefits among group members (Cognition and Technology Group at Vanderbilt, 1993). Concerns have also been raised on whether less motivated teachers can harness the advantages of project-based learning techniques, which is recognised as one form of constructivist model of learning (Panel On Education Technology, 1997).

The concerns have, in part, been due to the lack of constructivist teaching theory (Richardson, 2003) or constructivist teaching model (Windschitl, 2002). In translating the constructivist learning theory into practice, teachers have only been given a set of prescriptions about what actions should not be adopted from the transmission model that was traditionally used in schools (Richardson, 2003). Windschitl (2002) have used four dimensions to illustrate the dilemmas in relation to constructivism in practice, "conceptual dilemmas rooted in teachers' attempts to understand constructivism, pedagogical dilemmas for teachers arising from the more complex approaches to designing curriculum and fashioning learning experiences, cultural dilemmas emerging between teachers and students during the radical reorientation of classroom roles and expectations, and political dilemmas associated with resistance from various stakeholders in school communities when institutional norms are questioned and routines of privilege and authority are disturbed" (p. 132). Besides, some educators think that linearly sequenced instructional content of well-structured presentation order still has its pedagogical value in facilitating understanding (Panel On Education Technology, 1997). In view of the above concerns, it is recommended that more researches should be conducted to study whether students learning in classrooms that engage in constructivist pedagogy really make a difference (Richardson, 2003). Nevertheless, constructivism is a common framework that people adopt in examining the effectiveness of using ICT in teaching and learning. In this present study, constructivism will therefore be used as a lens to examine whether ICT implementation will gear teaching and learning towards a more constructivist paradigm.

Organisational Approach to Managing Educational Change

While acknowledging the core of educational change was best manifested by changes in the daily teaching and learning activities (Hall, 1995), education policymakers have begun to view educational change in a wider institutional context (McLaughlin, 1991). In adopting a coherent approach, it has been believed that attention to second-order changes like school restructuring is essential to the first-order changes related to classroom practices (Leithwood, 1994). Impressive accumulation of evidence has suggested that in the absence of second order changes in the school's organisation and culture, first-order changes related to classroom practices soon disappear (Fullan, 1993; Fullan & Hargreaves, 1992; Louis & Miles, 1990). Because of this, the focus of educational change has been shifted from the classroom level to the organisational level since 1980s (Cheng, 2002a). This shift has enormous consequences for the role of school organisation and leaders during the implementation of innovations (Sleegers et al., 2002). The focus of implementation research has been shifted gradually from describing the range of barriers to addressing the kinds of policies or practices that support effective translation of policy to practice (Levin, 2001). At the same time, increasing attention has also been given to conceptualising educational policy implementation as a form of individual and organisational learning (Cheng, 2002b; Fullan, 1995b; Hall & Carter, 1995).

Implementation of Educational Change at School Level

Despite facing repeated educational reforms, policymakers and educators have only come to acknowledge the importance of change at the school level since 1970s (Firestone & Corbett, 1988). Although external efforts outside schools contribute to their change, the present literature review is going to focus on the internal issues that affect changes in schools since internal issues are more central to the change process at school level.

Among the emerging perspectives on understanding dynamics of change at school level, the three perspectives proposed by House (1981) has gained widespread attention, namely the technological, the political, and the cultural perspectives (House, 1981). While the technological and political perspective dominated in the 1970s, the cultural perspective has now been gaining increasing attention (Firestone & Corbett, 1988). Viewing implementation of change as a technical task subject to rational analysis, the technological analyses have been criticised as failing to capture many complexities of implementation. While seeing rational analysis of change as impossible because of the divergent interests of many actors, the political perspective has still been found to be pessimistic. Among the three perspectives, the cultural perspective has been considered as a relatively more optimistic one as it attends to the cultural features of organisations that may help to promote innovation and effectiveness (Firestone & Corbett, 1988).

In understanding dynamics of change in schools, Ogawa (2005) has also identified four evolving perspectives focusing on school organisations instead of innovations. They are the rational perspective, the human relations school, schools as loosely coupled systems, and organisational culture. Before 1960s, the rational perspective dominated, with emphasis on organisational goals and formal structures. After that, the human relations school evolved, the focus of which was shifted to individual's attributes and informal structures. In 1970s, the perspective of schools as loosely coupled systems dominated, which uncovered the bases other than norms of rationality for connecting organisational elements. Specifically, institutional theory explained that organisations tend to adopt structures that reflect institutions in the environment to gain social legitimacy and shared meaning and values. Since 1980s, the application of culture as a metaphor for organisation has emerged, in which deep and abiding organisational values operate in shaping actions and relationships.

In the above analysis, both House (1981) and Ogawa (2005) have identified in their perspectives, a shift from the characteristics or component part, to the relationships between people, to the context itself. As indicated by House (1981), the cultural perspective emphasises on the context, in which meanings and values are the focal points. In understanding the change process, Firestone & Corbett (1988) have also believed that the most profitable direction is to apply more systematically the cultural perspective, which transcends the political perspective of personal interests and looks to deeply ingrained sets of values, beliefs, and norms.

Organisational Culture

Organisational culture is an abstraction, the forces of which are invisible and to a considerable degree unconscious, yet culture creates powerful and concrete impact to an organisation (Schein, 2004). Organisational culture has been a central concern in management since 1980s (Owens, 2001). The concept of organisational culture has been the subject of considerable academic debate in the last few decades (Schein, 2004). Organisational culture can be viewed as simple as the way people do things around the organisation (Deal & Kennedy, 1982). However, organisational culture can be further seen as a set of implicit rules (Schein, 1968), a set of observed behavioural regularities (Goffman, 1959), or a set of basic assumptions and beliefs (Schein, 1985) that are shared by members of an organisation. Using structural stability, depth, breadth, and patterning to characterise organisational culture, Schein (2004) went on to define culture of a group as "a pattern of shared basic assumptions that was learned by a group as it solved its problems of external adaptation and internal integration, that has worked well enough to be considered valid and, therefore, to be taught to new members as the correct way to perceive, think, and feel in relation to those problems", and culture formation as "a striving toward patterning and integration" (p. 17).

Owens (2001) defined organisational culture as a system of shared values and beliefs that interact with an organisation's people, organisational structures, and control systems to produce behavioural norms. While organisational culture and organisational climate are two distinct concepts, they are related. In distinguishing organisational culture from organisational climate, Owens (2001) pointed out, "Culture refers to the behavioural norms, assumptions, and beliefs of an organisation, whereas climate refers to perceptions of persons in the organisation that reflect those norms, assumptions, and beliefs (p. 145)."

Two themes have consistently pervaded the literature describing organisational culture: norms and assumptions. Norms are referred as rules of behaviour, written or unwritten, which have been accepted as legitimate by members of a group. Assumptions are thought as bedrock on which norms are built, and they are tacit, unconsciously taken for granted, rarely considered or talked about, and accepted as true and nonnegotiable (Owens, 2001). In the analysis of organisational culture, scholars have also highlighted the evolving and learning nature of unconscious thoughts: "organisational culture develops over a period of time and, in the process of developing, acquires significantly deeper meaning" (Owens, 2001, p. 145), "culture is the result of a complex group learning process that is only partially influenced by leader behaviour" (Schein, 2004, p.11).

Impact of school culture on organisational effectiveness. School culture is influenced in various ways by the social and cultural context within which education are located (Angus, 1996). The context influencing school culture may be characterised by symbolic elements of the school such as traditions and rituals, heroes and heroines, history, stories and myths, values and beliefs, and behavioural norms (Owens, 2001). Culture can exist as school-wide culture or subunit culture. The adult subculture of schools is the focus of attention among researchers, an example of which is the interactions among teachers in the school (Firestone & Louis, 1999).

Studies of schools have strongly supported the belief that organisational culture is a fundamental factor in determining the quality of educational organisations (Owens, 2001). Culture was believed to offer ways to address issues of togetherness and community (Firestone & Louis, 1999). In adopting a cultural perspective to view organisational change, it was suggested that school improvement is parallel to cultural transformation (Firestone & Corbett, 1988). Traditionally, educational administration literature tended to affirm the role of leadership in shaping school culture (Angus, 1996). Seeing culture as a dynamic phenomenon that are constantly enacted and created by our interactions with others and shaped by leadership, Schein (2004) has thought that leadership and culture are two sides of the same coin. He has further affirmed the role of leadership in creating culture in an era of changes. At times facing environment changes to the point where some of the assumptions in an organisation are no longer valid, leadership is the ability to step outside the established culture and to adaptively evolve a new culture (Schein, 2004). As indicated by Sergiovanni (1995), strong and functional cultures are deliberately nurtured and built by school leadership and membership. While strong and functional culture is capable of acting as a powerful socialiser of thought and programmer of behaviour, it is not necessarily characterised by harmony. Successful schools are both tightly coupled and loosely coupled, in a sense that there exists a strong culture and clear sense of purpose, but a great deal of freedom is given to teachers and others as to how these essential core values are to be realised (Sergiovanni, 1995).

However, alternative views have been offered by research traditions rooted in anthropologists who use conceptions of culture that emphasise conflict and complexity of cultural dynamics (Angus, 1996; Firestone & Louis, 1999). In acknowledging culture as an organisational variable that is shifting and contested,

Angus (1996), for instance, has argued against the prevailing thoughts of shaping organisational culture by leadership. Instead, he advocated a more complex notion of organisational dynamics, in which organisational culture is linked with subjectivity, cultural politics and social context.

While identifying limits to leaders' discretion to influence school culture, it must be pointed out that case study evidence has supported principals' effects on culture, though there are still debates about the extent which principals really affect the culture of the classroom (Firestone & Louis, 1999). Literature has explicitly addressed the role of principals in transforming culture and in maintaining culture. As mentioned by Firestone & Louis (1999), transformational leadership changes the school culture to support high performance, and a key element of transformational leadership is the need to manage the relationship between internal culture of the school and the external environment that is often unpredictable. The model of transformational leadership will be discussed in more detail in the section of school leadership. Two concepts in relation to building a particular kind of culture in school/organisation will be discussed here: learning organisation (Fullan, 1993; Senge, 1990) and knowledge management (Jensen & Frederiksen, 2001).

School as a learning organisation. According to Fullan (1993), building a learning organisation is the core solution to the current education problems since it is able to address the core of teaching and learning. Specifically, learning organisation is about changes in instructional practices and in the culture of teaching toward greater collaborative relationships among students, teachers and other potential partners (Fullan, 1993). Change is a complex process, and building a learning organisation is about changing norms, habits, skills and beliefs. Fullan (1993) has believed that effective change starts in the organisational culture first, which in turn spreads and leads to changes in structure. Fullan (1999) has further viewed learning organisations as living systems where people and relationships evolve and adapt to new situations by continuously creating new and better knowledge. The learning strategy is termed "living on the edge of chaos", which is about getting used to conflict, uncertainty and anxiety during the process of change. Conflict is seen as collaborative diversity, and is in fact a good source for learning as well as generating creative breakthroughs. Getting used to uncertainty refers to the needs to set up an adaptive structure in order to grow in uncertainty, which involves a system of people-based learning framed by a few key priorities and structures only. Such a system, having both structure and openendedness, is believed to foster intense communication and knowledge creation in relation to selected problems. In addition, members in a learning organisation should deal with anxiety by understanding that it always comes with change, and together they create a collaborative culture that can contain anxiety by supporting each other (Fullan, 1999).

In relating the disciplines of learning organisations in business to schools, Senge and colleagues (2000) have suggested that people need to master the five disciplines for learning organisations in order to foster learning at classroom, school as well as community levels. They are personal mastery, mental models, shared visions, team learning and systems thinking (Senge, 1990). Personal mastery is characterised by continually clarifying and deepening one's personal vision as well as seeing reality objectively in order to develop commitment and sense of mission in one's career. The mastery of mental models are about digging up one's ingrained generalisations and holding them rigorously to scrutiny so as to allow oneself clearly and honestly define current reality. By building shared visions, an organisation is able to foster genuine commitment and enrolment rather than compliance. The discipline of team building starts with the capacity of members of a team to suspend assumptions and enter into a genuine collective thinking. System thinking has been termed the fifth discipline by Senge (1990), which is about learning to better understand interdependency and change so as to trace how one creates the problems in a wider perspective rather than seeing problems as caused by someone else. This represents an approach that galvanises hope in meeting challenges in this fast changing world (Senge et al., 2000). As every organisation including a school is a product of how its members think and interact, the essence of the disciplines of learning organisations are basically concerned with continually shifting our points of orientation to foster learning. Senge (1990) pointed out that learning organisations would truly excel in the future since they discover how to tap people's commitment and capacity to learn at all levels in an organisation.

Knowledge management in schools. As the intellectual capital of the education system is the competence-base of the economy, the concept of knowledge management in schools is becoming important. It is about creating a culture of making tacit knowledge in the education system explicit (Jensen & Frederiksen, 2001). By tacit knowledge, Nonaka and Takeuchi (1995) have referred to the knowledge of experience, which is subjective, practical and analogue. By contrast, explicit knowledge is the knowledge of rationality, which is objective, theoretical and digital. It has been believed that the professional knowledge and skills gained by teachers are largely tacit as they are acquired through experience of working alone in classrooms with students, and transforming schools to learning communities will be a solution to unlock the tacit knowledge in the heads of teachers (D. Hargreaves, 2000). This actually echoes the idea of creating a workplace for learning so as to facilitate knowledge creation. In the first place, knowledge creation is considered to be a dynamic social process between individuals in which tacit knowledge and explicit knowledge interact with each other (Nonaka & Takeuchi, 1995). Along with this is the idea that learning by doing is the most effective method of transferring competence (Sveiby, 1997). In fact, the idea of learning communities is a gentler notion of learning organisations when applied to schools (Henri, 1999). Research has revealed that teachers can benefit from a strong professional community, which has a significant bearing on how they respond to students in the classroom and therefore on student learning outcomes (McLaughlin, 1993; Silins & Rosalind, 1999). Inside learning communities, teachers are actually practising workplace learning, which involves the transformation of knowledge, values and beliefs into classroom practice (Retallick, 1999). On such basis, workplace learning is job-embedded, and is in fact a process of knowledge management, in which tacit knowledge is converted to explicit knowledge and vice versa.

School Leadership

The role of school leadership has been gaining increasing importance in the context of educational change (Leithwood, 1994). A review of literature has indicated that most researches on school leadership focused predominantly on the principal's role (Heck & Hallinger, 1999) and conceptualisation of leadership (Leithwood & Duke, 1999). These studies are generally believed to be of use to guide future and incumbent administrators, and also to help explain school outcomes (Leithwood &

Duke, 1999). In another review of literature, Hallinger & Heck (1998) have also found that there are a substantial number of empirical studies to support the belief that principals exercise a measurable though indirect effect on school effectiveness and student achievement.

Conceptualising Leadership

Despite having no clear and agreed-upon definition of school leadership over the century, Liethwood & Duke (1999) have thought that it is counterproductive to persevere on the development of a precise definition of leadership because complex concepts are usually defined vaguely. Others have suggested that the differences in expression of cultural ideal or aspirations relating to leadership may explain the absence of an agreed-upon definition (Ogawa, 2005). Leadership has been described and defined in a number of ways. Leadership has been viewed as the process by which an individual induces a group to pursue the objectives of the individual (Gardner, 2000). Sergiovanni (2001) has argued that leadership is about a struggle to do the right thing, which is about helping people to understand and manage the problems they face, and even with helping them to live with problems. Likewise, Heifetz (1994) has referred leadership as an activity getting people to clarify what matters most, in what balance, with what trade-offs. Others have thought that leadership is not a quality that comes with an office or a person but it can spring from anywhere (Foster, 1989). Thus, a principal in school may at times be a leader and at other times a follower. It seems that many of the above definitions for leadership are guided by idealised conceptions of the kinds of qualities and relationships that leaders have with followers rather than on-the-job performance within given circumstances.

There is no absolute right definition of leadership, but it is only a matter of usefulness for conceptualisation. Definitions can be used to serve a variety of purposes of the study, but it was indicated that the definition of leadership has followed a progression of thought from as a focus of group process, to as the art of inducing compliance, to the more recent definitions in terms of influence relationships and the like (Bass, 1990). Anyway, Ogawa (2005) was of the opinion that "most definitions of leadership emphasise individuals and their relationship to organisational structures both as constraints on and as the means of exerting agency, which typically takes the form of influencing the actions and interactions of other participants" (p. 93). A main line of thought is that it is useful to study leadership as a process or an activity rather than as individual traits that come with a leader since this view allows for leadership from multiple positions in a social structure (Heifetz, 1994), and it accounts for the maintenance and continuation of leadership (Bass, 1990). At the same time, it is common to think that leadership involves intentional influence that is exerted by one person/ group over other people (Yukl, 1998). In this way, leadership involves a social influence process exerted by one person/ group over other people/ groups that often involves a structuring and restructuring of the situation and the perceptions and expectation of the members (Bass, 1990).

Theories of Leadership

It was suggested that theories of leadership have been shaped by the conceptualisation of organisation, following the progression from the rational perspective, to the institutionalism, to the cultural metaphor (Ogawa, 2005). It was also believed that democratically-oriented educational philosophy, social psychology, and critical theory are important in giving rise to and sustaining interest in participative and democratic forms of school leadership (Leithwood & Duke, 1999). Critical theory approaches to administration have contributed significantly to conceptualising leadership, serving as alternatives to challenge the structural-functional assumptions about organisations, and a control-oriented as well as managerial view of leadership (Leithwood & Duke, 1999). In another analysis, development of leadership theories was seen as reflecting the progression from trait theory, to behavioural theory, to contingency theory, to transformational theories (Crow & Grogan, 2005).

Owens (2001) distinguished traditional theories of leadership from contemporary theories of leadership. He viewed traditional theories of leadership under two dimensions, namely task-focused dimension and follower-focused dimension. Task-focused leaders give emphasis to rules and procedures for getting the job done, while follower-focused leaders place emphasis to concerns for people. Traditional leadership was then understood as a mix of these two kinds of behaviours, and the effectiveness of a leader depends on choosing the right blend in different situations. The focus of traditional leadership theories is therefore on prescriptive leadership styles (Owens, 2001). On the other hand, contemporary theories of leadership have shifted the emphasis from personality and rules to quality and kinds of relationships that leaders have with followers. As mentioned by Owens (2001), contemporary leadership theories recognise the importance of members of an organisation as valuable resources of knowledge, creativity, and energy for improving the organisation. At the same time, they favour the development of environments that are motivating, caring, inclusionary and empowering of members as forms of leadership.

Contemporary Models of School Leadership

From an analysis of a representative sample of literature, Leithwood & Duke (1999) identified six distinctly different models of leadership in schools: instructional, transformational, moral, participative, managerial, and contingent leadership. Studies from early 1980s were dominated by an instructional leadership conceptualisation drawn from the effective school literature (Hallinger & Heck, 1998), with a typical focus on teachers' behaviours that related to activities directly affecting the growth of students (Leithwood & Duke, 1999). Instructional leadership was also defined as leadership functions directly related to teaching and learning (Murphy, 1988). In the context of educational change where school restructuring and teacher professionalisation dominate, it has been argued that the heavily classroom-focused instructional leadership becomes inadequate (Leithwood, 1994; Marks & Printy, 2003). Since the 1990s, many researchers have adopted new perspectives in understanding school leadership other than an instructional leadership conceptualisation (Hallinger & Heck, 1998). Leadership has been conceptualised as an organisational quality that is embedded in a social and cultural context (Ogawa & Bossert, 1995), as an integration of transformational and instructional leadership (Marks & Printy, 2003), as cognitive leadership which is based less and less on positions, less and less on mandates, and more and more on ideas (Sergiovanni, 2001). Anyway, the dominant model has been transformational leadership that focuses on increasing the organisation's capacity to innovate (Hallinger & Heck, 1998), and adopting a commitment-building strategy for school restructuring (Leithwood, 1994).

Transformational Leadership

The idea of transformational leadership was proposed by James McGregor Burns (Burns, 1978), and it has directly influenced the thinking of scholars ever since (Owens, 2001). Comparing and contrasting with transactional leadership which is mainly based on transactions between leader and followers, transformational leadership was identified as a process in which leaders and followers raise one another to higher levels of morality and motivation (Burns, 1978). The focus of Burns's (1978) concept is on the relationships between leaders and followers (Leithwood, Tomlinson, & Genge, 1996), which is illustrated by the saying: "Burns's distinction between transactional and transformational types of leadership hinges on this appreciation of power-as-relationships" (p. 786).

Based on Burns's (1978) idea, many modifications of transformational leadership have been proposed. With a modification of Burns's (1978) conception of transactional and transformational leadership as opposite ends of a leadership continuum, Bass (1985) proposed a two-factor theory in which transactional and transformational leadership are complementary. While the concept of transformational leadership has been subject to varying interpretations in the literature, they share a common focus with emphasis on the commitments and capacities of organisational members, and a common concern for organisational culture (Leithwood & Duke, 1999). In comparison with instructional leadership, transformational leadership focuses on increasing the organisation's capacity to innovate instead of focusing specifically on curriculum and instruction (Hallinger & Heck, 1998). Being sensitive to organisation building: developing shared vision, creating productive work cultures, distributing leadership to others and the like, transformational leadership has been thought as crucial in the context of uncertainty and change as it calls for "commitment" rather than "control" strategies (Leithwood, 1994).

Of the developed models of transformational leadership, there is the two-factor theory proposed by Bass & Avolio (1994). The two-factor theory was constructed along seven dimensions, four of which come from transformational leadership: idealised influence, inspirational motivation, intellectual stimulation, and individualised consideration; and three come from transactional leadership: contingent reward, management by exception, and laissez-faire leadership (Bass & Avolio, 1994). While considerable empirical evidence was established for the two-factor theory (Bass, 1985), subsequent empirical studies have cast doubts on some dimensions, in particular the dimensions of management by exception and contingent reward (Leithwood, 1994; Leithwood et al., 1996). The most fully developed model of transformational leadership in schools has been provided by Leithwood and his colleagues (Leithwood & Duke, 1999), in which transformational leadership was conceptualised along eight dimensions: building school vision, establishing school goals, providing intellectual stimulation, offering individualised support, modelling best practices and important organisational values, demonstrating high performance expectations, creating a productive school culture, and developing structures to foster participation in school decisions (Leithwood, 1994). The eight dimensions of the model were also summarised into three areas: mission centred, performance centred, and culture centred (Marks & Printy, 2003).

A review of empirical research on transformational school leadership has offered modest amount of evidence for the contributions of such leadership to student participation, organisational learning, and the development of productive school climate (Leithwood et al., 1996). Specifically, substantial evidence of contributions to school outcome has been established in three dimensions of transformational school leadership: building school vision, providing intellectual stimulation, and offering individualised support. In addition, the empirical evidence has revealed that the effects of transformational school leadership on students were likely to be mediated by teachers (Leithwood et al., 1996).

Human and Social Capital Perspective to Managing Educational Change

Early school reforms in the Western countries, as in 1980s and early 1990s, were criticised as having a focus on the structural dimensions of schools (Elmore, Peterson, & McCarthey, 1996). In response to this, the reform research has strongly suggested that something on top of restructuring are needed to support productive teaching and learning (Smylie & Hart, 1999). Apart from the cultural perspective arising from the organisation and management theory mentioned in the previous sections, there is also the emergence of the human and social capital perspective to manage educational change, in particular applying the concepts to teachers and their work in schools (Smylie & Hart, 1999).

Concepts of Human and Social Capital Perspective

Although wanting to improve schools for student learning, educators have only recently come to adopt a human and social capital perspective to relate student learning to the extent to which schools support the ongoing development of teachers' knowledge and skills (Smylie & Hart, 1999). In believing human capital as in the form of knowledge, skills, and other attributes from humans, and in believing social capital as derived from relationships among individuals and from the social structures that frame the relationships, the human and social capital perspective affirms the values of social relationships in teacher learning that contribute to organisational effectiveness (Smylie & Hart, 1999). Building human capital means investing in teachers who are seen by many as the primary source of the problems we wish to solve in our schools (Smylie, 1997; Smylie, Miretzky, & Konkol, 2004), and social capital in the form of social relations in schools are considered to be crucial in building such human capital (Smylie & Hart, 1999).

Roles of Human and Social Capital Perspective in Educational Change

The instrumental role of capacity-building in educational change can be highlighted from the fact that it has been proposed as one of the alternative policy instruments in the field of educational policy analysis (McDonnell & Elmore, 1991), in which policymakers expect to reap long term returns like skill enhancement and competence. At the implementation level, recent implementation studies have revealed that building teacher capacity is considered to be conducive to successful policy implementation as teachers are found to play a more central role in educational change processes (Fung, 1995; Odden, 1991b). Capacity building has been suggested to be achieved in direct ways like staff development (Hord, 1995), and in indirect way like fostering collegial culture, teacher empowerment as well as professionalism (Odden, 1991b), and becoming learning organisations (Fullan, 1995b). In the study of implementing educational policy at the school level, capacity-building has been identified as one of the key organisational issues in the process of implementing change, which represents the ability to deliver the promised change (Smylie et al., 2004; Van der Vegt, Smyth, & Vandenberghe, 2001).

Human Capital

Framing teacher development as fundamentally a function of school organisation that works together to enhance teaching and student learning, Smylie et al (2004) have lent from the organisation and management literature the strategic human resource management perspective to manage educational change. The approach calls for a comprehensive, systemic, and strategic approach to teacher workforce development, and it is something more than developing the knowledge and skills of individual teachers, but is a collective and organisational issue encompassing a broader range of practices as arranged in an interrelated system (Smylie et al., 2004). As an expansion to the traditional concept of staff development, a similar concept of developing human capital has been proposed as to covering the needs of teachers at the individual, group and school level (Tam & Cheng, 1996). In comparison with the regulatory policy, the human capital perspective attaches high importance to teacher learning as well as to professional autonomy and discretion of teachers, which calls for social and normative accountability instead of bureaucratic work rules and procedures (Smylie, 1997). Along this line of thought, Sergiovanni (1995) has also proposed the professional development model and renewal model for teacher development. Valuing teacher's capacities and needs, the models encourage the building of a learning and caring community that encourages teachers to reflect and to engage in conversation and discourse. While there has been emerging, mostly descriptive body of research that points to the efficacy of a strategic human resource approach, its application to teacher workforce development is now largely theoretical and intuitive (Smylie et al., 2004).

Social Capital

Social capital is viewed as an important source to promote the development of human capital among teachers for instructional improvement since social capital is grounded in relationships and interactions that are educative (Smylie & Hart, 1999). Social learning, incidental learning, and organisational socialisation theories suggest that social relations characterised by open exchange of ideas are most conducive to creative, innovative and reflective learning (Smylie & Hart, 1999). Reasoning along this line, the self-management concept believes that teachers operating in a high degree of autonomy within the boundaries framed by the school and their work groups may undergo continuous self-learning and development to ensure quality of work in a changing environment (Cheung & Cheng, 1997). Social capital is also considered as an important source to enhance the development of human capital. In the process of continuous self-learning acquired through action and experience, teachers need to build a strong network of close affiliation with significant others in work, such as superiors, colleagues, students, parents and other work related parties in the community (Cheung & Cheng, 1997; Kochan, 2000). It is emphasised that frequent exchanges with colleagues may trigger reflection in teachers (Cheung & Cheng, 1997; Kochan, 2000).

While research has revealed that social relations among teachers and administrators in schools are crucial in promoting student learning and teachers' professional learning, there is relatively little empirical evidence as to show what kinds of social relations that are conducive to teacher learning and change (Smylie & Hart, 1999). Specifically, Smylie & Hart (1999) identified three types of such empirical studies: studies of teachers' collegial relations in schools, studies of teacher collaboration in improvement initiatives, and studies of teacher professional communities. The empirical studies have demonstrated some evidence that the goals, norms, expectations, and mutual obligations and accountability which define teachers' social relations also form a system of professional social controls that govern the activities of individual teachers (Smylie & Hart, 1999).

Perspectives on ICT and Education Research

Introduction

Entering into a new millennium of expecting schooling and society to produce lifelong learners with continuous development of new knowledge and skills (A. Hargreaves, 1999), initiatives for restructuring education follow. While the restructuring targets are not ICT, the implementation of these ideas of change often assumes the availability of ICT (Riel, 1994). This is in part due to the common belief that ICT has the potential to support certain fundamental changes in learning (Panel On Education Technology, 1997), and also the prevalent expectations that ICT will revolutionise education (Riel, 1994). Currently, the ongoing and unprecedented development of ICT has led to a widespread intention of using ICT to advance educational goals, and many national leaders have included ICT as a necessary component in the educational change (Cheng & Townsend, 2000). Many countries have already drawn up their master plans for the development of ICT in education (Cheng, 2002a). In Hong Kong, the commitment of promoting the application of ICT across the school curriculum was confirmed with the introduction of "Information Technology for Learning in a New Era: Five-year Strategy – 1998/99 to 2002/03" in 1998 (Education and Manpower Bureau, 1998).

Despite all these positive views on ICT, how may ICT advance changes in student learning? How may ICT revolutionise education? While the previous sections have provided us with a review on educational change, this section is going to link the concepts of educational change with the use of ICT in education. In a review of studies on ICT in education, it was found out that the focus has been shifted from studying particular technologies and their impact on student learning, to studying how technology is integrated into education settings (Honey et al., 2000). While such progression indicates an increased awareness of viewing the use of ICT in context, conceptualisation of ICT implementation is still inadequate, in a sense that there is a lack of a holistic conceptualisation of ICT implementation that links the use of ICT to educational change. We suggest that answering these questions requires the re-conceptualisation of a range of concepts relevant to educational change: (1) effective use of ICT cannot be accomplished by technological fixes alone, instead it is embedded in the larger process of educational change; (2) effective use of ICT requires pedagogical intervention from teachers, which involves a conceptual change of teaching and learning from a teacher-centred approach to a student-centred approach; and (3) effective use of ICT requires organisational intervention, which involves a conceptual change of leadership from instructional leadership to transformational leadership, and a conceptual change of professionalisation from teacher training to human and social capital development of teachers. Along these directions, the following sub-sections will lead us to understand the concept of learning with ICT as well as the concept of using ICT in context, and subsequently to explore how ICT may advance changes in student learning by ways of pedagogical intervention and organisational intervention that facilitate constructivist learning.

The Evolution of Educational Computing

The pace of technological development has dramatically accelerated during the past decade, which has brought increasingly diverse and more powerful technological tools and computer resources into schools (Honey et al., 2000). At the same time, increased performance have been matched by declining costs, and this also enables more and more schools to have access to these new technologies (Knapp & Glenn, 1996). Despite its diversity and power, each time when a technological innovation is brought into an education setting, positive changes in learning from its use is proposed to progress over time, and specifically through four stages, with each new stage building onto, rather than replacing, existing practice (Heppell, 1993). At the topicality stage, the focus is on learning about the technology. With the progression onto the surrogacy stage, ICT is supposed to act as a surrogate teacher having a discrete and relatively small body of expertise. At the progression stage, the focus is on the use of generic tools. At the pedagogic evolution stage, ICT alters the learning environment and the learners. According to Heppell (1993), the evolution to a final stage of educational computing requires an emphasis on participative media, radical change in pedagogy enabled by ICT, and recognition of the emergent capabilities of learners. Heppell's (1993) model indicates changes in assumptions about how ICT should be used in education when a technological innovation is brought into the education setting.

Emerging Use of ICT in Education

In and of itself, ICT contains neither pedagogical philosophy nor content basis (Means, 1994). Heppell's (1993) model distinguishes the concept of learning about ICT from the concept of learning with ICT. This also suggests that pedagogical and organisational intervention needs to be addressed if ICT use is to go beyond the early stages characterised by learning about ICT and proceed to the final stage of learning with ICT. In the process of learning with ICT, it has been reminded that ICT is a means for constructing knowledge but not an end in itself (Eib & Mehlinger, 1998; McCombs, 2000; Means, 1994). Apart from that, educators has begun to identify ICT as a tool for change, stimulation and restructuring of education (Collis & Carleer, 1992; Means, 1994). In this process, ICT is expected to facilitate the transformation of existing classroom practices, which requires qualitative changes in the teaching and learning process (Itzkan, 1994; Thomas et al., 1998; Twining, 2002a). Basically, it involves changes in the curriculum content, the instructional process, and the teacher-student relationships (Plomp & Brummelhuis, 1999; Squires & McDougall, 1994). It has been emphasised that students must be seen as knowledge generators and active participants in their own learning, and teachers are seen as facilitators (Jonassen, Peck, & Wilson, 1999; McCombs, 2000; Means, 1994).

ICT as a Tool for Learning

ICT as tutor, tool and tutee. The increasing impact of ICT on our society is influencing teaching and learning in a variety of ways. Contemporary views of a large part of the educational research community generally suggest that the way we use ICT in schools should change from the traditional roles of technology as a tutor to the role of technology as a tool in the learning process. In 1970s, when personal computers and educational software were not common, roles of ICT in education were seen as tutor, tool and tutee (Taylor, 1980). The intelligent tutoring system served the role of tutor, and the generic tools like statistical analysis program served the role of tool, and lastly computer programming served the role of tutee. Taylor (1980) favoured the tutee role of ICT as it qualitatively changed the learning experience by shifting the focus from end product to process, from acquiring facts to understanding them.

ICT as a mindtool. While looking at the role of ICT as a tool, Jonassen (1996) recently has taken a rather different perspective with an emphasis on the function of ICT as intellectual partners with the learners. He has chosen to use the term mindtools to include databases, spreadsheets, semantic networks, expert systems, computer conferencing, multimedia and hypermedia construction (Jonassen, 1996). It has been further pointed out that technology integration into education is a process whereby education technology is no longer the object of study, it is the means for conducting inquiry (Eib & Mehlinger, 1998). As a result, ICT has been considered as a tool that is used for inquiry, knowledge building, and collaboration, but not as an end in itself (McCombs, 2000; Means, 1994).

ICT as a Tool to Catalyse Educational Change

ICT as a tool for transformation of existing practice. Apart from seeing ICT as a tool for learning, some researchers has gone further to identify ICT as a tool for transformation of existing practices (Itzkan, 1994; Thomas et al., 1998; Twining, 2002a), and to support the kinds of sustained and substantial inquiry and analysis that is proposed by educational reforms (Honey et al., 2000). This involves how ICT acts as a tool or medium to facilitate fundamental and qualitative changes in the nature of teaching and learning (Panel On Education Technology, 1997). If such potential is to be realised, it has been suggested that ICT should not only be used for replacing existing tasks or reinforcing old pedagogy (Lankshear et al., 2000; Papert, 1997), but it should enter what is termed a transformation stage, which requires the process of instructional practices and the curriculum content to be changed at the same time (Itzkan, 1994; Twining, 2002a). In elaborating such kind of educational changes, Twining (2002) has used three different modes to describe the impact of ICT on the curriculum: support, extend and transform. When use of ICT is to support the curriculum, the instructional process becomes more efficient but the content remains unchanged. If use of ICT is to extend the curriculum, the content changes and/or the instructional process changes but they could have been achieved without the use of ICT. When entering into the transformation stage, the curriculum content and/or instructional process changes and they could not have been achieved without the use of ICT (Twining, 2002a).

ICT use in classroom context. Apart from focusing on ICT itself, some researchers have attended to the ways in which ICT is used in different context. In the use of ICT, the learning process has been construed as the result of the interplay among four forces, namely teachers, learners, content, and materials & technical infrastructure (Plomp & Brummelhuis, 1999). In a similar vein, some researchers have focused on the interactions between three key actors in the use of ICT: students, teacher and designer of software (Squires & McDougall, 1994). All these have helped educators to move the focus from the ICT itself to the issues of educational change, in particular the learning process in relation to roles of teachers, curriculum issues, and student responsibility for learning. Along this line of reasoning, a learner-centred perspective has been proposed to assess the role of ICT in the teaching and learning process (McCombs, 2000). Being a reflection in practice of the Learner-Centred Psychological Principles of the American Psychological Association, the learner-centred model focuses on human needs and teacher-student relationships. It is emphasised that students must be seen as knowledge generators and active participants in their own learning.

ICT to support constructivist learning. In this regard, ICT applications not only structure the learning process, but students themselves need to increasingly structure their own learning process. This echoes the belief that ICT can be used to support constructivist learning or constructivism, in which learners construct knowledge with technology, and technology is a tool to think and learn with (Jonassen et al., 1999). In this sense, ICT is viewed as a partner and tool for change, stimulation and restructuring of education (Collis & Carleer, 1992; Means, 1994). Specifically, in the teaching and learning process, ICT can be used as tools to catalyse the process of change that is required for constructivism, but the key to success still depends on whether teachers and learners can have a clear understanding of their roles in the interplay (Jonassen et al., 1999).

Looking at ICT Use in a Larger Process of School Change

Some researchers have suggested that effective ICT use is embedded in a larger process of school change (Honey et al., 2000), and effort of ICT implementation needs to be examined as part of multifaceted school reform rather than in isolation (Bober, 2002). Along this line of reasoning, effective use of ICT in teaching and learning largely depends on pedagogical interventions as well as organisational interventions.

Pedagogical intervention. In the process of using ICT in teaching and learning, it has become clear through research evidence that instructional strategy underlying the use of ICT determines learning effectiveness (Honey et al., 2000; McCombs, 2000;

51

Means, 1994; Mehlinger, 1995). As new technologies continue to evolve into more powerful and sophisticated application, it provides the opportunity for educators to explore in different ways to teach and design instruction to realise constructivist teaching and learning (Knapp & Glenn, 1996). As remarked by Mehlinger (1995), the effort to infuse ICT into teaching and learning may encourage teachers to use more powerful instructional strategies that they felt incapable of employing in the past. In Means' (1994) view, the move to constructivism is characterised by centring instruction on authentic and challenging tasks, which are essentially multidisciplinary in context, with interactive instruction and collaborative work.

Organisational intervention. It has been pointed out that effective ICT integration into curriculum requires organisational intervention since the benefits of ICT cannot be adequately separated from other variables that impact learning in the larger instructional context (Honey et al., 2000; McCombs, 2000). This alerts educators to move the focus from ICT itself, to ICT use in classroom context, and further to ICT use in an organisational context. It has been suggested that in exploring how best to use ICT to advance educational goals, in addition to pedagogical methods of the teachers, issues like organisation of the classroom and socio-cultural setting of the schools have to be considered (Honey et al., 2000). In the comprehensive program of reform that has taken place in the Union City New Jersey schools, research results identified a number of organisational factors that have positive impact on students' learning, such as leadership, professional development of teachers, school culture (Honey et al., 2000).

Educational Research Direction Concerning ICT

It must be pointed out that early studies sought to focus specifically on particular technologies and their impact on student learning (Honey et al., 2000). As the value of these studies was tied to the specific technologies, they tended to become outdated very fast under the rapid advancement of technology. Studies focusing specifically on text-based, stand-alone computer assisted instruction applications were examples (Honey et al., 2000). Along with this, it has been realised that impact of ICT on teaching and learning needs to be construed in classroom context and organisational context. To this end, the research direction has begun to move from studying the impact of particular kinds of ICT on student learning to studying how use of ICT is embedded in the larger process of school change (Honey et al., 2000). One direction is to gain an understanding of how ICT use is mediated by pedagogical and organisational factors mentioned above.

Theoretical Framework

Introduction

While policymakers and educators have already acknowledged the importance of implementation at school level since 1970s (Firestone & Corbett, 1988), a considerable body of research in education and other policy fields also lays out the difficulties of moving from policy to practice (Levin, 2001). Early implementation research findings showed that the educational change process at schools was inherently at odds with state program initiative in the United States (Odden, 1991a). An understanding of how implementation takes place in schools is therefore vital to our understanding of success with policy initiatives. Recently, implementation research has also been shifting gradually from describing the range of barriers to addressing the kinds of policies or practices that support effective translation of policy to practice. Nevertheless, there has been less deliberation on models of implementation (Levin, 2001).

In light of the above findings and taking into account the fast development of ICT that will possibly diminish the value of studying the impact of particular kinds of ICT on student learning, the present study is going to build a theoretical model of ICT implementation that helps to examine how ICT use is mediated by pedagogical and organisational factors. Past research on ICT implementation in schools generally focused on individual issues like: technical infrastructure such as access and technical support; institutional factors such as scheduling, school leadership, school culture, sense of support and collegiality; personal factors such as willingness, ability, and confidence to use ICT; pedagogical factors such as pedagogical style and classroom organisation (Frank, Zhao, & Borman, 2004; Twining, 2002b). Nevertheless, few studies provided a framework to unfold ICT implementation in the way of mapping interaction of the use of ICT with changes in student learning, and with pedagogical and organisational issues. We believe that a model of ICT implementation is useful in understanding how effective ICT use is embedded in the change process that is targeted to advance student learning in the context of schooling. This study attempts to bridge this gap by constructing a model to inspect the whole-school ICT implementation in the context of managing change in schools. The conceptual framework is constructed under the following core ideas:

First, effective use of ICT in teaching and learning is embedded in the larger context of managing change in schools, and therefore the benefits of ICT cannot be adequately separated from other variables that impact learning in the larger instructional context (Honey et al., 2000; McCombs, 2000).

Second, in adopting a situative perspective, ICT implementation practice is constituted in the interaction of individuals and their situation in the execution of particular tasks (Greeno, 1998; Spillane et al., 2002). The focus of analysis is shifted from isolated individuals to the social collective or activity system, the purpose is to locate the isolated individuals in relation to functions of the larger systems in which individuals participate (Greeno, 1998).

Third, in adopting an approach of ecological metaphor and sociocultural approach, the framework highlights the aspect of social dynamics in the ICT implementation process. Accordingly, the framework unfolds the interactions of contextual factors in the implementation of change. In and of itself, ICT contains neither pedagogical philosophy nor content basis (Means, 1994). Therefore, contextual factors will be conceptualised along the direction of pedagogical intervention as well as organisational intervention.

Frameworks for Conceptualising Implementation in Education

A review of literature has suggested that there are many general models for policy analysis but very few have been developed specifically for the analysis of education policies (Cheng & Cheung, 1995), and there is even fewer models of implementation specially designed for educational settings (Levin, 2001). As suggested by MacLaughlin (1991), implementation models need to take into consideration its institutional context since policy effect is complex and transitory.

The following brief review of implementation models are those designed specifically for the analysis of education policies, and at the same time they are also relevant to the present study.

Micro- and macro-implementation models. Literature review has suggested that there are two distinct approaches to constructing implementation models. Macro-level analyses operate at the level of the system, which stress regularities of process and organisational structures. Micro-level analyses operate at the individual level, which interpret organisational action as the outcome of autonomous actors, motivated by self-interest (McLaughlin, 1991). While macro-analyses generally provide insufficient guidance in developing models of how policies operate in practice, micro-analyses ignore systemic attainments for the institutional setting as a whole so that they provide limited guidance to policymakers faced with system-wide decisions (McLaughlin, 1991). McLaughlin (1991) construed model of micro-implementation as a process of bargaining, and framed implementation issues in terms of perspectives at individual level like motivation, beliefs and capacity. This perspective on the implementation has focused on individuals rather than organisations, and is helpful in understanding how policies operate in practice. On the other hand, McDonnell & Elmore (1991) suggested that new research should focus less on specific programs but more on developing policy instruments that were able to analyse education policies that targeted at all students and aimed at the core of schooling. They ended up with conceptualising a range of alternative policy instruments (mandates, inducements, capacity-building, system-changing) and hypothesising why policymakers select different instruments. The instrument has been for macro-level analysis and it helps to inform policymakers in formulating policy initiatives.

Organisational dynamics model. Acknowledging that organisation and individual phenomena in implementation have not usually been explicitly related conceptually, Van der Vegt et al. (2001) have built a framework that maps the interplay of organisational issues and personal concerns, which serves as a reflection on the dynamics of change. The important organisational issues identified are amending organisation's identity, capacity building, redesigning structure and controls, allocating incentives; and personal concerns are represented by personal identity, competence concern, power-and-influence concern, fairness concern. Implementation has been seen as generating of specific organisational issues, which in turn activating personal concerns. Likewise, Levin (2001) has used an organisational perspective to view implementation, with particular reference to organisational learning. He has thought that acceptance of change is strongly affected by the degree to which a policy is consistent with the existing culture in the process of implementation.

Perspectives and Approaches for Conceptualising ICT Implementation

Situative perspective. The situative perspective has been widely applied to knowing, thinking and learning (Brown et al., 1989; Greeno, 1998; Lave & Wenger, 1991), teacher education (Borko, Romagnano, Knuth, & Willis-Yorker, 2000; Peressini, Borko, Romagnano, Knuth, & Willis, 2004; Putnam & Borko, 2000), and has also been applied to policy implementation (Spillane et al., 2002). Situative perspective argues that a phenomenon (e.g. knowing, thinking, learning, and policy implementation) does not exist in isolation, and therefore one must study the phenomenon within the contexts in which they occur, in particular the physical and social systems in which individuals participate. The situative perspective helps researchers to shift the unit of analysis from isolated individuals to the social collective or activity system, the purpose is to locate the isolated individuals in relation to functions of the larger systems in which individuals participate (Greeno, 1998). Working along a similar direction, instruction capacity has been conceptualised as a function of instructional unit as a result of interactions between teachers and students around educational materials, and it follows that improvement in capacity can be understood as a function of how well the teachers, students, and educational materials interact with one another (Cohen & Ball, 1999). Likewise, use of ICT in a classroom lesson has also been conceptualised as an activity system in which students, teachers, ICT tools situated, and is mediated by rules and division of labour (Lim, 2002; Lim & Hang, 2003).

Sociocultural approach. Apart from viewing the unit of analysis as an activity system, many researchers have begun to attach high importance to the way the activity is organised: a socially organised activity (Greeno, 1998; Lave & Wenger, 1991; Lim, 2002). However, in studying ICT implementation in schools, past research tended to ignore both the social contexts and institutional culture in which teachers situated (Windschitl & Sahl, 2002). In contrast, the platform of teacher learning has been changing from isolated individuals to communities of practice (Stein & Brown, 1997). The constructivist view also takes a social and cultural perspective to analyse educational issues, giving special importance to collegial supervision, teacher development as inquiry and reflection, and leadership as community building (Sergiovanni, 1995).

Ecological approach. To unfold implementation dynamics at an organisational level, it is vital to study the interplay of contextual variables (Van der Vegt et al., 2001). Ecology is a branch of biology that is concerned with studying the interrelationships between organisms and their environment, and adopting an ecological approach in studying implementation signifies its focus on the interactions of contextual factors and its context. Actually, the approach of adopting an ecological metaphor has been found to be a powerful tool in understanding the use of ICT in schools, in a sense that the ecological model transcends the boundary of simply identifying factors affecting use of ICT in schools and takes a further step to study the interactions among the factors (Zhao & Frank, 2003).

Perspective of learning organisations. It has been argued that typical implementation strategies with a conventional answer-driven approach are likely to result in failure of school change. On the contrary, when presented with questions rather than answers, teachers are challenged deeply to work out solutions in a collaborative manner (Wagner, 1998). Along this line of reasoning, successful implementation of change requires reculturing of the school into a learning organisation (Fullan, 1995b) that taps people's commitment and capacity to learn at all levels in an organisation (Senge, 1990) with an inquiry approach (Fullan, 1995b; Wagner, 1998). The learning organisation has the capacity to adapt to new situations on a continuous basis (Fullan, 1999), which is termed the innovative capacity (van den Berg & Sleegers, 1996). The widespread support for learning organisations indicates a shift from the structural-functional perspective to a cultural-individual perspective on educational innovation (Sleegers et al., 2002). Studies from this perspective favour the human and social capital theory (Smylie & Hart, 1999) mentioned in the previous sections, which places great emphasis on professional development of teachers and their opportunities to learn. According to this perspective, teachers' commitment to change, teachers' empowerment, and collegiality and collaboration among teachers are important issues to implementation of change (Sleegers et al., 2002).

Conceptualisation of the Theoretical Framework

In constructing the theoretical framework, the present study has made reference to the aforementioned implementation models, in particular the concept of viewing implementation dynamics as the interplay between organisational issues and personal concerns (Van der Vegt et al., 2001). Adopting a situative perspective with the unit of analysis as the activity system to frame schoolwide ICT implementation, the present theoretical framework has also borrowed the concept of viewing implementation from the perspective of learning organisation (Levin, 2001). At the same time, the micro-implementation model (McLaughlin, 1991) and macro-implementation model (McDonnell & Elmore, 1991) have offered insight to building the contextual variables for the present theoretical framework, in particular the variable "collegial capacity of ICT implementation strategies" (for details, please refer to the later section for contextual factors). The contributions are summarised as follows:

First, both the micro- and macro-implementation models (McDonnell & Elmore, 1991; McLaughlin, 1991) highlight the concept of capacity building in the form of a workforce unit, which has yielded important insights in conceptualising ICT implementation strategies in the present theoretical framework as the collegial capacity of ICT implementation strategies.

Second, the concept of motivation and beliefs (McLaughlin, 1991) and the concept of learning organisation (Levin, 2001) have steered the present framework in adopting a sociocultural approach in conceptualising the contextual variables, in a way of taking into account the often overlooked but deep-seated social problems in implementing ICT practices in schools.

Third, in the present study, the unit of analysis is the activity system for schoolwide ICT implementation. With its emphasis on the larger systems in which the principal, teachers, and students interacting with each other, the situative perspective has guided the researcher to understand schoolwide ICT implementation as a result of activity between pedagogical intervention and organisational intervention to achieve changes in student learning. Implementation dynamics emerges from the elements interacting in the activity system. Accordingly, ICT implementation is framed in an activity system characterised by the interplay between contextual variables that are derived from interaction of the principal, teachers, and students.

Fourth, the present study has used an ecological approach to conceptualise schoolwide ICT implementation as "the social ecology of ICT implementation". With an ecological approach, the study of ICT implementation will focus on the interactions of the factors within its context, and therefore "the social ecology of ICT implementation" is conceptualised as the interplay of contextual variables as a result of interaction of individuals and their situation to advance changes in student learning in a sociocultural context of schooling.

Identification of Contextual Factors for the Theoretical Framework

Concept of implementation. In attempting to identify the contextual factors interacting with each other in the process of ICT implementation, it is necessary to understand the concept of implementation first. Implementation focuses on what happens in practice, and with the nature and extent of actual change, as well as the factors and processes that influence how and what changes are achieved (Fullan, 1992). Specifically, implementation concerns with strategies used to translate policies into practice (Levin, 2001). In a way, implementation is relevant to the interaction of implementation strategies, the changes achieved, and also the factors that influence these changes. Guided by the above theoretical underpinning, the contextual factors comprising the social ecology of ICT implementation in a context of managing change in schools will include: ICT implementation strategies, changes in student learning, and some key contextual factors.

Perspective of learning organisations. In identifying the contextual factors, the social aspect of the ecology will be emphasised in the way of taking into account the often overlooked but deep-seated social problems in implementing ICT strategies in schools. Borrowing the idea of viewing implementation from an organisational learning perspective (Levin, 2001), we have chosen to include organisational variables that are important to teachers' commitment and capacity to learn (Sleegers et al., 2002). As mentioned in the literature review section, the idea of learning organisation is about fostering a culture of learning (Fullan, 1993; Senge, 1990), and it seems plausible to put the "climate for collaboration and experimentation" as a pillar of the theoretical framework. As there are a substantial number of empirical studies supporting the belief that principals exercised a measurable though indirect effect on school effectiveness and student achievement (Hallinger & Heck, 1998), and educational administration literature tended to affirm the role of leadership in shaping school culture (Schein, 2004; Sergiovanni, 1995), "school leadership with particular reference to principals" is selected as another contextual factor. Apart from organisational interventions represented by school climate and leadership, effective ICT implementation depends on pedagogical interventions as well. Therefore, "changes in teacher pedagogy" is proposed as one of the contextual factors comprising the social ecology of ICT implementation in schools.

External input - government ICT policy. In adopting the ecological approach and situative approach to construct the theoretical framework, the platform of ICT implementation is framed as the school system in which individuals interact with themselves and with the situations to produce contextual factors. However, both approaches recognise that the activity system or ecosystem is not purely isolated. While the ecosystem has the tendency to maintain internal equilibrium, it is an open system with things constantly entering and leaving (Zhao & Frank, 2003). In a similar vein, the activity system as a unit of analysis is considered as being surrounded by different level of circles made up of broader context (Lim, 2002). In the context of the social ecology of ICT implementation in schools, "Government ICT policies" is identified as a relevant external input that will influence the dynamics of ICT implementation, it is therefore proposed as one of the contextual factors.

Conceptualisation of Contextual Factors

School climate. In building a learning organisation that ultimately uses ICT to advance changes in students' learning, a collaborative school culture with innovative

63

capacity is thought to be important. The concept of organisational culture has emerged as central in the analysis of organisational effectiveness and organisational change, and studies of schools have strongly supported the belief that organisational culture is a fundamental factor in determining the quality of educational organisations (Owens, 2001). Specifically, school culture has been considered as playing a critical role in promoting and sustaining the implementation of ICT in education (Dexter, Anderson, & Becker, 2000; Fullan, 1993, 1995a, 1999, 2001; Senge et al., 2000). While school culture and school climate are two distinct concepts, they are related. School culture refers to deeply embedded beliefs, and is difficult to assess. On the other hand, school climate refers to current feelings and attitudes, and is easier to assess (Gonder, 1994). To facilitate measurement, school climate is used in this study. Conceptualisation of school climate is along the direction of collaboration and experimentation.

School leadership. In the context of managing change in schools to implement ICT strategies to advance changes in student learning, it is appropriate to propose transformational leadership as one of the key contextual factors. First, transformational leadership focuses on increasing the organisation's capacity to innovate (Hallinger & Heck, 1998), and it is also crucial in the context of uncertainty and change as it calls for "commitment" rather than "control" strategies (Leithwood, 1994). Second, while the concept of transformational leadership was subject to varying interpretations in the literature, they have shared a common focus with emphasis on a common concern for school culture (Leithwood & Duke, 1999), which is the pillar of the present theoretical framework. Conceptualisation of transformational leadership is along the direction of establishing shared visions and building innovative capacity.

Government ICT policy. Acknowledging that ICT in education is part and parcel of the overall plan of educational change, the initiative "Information Technology for Learning in a New Era: Five-year Strategy – 1998/99 to 2002/03" explicitly stated its aim as enabling a paradigm shift from a largely textbook-based and teacher-centred approach to a more interactive and learner-centred approach (Education and Manpower Bureau, 1998). Viewing the government initiative as an external input that is going to influence ICT implementation in schools, conceptualisation of government ICT policy is along the direction of how government supports use of ICT in learning and teaching.

ICT implementation strategies. As mentioned in the previous sections, both the micro- and macro-implementation models (McDonnell & Elmore, 1991; McLaughlin, 1991) highlight the concept of capacity building in the form of a workforce unit, which has yielded important insights in conceptualising ICT implementation strategies in the present theoretical framework as the collegial capacity of ICT implementation strategies. In the process of ICT implementation, it has been proposed that a supportive context in schools may offer collegial exchange of professional-development experiences that enhance changes in pedagogical practices (Dexter et al., 2000). Specifically, it has been argued that new pedagogies enter a receptive teacher through collegial ties within the school (Zhao & Frank, 2003). Therefore, conceptualisation of ICT implementation strategies is along the direction of enhancing a collegial exchange of ICT knowledge and of mobilising resources.

Changes in pedagogy and changes in learning. In conceptualising changes in pedagogy and learning, we have taken into consideration some important concepts. First, implementation is actually an ongoing developmental process, in which the outcome will differ depending on the point in the process under study (McLaughlin, 1991). As such, changes in pedagogy and changes in learning are defined along a continuum rather than on an absolute measure. Second, as discussed in the previous sections, learning outcomes are defined more in terms of teaching for understanding and performance (Fullan, 1996), and constructivism is the common lens that people adopt in examining the effectiveness of using ICT in teaching and learning. Hence, conceptualisation of changes in pedagogy and learning will lend its support from the dominant theory of learning: constructivism. Thus, it would be interesting to see whether ICT implementation will gear teaching and learning towards a more constructivist paradigm. Third, constructivist learning and teaching are theories that are difficult to measure. This is partly due to the lack of constructivist teaching theory (Richardson, 2003) or constructivist teaching model (Windschitl, 2002) for translating the constructivist learning theory into practice. As constructivism stresses the importance of each individual's autonomy as a thinker and the importance of the social context of learning, and thereby emphasising the importance of the learner's initiative (Tiene & Ingram, 2001), changes in pedagogy and changes in learning is conceptualised as changes from a teacher-centred approach to a student-centred approach. As a whole, conceptualisation of changes in pedagogy and changes in learning is along the direction of measuring the degree of paradigm shift in pedagogy and learning from a teacher-centred approach to a student-centred approach over a period of two years' time. A period of two years is chosen on consideration of several reasons. First, educational change is a process but not an event (Whitaker, 1993), and we have proposed that a time of at least two years is needed for the ICT implementation to effect a change. Second, it will be practical to request teachers to recall a change not more than two years' time. Third, since the primary purpose of the study is to model the pattern of relationships between the various sociocultural factors pertaining to ICT implementation and changes in student learning, two years' time will suffice to enable the differentiation of the changes in student learning.

Summary

ICT implementation in education is understood as a dynamic process with use of ICT being embedded in a larger context of managing change in schools that requires pedagogical as well as organisational interventions. Implementation is conceptualised along the direction of learning organisation, with the adoption of a situative perspective, a sociocultural approach, and an ecological approach. As such, ICT implementation is framed in an activity system characterised by the interplay between contextual variables that are derived from interaction of the principal, teachers, and students. It is a result of activity between pedagogical intervention and organisational intervention to achieve changes in student learning, and therefore changes in student learning is an outcome variable. Implementation dynamics emerges from the elements interacting in the activity system, in which the collegial capacity of ICT implementation strategies is the central focus. Figure 2.1 helps to depict the theoretical framework.

Activity System: School

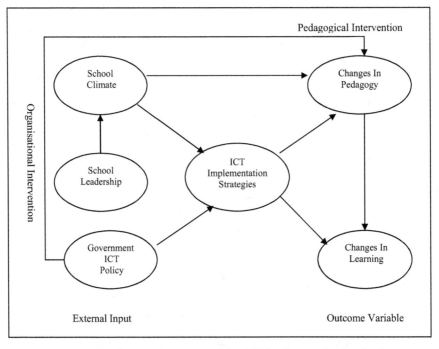

Figure 2.1. Theoretical framework conceptualising ICT implementation in schools.

Multilevel Analysis

While the theoretical framework helps us to conceptualise the dynamics of ICT implementation in schools, it needs to be empirically tested with the help of data. The data set in the present study is in the form of teachers nested within schools. After defining the conceptual variables, they are operationalised as survey instrument measuring perceptions of teachers. With the use of the promising structural equation modelling (SEM) technique, the present study is able to model adequately the complex relationships in the social ecology of ICT implementation in schools given that SEM has the advantage of allowing the researcher to accommodate multiple interrelated dependence relationships in a single model that are not possible with other popular multivariate techniques (Hair et al., 1998). While pooling teacher data across all schools in the structural analysis enables us to examine how the contextual variables and the outcome variable interact with each other, it fails to take into consideration the multilevel nature of such typical kind of educational data (Raudenbush & Bryk, 2002). In other words, if we regard school as a single unit and do a regression analysis for each individual school, we may obtain a very different set of regression coefficients as compared to those obtained from the regression over the data pooled from the entire sample. To overcome the limitation, the study will employ hierarchical linear modelling (HLM) technique to build multi-level models so as to examine the multilevel effects on the outcome variable, to see (1) whether the regression results obtained from each individual school are consistent with those in the SEM analysis and (2) how much variation found in the outcome variable can be attributed to school-level variables.

In building a 2-level model, the present study is able to combine teacher-level variables with school-level variables. In the 2-level model of the present study, "changes in student learning" is viewed as a function of both teacher-level (level-1) variables and school-level (level-2) variables. In other words, variation in "changes in student learning" will be explained in terms of teacher-level variables (perception of teachers) as well as school-level variables (teacher aggregate of the same school and structural characteristics from other sources). At level-1, the units are teachers within the same school, and each teacher's outcome is represented as function of a set of teacher-level variables. At level-2, the units are schools. The regression coefficients (slopes and intercepts) in level-1 model for each school are conceived as outcome variables that are hypothesised to depend on specific school-level variables (Raudenbush & Bryk, 2002).

In sum, multilevel analysis will enable the study to examine, in the social ecology of ICT implementation in schools, how school-level variables account for the variation of changes in student learning among schools. The following 2-level model helps to depict the 2-level analysis on the social ecology of ICT implementation in schools:

$$\text{Level 1: (learning)ij} = \beta_{0j} + \beta_{1j}(\text{age}) + \beta_{2j}((\text{changes in pedagogy})$$
$$+ \beta_{3j}(\text{ICT implementation strategies}) + r_{ij}$$
$$\text{Level 2: } \beta_{0j} = \gamma_{00} + \gamma_{01}(\text{school type}) + \gamma_{02}(\text{mean climate})$$
$$+ \gamma_{03}(\text{mean govt ICT policy})$$
$$+ \gamma_{04}(\text{mean changes in pedagogy})$$
$$+ \mu_{0j}$$
$$\beta_{2j} = \gamma_{20} + \gamma_{21}(\text{school type}) + \mu_{2j}$$

β_{0j} is the school mean for changes in learning for school j,

γ_{00} is the average school mean (grand mean) for changes in student learning,

β_{2j} is the mean slope for changes in pedagogy for school j,

γ_{20} is the grand mean slope for changes in pedagogy across schools,

r_{ij} is the residual error at level-1,

μ_{0j} and μ_{2j} are the residual error at level-2.

Research Questions

The proposed theoretical framework assumes that (1) ICT implementation is embedded in the larger process of school change that aims at advancing changes in student learning; (2) ICT implementation is a dynamic process that needs to be examined in an activity system as the interplay between contextual factors as a result of interactions among students, teachers, and school principal; (3) effective ICT use requires pedagogical as well as organisational interventions; (4) the theoretical framework is limited to verification as from the perception of teachers, that is at the teacher-level. It needs to be modified to a 2-level model that may examine how school-level variables may bring about a paradigm shift in learning among schools. Alternative models based on the theoretical framework will be proposed for empirical testing, and the purpose of which is to address the following research questions:

1. In a context of managing change in schools with ICT implementation, what are the factors affecting changes in student learning from a teacher-centred approach to a student-centred approach?

2. In an overall picture (the social ecology of ICT implementation), how do the above factors interact with one another to bring about the paradigm shift in student learning?

3. In the social ecology of ICT implementation within schools, how does ICT act as a lever to bring about the paradigm shift in student learning?

4. Among schools, what are the organisational features of schooling facilitating the paradigm shift in student learning?

Chapter 3: Research Design and Methodology

Introduction

The goal of this chapter is to examine how data was collected with reference to the sampling method, measures to be used to get the results, and the analysis to be performed on the resulting data. In brief, the sample consisted of Hong Kong primary and secondary school teachers coming from about one tenth of Hong Kong public schools. Data were collected using a mail-out questionnaire survey that consisted of 6 scales measuring changes in student learning, changes in pedagogy, school leadership, school climate, government ICT policy, ICT implementation strategies, as well as several other questions to collect supplementary data. In light of the fast development of ICT that possibly diminishes the value of studying the impact of particular kinds of ICT on student learning, with the help of structural equation modelling (SEM) and hierarchical linear modelling (HLM) techniques, the present study has built a theoretical model of ICT implementation that helped to examine how ICT use was mediated by pedagogical and organisational factors within a context of managing change in schools, with a view to understanding how ICT acted as a lever to bring about a paradigm shift in student learning in terms of teacher's perceptions as well as organisational features of schools. The special feature of the present framework was that it unfolded ICT implementation in the way of mapping the interaction of ICT use with changes in student learning, and with pedagogical and organisational issues.

Research Design

The research was exploratory in nature to empirically test alternative theoretical models that conceptualised and unfolded the dynamics of ICT implementation in schools. In searching for a theoretical model, the research also sought to explain how ICT implementation strategies interacted with organisational intervention and pedagogical intervention to enhance changes in student learning from a teacher-centred approach to a student-centred approach. In this sense, the research objective was exploratory and also explanatory. Quantitative approach was used for the research, with a cross-sectional questionnaire survey administered to teachers, school principals, and heads of IT team from a convenient sample of schools located in various districts in Hong Kong. The convenient sample would suffice to serve the primary aim of the present study, which was to deepen our understanding of the sociocultural factors affecting ICT implementation in and across schools rather than to make generalisation over the entire Hong Kong school population.

In building the models, we relied on teacher perceptions on the contextual factors and we also used teacher judgements of student learning as an outcome variable. While studies based on perceptions of teachers have its limitations with regard to the actual impact on teaching and learning, we believe perceived impacts are important to examine in school studies. First and foremost, the gist of the study is not on measuring improvements in traditional processes and knowledge as measured by attainment tests, but rather on discerning whether there are changes in teaching and learning towards a constructivist approach or student-centred approach as a result of ICT implementation. While many education systems recognise the constructivist approach in teaching and learning, few have formally adopted new modes of assessment with greater emphasis in measuring learning attitude and learning ability in terms of active construction of knowledge and collaborative work (Balanskat et al., 2006; Valli, Cooper, & Frankes, 1997). To overcome the inherent difficulties in measuring changes in teaching and learning towards a student-centred approach as a result of ICT use, some researchers resort to gathering evidence from perceptions of teachers. Such opinion-based approach has been regarded as informative and was used by a number of Nordic studies in investigating perceived impacts of ICT on student learning such as team work, independent learning and critical thinking (Balanskat et al., 2006). In fact, the practice of eliciting perceptions of participants is quite common and has been reflected by a huge body of literature on leadership, school improvement and managing change (Owens, 2001; Smylie & Hart, 1999). According to Owens (2001), while perceptions themselves are not objective reflections of "reality", the point is that whatever people in the organisation perceive as their experience is the reality to be described.

Along this direction, the key issue was to obtain an objective description of those perceived behaviour. Therefore, construct validity was central to the present study as it relied heavily on the use of questionnaires as the primary means of data collection (Hinkin, 1998), and also on consideration that construct validity is a necessary condition for theory development and testing (Peter, 1981). The researcher would therefore use a whole section of survey instrument to address content validity, and a whole chapter (chapter 4) to address reliablilty, unidimensionality, convergent validity, and discriminant validity of the constructs. In the HLM analysis, apart from expanding the data pool from teachers to school principals as well as heads of school IT team, school has also been used as a unit of analysis rather than individual teachers. This would help to reduce the possible spurious results arising from individuals. Specifically concerning the outcome variable "changes in student learning", as the primary purpose of the study is not to have fine differentiation of students' learning outcomes but to draw a demarcation line for discerning changes exhibited in student learning, probing teachers' perceived changes in student learning is thus sufficient for the present analysis.

The major concerns that guided the present research design were considerations of (1) adopting a situative perspective to conceptualise schoolwide ICT implementation as a result of social and ecological interaction between pedagogical intervention and organisational intervention to achieve changes in student learning; (2) adopting SEM technique to construct a theoretical model delineating the relationships between the outcome variable "changes in student learning" and other interacting factors; (3) adopting HLM technique to reveal the effect of various organisational setting (school-level variables) on changes in student learning.

A combination of cross-sectional survey with statistical analysis was considered as appropriate in the present study because of the considerations of: (1) the feasibility issue: questionnaire survey allowed the collection of large amount of data from a selected sample within a relatively shorter period of time; (2) the methodological power of using the statistical technique of SEM. SEM allowed the researcher to model not only direct relationship but complex interrelationships among variables as it could provide estimates of the strength of all the hypothesised relationships between variables in a theoretical model. In other words, SEM could provide information about hypothesised impact, both directly from one variable to another and via intervening or mediating variables (Maruyama, 1998). In a theory-driven way of testing alternatively proposed models, SEM allowed the determination of whether the model might be considered a plausible representation of data; (3) the methodological power of using the statistical technique of HLM to address the unit of analysis problem. With the use of HLM, multi-level models would be constructed so as to obtain estimation of effects within schools without pooling data across all schools, and estimation of effect across schools without aggregating data from individuals within each school by obtaining a mean (Heck & Thomas, 2000).

The study could be summarised into four stages: instrument development, instrument validation, structural equation model building, and multilevel model building. First, it began with conceptualisation of the variables that build up the model of social ecology of ICT implementation, and then operationalised the concepts by generating an initial set of items for the survey instrument. Second, after collection of data, the instrument's reliability and validity were examined and refined by tests specifically designed for factor analysis, reliability, convergent validity, and discriminant validity. Third, pooling all the teacher data in an analysis, a structural equation model was constructed to delineate the multiple interrelated dependence relationships among the variables so as to uncover how ICT implementation strategies and pedagogical as well as organisational variables affected changes in student learning. Fourth, as an extension to the SEM model, a HLM analysis was carried out to examine whether the regression results obtained from each individual school are consistent with those in the SEM analysis, and also how much variation found in changes in student learning (outcome variable) could be attributed to school-level variables.

Sample and Procedure

The current study worked on part of the data collected from a large scale questionnaire survey that covered multiple sources including school principals, heads of ICT team, subject panel chairmen, teachers and students. In this study, most of the analysis was conducted on the part of the quantitative data collected from teachers. Therefore, the data analysis largely represented the perception from teachers. However, in the HLM analysis, the source of data was expanded to school principals and heads of school IT team as well.

The working population was defined as the public primary and secondary schools in HKSAR, and all of them were invited to participate in the survey. Out of the working population, a total of 1076 teachers from 130 schools were recruited on voluntary basis, which was about one tenth of the school population. The sample of recruited schools therefore was a convenient sample located in all 18 school districts in Hong Kong. Convenient sampling suffices to address the research questions posed since the gist of the entire study is not to make generalisation over the entire population in Hong Kong but to examine the dynamics of interplay of various constructs pertaining to ICT implementation and learning in schools. In terms of distribution according to finance types (government, direct-subsidy scheme, and aided), the sample population closely resembled the true population. Primary schools made up 53% of the sample, and secondary schools 47%, the distribution of which was quite similar to the true population (60% primary schools and 40% secondary schools). To increase the representativeness, teachers in each school were selected from diversified subject domains including languages, mathematics, humanities, science, technology, art, and physical education. In each primary school, 10 teachers coming from specified subject domains were selected by schools to fill out the questionnaires; and in each secondary school, 9 teachers from specified subject domains were selected by schools to fill out the questionnaires. Of the 1076 teachers, 56.3% were primary school teachers, while 43.7% were secondary school teachers. The sample consisted of relatively more female participants (61.8%). Ages were in the range from 18-25 years old to over 50 years old, with the majority (35.4%) at the age of 31-40, and the minority (7.8%) over 50. Teaching experience of the sample was of the range of below 3 years to over 20 years, and the majority (39.1%) had 4-10 years of teaching experience. Overall, nearly half of the sample population (42.8%) had a teaching experience of over 10 years. Details were shown in Table 3.1.

Mail-out survey was used in this study as an instrument to collect data, and the questionnaire adopted 4-point Likert scale questions having close-ended response choices. Letters of invitation were sent by fax to target school principals, and they were requested to indicate their acceptance by returning a fax or registering via Internet. After securing confirmation, questionnaires were delivered to each school by mail with an enclosed business return envelope for returning the completed questionnaires in one month's time. Informed consent was sought throughout the whole process, firstly via the letter of invitation sent to school principals, and then via an opening paragraph in the questionnaires to teachers. Participants were informed of the purpose of the research, and were reassured that anonymity could be kept throughout the whole process. In return, the research findings would be shared with the schools.

Table 3.1
Demographic Data of the Sample Teachers

		No. of participants	Percent (%)	Cumulative %
Age	18 - 25	134	12.5	12.5
	26 - 30	309	28.8	41.3
	31 - 40	380	35.4	76.7
	41 - 50	166	15.5	92.2
	Over 50	84	7.8	100.0
	Total	1073		
Teaching experience	0 – 3 years	195	18.1	18.1
	4 – 10 years	421	39.1	57.2
	11-15 years	209	19.4	76.7
	16 – 20 years	97	9	85.7
	Over 20 years	154	14.3	100.0
	Total	1076		
Sex	Male	411	38.2	38.2
	Female	664	61.8	100
	Total	1075		

Survey Instrument

Overview. Generating an initial set of items for the survey instrument is an important first step to attaining construct validity, and therefore considerably more attention was given to the conceptual definition of the instrument (Peter, 1981). Deductive scale development has been a common approach for creating preliminary items to assess the constructs under examination, in which theoretical foundation is expected to provide enough information to generate the initial set of items (Hinkin, 1998). With the approach of deductive scale development guided by a thorough literature review and expert opinions, the initial set of items for the survey instrument was conceptually defined on theoretical basis that would indicate the content domains for each construct. As it was not possible to measure the complete domain of interest, the domain sampling theory was adopted, which aimed at generating sample of items that adequately represented the construct under examination (Hinkin, 1998). The set of items was then subjectively assessed by experts to ensure content validity (Hair et al., 1998; Hinkin, 1998). The deductive approach was considered to be appropriate for the present study as there being relevant contemporary education theories as a starting point to generate the scales. A small-scale pilot test was then conducted to refine the instrument before the questionnaire administration, mainly in terms of clarity in wording. The survey instrument for studying the social ecology of ICT implementation in schools consisted of the following six scales and some other supplementary data:

Scales

1. Perceived school leadership

2. Perceived school climate for collaboration and experimentation

3. Perceived effectiveness of government ICT policy

4. Perceived collegial capacity of school's ICT implementation strategies

5. Perceived changes in teacher pedagogy

6. Perceived changes in student learning

Data

1. Sex

2. Age

3. Teaching experience

4. School type (primary school or secondary school)

5. School's ICT infrastructure

6. School's ICT competency

Questions collecting supplementary data provided categorical options that collected information on teachers' social background and school's structural characteristics, as shown in Appendix 2. On the other hand, on average, each of the six scales was made up of about 6 items, with a total of 42 items (ultimately reduced to 33 items) on the whole, as shown in Appendix 1. All items were measured on a 4-point Likert scale ranging from "strongly disagree" to "strongly agree". This type of ordinal scales measuring the intensity of feeling toward the item has been regarded as generating more information than dichotomous scoring, which facilitates statistical analysis by more faithfully reflecting the individual differences on the attribute (Nunnally & Bernstein, 1994). On the other hand, statistical analyses are usually based on Pearson product-moment correlation (PPM) matrix that assumes the input data as variables of continuous scale. In fact, variables based on ordinal response modes from Likert scale measurements actually depart from the representation of a continuous scale (Joreskog & Sorbom, 2001). To overcome the limitation, the present study used a polychoric correlation matrix as the input matrix for confirmatory factor analysis and structural equation modelling (Flora & Curran, 2004; Joreskog & Sorbom, 2001). Assuming that there was a continuous variable underlying each ordinal variable, polychoric correlations were not correlations computed from actual scores but were rather estimated theoretical correlations of the underlying continuous variables (Joreskog & Sorbom, 2001). The next section on data analysis and the method section in Chapter 4 will give more details on the issue.

School leadership scale. As mentioned in the theoretical framework section of Chapter 2, conceptualisation of leadership was along the direction of transformational leadership in establishing shared visions and building innovative capacity. In this way, the leadership scale in this study measured the perceived school leadership with a focus on measuring the school principal's transformational leadership in four domains: (1) the capability of establishing shared visions, (2) the capability of trusting and empowering teachers, (3) the capability of supporting innovations in classroom practices, and (4) the capability of supporting staff development (see Table 3.2).

Table 3.2
School Leadership Scale

Domain	Item	Questions
Establishing shared visions	Ld1	My educational beliefs are reflected in the school goals.
	Ld2*	The school goals are reflected in school plans.
Trust and Empowerment	Ld3	The principal trusts me with school matters.
	Ld4	Teachers have autonomy to make decisions relevant to their teaching.
Support for innovation	Ld5	The principal encourages me to experiment with new ideas in classroom practice.
Support for staff development	Ld6	My school provides adequate resources to support staff professional development.
	Ld7	The principal values staff professional development.

Items marked with * were eventually discarded

School climate scale. School climate referred to the atmosphere in a school, and was generally considered to be positive or negative (Gonder, 1994). Specifically, school climate has been described as the quality of a school that stimulates teachers' creativity, enthusiasm, and sense of belonging (Freiberg & Stein, 1999), in particular the collegial relationships (Reinhartz & Beach, 2004). As conceptualisation of school climate in the present study was along the direction of collaboration and experimentation, it also included the domains of striving for progress among teachers and flow of information within the school, in addition to the traditional domains of collegial relationships. In a context of transformational leadership that focuses on increasing the organisation's capacity to innovate (Hallinger & Heck, 1998), it has been believed that a great deal of freedom is given to teachers and others as to how essential core values are to be realised (Sergiovanni, 1995). Thus, an open and efficient communication between school principals and teachers may help to establish a collaborative school climate with innovative capacity, and that's why the domain of flow of information within school was included here. With a socio-cultural perspective, the school climate scale of the present study measured among teachers their (1) collegial relationships, (2) their spirit to strive for progress, and also (3) the flow of information within school (see Table 3.3). The domain flow of information was eventually removed after construct validation (please refer to Chapter 4 for details).

Table 3.3
School Climate Scale

Domain	Item	Questions
Flow of information	Cm1#	I feel comfortable to talk to the principal about school matters.
	Cm2*	My opinions can be conveyed to policy-making units (e.g. school management board/committees/departments) effectively within school.
Collegiality	Cm3	I have a sense of belonging to my school.
	Cm4	Teachers in our school work in a collegial manner.
Striving for progress	Cm5	Teachers in our school are willing to experiment with new ideas in classroom practice.
	Cm6*	I share with colleagues my experiences of improving classroom practice.

Items marked with * were eventually discarded, with # was moved to leadership.

Government ICT policy scale. The government ICT policy was viewed as an external input influenced ICT implementation in schools, and its conceptualisation was along the direction of enhancing learning and teaching (Chapter 2). While the government ICT policy might directly affect learning and teaching, the influence might also be indirect through, for instance, curriculum and resource support and ICT access and connectivity. In this way, the scale for government ICT policy measured the perceived effectiveness of the initiative "Information Technology for Learning in a New Era: Five-year Strategy – 1998/99 to 2002/03" (Education and Manpower Bureau, 1998), with particular reference to effectiveness in enhancing four domains: (1) teaching, (2) curriculum and resource support, (3) ICT access and connectivity, and (4) learning (see Table 3.4).

Table 3.4
Government ICT Policy Scale

Domains	Item	Questions
Teaching	Gp1	The initiative provides adequate professional development for teachers.
	Gp8	The initiative helps to strengthen teachers' quality of teaching.
Curriculum and resource support	Gp3	The initiative enables schools to have autonomy in allocating or recruiting resources relevant to ICT.
	Gp5	The initiative provides adequate support on curriculum resources for teachers.
ICT access and connectivity	Gp4	The initiative provides adequate hardware facilities for schools.
	Gp6	The initiative provides sound network infrastructure for schools.
	Gp2	The initiative provides adequate technical support for schools.
Learning	Gp7	The initiative enables students to learn more effectively.

Collegial capacity of ICT implementation strategies scale. As mentioned in the theoretical framework section of Chapter 2, conceptualisation of ICT implementation strategies was along the direction of providing a supportive ICT context in schools that might offer collegial exchange of professional-development experiences that enhanced changes in pedagogical practices (Dexter et al., 2000; Zhao & Frank, 2003). While the supportive context might be in a direct manner of providing formal and informal channels of exchanging knowledge and ideas, it might also be an indirect one of capacity building in the form of pulling together human resources and experiences. Along this line of reasoning, the scale measured the capacity of the school's ICT implementation strategies in two domains: (1) enhancing a collegial exchange of ICT knowledge, and (2) enhancing the mobilisation of resources (see Table 3.5).

Table 3.5

Collegial Capacity of ICT Implementation Strategies Scale

Domains	Item	Questions
Mobilising resources	Is1	My school has a mechanism to disseminate the experiences of using ICT resources for teaching and learning.
	Is2*	The principal encourages teachers to experiment with new ICT practices.
	Is3	My school has mobilised resources from external parties (e.g. parents/ alumni/ other schools/ organisations) to help to implement ICT in teaching and learning.
	Is4*	The principal is willing to release teachers to attend training/workshops/conferences on using ICT in teaching and learning.
Collegial exchange of ICT knowledge	Is5	Teachers' opinions can be conveyed to ICT policy-making bodies effectively within school.
	Is6	Colleagues in my school exchange experiences of using ICT to enhance teaching and learning.
	Is7	I participate in sharing sessions for exchanging experiences (of using ICT to enhance teaching and learning) with teachers from other schools.
	Is8	I participate in sharing sessions for exchanging experiences (of using ICT to enhance teaching and learning) with educators from tertiary institutions.

Items marked with * were eventually discarded

Changes in teacher pedagogy scale and changes in student learning scale. As mentioned in the theoretical framework section of Chapter 2, changes in pedagogy and changes in learning were conceptualised with reference to the HKSAR official document "Learning to Learn: Life-long Learning and Whole-person Development" (Curriculum Development Council, 2001). The scales measured the degree of paradigm shift in pedagogy and learning from a teacher-centred approach to a student-centred approach over a period of two years' time, which was parallel to changes in classroom practices from a traditional approach to a constructivist approach. Apart from changes in classroom practice, it was believed that changes in curriculum content as well as changes towards a collaborative kind of teaching would enhance changes in pedagogy from a teacher-centred approach to a student-centred approach. Hence there were three domains indicating changes in pedagogy: (1) curriculum, (2) collaborative teaching, and (3) classroom practice (see Table 3.6). The domains curriculum and collaborative teaching were eventually removed after construct validation (please refer to Chapter 4 for details).

Table 3.6
Changes in Teacher Pedagogy Scale

Domains	Item	Questions
Curriculum	Pd1*	Compared to the past two academic years, our school curriculum is putting more emphasis on infusing generic skills (e.g. critical thinking/creativity/ communication skills) into teaching and learning of various subjects.
Collaborative teaching	Pd2*	Compared to the past two academic years, I am more willing to have collaborative lesson preparation with my colleagues.
	Pd3*	Compared to the past two academic years, I am more willing to participate in peer classroom observation.
Classroom practice	Pd4	Compared to the past two academic years, I am creating more opportunities for discussions to develop students' expressive and analytical abilities.
	Pd5	Compared to the past two academic years, I am encouraging students to explore and to inquire in learning more.
	Pd6	Compared to the past two academic years, I am providing more opportunities for students to determine their learning activities.

Items marked with * were eventually discarded

In addition to learning ability, it was suggested that students' attitude in learning as well as the learning process were also important (Curriculum Development Council, 2001). Hence the changes in learning scale focused in measuring the process of learning, in which the learning ability and attitude were both considered. Learning ability was concerned with competency in generic skills and active learning, whereas learning attitude was concerned with the demonstrating motivation and enthusiasm in the process of learning. In this way, perceived changes in student learning captured domains of (1) learning ability and, (2) learning attitude (see Table 3.7).

Table 3.7

Changes in Student Learning Scale

Domains	Item	Questions
Learning ability: generic skills and active learning	Ln1*	Compared to students of the same level in the past two academic years, my students are more competent in mastering various generic skills (e.g. collaboration skills, communication skills, creativity, critical thinking skills, information technology skills, numeracy skills, problem-solving skills, self-management skills, study skills).
	Ln2	Compared to students of the same level in the past two academic years, my students are more independent in their learning.
	Ln3	Compared to students of the same level in the past two academic years, my students are more active in constructing knowledge.
	Ln4	Compared to students of the same level in the past two academic years, my students are able to make better use of collaborative work to facilitate learning.
Learning attitude: motivation and enthusiasm	Ln5	Compared to students of the same level in the past two academic years, my students have more courage to express ideas in class.
	Ln6	Compared to students of the same level in the past two academic years, my students are more motivated in their learning.
	Ln7	Compared to students of the same level in the past two academic years, my students enjoy learning more.

Items marked with * were eventually discarded

Data Analysis

Data analysis fell into 3 major parts. The first part was to validate the instrument (Chapter 4), the second part was to test alternative proposed theoretical models using SEM (Chapter 5), and the last part was to build a 2-level model to investigate the within-school and between-school effects (Chapter 6). The following sections will give a brief account on the methodology used, while most of the remaining details will be discussed in the method section in Chapter 4, 5 and 6.

Instrument validation. The survey instrument contained items that represented abstract concepts termed constructs. Constructs are variables that are abstract and latent rather than concrete and observable (Nunnally & Bernstein, 1994). As suggested by Nunnally & Bernstein (1994), there are three major aspects of construct validation: (1) specifying the domain of the construct, (2) empirically determining the extent to which items measure that domain, and (3) examining the extent to which the measure produces results that are predictable from theoretical hypotheses. The first aspect is concerned with content validity issue, which has already been discussed in the section of theoretical framework in Chapter 2 and in the above section of survey instrument. The last aspect of examining the extent to which the measure produces results that are predictable from theoretical hypotheses is concerned with nomological validity, which will be discussed in the section below and Chapter 5 concerning model building using SEM. The rest of this section and Chapter 4 will concentrate on the second aspect of empirically determining the extent to which items measure the domain. Overall, instrument validation covers four major aspects: (1) unidimensionality, (2) reliability, (3) convergent validity, and (4) discriminant validity. Achieving unidimensionality is a crucial undertaking in scale development, which refers to the condition that each set of alternate indicators has only one underlying construct in common (Anderson & Gerbing, 1988). Other than unidimensionality, validity and reliability are two basic but different characteristics of measures, with validity addressing the issue of the direct correspondence between a measure and a concept, and reliability assessing the consistency of a measure regardless of whether it is valid (Bollen, 1989). Of construct validity, the most widely accepted forms of validity will be examined: convergent validity and discriminant validity (Hair et al., 1998).

As a preliminary analysis, the factor structure of the instrument was first examined by exploratory factor analysis using SPSS. Reliability of the constructs was then examined by Cronbach alpha test using SPSS and structural equation-based approach of composite reliability from confirmatory factor analysis (CFA) using LISREL. Unidimensionality, convergent validity and discriminant validity was then examined by structural equation-based approach using LISREL. As the structural equation-based approach has been considered as a powerful method for addressing construct validity (Anderson & Gerbing, 1988; Bagozzi et al., 1991; Bollen, 1989), the present study relied mainly on the structural equation-based approach and its complementary methods for construct validation.

The basic assumption underlying the standard use of CFA in structural equation modelling is that the observations are drawn from a continuous and multivariate normal population, yet researchers often ignore the assumptions (Flora & Curran, 2004; West, Finch, & Curran, 1995). With the use of Pearson's product-moment (PPM) correlation matrix as an input matrix for CFA, it assumes that the data is continuous. As the data obtained in the present study was based on ordinal response modes from Likert scale measurements, the variables actually departed from the representation of a continuous scale (Joreskog & Sorbom, 2001). To overcome the limitation, the present study used a polychoric correlation matrix as the input matrix (Flora & Curran, 2004; Joreskog & Sorbom, 2001; Wang & Cunningham, 2005). Assuming that there was a continuous variable underlying each ordinal variable, polychoric correlation were not correlations computed from actual scores but were rather estimated theoretical correlations of the underlying continuous variables (Joreskog & Sorbom, 2001).

Regarding the issue of estimation methods in structural equation modelling, the commonly used estimation methods of maximum likelihood (ML) and generalised least squares (GLS) are developed under multivariate normality assumption (Chou & Bentler, 1995; Raykov & Marcoulides, 2000). Nevertheless, real social science data may depart from the normality assumption, as revealed in a study of over 400 large data sets. Findings revealed that the great majority of data collected in behavioural research were not normally distributed (West et al., 1995). While ML estimates have been found to be quite robust to the violation of normality, simulation studies showed that ML yielded very unsatisfactory results under extreme nonnormality (Chou & Bentler, 1995; West et al., 1995). On the other hand, two alternative estimation methods, namely weighted least squares (WLS) and diagonally weighted least squares (DWLS), do not make assumptions about the distribution of the observed variables (Diamantopoulos & Siguaw, 2000; Flora & Curran, 2004). Despite the advantage of distribution-free nature for observed variables, WLS has been shown to behave poorly in small sample sizes, particularly in large models (Muthen, 1993; Wang & Cunningham, 2005). In contrast, support in terms of parameter estimates and goodness-of-fit statistics has been gained from CFA analyses using polychoric correlations with DWLS estimation when analysing ordinal nornormal data (Wang & Cunningham, 2005).

As mentioned above, the instrument of the present study adopted a Likert scale measurements that were ordinal in nature, and on consideration of the usually non-normality nature of the real social science data, the present study employed DWLS estimation with polcychoric correlations for CFA. In evaluating the CFA models, parameter estimates related to indicators of the CFA model were examined, and the higher the value with respect to factor loading, t-value, and the squared multiple correlation (R^2) on its target construct, the better the indicator representing the latent variable concerned (Bollen, 1989; Diamantopoulos & Siguaw, 2000). In evaluating the overall fitness of the model for CFA, in addition to the traditional inferential goodness-of-fit index of chi-square value, a number of goodness of fit indexes were used (Browne & Cudeck, 1993; Diamantopoulos & Siguaw, 2000; Kelloway, 1998; Marsh, Balla, & Hau, 1996), in particular the root mean square error of approximation (RMSEA), comparative fit index (CFI), and non-normed fit index (NNFI).

Structural model building. SEM was used to model the complex relationships in the social ecology of ICT implementation in schools given that SEM has the advantage of allowing the researcher to accommodate multiple interrelated dependence relationships in a single model that are not possible with other popular multivariate techniques (Hair et al., 1998). In addition, structural equation modelling technique has advantage over traditional regression analysis as it provides a mechanism for explicitly taking into account measurement error in the variables (Hoyle, 1995; Jaccard & Wan, 1996; Raykov & Marcoulides, 2000). Another compelling characteristic of the SEM technique is its capacity to estimate and test relations between latent variables that are implied by covariances among multiple indicators (Hoyle, 1995).

A basic model was proposed and examined with SEM techniques using LISREL 8.54 (Joreskog & Sorbom, 2001). The analysis adopted the two-step approach for analysing the postulated model (Anderson & Gerbing, 1988), where a confirmatory measurement model was specified prior to the simultaneous estimation of the measurement and the structural model. In addition, the present study adopted the model comparison strategy as a means for evaluating the proposed models, in which competing models were generated on the basis of alternative formulations of the underlying theory (Browne & Cudeck, 1993; Hair et al., 1998; MacCallum, 1995). In addition, after identifying the most plausible model among the alternative models, the most plausible model was further tested using the method of nested model comparison.

In assessing alternative models, the present study adopted an approach of considering criteria that took fit as well as parsimony (in the sense of number of parameters) into account, of which three strongly related criteria were used: the Akaike Information Criterion (AIC), the Consistent Akaike Information Criterion (CAIC), and the Expected Cross-Validation Index (ECVI) (Joreskog, 1993). The essence of the indexes was to examine measures of fit that took the number of parameters in the model into account, thereby the fit of model was not necessarily improved as parameters were added to the model (Browne & Cudeck, 1993). It must be however reminded that the indexes should be assessed together with other considerations such as the plausibility of the model and the meaningfulness of its parameters (Bollen, 1989; Browne & Cudeck, 1989; Hu & Bentler, 1995). For this reason, parameter estimates were examined, which included the squared multiple correlation (R^2) for each pair of relationships and the t-value of the path coefficients.

Multilevel model building. HLM (Raudenbush, Bryk, Cheong, & Congdon, 2001), a multilevel analytic method, was used to build a 2-level model to deepen our understanding on the SEM model, with particular emphasis on the school-level effect. To address the school-level effect, the HLM analysis had the advantage of providing a means of partitioning the outcome variable's variance into within-units (teacher level: variance among teachers within the same school) and between-units (school level: variance among schools); and within the analysis, a means of assigning variables to different levels (Heck & Thomas, 2000).

To examine the school effect on the outcome variable, we adopted an exploratory procedure to select a hierarchical model in a systemic manner as suggested by Hox (2000). Exploratory in a sense that we found a plausible model through the means of constructing alternative level-2 models based on different proposition for comparison. A "step-up" strategy was adopted in the exploratory process of model building, with theoretical guidance to define a small set of variables. The "step-up" strategy referred to starting with the simplest possible model, and then adding various types of parameters step by step (Hox, 2002). The SEM model derived in Chapter 5 was used as the cornerstone for formulating the level-1 (teacher-level) model for the HLM analysis. In addition, teacher-level variables should also include some demographic data of teachers to adjust for the differences among schools. According to Raudenbush & Bryk (2002), such kind of statistical adjustments for individual background are very important for two reasons. First, because teachers are not usually assigned at random to schools, controlling for background may reduce its bias on estimation of school effects. Second, if variables of individual background are strongly related to the outcome variable (changes in student learning), controlling for them will increase the precision of any estimates of school effect and the power of hypothesis tests by reducing unexplained level-1 error variance.

In selecting school-level (2-level) variables, in addition to those aggregate as derived from variables used in the SEM model building, the researcher decided to include some relevant structural characteristics of the school that were not derived from teacher aggregate but from sources like heads of ICT team and school principals. The rationale behind was to widen the context and source of school-level factors. Ultimately, we had two types of school-level variables coming from a variety of sources, those measuring the sociocultural factors relating to ICT implementation as identified in the SEM model, and those measuring structural characteristics of schools.

With the intention of identifying the source of variations within the outcome variable "changes in learning", the first step was to build a fully unconditional model to partition the total variance in the outcome variable into its within- and between-school components (Raudenbush & Bryk, 2002). The exploration then started with school-level variables being held aside first, and data analysis focused on adding teacher-level variables to the fully unconditional model (Raudenbush & Bryk, 2002). At this juncture, it was also necessary to decide whether a teacher-level coefficient should be conceived as fixed or random. As suggested by Raudenbush & Bryk (2002), the decision would be based on theoretical considerations as well as the reliability of the random level-1 coefficient as well as the p-value of its estimated level-2 variances. The reliability of the estimated level-1 coefficient measured the ratio of the true score relative to the observed score, and the p-value of its estimated level-2 variances indicated the probability of rejecting the hypothesis of slope homogeneity (chi-square and likelihood-ratio tests). Afterwards, school-level variables were introduced into level-2 of the model, in which the random level-1 coefficients were regressed on the school-level variables. Alternative models were built on theoretical consideration. In the process of model comparison, the proportion of reduction in variance as accounted

for by the model would serve as the basis for making judgement about the importance

of variables (Raudenbush & Bryk, 2002).

Chapter 4: Results (Part 1) – Instrument Reliability and Validity

Introduction

This chapter is going to demonstrate the unidimensionality, reliability and validity of the instrument for measuring the constructs of school leadership, school climate, effectiveness of government ICT policy, collegial capacity of ICT implementation strategies, changes in pedagogy, and changes in student learning. These abstract concepts are termed constructs as they are variables that are abstract and latent rather than concrete and observable (Nunnally & Bernstein, 1994). Achieving unidimensionality is a crucial undertaking in scale development, which refers to the condition that each set of alternate indicators has only one underlying construct in common (Anderson & Gerbing, 1988). Other than unidimensionality, validity and reliability are two basic but different characteristics of measures, with validity addressing the issue of the direct correspondence between a measure and a concept, and reliability assessing the consistency of a measure regardless of whether it is valid (Bollen, 1989).

Construct validity was a core issue here as the present study relied heavily on the use of questionnaires as the primary means of data collection (Hinkin, 1998), and the present study also aimed at theory development and testing (Peter, 1981). As suggested by Nunnally & Bernstein (1994), there are three major aspects of construct validation: (1) specifying the domain of the construct, (2) empirically determining the extent to which items measure that domain, and (3) examining the extent to which the measure produces results that are predictable from theoretical hypotheses. The first aspect is concerned with content validity issue, which has already been discussed in the section of theoretical framework in Chapter 2 and the section of survey instrument in Chapter 3. The last aspect of examining the extent to which the measure produces results that are predictable from theoretical hypotheses is concerned with nomological validity, which will be discussed in Chapter 5 concerning model building using structural equation modelling (SEM). The present chapter is going to concentrate on discussing the second aspect of empirically determining the extent to which items measure the domain, and the most widely accepted forms of validity will be examined: convergent validity and discriminant validity (Hair et al., 1998).

As a preliminary analysis, the factor structure of the instrument was first examined by exploratory factor analysis using SPSS. Reliability of the constructs was then examined by Cronbach alpha test using SPSS. Unidimensionality, convergent validity and discriminant validity was then examined by structural equation-based approach using LISREL. As the structural equation-based approach has been considered as a powerful method for addressing construct validity (Anderson & Gerbing, 1988; Bagozzi et al., 1991, Bollen, 1989), the present study relied mainly on the structural equation-based approach and its complementary methods for construct validation.

Methods and Results

Exploratory Factor Analysis Examining Factor Structure

Exploratory factor analysis (EFA) was used as a preliminary technique to factor the overall set of items, and the resulting factor loadings were then used as a reference, in addition to theoretical basis, to construct the scales for studying the social ecology of ICT implementation in schools (Gerbing & Anderson, 1988).

Exploratory factor analysis (EFA) is a method that attempts to determine the number of latent variables (factors or constructs) that adequately explain the variation and covariation in a set of observed variables (Preacher & MacCallum, 2003). The technique is exploratory in a sense that it does not impose a structure on the relationship between the observed variables and the latent variables (Muthen & Muthen, 2005). However, the EFA model that is tested is underidentified, in other words, there is no unique solution but rather an infinite number of possible solutions that fit the data equally well. Despite this, EFA is still valuable when used in anticipation of use of the hypothesis testing technique confirmatory factor analysis (CFA) (Maruyama, 1998). As an initial step to refine the measurement scales and thereby providing evidence of construct validity, SPSS exploratory factor analysis was thus conducted to assess the factor structure of the variables comprising the scales (Hair et al., 1998; Hinkin, 1998).

In choosing the method of analysis, maximum likelihood method was used instead of principal component method of analysis since the principal component method assumed that there was no unique variance and therefore the error part of the model disappeared (Maruyama, 1998). About rotation methods, orthogonal rotation methods restricted the factors to be uncorrelated, while oblique methods allowed correlated factors (Preacher & MacCallum, 2003). The aim of the present study was to trace the relationship among the theoretical constructs (factors), and factors were assumed correlating with each other. As a result, it was safe to use oblique rotation instead of orthogonal one.

Prior to conducting EFA, the suitability of the data for factor analysis was assessed by sample size requirement, case-to-item ratio, Kaiser-Mayer-Olkin (KMO) measure of sampling adequacy, and the Bartlett's test of sphericity (Pallant, 2001).

For sample size requirement, the present sample size of 1076 was greater than the 300

as recommended by Tabachnick and Fidell, and the case-to-item ratio of 24:1 was

also greater than the 10:1 ratio recommended by Nunnally (Pallant, 2001). For

strength of the relationship among the items, the KMO measure of sampling adequacy

was 0.9, which was greater than the minimum requirement of 0.6 as suggested by

Tabachnick and Fidell, and the Bartlett's test of sphericity was also statistically

significant at p = .000 (Pallant, 2001). The results indicated that the data set was

suitable for conducting EFA.

Using SPSS 11.0 for Windows, the 42 items of the 6 scales (Appendix 1) were

subjected to maximum likelihood analysis with oblique rotation method. In

identifying the number of interpretable factors among the set of variables, theoretical

basis in addition to fitness of model were taken into account (Muthen & Muthen,

2005). In view of that, multiple criteria were used to retain the number of factors, such

as eigenvalues, scree plot, factor loadings as well as the underlying theoretical

justifications (Hair et al., 1998). Initially, the analysis revealed the presence of 9

components with eigenvalues exceeding 1. On consideration of the underlying

theoretical justifications (Hinkin, 1998; Muthen & Muthen, 2005), an attempt for

extraction of 6 factors was conducted. The results however indicated that the

leadership scale and climate scale tended to mingle in a single factor.

An inspection of the scree plot revealed a clear break after the fifth component.

Therefore, another attempt for extraction of 5 factors was conducted. The preliminary

results of EFA with extraction of 5 factors were satisfactory, despite items for scales of climate and leadership mingling in a single factor. Subsequently, factor loadings were evaluated on the basis of obtaining a statistical power of 80 percent and at the significance level of .05. For a sample size of 350 or greater (the current sample size was 1076), a factor loading of .30 and above could be considered significant (Hair et al., 1998). As indicated in Table 4.1, all factor loadings were higher than .30 except Cm6, which was therefore discarded. Cross loading occurred in some items, in particular for those belonging to the construct of changes in pedagogy and the construct of changes in learning. As EFA was a preliminary screening process, items with cross-loadings were retained on the theoretical basis.

On the other hand, three items (Is2, Is4, and Is5) loaded on a factor that was not consistent with the underlying theory. Is2, Is4, and Is5 were supposed to measure the collegial capacity of ICT implementation strategies, but they loaded on the factor of leadership and climate. Is2 stated that the principal encourages teachers to experiment with new ICT practices, and Is4 stated that the principal is willing to release teachers to attend training/ workshops /conferences on using ICT in teaching and learning. On careful consideration, Is2 and Is4 were discarded because they were conceptually ambiguous, in a sense that they overlapped with the concept of school leadership. Is5 stated that teachers' opinions can be conveyed to ICT policy-making bodies effectively within school, and it was retained on theoretical consideration, in a sense that Is5 was important in measuring the collegial capacity of ICT implementation strategies that has been conceptualised along the direction of enhancing the collegial exchange of ICT knowledge. In the process of ICT implementation, staff involvement in ICT policy making was regarded as an important move in creating a supportive context in schools that might offer collegial exchange of professional-development experiences. In the end, with Is5 reassigned to ICT implementation strategies and the removal of 3 items (Cm6, Is2, and Is4), the emerged factors corresponded to the items of the scales developed except the mingling of leadership and climate in a single factor.

While leadership and climate were highly related to each other, they were two distinct concepts on theoretical basis (Hall & George, 1999; Reinhartz & Beach, 2004; Sergiovanni, 1995). Despite the EFA results indicating the mingling of leadership and climate in a single factor, it was decided on theoretical consideration to retain leadership and climate as two separate factors in subsequent analyses, but they would be subject to CFA assessment using LISREL 8.54.

Table 4.1
EFA Results with Factor Loadings (Factor Loadings < .30 were Suppressed)

Scales	Items	Factors				
		1	2	3	4	5
Changes in Learning	Ln6	.814				.333
	Ln2	.780				.376
	Ln7	.742				.376
	Ln3	.734				.413
	Ln4	.643				.395
	Ln5	.633				.371
	Ln1**	.630				.427
Govt. ICT Policy	Gp5		.755			
	Gp2		.743			
	Gp4		.687			
	Gp6		.666			
	Gp1		.660			
	Gp3		.634			
	Gp7		.612			
	Gp8		.599			
Leadership & Climate	Cm2**	.318		.701		
	Cm1 → Ld8			.694		
	Ld6			.657		
	Cm3			.641		
	Ld2**			.632		
	Ld7			.618		
	Ld1			.614		
	Ld3			.604		
	Cm4			.538		
	Ld4			.513		
	Ld5			.497		
	Cm5			.484		
	Is2*			.448		
	Is4*			.417		
	Is5			.379		
	Cm6*			<.300		
ICT Implement Strategies	Is7				.850	
	Is8				.828	
	Is6				.427	
	Is3				.360	
	Is1				.315	
Changes in Pedagogy	Pd4	.349				.705
	Pd5	.341				.634
	Pd6	.409			.338	.527
	Pd3**	.365				.419
	Pd2**	.361				.401
	Pd1**	.343				.399
	Cronbach α	.8785	.8677	.8131/ .7764	.7203	.7001
		.8678^	---	.8148/ .7035^	---	.6822^

Notes: * Items being discarded after running EFA
** Items being discarded after running discriminant validity checks
^ Values for the modified measure after running EFA and discriminant validity checks

Internal Consistency Measuring Reliability

Reliability is the degree of consistency between multiple measurements of a variable (Hair et al., 1998), and it is a necessary but not sufficient condition for scale development (Bollen, 1989; Hinkin, 1998). The main techniques of estimating reliability are the test-retest, alternative forms, split-halves, and Cronbach's alpha (Bollen, 1989). A commonly used measure of reliability is internal consistency using Cronbach's alpha (Hair et al., 1998; Hinkin, 1998), which examines whether the individual items of the scale should all be measuring the same construct and thus be highly intercorrelated (Churchill, 1979; Nunnally & Bernstein, 1994). A Cronbach's alpha higher than the accepted value of .70 for newly developed measures is considered as demonstrating internal consistency (Hinkin, 1998). Therefore, after conducting exploratory factor analysis, the reliability of the scales was assessed by internal consistency using Cronbach's alpha test. As shown in Table 4.1, all the reliability values were higher than the accepted value of .70 for newly developed measures, indicating that the scales were internally consistent (Hinkin, 1998).

CFA Assessing Unidimensionality

Unidimensionality refers to the condition that each set of alternate indicators has only one underlying construct in common. In other words, measurement models that have indicators loading on more than one estimated construct do not represent unidimensionality (Anderson & Gerbing, 1988). Achieving unidimensionality is a crucial undertaking in scale development. As EFA typically does not provide an explicit test of unidimensionality, unidimensionality of the scales was assessed using structural equation-based approach of CFA (Gerbing & Anderson, 1988; Hair et al., 1998). CFA affords a stricter interpretation of unidimensionality by making possible an assessment of the internal consistency and external consistency implied by the multiple-indicator measurement (Gerbing & Anderson, 1988). Using LISREL 8.54 program, the resulting scales was subjected to CFA analysis, with leadership and climate retained as two separate factors in the CFA model (Table 4.2).

Table 4.2

CFA Model for the Scales Measuring the Social Ecology of ICT Implementation

Constructs	Indicators	Coefficients	t-value	R^2	Error Variance
Leadership	Ld1	.82	24.02	.67	.33
	Ld2	.81	25.82	.66	.34
	Ld3	.74	26.98	.55	.45
	Ld4	.56	13.13	.32	.68
	Ld5	.61	17.84	.37	.63
	Ld6	.80	35.48	.63	.37
	Ld7	.73	25.91	.54	.46
Climate	Cm1	.80	34.58	.63	.37
	Cm2	.83	40.02	.68	.32
	Cm3	.75	27.11	.56	.44
	Cm4	.65	21.28	.43	.57
	Cm5	.64	18.24	.41	.59
Government ICT policy	Gp1	.78	26.65	.61	.39
	Gp2	.83	34.44	.70	.30
	Gp3	.68	18.60	.47	.53
	Gp4	.66	19.03	.44	.56
	Gp5	.85	34.77	.72	.28
	Gp6	.68	18.07	.46	.54
	Gp7	.81	29.36	.66	.34
	Gp8	.78	24.52	.61	.39
ICT implementation strategies	Is1	.66	17.05	.43	.57
	Is3	.64	18.19	.41	.59
	Is5	.71	19.45	.50	.50
	Is6	.65	18.43	.42	.58
	Is7	.69	17.90	.47	.53
	Is8	.69	18.85	.47	.53
Changes in pedagogy	Pd1	.66	15.62	.43	.57
	Pd2	.58	14.02	.33	.67
	Pd3	.64	16.53	.41	.59
	Pd4	.60	15.23	.36	.64
	Pd5	.57	13.73	.33	.67
	Pd6	.65	19.56	.43	.57
Changes in learning	Ln1	.79	33.33	.63	.37
	Ln2	.87	37.55	.75	.25
	Ln3	.83	38.01	.72	.28
	Ln4	.77	29.15	.59	.41
	Ln5	.72	27.53	.51	.49
	Ln6	.81	34.56	.65	.35
	Ln7	.80	28.86	.64	.36

Basic Assumption Underlying Use of CFA

The basic assumption underlying the standard use of CFA in SEM is that the observations are drawn from a continuous and multivariate normal population, yet researchers often ignore the assumption (Flora & Curran, 2004; West et al., 1995). The assumption also applies to the common practice of using the Pearson's product-moment (PPM) correlation matrix as an input matrix for CFA analysis. As the data obtained in the present study was based on ordinal response modes from Likert scale measurements, the variables actually departed from the representation of an interval scale (Joreskog & Sorbom, 2001). In the large-scale simulation study in which Likert scale data with ordered categories was used, it has been demonstrated that the use of PPM coefficients instead of polychoric coefficients generated underestimated parameters with overestimated standard errors (Babkus, Ferguson, & Joreskog, 1987). To overcome the limitation, the present study would use a polychoric correlation matrix as the input matrix (Flora & Curran, 2004; Joreskog & Sorbom, 2001; Wang & Cunningham, 2005). Assuming that there was a continuous variable underlying each ordinal variable, polychoric correlation were not correlations computed from actual scores but were rather estimated theoretical correlations of the underlying continuous variables (Joreskog & Sorbom, 2001).

Regarding the issue of estimation methods in SEM, the commonly used estimation methods of maximum likelihood (ML) and generalised least squares (GLS) are developed under multivariate normality assumption (Chou & Bentler, 1995; Raykov & Marcoulides, 2000). Nevertheless, real social science data may depart from the normality assumption. As revealed in a study of over 400 large data sets, the great majority of data collected in behavioural research were not normally distributed (West et al., 1995). Although ML estimates have been found to be quite robust to the violation of normality, simulation studies showed that ML yielded very unsatisfactory results under extreme nonnormality (Chou & Bentler, 1995; West et al., 1995). On the other hand, two alternative estimation methods, namely weighted least squares (WLS) and diagonally weighted least squares (DWLS), do not make assumptions about the distribution of the observed variables (Diamantopoulos & Siguaw, 2000; Flora & Curran, 2004). Despite the advantage of distribution-free nature for observed variables, WLS has been shown to behave poorly in small sample sizes, particularly in large models (Muthen, 1993). Holemes-Smith, Coote, & Cunningham (2004) have pointed out that the minimum sample size required to compute a stable weight matrix for WLS estimation is as large as 1.5K (K + 1), where K is the number of variables (Wang, 2005). This meant that for an instrument containing 39 items as in the present study, a minimum of 2340 cases was required, which far exceeded the present sample size of 1076.

An alternative to WLS estimation is DWLS estimation. In a simulation study, it was found out that DWLS performed better than WLS for a medium size model comprising 21 dichotomous variables and with sample sizes down to 100 cases (Maydeu-Olivares, 2001). In another simulation study using polychoric correlation matrix, Flora & Curran (2004) showed that DWLS (robust WLS), in comparison to WLS, produced less biased chi-square statistics, parameter estimates, and standard errors. Specifically, the biases in test statistics and standard errors for WLS was demonstrated to increase as a function of decreasing sample size and increasing model complexity (Flora & Curran, 2004). In a study of CFA models using ordinal and nonnormal data with various combinations of input matrices (PPM matrix, polychoric correlation matrix) and estimation methods (ML, WLS and DWLS), it was shown that DWLS estimation using a polychoric correlation as an input matrix produced better outcomes on parameter estimates and goodness-of-fit statistics (Wang & Cunningham, 2005).

As mentioned before, the instrument of the present study adopted a Likert scale measurements that were ordinal in nature, and on consideration of the usually non-normality nature of the real social science data, the present study employed DWLS estimation with polychoric correlations for CFA.

Evaluating CFA Model

In evaluating CFA models, parameter estimates related to indicators of the CFA model were examined, of which the higher the value with respect to factor loading, t-value, and the squared multiple correlation (R^2) on its target construct, the better the indicator representing the latent variable concerned (Bollen, 1989; Diamantopoulos & Siguaw, 2000). A t-value was obtained from dividing the estimated factor loading by its standard error, and a t-value of less than the minimum value of 1.96 indicated that the corresponding parameter was not statistically significantly at the 5% level. In addition, R^2 was indicative of the degree to which the indicators were free from measurement error (Diamantopoulos & Siguaw, 2000).

In evaluating the overall fitness of the model for CFA, in addition to the traditional inferential goodness-of-fit index of chi-square value, a number of goodness of fit indexes were used (Browne & Cudeck, 1993; Diamantopoulos & Siguaw, 2000; Kelloway, 1998; Marsh et al., 1996), in particular the root mean square error of approximation (RMSEA), comparative fit index (CFI), and non-normed fit index (NNFI).

The chi-square test is a traditional measure for evaluating overall model fit. The chi-square statistic for a just-identified model is zero with no degrees of freedom. Within a large sample and under the assumption of multivariate normality, the chi-square statistic for an overidentified model is interpreted as a test of significance of the difference in fit between that model and a just-identified model. Thus, a low and non-significant value (usually for $\alpha = .05$, $p > .05$) of the chi-square index indicates a good model fit (Kline, 1998). Despite this, caution needs to be exercised in chi-square test application due to its sensitivity to sample size and the highly restrictive assumption of perfect fit (Diamantopoulos & Siguaw, 2000). Chi-square test is highly restrictive in a sense that it provides a test of perfect fit, in which the null hypothesis is that the model fits the population data perfectly. A statistically significant chi-square causes rejection of the null hypothesis (Diamantopoulos & Siguaw, 2000). In other words, if the probability of achieving the chi-square value (p value) is greater than .05, it indicates that there is no difference between the implied sample covariance matrix and the fitted covariance matrix. Along with this restrictive assumption of perfect fit, the chi-square statistics is also known to be sensitive to departures from large sample size. In other words, the value of chi-square statistics for a particular model and its data will increase simply by increasing the sample size (Kline, 1998). With very large samples, there is an artificial tendency to obtain a small p value associated with chi-square statistics, and therefore to reject the model even if it is only marginally inconsistent with the data (Raykov & Marcoulides, 2000). This is akin to a Type I error, of which one rejects the null hypothesis of perfect fit when it is true.

In addition, chi-square test has the multivariate normality assumption on the data. When the assumption of multivariate normality is violated, Satorra-Bentler chi-square statistic (Satorra & Bentler, 1994) is introduced to adjust the value of the standard chi-square downward by a constant that reflects the degree of kurtosis (Kline, 1998). To reduce the sensitivity of the chi-square statistic to sample size, some researchers divide its value by the degree of freedom, and a frequent suggestion is that this ratio be less than three (Kline, 1998). On consideration of the usually non-normality nature of the real social science data, Satorra-Bentler chi-square was used in the present study, and as an additional piece of information for reference, the value of Satorra-Bentler chi-square per degree of freedom was also shown.

Unlike the chi-square value, RMSEA is not sample-dependent. Instead, the RMSEA is a measure used to test the null hypothesis of close fit (Browne & Cudeck, 1993), and is primarily concerned with evaluating the extent to which the model fails to fit the data (Raykov & Marcoulides, 2000). In addition, there is also the value for the 90% confidence level of the RMSEA, which offers further indication to the precision of estimation. As such, it is generally regarded as one of the most informative fit indexes (Diamantopoulos & Siguaw, 2000), and has recently become quite a popular index of model fit (Raykov & Marcoulides, 2000). An index value of RMSEA below .05 and a p-value for close fit greater than .05 indicate a close fit of the model (Browne & Cudeck, 1993; Kelloway, 1998). However, in a recent research examining the conventional cutoff criteria for fit indexes, it has been suggested that RMSEA values of .06 or less indicate a close-fitting model (Hu & Bentler, 1999).

The NNFI and CFI measure the percentage of improvement in fit over the null (independence) model in which the observed variables are assumed to be uncorrelated (Kline, 1998). The indexes are recommended because of their relative independence to sample size, with NNFI further incorporating a penalty for model complexity and a reward for model parsimony (Marsh et al., 1996). In a large simulation study evaluating the behaviour of a number of incremental fit indexes according to criteria such as bias due to sample size, penalties for model complexity, rewards for model parsimony, and estimation reliability, NNFI was found to be the most successful index. While CFI was unrelated to sample size, it had no penalty for model complexity and reward for model parsimony (Marsh et al., 1996). Therefore, Marsh et al. (1996) recommended researchers to use both CFI and NNFI. An index value of above .90 for CFI and NNFI is interpreted as an indication of reasonably good fit of model data (Kelloway, 1998; Kline, 1998; Marsh et al., 1996). In a recent research examining the conventional cutoff criteria for fit indexes, it has been suggested that CFI values greater than .95 are indicative of good fitting model (Hu & Bentler, 1999).

CFA Results Indicating Unidimensionality

Table 4.4 shows the results of the CFA measurement model for the scales measuring the social ecology of ICT implementation. The results indicated that all factors were clearly defined, in a sense that each set of alternate indicators had only one underlying construct in common, and each indicator had a substantially high factor loading (ranging from .61 to .83), t-value (ranging from 13.13 to 40.02), and R^2 (ranging from .32 to .75) on its target construct. In addition, Table 4.3 (original measure) shows the summary of goodness of fit statistics for the CFA model, in which the value of RMSEA, CFI and NNFI all fell within an acceptable range, except the p-value of chi-square test. As mentioned above, the chi-square statistics is known to be sensitive to departures from large sample size and therefore was considered not indicative in the present study. On the other hand, the values of RMSEA index (.045) and its endpoint of the 90% confidence interval (.043 and .047) were reasonably small, with the p-value for close fit (.99) being larger than .05. These indicated a close fit of the CFA model. In addition, the values of CFI (.98) and NNFI (.98) were substantially high, which also pointed to an indication of reasonably good fit of model data. The overall CFA results provided support for unidimensionality of the 6 scales for measuring the social ecology of ICT implementation.

Table 4.3
A Summary of Fitting Index Results from CFA

Goodness of fit index	Values obtained from CFA		
	Original Measure	Modified Measure	Recommended good fit values
Satorra-Bentler chi-square (degree of freedom)	2209.32 (687)	1363.31 (480)	---
P-value of chi-square test	.00	.00	>.05
Chi-square per degree of freedom	3.22	2.84	<3.0
Root Mean Square Error of Approximation (RMSEA)	.045	.047	<.06
90% Confidence Interval for RMSEA	.043, .047	.044, .049	<.06
P-Value for Test of Close Fit (RMSEA < .05)	.99	.99	>.05
Non-Normed Fit Index (NNFI)	.98	.98	>.95
Comparative Fit Index (CFI)	.98	.98	>.95

Table 4.4

Parameter Estimates for the CFA Measurement Model of the 6 Constructs (with Values of the Modified Measure in Parentheses)

Constructs	Indicators	Factor loading	t-value	R^2	Composite reliability	Variance extracted
Leadership	Ld1	.82 (.82)	24.02 (23.45)	.67 (.67)	.89 (.89)	.53 (.54)
	Ld2	.81 (---)	25.82 (---)	.66 (---)		
	Ld3	.74 (.74)	26.98 (26.29)	.55 (.55)		
	Ld4	.56 (.58)	13.13 (14.26)	.32 (.34)		
	Ld5	.61 (.62)	17.84 (18.18)	.37 (.39)		
	Ld6	.80 (.80)	35.48 (32.18)	.63 (.64)		
	Ld7	.73 (.75)	25.91 (25.67)	.54 (.56)		
	Cm1/ Ld8	.80 (.79)	34.58 (35.46)	.63 (.63)		
Climate	Cm2	.83 (---)	40.02 (---)	.68 (---)	.85 (.85)	.54 (.59)
	Cm3	.75 (.84)	27.11 (30.78)	.56 (.70)		
	Cm4	.65 (.73)	21.28 (22.05)	.43 (.53)		
	Cm5	.64 (.72)	18.24 (18.78)	.41 (.51)		
Government ICT policy	Gp1	.78 (.78)	26.65 (27.37)	.61 (.60)	.92 (.92)	.58 (.58)
	Gp2	.83 (.83)	34.44 (35.80)	.70 (.70)		
	Gp3	.68 (.69)	18.60 (19.39)	.47 (.47)		
	Gp4	.66 (.68)	19.03 (21.15)	.44 (.46)		
	Gp5	.85 (.85)	34.77 (36.47)	.72 (.71)		
	Gp6	.68 (.70)	18.07 (19.25)	.46 (.48)		
	Gp7	.81 (.80)	29.36 (29.45)	.66 (.64)		
	Gp8	.78 (.77)	24.52 (24.97)	.61 (.60)		
ICT implement strategies	Is1	.66 (.63)	17.05 (16.23)	.43 (.40)	.83 (.83)	.45 (.45)
	Is3	.64 (.64)	18.19 (18.03)	.41 (.40)		
	Is5	.71 (.68)	19.45 (18.60)	.50 (.46)		
	Is6	.65 (.64)	18.43 (18.50)	.42 (.42)		
	Is7	.69 (.72)	17.90 (20.31)	.47 (.52)		
	Is8	.69 (.72)	18.85 (22.22)	.47 (.51)		
Changes in pedagogy	Pd1	.66 (---)	15.62 (---)	.43 (---)	.79 (.78)	.38 (.54)
	Pd2	.58 (---)	14.02 (---)	.33 (---)		
	Pd3	.64 (---)	16.53 (---)	.41 (---)		
	Pd4	.60 (.70)	15.23 (15.61)	.36 (.49)		
	Pd5	.57 (.69)	13.73 (14.04)	.33 (.47)		
	Pd6	.65 (.81)	19.56 (21.41)	.43 (.65)		
Changes in learning	Ln1	.79 (---)	33.33 (---)	.63 (---)	.93 (.92)	.64 (.66)
	Ln2	.87 (.87)	37.55 (36.50)	.75 (.76)		
	Ln3	.83 (.86)	38.01 (37.42)	.72 (.74)		
	Ln4	.77 (.77)	29.15 (28.69)	.59 (.59)		
	Ln5	.72 (.72)	27.53 (26.55)	.51 (.51)		
	Ln6	.81 (.83)	34.56 (33.77)	.65 (.68)		
	Ln7	.80 (.81)	28.86 (28.62)	.64 (.66)		

Construct Validity

In addition to ensuring unidimensionality and reliability of the scales, attainment of construct validity was also necessary. Notably, validity usually is a matter of degree rather than an all-or-none property, and therefore scales are often validated independently of their development and well after development is complete (Nunnally & Bernstein, 1994). Along this line of reasoning, validation of the present developed scales was considered as a preliminary one. As the structural equation-based approach has been considered as a powerful method for addressing construct validity (Anderson & Gerbing, 1988; Bagozzi et al., 1991), the present study relied mainly on the structural equation-based approach and its complementary methods for construct validation.

Construct validity is the extent to which a scale accurately represents the concept of interest, and are widely represented by content validity, convergent validity, discriminant validity, and nomological validity (Hair et al., 1998; Peter, 1981). Content validity has already been discussed in the section of theoretical framework in Chapter 2 and the section of survey instrument in Chapter 3. Nomological validity measures the degree that the scale makes accurate predictions of other concepts that are in the theoretical model (Hinkin, 1998), and according to Anderson & Gerbing (1988)'s suggestions, "The measurement model provides a confirmatory assessment of convergent and discriminant validity. Given an acceptable convergent and discriminant validities, the test of the structural model will then constitute a confirmatory assessment of nomological validity" (p.411), nomological validity will be demonstrated in the next chapter of structural model building (Chapter 5). In view of the above, the following sections will discuss construct validity by examining convergent validity and discriminant validity of the constructs.

Convergent Validity

Convergent validity assesses the degree to which two measures of the same concepts are correlated (Hair et al., 1998). One way of demonstrating convergent validity is to correlate a new measure with an existing measure that is measuring the same concept (Hair et al., 1998). However, some of the constructs of the present study were rather new, which imposed difficulties on finding alternative measures of the same concept for validation. Therefore, an alternative method of using SEM, as suggested by Anderson & Gerbing (1988) and Kline (1998), was selected to demonstrate convergent validity. Convergent validity of a construct was assessed by examining the estimated loading for each indicator from the measurement model obtained by CFA using LISREL 8.54. High and statistically significant loadings of sets of indicators on their common factor provided evidence for convergent validity (Anderson & Gerbing, 1988; Bollen, 1989; Kline, 1998).

As shown in Table 4.4, each indicator had a substantially high factor loadings (ranging from .61 to .83) and t-value (ranging from 13.13 to 40.02) on its target construct. T-values of all the indicators' estimated coefficient was greater than 2 at $p<.05$, indicating that all the estimated coefficients were significantly different from zero. In addition, the values of the squared multiple correlations (R^2) were moderate to high, ranging from .32 to .75, suggesting that the indicators were reasonably successful as measures of the six latent variables (constructs) in the model (Diamantopoulos & Siguaw, 2000). Such strong relationships between indicators and their corresponding latent construct also demonstrated convergent validity (Bollen, 1989).

Composite Reliability and Average Variance Extracted

Apart from the Cronbach's alpha measurement, the composite reliability and the average variance extracted derived from confirmatory factor analysis are considered as alternative reliability measures (Diamantopoulos & Siguaw, 2000; Hair et al., 1998). Composite reliability of a construct was computed from dividing the square of sum of standardised loadings of indicators by the total value obtained from the square of sum of standardised loadings of indicators plus sum of indicator measurement error. Variance extracted for a construct was computed from dividing the sum of squared standardised loadings of indicators by the total value obtained from the sum of squared standardised loading of indicators plus sum of indicator measurement error (Diamantopoulos & Siguaw, 2000; Hair et al., 1998). Reliability was demonstrated if the value for composite reliability of a construct exceeded the recommended level of .70 (Hair et al., 1998), and the value of the variance-extracted measure exceeded the recommended value of .50 (Fornell & Larcker, 1981). Thus, in addition to convergent validity, the composite reliability and variance extracted for each construct of the CFA measurement model were also examined in the present study.

As indicated from Table 4.4, the composite reliability for all 6 constructs exceeded the recommended level of .70 (Hair et al., 1998). For the variance-extracted measures, other than two constructs, the average variance extracted for all other four constructs exceeded the recommended value of .50 (Fornell & Larcker, 1981). The construct for ICT implementation strategies and the construct for changes in pedagogy had a value for average variance of .45 and .38 respectively, falling a bit short of the recommended value of .50.

For the construct of pedagogy, apart from its relatively lower value of average variance extracted (.38), results from discriminant validity checks also indicated that it was not clearly distinct from the construct of changes in learning (Table 4.6). For this reason, the construct of changes in pedagogy was modified with a subsequent removal of three items. The details will be discussed in the section below reporting construct refinement. It was noteworthy to mention here that the t-values, R^2, and composite reliability for the modified measure remained more or less the same when compared with the original measure; on the other hand, the variance extracted for the construct of pedagogy after modification had its value improved from .38 to .54 (Table 4.4). In brief, upon modification of the constructs, the overall results provided evidence of convergent validity and discriminnat validity.

Discriminant Validity

Discriminant validity is the degree to which two conceptually similar cosntructs are distinct, and evidence of discriminant validity is therefore to ensure that similar constructs are sufficiently different from one another (Hair et al., 1998). Discriminant validity of the constructs was checked by two structural equation-based methods. The first one was the chi-square difference test (Joreskog, 1971) and its complementary assessment (Anderson & Gerbing, 1988), and the other one was the variance extracted against squared correlation test (Fornell & Larcker, 1981).

For the chi-square difference test, discriminant validity was assessed using structural equation-based approach to construct a constrained and an unconstrained CFA measurement model with one pair of target constructs each time (Anderson & Gerbing, 1988). In this study, a total of 15 target pairs out of the 6 constructs were constructed (Table 4.5). Specifically, for each pair of target constructs, one pair at a time, a constrained measurement model (correlation Φ fixed to 1) and an unconstrained measurement model (free correlation) were constructed using LISREL 8.54. The constrained measurement model was obtained by constraining the estimated correlation between the pair of constructs to 1.0 (Anderson & Gerbing, 1988). After obtaining the constrained and unconstrained models, a chi-square difference test was performed between the constrained model and the unconstrained model. A significant lower chi-square value for the unconstrained model would indicate discriminant validity (Bagozzi & Phillips, 1982). According to Hair et al. (1998), a value of 3.84 or greater suggested a statistically significant reduction in chi-square. In this study, the 15 pairs of constructs out of the combination of the 6 constructs were subjected to chi-square difference tests. As shown in Table 4.5, other than the pair of constructs for leadership and climate, all the chi-square differences were highly significant. In any case, the chi-square difference for leadership and climate was still significant, though not highly significant.

The complementary assessment of the chi-square difference test was derived from the unconstrained CFA model for the pair of target constructs. In getting the unconstrained CFA model, a correlation estimate between the pair of constructs was obtained when they were allowed to correlate freely. Discriminant validity was obtained if the correlation estimate between the two constructs (+ and – two standard errors) did not include the value of 1.0 (Anderson & Gerbing, 1988). As shown in Table 4.5, results for the complementary assessment of chi-square difference test in this study also supported the presence of discriminant validity in the constructs. The results indicated that all the confidence interval (+ and − two standard errors) around the correlation estimate between any two constructs did not include 1.0.

Discriminant validity was further assessed by variance extracted against squared correlation test recommended by Fornell & Larcker (1981). The test started with using LISREL 8.54 in running CFA for each pair of constructs at a time, analysis of which focused on comparing the average variance extracted for the two constructs with the correlation estimate between the pair of constructs. The average variance extracted for a construct was computed from dividing the sum of squared standardised loadings of indicators by the total value obtained from the sum of squared standardised loading of indicators plus sum of indicator measurement error (Diamantopoulos & Siguaw, 2000; Hair et al., 1998). Discriminant validity for a pair of constructs was achieved if both their values of average variance extracted were larger than the squared correlation between the two constructs (Fornell & Larcker, 1981). The results from the variance extracted against squared correlation test did not indicate full discriminant validity of the scales. When the average variance extracted for each pair of constructs were compared with its square of correlation, average variance extracted for all 13 pairs of constructs were larger than their corresponding correlation, except for 2 pairs of constructs. As shown in Table 4.6, the two exceptions were leadership vs. climate, and changes in pedagogy vs. changes in learning. Although the test offered by Fornell and Larcker (1981) had been regarded as having some limitations (Bagozzi, 1981), and the chi-square difference test and its complementary method had been considered as a powerful method for addressing construct validity (Anderson & Gerbing, 1988; Bagozzi et al., 1991), the researcher still felt obliged to strive for improvement of the scales on consideration of mingling of the constructs for leadership and climate in EFA and the low value of variance extracted for the construct of changes in pedagogy (Table 4.4).

Table 4.5
Discriminant Validity Checks Using Chi-Square Difference Test (with Values of the Modified Measure in the Lower Row)

Pair of constructs	χ^2 (correlation fixed to 1)	χ^2 (unconstrained correlation)	χ^2 difference	Standard error for correlation estimate	Correlation estimate +/- 2 standard error
Leadership vs. climate	336.53 303.22	325.09 (.89) 214.00 (.72)	11.4 89.22	.018 .031	.93 - .85 .78 - .66
Leadership vs. govt ICT policy	2649.45 3325.55	428.28 (.27) 463.28 (.25)	2221.17 2862.27	.044 .044	.36 - .18 .34 - .16
Leadership vs. ICT strategies	1419.87 1493.12	582.29 (.34) 691.15 (.35)	837.58 801.97	.048 .046	.44 - .24 .44 - .25
Leadership vs. change in pedagogy	694.38 383.55	244.70 (.43) 135.03 (.31)	449.68 248.52	.042 .047	.51 - .35 .41 - .22
Leadership vs. change in learning	1510.94 1820.07	153.33 (.43) 113.71 (.41)	1357.61 1706.36	.040 .039	.51 - .35 .48 - .33
Climate vs. govt ICT policy	1666.62 587.13	487.73 (.28) 289.51 (.27)	1178.89 297.62	.042 .041	.36 - .20 .35 - .19
Climate vs. ICT strategies	1539.85 1157.47	878.24 (.39) 609.69 (.34)	661.61 547.78	.040 .045	.47 - .31 .43 - .25
Climate vs. change in pedagogy	866.91 433.55	382.39 (.46) 47.56 (.28)	484.52 385.99	.042 .052	.54 - .38 .39 - .18
Climate vs. change in learning	766.09 150.40	135.86 (.48) 17.44 (.43)	630.23 132.96	.039 .043	.52 - .44 .48 - .33
Govt ICT policy vs. ICT strategies	1455.05	780.12 (.30)	647.93	.048	.40 - .20
Govt ICT policy vs. change in pedagogy	1050.49 539.77	435.39 (.29) 293.79 (.22)	615.10 299.98	.044 .047	.38 - .20 .32 - .13
Govt ICT policy vs. change in learning	3676.08 2992.02	278.23 (.32) 242.95 (.31)	3397.85 2749.07	.043 .044	.41 - .23 .40 - .22
ICT strategies vs. change in pedagogy	1210.82 814.64	696.49 (.49) 525.67 (.43)	514.33 288.97	.049 .041	.59 - .39 .52 - .35
ICT strategies vs. change in learning	802.58 884.79	324.63 (.44) 273.55 (.42)	477.95 611.24	.044 .037	.53 - .35 .50 - .35
Change in pedagogy vs. change in learning	259.70 95.11	139.12 (.75) 22.57 (.62)	120.58 72.54	.032 .039	.81 - .67 .70 - .55

Table 4.6

Discriminant Validity Checks by Comparing the Average Variance Extracted and the Square of Correlation

Construct	Square of correlation (average variance extracted for the pair of constructs)				
	Climate	Govt. ICT policy	ICT Strategies	Changes in Pedagogy	Changes in Learning
Leader-ship	.79 (.53, .54) .52 (.54, .61)*	.073 (.53, .59) .062 (.54, .58)*	.12 (.53, .44) .12 (.54, .44)*	.18 (.53, .39) .10 (.54, .56)*	.18 (.53, .64) .16 (.54, .66)*
Climate		.079 (.55, .59) .072 (.59, .59)*	.15 (.55, .45) .11 (.58, .47)*	.21 (.55, .39) .08 (.59, .56)*	.23 (.55, .65) .16 (.58, .66)*
Govt. ICT Policy			.09 (.59, .44)	.084 (.59, .39) .049 (.59, .55)*	.10 (.59, .65) .10 (.59, .66)*
ICT Strategies				.24 (.44, .39) .19 (.47, .54)*	.19 (.44, .64) .18 (.47, .66)*
Changes in Pedagogy					.56 (.39, .64) .39 (.55, .66)*

* Values for the modified measure

Construct Refinement

Construct refinement was conducted in response mainly to the discriminant validity checks indicating that two pairs of constructs were not clearly distinct from one another, which were leadership vs. climate, and changes in pedagogy vs. changes in learning. As an initial attempt in refining the scales to improve discriminant validity, the pair of constructs was subject to EFA with an extraction of two factors. The EFA results served as a preliminary reference for discarding items with cross loading on the pair of not clearly distinct constructs. Yet, structural equation-based approach pertinent to CFA and discriminant validity checks served as the main source in guiding the refinement process.

Leadership and climate. In an attempt to extract two factors using EFA with oblique rotation, results indicated that the majority of items cross-loaded on both factors of leadership and climate, and also three items for climate loaded on leadership (Table 4.7). The strategy for refinement was to discard items with high cross loading based on theoretical considerations, of which Ld2 "The school goals are reflected in school plans" and Cm2 "My opinions can be conveyed to policy-making units effectively within school" were discarded. While it was likely that the items were not good enough to measure the concept for establishing shared vision (a domain of the construct of leadership, Chapter 3) and the concept for flow of information within school (a domain of the construct of climate, Chapter 3), there was also the possibility that the items measured another common concept behind, such as participative decision, and therefore rendering the two scales not clearly separable from one another. At the same time, attempts were also made to re-assign an item originally belonging to climate but loading on leadership, Cm1 "I feel comfortable to talk to the principal about school matters" was re-assigned to leadership. The decision was based on considerations that Cm1 actually measured the domain of trust and empowerment as specified by leadership. Finally, the modified measures (Table 4.1 and Appendix 1) were subject to structural equation-based approach of discriminant validity checks. The discriminant validity for the modified leadership and climate scales was improved (Table 4.5, 4.6). Specifically, the average variance extracted for the pair of constructs of leadership and climate (.54, and .61) were now higher than the square of correlation (.52), indicating that discriminant validity was obtained (Table 4.6).

Changes in pedagogy and changes in learning. In an attempt to extract two factors using EFA with oblique rotation, results indicated that the majority of items cross-loaded on both factors of changes in pedagogy and changes in learning, and one item for changes in pedagogy (Pd2) loaded on changes in learning (Table 4.8). Again, the strategy for refinement was to discard items with high cross loading based on theoretical considerations. The item Pd2 "Compared to the past two academic years, I am more willing to have collaborative lesson preparation with my colleagues" was removed on consideration that it loaded on the wrong factor. Ln1 "Compared to students of the same level in the past two academic years, my students are more competent in mastering various generic skills" was also removed on consideration of its non-specific nature.

Despite removing the two items (Pd2, Ln1) with high cross-loadings from the scales, the average variance extracted for the pair of construct (.41, .66) was still smaller than the square of its correlation (.53) (Table 4.9). As Pd2 and Pd3 measured the same domain of collaborative teaching, therefore, an attempt of further removing the item Pd3 "Compared to the past two academic years, I am more willing to participate in peer classroom observation" from the construct changes in pedagogy made the two constructs barely discriminant from one another, but the average variance extracted for changes in pedagogy (.45) still slightly fell short of .50, and its Cronbach α dropped from .7001 to .6583 (Table 4.9). In developing the construct for changes in pedagogy, the researcher considered that changes in curriculum content (Pd1) as well as changes towards a collaborative kind of teaching (Pd2 and Pd3) would enhance changes of pedagogy from a teacher-centred approach to a student-centred approach (Chapter 3). However, the changes might only represent the school policy behind necessitating curriculum change and collaboration among teachers, which did not necessarily have association with an actual change in classroom practices. With a further removal of Pd1 "Compared to the past two academic years, our school curriculum is putting more emphasis on infusing generic skills into teaching and learning of various subjects" from the construct changes in pedagogy, the values with respect to discriminant validity (.55, .66 > .39), average variance extracted from the CFA measurement model (.54) as well as the Cronbach α (.6822) were all improved (Table 4.9). Upon refinement, the modified scales of changes in pedagogy (with removal of Pd1, Pd2, and Pd3) and changes in learning (with removal of Ln1) demonstrated discriminant validity and convergent validity, with Cronbach α (.6822) slightly below the recommended value of .70 for newly developed measures (Hinkin, 1998). However, reliability of the construct could still be demonstrated in the structural equation-based approach of measuring composite reliability and average variance extracted (Hair et al., 1998) (Table 4.4). A value of composite reliability of .78 of the modified construct for changes in pedagogy exceeded the recommended level of .70 (Hair et al., 1998), and a value of average variance extracted of .54 exceeded the recommended level of .50 (Fornell & Larcker, 1981).

Overall validity. At the outset, while the Cronbach's alpha reliability test identified the 6 constructs as internally consistent, a preliminary EFA has found out that the factors of leadership and climate were not clearly separable. However, a more rigorous structural equation-based approach of CFA subsequently provided strong evidence in supporting the constructs as unidimensional in representing a single concept. Of the CFA model, the composite reliability and variance-extracted for each construct provided further support of convergent validity. The chi-square difference

126

test and its complementary assessment method as suggested by Anderson & Gerbing (1988) provided strong evidence of discriminant validity. Despite this, the variance extracted against squared correlation test as suggested by Fornell and Larcker (1981) found that leadership against climate as well as changes in pedagogy against changes in learning were not clearly separable from one another. On such considerations, modifications were subsequently made to the constructs for leadership, climate, changes in pedagogy, and changes in learning. With a total deletion of 9 items, 3 of which were deleted after conducting EFA, and 6 of which were deleted after conducting discriminant validity checks, the modified measure subsequently contained of 33 items (Table 4.1).

The modified measure for the most part demonstrated unidimensionality, reliability, convergent validity, and discriminant validity. Results from the structural equation-based approach of CFA and discriminant validity checks indicated that the values all reached an acceptable level indicating construct validity (Table 4.3, 4.4, 4.5, and 4.6). However, the Cronbach α for the construct of changes in pedagogy (.6822) was slightly below the value of .70 as recommended for newly developed measures (Hinkin, 1998). In addition, the average variance extracted for the construct of ICT implementation strategies (.45) was also slightly below the recommended value of .50 (Fornell & Larcker, 1981). These limitations were, however, compensated by the validity of the scales strongly demonstrated in the dimensions of unidimensionality, composite reliability, convergent validity, and discriminant validity.

Summary

In the validation process, the scales were subject to two major modifications after conducting EFA and discriminant validity checks. As an initial refinement, results of EFA guided the removal of three items loading on other factors (Table 4.1). In addition, EFA alerted us that the constructs of leadership and climate tended to mingle in one composite factor, and the items for the constructs of changes in pedagogy and changes in learning cross-loaded with each other. The discriminant checks using variance extracted against squared correlation test as recommended by Fornell & Larcker (1981) further revealed that the constructs for leadership and climate were not sufficiently different from one another, and that the constructs for changes in pedagogy and changes in learning could be barely considered distinct from one another. To increase discriminant validity, the scales of leadership, climate, changes in pedagogy, and changes in learning were modified. In the process of refinement, EFA served as a preliminary tool in guiding the researcher to discard items with cross loading on the pair of not clearly distinct factors. Yet, structural equation-based approach as well as theoretical considerations served as the main source in guiding the refinement process. Eventually, in addition to the three items, a further six items were discarded, with a total removal of nine items. With such modifications, construct validity was on the whole improved in the dimensions of factor structure, reliability, convergent validity, and discriminant validity, but at the expense of a slight decrease in the index of internal consistency in terms of Cronbach's alpha measurement. Yet, reliability of the construct was still demonstrated in the structural-equation based approach of measuring composite reliability and average variance extracted (Hair et al., 1998).

Table 4.7

EFA Results for the Constructs of Leadership and Climate (Factor Loadings < .30 were Suppressed)

Construct	Items	Factors	
		1	2
Leadership	Cm1 → Ld8	.730	-.366
	Cm2*	.685	-.497
	Ld3	.664	
	Ld6	.645	-.434
	Ld7	.641	-.326
	Ld2*	.625	-.526
	Ld1	.614	-.486
	Cm3	.611	-.554
	Ld4	.542	
	Ld5	.512	
Climate	Cm4	.449	-.841
	Cm5	.403	-.612

* Items being discarded eventually

Table 4.8

EFA Results for the Constructs of Changes in Pedagogy and Changes in Learning (Factor Loadings < .30 were Suppressed)

Construct	Items	Factors	
		1	2
Changes in Learning	Ln6	.802	.442
	Ln2	.771	.450
	Ln3	.740	.442
	Ln7	.736	.429
	Ln4	.655	.433
	Ln1*	.653	.485
	Ln5	.642	.417
	Pd2*	.411	.397
Change in Pedagogy	Pd4	.388	.711
	Pd5	.376	.680
	Pd6	.431	.558
	Pd3*	.386	.435
	Pd1*	.355	.374

* Items being discarded eventually

Table 4.9

Reliability and Validity Checks for the Modified Scale of Changes in Pedagogy

Construct of Changes in Pedagogy	Discriminant Validity Checks on Changes in Pedagogy vs Changes in Learning		Reliability		
	Square of Correlation	Average Variance Extracted	Composite Reliability	Variance Extracted	Cronbach α
Deleting item Pd2	.527	.407, .656	---	---	---
Deleting item Pd2, Pd3	.457	.461, .655	.77	.45	.6583
Deleting Pd2, Pd3, Pd1	.388	.546, .656	.78	.54	.6822

Discussions

The present chapter has mainly demonstrated empirically the extent to which items of the constructs measured the abstract concepts concerning leadership, climate, effectiveness of government ICT policy, collegial capacity of ICT implementation strategies, changes in pedagogy, and changes in student learning. While the proposed constructs strongly demonstrated unidimensionality, reliability, and convergent validity, it did not clearly demonstrate discriminant validity with respect to two pairs of constructs: leadership vs climate, changes in pedagogy vs changes in learning. After refinement, the modified measure for the most part strongly demonstrated unidimensionality, reliability, convergent validity, and discriminant validity.

In the process of construct validation, on theoretical basis, it had been assumed that the constructs of leadership and climate were conceptually similar but were still distinct from one another. Despite this, when coming to the point of empirically translating the concepts into items, the constructs were not clearly distinct from one another. In fact, the constructs for leadership and climate in the present study were modified several times before they were clearly distinct from one another. On theoretical basis, the leadership scale in this study was supposed to measure how teachers perceived school principal's transformational leadership. The domains measuring transformational leadership included the capability of establishing shared visions, of trusting and empowering teachers, of supporting innovations in classroom practices, and of supporting staff development. On the other hand, the school climate scale of the present study was supposed to measure among teachers their collegial relationships, their spirit to strive for progress, and also the flow of information within school. To increase the level of discrimination of the scales between leadership and climate, the item Cm2 "My opinions can be conveyed to policy-making units effectively within school" had to be removed from the climate scale, and the item Ld2 "The school goals are reflected in school plans" had to be removed from the leadership scale. While it was likely that the items were not good enough to measure the concept for flow of information within school and the concept for establishing shared vision, there was also the possibility that the items measured another common concept behind, such as participative decision, and therefore rendering the two scales not clearly separable from one another.

In the process of improving the discriminant validity of the constructs for changes in pedagogy and changes in learning, it was interesting to find out that the items deleted were in connection with school policy concerning curriculum and teacher collaboration. The items deleted were Pd1 "Compared to the past two academic years, our school curriculum is putting more emphasis on infusing generic skills into teaching and learning of various subjects", Pd2 "Compared to the past two academic years, I am more willing to have collaborative lesson preparation with my colleagues", Pd3 "Compared to the past two academic years, I am more willing to participate in peer classroom observation", and Ln1 "Compared to students of the same level in the past two academic years, my students are more competent in mastering various generic skills". It seemed likely to postulate that participations in collaborative lesson preparation and peer classroom observation did not necessarily have association with an actual change in classroom practices. Such phenomenon might be attributed to the school policy behind necessitating curriculum change and collaboration among teachers.

As mentioned in the method and results section, construct validity was on the whole improved in the dimensions of factor structure, convergent validity, and discriminant validity, but at the expense of a slight decrease in the index of internal consistency in terms of Cronbach's alpha measurement. However, also available were reliability measures derived from CFA which included the composite reliability and the average variance extracted (Hair et al., 1998). With a decrease in Cronbach alpha value to .6822, yet the average variance extracted for the modified construct of changes in pedagogy increased from .38 to .54. As the structural equation-based approach has been considered as a powerful method for addressing construct validity (Anderson & Gerbing, 1988; Bagozzi et al., 1991), the present study used structural equation-based approach as the main source in guiding the refinement process. Ultimately, with a total deletion of nine items, the modified measure was left with thirty-three items on the whole, with climate and changes in pedagogy each having only three items. Although the suggested number of items for most constructs was four to six items, it was finally determined to keep three items for the scales of climate and changes in pedagogy. The decision was made on consideration that gaining accumulated evidence in support of construct validity was far more important (Hinkin, 1998). Nevertheless, the scales met the recommendation of at least three indicators per factor for CFA (Bollen, 1989).

Chapter 5: Results (Part 2) – Model Building

Introduction

This chapter discussed the issue of finding out a plausible model to depict the social ecology of ICT implementation in schools, and conceptualisation of which was based on the theoretical framework constructed in Chapter 2 of the present study. The social ecology of ICT implementation was conceptualised as the interplay of contextual variables to advance changes in student learning in a sociocultural context of schooling. Structural equation modelling (SEM) technique was used to model the complex relationships in the social ecology of ICT implementation in schools since SEM has the advantage of allowing the researcher to accommodate multiple interrelated dependence relationships in a single model that are not possible with other popular multivariate techniques (Hair et al., 1998). As a way of empirically testing the theoretical framework about relationships between the various contextual variables (Joreskog, 1993), the process of structural model building at the same time also helped to examine the extent to which the constructs produced results that were posited from theories and thereby providing nomological validity (Anderson & Gerbing, 1988).

It should be emphasised that the primary purpose of the SEM analysis was to examine, in the social ecology of ICT implementation in schools, the pattern of relationships among the contextual variables rather than about the predictive ability of the contextual variables. At the same time, the evaluation of the structural model was not considered as entirely a statistical matter, in a way its fit to the data was assessed in relation to the substantive theory. Accordingly, the goal was to find a model that not only fit the data well from a statistical point of view, but also gave meaningful interpretation (Joreskog, 1993). The aim was to achieve a balance between the two desirable but conflicting characteristics of a model: interpretability and goodness of fit (Browne & Cudeck, 1993). Guided by the theoretical framework of the present study, the social ecology of ICT implementation was modelled with "changes in learning" as an outcome variable, and a set of competing models were generated on the basis of alternative formulations of the underlying theory (Browne & Cudeck, 1993; Hair et al., 1998).

Methods

SEM technique was used to model the complex relationships in the social ecology of ICT implementation in schools given that SEM has the advantage of allowing the researcher to accommodate multiple interrelated dependence relationships in a single model (Hair et al., 1998). In addition, SEM technique has advantage over traditional regression analysis as it provides a mechanism for explicitly taking into account measurement error in the variables (Hoyle, 1995; Jaccard & Wan, 1996; Raykov & Marcoulides, 2000). Another compelling characteristic of the SEM technique is its capacity to estimate and test relations between latent variables that are implied by covariances among multiple indicators (Hoyle, 1995). In research across the social sciences where experimentation is not readily available, SEM technique is particularly valuable as it provides an alternative and complementary methodology to experimentation for examining plausibility of hypothesised models (Maruyama, 1998).

Strategies

The model building task could be thought of as the analysis of two conceptually distinct models, a confirmatory measurement (factor analysis) model that specified the relations of the indicators to their underlying constructs, and a structural model that specified the relations of the constructs to one another (Anderson & Gerbing, 1988). While the measurement and structural models could be estimated simultaneously, Anderson & Gerbing (1988) recommended researchers to use a two-step modelling approach for analysing a postulated model. In the two-step approach, a confirmatory measurement model was specified prior to the simultaneous estimation of the measurement and the structural model (Anderson & Gerbing, 1988). According to Anderson & Gerbing (1988), interpretational confounding occurs when empirical meaning is assigned to a construct of little validity, and the two-step approach had the advantage of reducing the risk of interpretational confounding. With a one-step approach, the presence of interpretational confounding may not be detected, and it increases the risks of maximising fit at the expense of meaningful interpretability of the constructs (Anderson & Gerbing, 1988). The measurement model (CFA model) has already been established in Chapter 4 discussing construct validity. In this chapter, the 2-step approach was used, in which the measurement model together with the proposed structural model were examined simultaneously using LISREL 8.54 (Joreskog & Sorbom, 2001).

In the process of exploring the most plausible model, a theoretically-driven approach of model comparison strategy was adopted (Browne & Cudeck, 1993), in which competing models were generated on the basis of alternative formulations of the underlying theory. The most plausible model was then fine-tuned using the method of nested model comparison, in which the specific hypothesis posited was further tested by adding paths within the most plausible model. It must be emphasised that the construction of nested models in the present study was theoretical but not empirically driven. While relevant details would be discussed in the section of results and discussions, the following sub-sections state out the reasons why we adopted the model comparison strategy and the method of nested model comparison.

Model comparison strategy. There are three distinct strategies that may be employed for model construction and development, namely strictly confirmatory, model generation, and model comparison (Joreskog & Sorbom, 1993). We have not employed the strictly confirmatory strategy as the latter is generally considered as overly rigid, and it is also not used much since only one model of interest will be constructed (MacCallum, 1995). In the model generation strategy, the specified model is evaluated with empirical data for the purpose of modifying the model to improve its parsimony (in the sense of number of parameters) and/ or its fit to the observed data (MacCallum, 1995). While the model generation strategy has been used widely in practice, we considered it as a phenomenological approach of data-driven nature and it would therefore be inappropriate for the present study. In comparison with the model generation strategy, the model comparison strategy is a relatively more theoretically-driven approach, in which model testing is based on comparing a set of competing models which are generated on the basis of alternative formulations of the underlying theory (Browne & Cudeck, 1993; Hair et al., 1998; MacCallum, 1995). Other than that, the researcher regarded some of the competing models in the present study as approximations to reality. It therefore made sense to adopt the model comparison strategy in the present study to investigate similarities among seemingly different approximations to reality (Kumar & Sharma, 1999).

Nested model comparison. Nested model comparison is a common type of model testing method to compare nested models by imposing/releasing a constraint on a starting model (Kline, 1998). Two models are nested if one is the subset of other, an example of which is to drop/ add a path from one model to form its nested model. As recommended by Kline (1998), we used nested model comparison to test the specific and priori hypothesis posited by the most plausible model. Adding paths to the most plausible model according to theoretical criteria, the nested model comparison method supported theoretically guided re-specification of models. Given that all other criteria are equal, this method had the benefit of using of chi-square difference test in evaluating the significance of the improvement in overall fit as paths are added, and a significant difference between the chi-square values of the two nested models supports retention of the path that was just added (Kline, 1998). As will be discussed in the next section, nested model comparison was used to test whether the collegial capacity of ICT implementation strategies fully mediated the effect of organisational interventions on changes in classroom practices (pedagogy and learning).

Input Matrix and Estimation Method

The measure of the present study adopted a 4-point Likert scale. Although the responses to a set of ordered categories obtained from Likert scale measurements were assumed to be quantitative, the variables actually departed from the representation of an interval scale. Ignoring this problem will seriously bias results and conclusions (Joreskog, 1993). To overcome the limitation, an asymptotic covariance matrix of the polychoric correlations was generated (Joreskog & Sorbom,

2001). The present sample size of over 1000 was sufficient enough to produce an accurate measure of an asymptotic covariance matrix of polychoric correlations. Regarding the issue of estimation methods in SEM, the commonly used estimation methods of maximum likelihood (ML) and generalised least squares (GLS) are developed under multivariate normality assumption (Chou & Bentler, 1995; Raykov & Marcoulides, 2000). Nevertheless, real social science data may depart from the normality assumption. As revealed in a study of over 400 large data sets, the great majority of data collected in behavioural research were not normally distributed (West et al., 1995). On the other hand, two alternative estimation methods, namely weighted least squares (WLS) and diagonally weighted least squares (DWLS), do not make assumptions about the distribution of the observed variables (Diamantopoulos & Siguaw, 2000; Flora & Curran, 2004). The estimation method of diagonally weighted least squares (DWLS) was used instead of weighted least squares (WLS) since it was demonstrated that in small sample sizes, DWLS performed better than WLS in terms of chi-square statistics, parameter estimates, and standard errors (Flora & Curran, 2004; Maydeu-Olivares, 2001; Wang & Cunningham, 2005). In specifying the structural models, each construct was made scale invariant by fixing the loading of one indicator per construct to a value of 1.0 (Joreskog & Sorbom, 1993).

Model Evaluation

In evaluating the alternative models, consideration was given to three aspects: (1) the overall fit measures based on a number of fit indexes namely RMSEA (Browne & Cudeck, 1993), NNFI, CFI (Marsh et al., 1996); (2) the component fit measures came from parameter estimates (Bollen, 1989); and (3) the model parsimony based on fit indexes of AIC, CAIC, and ECVI (Joreskog, 1993; Kline,

1998). Before proceeding to model comparison, overall fit indexes of individual models and component fit measures (parameter estimates and R^2) were examined first to check whether any models would be rejected. In other words, overall fit and component fit evaluations were supposed to serve as a preliminary step to screen out unacceptable models. The acceptable models in terms of overall fit and component fit were then compared and evaluated in terms of parsimony indexes that take fit as well as parsimony (in the sense of number of parameters) into account.

Overall fit measures. In evaluating the overall fitness of the models, in addition to the traditional inferential goodness-of-fit index of chi-square value, a number of goodness of fit indexes were used (Browne & Cudeck, 1993; Diamantopoulos & Siguaw, 2000; Kelloway, 1998; Marsh et al., 1996), in particular the root mean square error of approximation (RMSEA), comparative fit index (CFI), and non-normed fit index (NNFI). The traditional chi-square test assumes that the model fits perfectly in the population, and the chi-square statistics is known to be sensitive to departures from large sample size. Models typically do not fit exactly in the population, and in such cases, the magnitude of the chi-square statistics becomes a function of sample size. With very large samples, there is an artificial tendency to obtain a small p value associated with chi-square statistics, and therefore to reject the model even if it is only marginally inconsistent with the data (Raykov & Marcoulides, 2000). This is akin to a Type I error, of which one rejects the null hypothesis of perfect fit when it is true.

Unlike the chi-square value, RMSEA is not sample-dependent (Joreskog & Sorbom, 1993). The RMSEA is a measure used to test the null hypothesis of close fit

(Browne & Cudeck, 1993), and is primarily concerned with evaluating the extent to which the model fails to fit the data (Raykov & Marcoulides, 2000). In addition, there is also the value for the 90% confidence level of the RMSEA, which offers further indication to the precision of estimation. As such, it is generally regarded as one of the most informative fit indexes (Diamantopoulos & Siguaw, 2000), and has recently become quite a popular index of model fit (Raykov & Marcoulides, 2000). It has been suggested that a value of the RMSEA of .05 or less would indicate a close fit of the model in relation to the degrees of freedom (Browne & Cudeck, 1993; Kelloway, 1998), and the null hypothesis of close fit will not be rejected at the 5 % level if the p-value is greater than .05 (Browne & Cudeck, 1993). On the other hand, in simulation studies, NNFI and CFI excelled in areas such as bias due to sample size, penalties for model complexity, rewards for model parsimony, and estimation reliability (Marsh et al., 1996). For CFI and NNFI, an index value of above .90 was interpreted as an indication of reasonably good fit of model data (Kelloway, 1998; Marsh et al., 1996). However, in a recent research examining the conventional cutoff criteria for fit indexes, it has been suggested that RMSEA values of .06 or less indicate a close-fitting model, and CFI values greater than .95 are indicative of good fitting model (Hu & Bentler, 1999).

Component fit measures. It must be however reminded that the indexes should be assessed together with other considerations such as the plausibility of the model and the meaningfulness of its parameters (Browne & Cudeck, 1989; Hu & Bentler, 1995). It has been pointed out that the overall fit of the model may be good, but parameter estimates may not be statistically significant or may have signs opposite to that predicted (Bollen, 1989). For this reason, parameter estimates for the models were examined, which included the signs, magnitude, and the t-value of the path coefficients. A t-value was obtained from dividing the parameter estimate by its standard error, and a t-value of the path coefficients of less than the minimum value of 1.96 indicated that the corresponding parameter was not statistically significantly at the 5% level (Diamantopoulos & Siguaw, 2000). In addition, the squared multiple correlation (R^2) for structural equations were also be examined, which indicated the amount of variance in each endogenous variable that was accounted for by the independent variables that were expected to impact upon it. A small R^2 indicated that the independent variables explain a small portion of variance in the endogenous variables concerned (Bollen, 1989; Diamantopoulos & Siguaw, 2000).

Model parsimony. When models are fitted to the data, the fit of the model can usually be improved by increasing the number of parameters (Browne & Cudeck, 1993; Kline, 1998). Hence, in assessing alternative models, the present study adopted an approach of considering criteria that take fit as well as parsimony (in the sense of number of parameters) into account, of which three strongly related criteria were used: the Akaike Information Criterion (AIC), the Consistent Akaike Information Criterion (CAIC), and the Expected Cross-Validation Index (ECVI) (Joreskog, 1993; Kline, 1998). The essence of the indexes was to examine measures of fit that took the number of parameters in the model into account, thereby the fit of model was not necessarily improved as parameters were added to the model (Browne & Cudeck, 1993). The AIC and CAIC are a modification of the standard goodness-of-fit of chi-square statistic that includes penalty for complexity (Kline, 1998). As single sample approximations to cross-validation coefficient, the ECVI (Browne & Cudeck, 1989) is a cross-validity index that measures the discrepancy between the fitted covariance matrix in the analysed sample and the expected covariance matrix that would be obtained in another sample of the same size (Joreskog, 1993). In applying measures of AIC, CAIC and ECVI to model comparison, a smaller value indicated better fit (Joreskog, 1993; Kline, 1998).

Nested model comparison. In addition to the parsimony fit indexes, the chi-square difference test can be used to evaluate nested models (Kelloway, 1998; Kline, 1998). As recommended by Kline (1998), we used nested model comparison to test the specific and priori hypothesis posited by the most plausible model. In generating a nested model by adding paths, the chi-square generally decreases as paths are added to the model, in other words, the overall fit is improved. In that case, a significant difference in chi-square (3.84) per degree of freedom supported the retention of the path that was just added (Kline, 1998).

Basic Model – Strategic Model

The strategic model represented the hypothesis that the collegial capacity of ICT implementation strategies mediated the effect of organisational interventions on changes in classroom practices (pedagogy and learning). Specifically, "ICT implementation strategies" was proposed as having a direct effect on "changes in learning", and an indirect effect through "changes in pedagogy" on "changes in learning". At the same time, "ICT implementation strategies" was proposed as a variable that mediated the effect of "school climate" and "government ICT policy" on "changes in learning", and "leadership" was proposed as an exogenous variable that related only to "school climate (Figure 5.1).

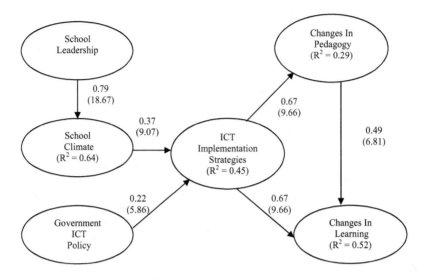

Notes: Path coefficients are shown with t-values in parentheses

Figure 5.1. The basic model (strategic model) examining the pattern of relationships among the constructs in the social ecology of ICT implementation in schools.

First, it has been acknowledged that effective ICT use largely depends on pedagogical interventions as well as organisational interventions. As mentioned in Chapter 2, it has become clear through research evidence that instructional strategy underlying the use of ICT determines learning effectiveness (Honey et al., 2000; McCombs, 2000; Means, 1994; Mehlinger, 1995). At the same time, it has been pointed out that effective ICT integration into curriculum requires organisational intervention since the benefits of ICT cannot be adequately separated from other variables that impact learning in the larger instructional context (Honey et al., 2000; McCombs, 2000).

Second, it was proposed that the collegial capacity of ICT implementation strategies had a direct and indirect effect on changes in student learning. As it has become clear through research evidence that instructional strategy underlying the use of ICT determines learning effectiveness (Honey et al., 2000; McCombs, 2000; Means, 1994; Mehlinger, 1995), an indirect effect of ICT implementation strategies may be through pedagogical intervention on changes in learning. Accordingly "ICT implementation strategies" was proposed as having a direct effect on "changes in learning", and an indirect effect through "changes in pedagogy" on "changes in learning".

Third, it was proposed that the collegial capacity of ICT implementation strategies played a central role in ICT implementation by mediating the effect of other organisational interventions on changes in pedagogy and learning. In building a learning organisation that ultimately uses ICT to advance changes in students' learning, a collaborative school climate with innovative capacity is thought to be important. While there is clear literature support on the role of transformational leadership in developing a positive school climate (Hall & George, 1999; Reinhartz & Beach, 2004; Sergiovanni, 1995), little research evidence has been established on the way how school climate interact with other contextual factors to enhance a paradigm shift in student learning. On the other hand, in a study investigating the organisational dynamics on use of ICT, it was suggested that the collegial capacity of ICT implementation strategies played an important role in offering collegial exchange of professional-development experiences that enhanced changes in pedagogical practices (Dexter et al., 2000; Zhao & Frank, 2003). As such, "ICT implementation strategies" was proposed as a variable that mediated the effect of "school climate" and "government ICT policy" on "changes in pedagogy" and "changes in learning". Taking into account the essential role of transformational leadership on shaping school climate, "leadership" was proposed as an exogenous variable that related only to "school climate".

Competing Model 1 - Non-mediating Model

The non-mediating model was supposed to test against the basic model and other competitive models that were constructed on the basis of theoretical justifications, which supported the mediating effect of organisational variables on changes in student learning. Therefore, in the non-mediating model, all contextual variables were simply proposed as directly related to the outcome variable "changes in learning" (Figure 5.2).

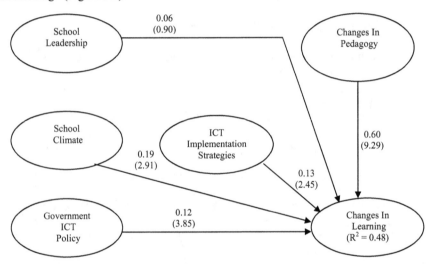

Notes: Path coefficients are shown with t-values in parentheses

Figure 5.2. The non-mediating model examining the pattern of relationships among the constructs in the social ecology of ICT implementation in schools.
Competing Model 2 – Climatic Model

The climatic model represented the hypothesis that the climate for collaboration and experimentation mediated the effect of organisational interventions on changes in classroom practices (pedagogy and learning). While acknowledging the importance of a collaborative school climate with innovative capacity in building a learning organisation that ultimately uses ICT to advance changes in students' learning, little research evidence has been established on how school climate interacts with other contextual factors to enhance a paradigm shift in student learning. As a competitive model based on alternative hypothesis, it was proposed that a climate for collaboration and experimentation played a central role in the social ecology of ICT implementation by mediating the effect of other organisational interventions on changes in pedagogy and learning. Specifically, "school climate" was proposed as having a direct effect on "changes in learning", and an indirect effect through "changes in pedagogy" on "changes in learning". At the same time, "school climate" was proposed as a variable that mediated the effect of "leadership", "ICT implementation strategies" and "government ICT policy" on "changes in learning" (Figure 5.3).

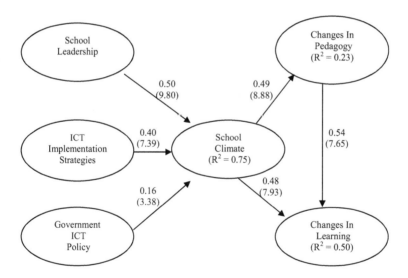

Notes: Path coefficients are shown with t-values in parentheses

Figure 5.3. The climatic model examining the pattern of relationships among the constructs in the social ecology of ICT implementation in schools.

Competing Model 3 – External Model

The external model represented the hypothesis that the external input in the form of government ICT policy mediated the effect of organisational interventions on changes in classroom practices (pedagogy and learning). As mentioned in Chapter 2, "Government ICT policies" was identified as a relevant external input that influenced the dynamics of ICT implementation. As a competitive model based on alternative hypothesis, it was proposed that the government ICT policy played a central role in ICT implementation by mediating the effect of other organisational interventions on changes in pedagogy and learning. Specifically, "government ICT policy" was proposed as having a direct effect on "changes in learning", and an indirect effect through "changes in pedagogy" on "changes in learning". At the same time, "government ICT policy" was proposed as a variable that mediated the effect of "climate" and "ICT implementation strategies" on "changes in learning". Taking into account the essential role of transformational leadership on shaping school climate, "leadership" was proposed as an exogenous variable that related only to "school climate" (Figure 5.4).

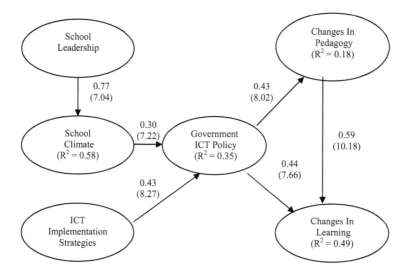

Notes: Path coefficients are shown with t-values in parentheses

Figure 5.4. The external model examining the pattern of relationships among the constructs in the social ecology of ICT implementation in schools.

Competing Model 4 – Pedagogical Model

The pedagogical model represented the hypothesis that pedagogical intervention mediated the effect of organisational interventions on changes in learning. As mentioned in Chapter 2, it has become clear through research evidence that instructional strategy underlying the use of ICT determines learning effectiveness (Honey et al., 2000; McCombs, 2000; Means, 1994; Mehlinger, 1995). As a competitive model based on alternative hypothesis, it was proposed that in the social ecology of ICT implementation, the effect of organisational interventions on changes in learning was mediated entirely by pedagogical intervention. Specifically, "changes in pedagogy" was proposed as the variable that mediated the effect of "climate", "ICT implementation strategies", and "government ICT policy" on "changes in learning". Taking into account the essential role of transformational leadership on shaping school climate, "leadership" was proposed as an exogenous variable that related only to "school climate" (Figure 5.5).

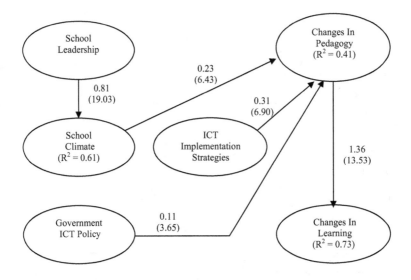

Notes: Path coefficients are shown with t-values in parentheses

Figure 5.5. The pedagogical model examining the pattern of relationships among the constructs in the social ecology of ICT implementation in schools.

Competing Model 5 – Strategic and Pedagogical Model

The strategic and pedagogical model was a simple modification of the strategic model, in which the path from "ICT implementation strategies" to "changes in learning" was trimmed from the model. The strategic model (Figure 5.2) represented the hypothesis that the collegial capacity of ICT implementation strategies mediated the effect of organisational interventions on changes in classroom practices (pedagogy and learning), and specifically "ICT implementation strategies" was proposed as having a direct effect on "changes in learning", and an indirect effect through "changes in pedagogy" on "changes in learning". For the strategic and pedagogical model, taking into account the research evidence supporting that instructional strategy underlying the use of ICT determined learning effectiveness

(Honey et al., 2000; McCombs, 2000; Means, 1994; Mehlinger, 1995), it was proposed that the effect of "ICT implementation strategies" on "changes in learning" was mediated entirely by "changes in pedagogy". As a result, the path from "ICT implementation strategies" to "changes in learning" was trimmed from the model (Figure 5.6).

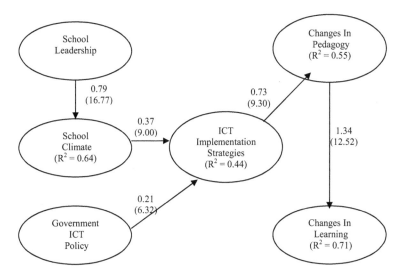

Notes: Path coefficients are shown with t-values in parentheses

Figure 5.6. The strategic and pedagogical model examining the pattern of relationships among the constructs in the social ecology of ICT implementation in schools.

Results and Interpretations

It should be emphasised that the primary purpose of the SEM analysis was to examine, in the social ecology of ICT implementation in schools, the pattern of relationships among the constructs rather than about the predictive ability of the constructs. At the same time, the evaluation of the structural model was not considered as entirely a statistical matter, in a way its fit to the data was to be assessed in relation to the substantive theory. Accordingly, the goal was to find a model that not only fit the data well from a statistical point of view, but also gave meaningful interpretation (Joreskog, 1993). A competing model strategy was adopted (Joreskog & Sorbom, 1993). In addition to the basic model (Figure 5.1), a competing model (non-mediating model of Figure 5.2) was generated for testing the proposition of mediation effect, and four other competing models (Figure 5.3, 5.4, 5.5 and 5.6) were generated on the basis of alternative formulations of the underlying theory (Browne & Cudeck, 1993; Hair et al., 1998; MacCallum, 1995).

Evaluation of Competitive Models

Results for model evaluations will be presented in the following sub-sections, with a focus on three aspects, namely overall fit, component fit, and parsimony fit. For easy reference, Table 5.7 shows the correlation matrix for the variables, Table 5.8 and 5.9 show the values of selected fit indexes and parameter estimates for each proposed model.

Table 5.7
Correlation Matrix for the 6 Variables in the Social Ecology of ICT Implementation

	1	2	3	4	5	6
1. Leadership	1.00					
2. Climate	.73	1.00				
3. Govt ICT Policy	.25	.27	1.00			
4. ICT Implementation Strategies	.37	.37	.31	1.00		
5. Changes in Pedagogy	.31	.30	.23	.43	1.00	
6. Changes in Learning	.41	.44	.32	.43	.62	1.00

Overall fit evaluation. Before proceeding to model comparison, overall fit indexes of individual models were examined first to check whether any models would be rejected. It can be seen that nearly all the selected fit indexes for the models fell

152

within an acceptable range, except the chi-square values for the models and the NNFI

(.94) for the external model (Table 5.8). As the chi-square statistic assumes that the

model fits perfectly in the population, and is known to be sensitive to departures from

the present large sample size, we therefore made reference to other indexes such as

RMSEA, NNFI and CFI as well. Overall, other than the external model, all models

had index values indicating reasonably good fit of model data.

Table 5.8
Indexes Comparing the Goodness of Fit and Parsimony for Alternative Models

Goodness of fit index	Recommended good fit values	Basic model: Strategic	Non-mediating model	Climatic model	External model	Pedagogical model	Strategic & pedagogical model
Satorra-Bentler chi-square	---	1597.98	1636.31	1689.54	1769.46	1714.38	1626.86
Degree of freedom	---	488	480	486	488	487	489
Chi-square per degree of freedom	<3.0	3.27	3.41	3.48	3.63	3.52	3.33
P-value for chi-square test	>.05	.00	.00	.00	.00	.00	.00
RMSEA	<.06	.045	.047	.047	.049	.048	.046
90% Confidence Interval for RMSEA	<.06	.043 .048	.044 .049	.045 .050	.046 .051	.045 .050	.043 .048
P-value for Test of Close Fit (RMSEA < .05)	>.05	1.00	.99	.96	.79	.93	1.00
NNFI	>.95	.97	.98	.97	.94	.97	.97
CFI	>.95	.97	.98	.97	.95	.98	.97
SRMR	<.08	.072	.067	.074	.10	.071	.074
ECVI	Smaller value	1.58	1.63	1.67	1.74	1.69	1.60
AIC	Smaller value	1743.98	1798.31	1839.54	1915.46	1862.38	1770.86
CAIC	Smaller value	2182.54	2284.92	2290.11	2354.02	2306.94	2203.40

Component fit evaluation. It must be however reminded that the indexes of overall fit should be assessed together with other considerations such as the plausibility of the model and the meaningfulness of its parameters (Browne & Cudeck, 1989; Hu & Bentler, 1995). Results of the parameter estimates were shown in Figure 5.1 to 5.6 and Table 5.9. Upon examination, it can be seen from Table 5.9 that the non-mediating model had some unreasonable results. The magnitude of the path coefficients from "leadership" to "changes in learning" was very low (.07) with t-value (.89) less than the minimum value of 1.96, indicating that the corresponding parameter was not statistically significantly at the 5% level (Figure 5.2 and Table 5.9). In addition, specific R^2 for the external model was relatively low, which indicated that the independent variables explained a small portion of variance in the endogenous variables concerned (Bollen, 1989; Diamantopoulos & Siguaw, 2000). Specifically, the effect of "government ICT policy" explained only 18% of variance in "changes in pedagogy" (Figure 5.4 and Table 5.9). Overall, results of parameter estimates supported the rejection of the non-mediating model and the external model.

Model parsimony evaluation. Before proceeding to model parsimony evaluation, the overall fit and component fit evaluations already screened out the external model and the non-mediating model. The external model was rejected owing to its low NNFI (.94) and low R^2 for pedagogy (.18), and the non-mediating model was rejected owing to an unreasonable parameter estimate (.07 with t-value of .89). On the other hand, the other four models (strategic model, climatic model, pedagogical model, strategic & pedagogical models) had a reasonably good fit of model data in terms of RMSEA (<.06), NNFI (>.95), and CFI (>.95). Magnitude and t-values of path coefficients (>1.96) as well as R^2 (ranging from .23 to .75) for the models were also within the acceptable range. The four models were further compared with reference to selected parsimony indexes (ECVI, AIC, and CAIC) that took fit as well as parsimony (in the sense of number of parameters) into account. From Table 5.8, it can be seen that the values of ECVI (1.58), AIC (1743.98), and CAIC (2182.54) all favoured the basic model (strategic model), which had the lowest values among the models. Nevertheless, its magnitude of difference in terms of parsimony indexes over the other models was not very significant. As the four competing models were supposed to be approximations to reality, it was worthwhile to investigate similarities among the seemingly different approximations to reality instead of finding the most plausible models.

Table 5.9
Parameter Estimates for Alternative Models

Model	Structural Equations	R^2	Unreasonable Parameter Estimates (t-value)
Basic (Strategic)	climate = .79 leadership	.64	---
	strategies = .37 climate + .22 policy	.45	---
	pedagogy = .67 strategies	.29	---
	learning = .49 pedagogy + .67 strategies	.52	---
Non-mediating	learning = .07 leadership + .19 climate + .12 policy + .13 strategies + .60 pedagogy	.48	Leadership $\rightarrow$ learning .07 (.89)
Climatic	climate = .50 leadership + .40 strategies + .16 policy	.75	---
	pedagogy = .49 climate	.23	---
	learning = .54 pedagogy + .48 climate	.50	---
External	climate = .77 leadership	.58	---
	policy = .58 climate + .43 strategies	.32	---
	pedagogy = .43 policy	.18	---
	learning = .59 pedagogy + .44 policy	.48	---
Pedagogical	climate = .81 leadership	.61	---
	pedagogy = .23 climate + .31 strategies + .11 policy	.41	---
	learn = 1.36 pedagogy	.73	---
Strategic & pedagogical	climate = .79 leadership	.64	---
	strategies = .37 climate + .21 policy	.44	---
	pedagogy = .73 strategies	.55	---
	learning = 1.34 pedagogy	.71	---

Investigating Similarities Among Plausible Models

The four plausible models shared similarities. In investigating similarities among the models, we expected to discover some patterns that might give insights into the social ecology of ICT implementation in schools. First, the models supported the proposition that changes in pedagogy mediated the effect of organisational intervention on changes in learning, though in different patterns. Second, results of comparison did not give a unanimous answer on the pattern of the mediation effect with regard to changes in pedagogy. The basic model and the climatic model supported a partial mediation effect of changes in pedagogy. In other words, organisational interventions might directly affect changes in learning, or indirectly through changes in pedagogy to affect changes in learning. However, the pedagogical model and the strategic & pedagogical model supported a full mediation effect of changes in pedagogy, of which organisational interventions only indirectly affected changes in learning through changes in pedagogy. Third, the models provided support for the important role of transformational leadership on shaping a climate for collaboration and experimentation, in which the path coefficients from transformational leadership to school climate were high (ranging from .50 to .81). Despite the above differences, the similarities shared among the four models of approximations to reality confirmed the critical role of pedagogical intervention in bringing about changes in learning as well as the critical role of transformational leadership in shaping school climate.

Dichotomy Comparison of Groups of Models

While the comparisons pointed out that changes in pedagogy assumed a mediation role in transferring the effect of organisational interventions to changes in learning, the pattern of mediation was not yet confirmed. Digging deeper into the analysis of parsimony fit indexes for the four plausible models, we discovered a dichotomy categorisation pattern. The dichotomy pattern pointed to two groups of models, of which the basic model and strategic & pedagogical model belonged to a group of lower magnitude of parsimony fit indexes that represented better fit (ECVI of 1.58 and 1.60), and the climatic model and pedagogical model represented another group of model of higher magnitude of parsimony fit indexes (ECVI of 1.67 and 1.69).

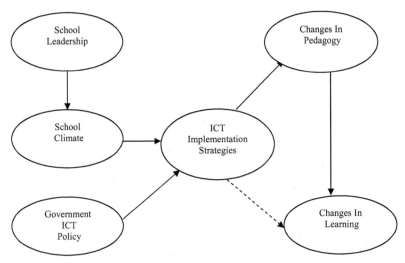

- - - -► Representing path that was present in the basic model but absent in the strategic & pedagogical model.

Figure 5.10. Models Belonging to the Group of Better Fit in Parsimony Indexes

The better-fit group of models (basic model and strategic & pedagogical model (Figure 5.10), shared the same proposition of placing the collegial capacity of ICT implementation strategies as a central focus, which mediated the effect of organisational interventions on changes in pedagogy or changes in classroom practices (pedagogy and learning). Alternatively, it was also plausible, but relatively less affirmative, to place the school climate as a central focus in mediating the effect of organisational interventions on changes in classroom practices (pedagogy and learning). And it was also plausible, again less affirmative, to place the changes in pedagogy as a central focus in mediating the effect of organisational interventions on changes in learning. In any case, results of parsimony index comparison affirmed the central role of the collegial capacity of ICT implementation strategies in the social ecology of ICT implementation.

Nested Model Comparison

Despite its magnitude of difference in terms of parsimony indexes over the other models was not very significant, the basic model excelled the other models in terms of parsimony indexes, and had the lowest values of ECVI (1.58), AIC (1743.98), and CAIC (2182.54). The basic (strategic) model represented the hypothesis that the collegial capacity of ICT implementation strategies mediated the effect of organisational interventions on changes in classroom practices (Figure 5.1). At this juncture, the effect of school climate and government ICT policy on changes in pedagogy was mediated fully by the collegial capacity of ICT implementation strategies. However, it was likely that school climate and/or government ICT policy might also have a direct effect on changes in pedagogy. The mediating hypothesis with regard to the collegial capacity of ICT implementation strategies was then tested using method of nested model comparison, in which two nested models were generated with the addition of a direct path from school climate and/or government ICT policy to changes in pedagogy.

Nested model 1 (Figure 5.11) represented the partially mediating role of collegial capacity of ICT implementation strategies with respect to school climate, in which school climate had a direct effect on changes in pedagogy, and an indirect effect through the collegial capacity of ICT implementation strategies on changes in pedagogy.

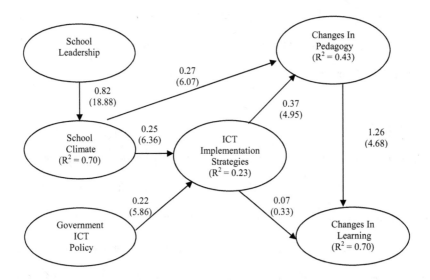

Notes: Path coefficients are shown with t-values in parentheses

Figure 5.11. Nested model 1 testing against the basic model.

Nested model 2 (Figure 5.12) represented the partially mediating role of collegial capacity of ICT implementation strategies with respect to school climate and government ICT policy, in which both school climate and government ICT policy had a direct effect on changes in pedagogy, and an indirect effect through the collegial capacity of ICT implementation strategies on changes in pedagogy.

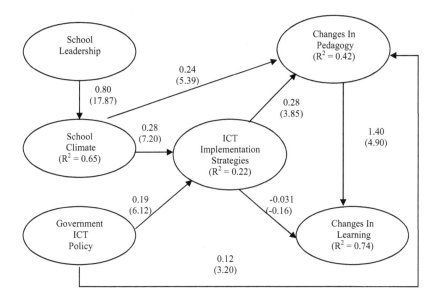

Notes: Path coefficients are shown with t-values in parentheses

Figure 5.12. Nested model 2 testing against the basic model

Table 5.13 presents the results of the nested model comparison. As shown from the chi-square difference test and indexes indicating overall fit and parsimony fit, the basic model was the best model in fitting the data. In nested models, the chi-square generally decreases as paths are added, in other words, the overall fit is improved (Kline, 1998). However, as paths were added in generating nested model 1 and 2, the chi-square increased. This indicated that the overall fit of the nested models worsened, and therefore the alternative hypotheses of adding paths were rejected. This further supported the hypothesis that the collegial capacity of ICT implementation strategies fully mediated the effect of organisational interventions (school climate and government ICT policy) on changes in classroom practices (pedagogy and learning).

Table 5.13

Indexes Comparing the Goodness of Fit and Parsimony for Nested Models Testing the Mediating Effect of ICT Implementation Strategies from Organisational Interventions to Changes in Pedagogy and Learning

Goodness of fit index	Recommended good fit values	Basic model: fully mediating	Nested model 1: fully & partially mediating	Nested model 2: partially mediating
Satorra-Bentler chi-square	---	1597.98	1661.70	1701.89
Degree of freedom	---	488	487	486
Chi-square difference with reference to basic model		---	63.72	103.91
RMSEA	<.06	.045	.047	.048
ECVI	Smaller value	1.58	1.64	1.68
AIC	Smaller value	1743.98	1809.70	1851.89
CAIC	Smaller value	2182.54	2254.26	2302.46

Summary

This paragraph summed up the overall results of model building. Table 5.8 and Table 5.9 show the values of selected fit indexes and parameter estimates for each model. Out of the six competitive models, the external model and the non-mediating model were rejected due to the poor overall fit and component fit. Of the four competing models, while the values of parsimony indexes all favoured the basic model (strategic model), its magnitude of difference over the other models was not very significant. Nevertheless, the similarities shared among the four models of approximations to reality confirmed the critical role of pedagogical intervention in the social ecology of ICT implementation in schools. The comparisons further pointed out that changes in pedagogy assumed a mediation role in transferring the effect of organisational interventions to changes in learning, yet the pattern of mediation was not yet confirmed. Digging deeper into the analysis of parsimony fit indexes for the four plausible models, we discovered a dichotomy categorisation pattern. The better-fit group of models (basic model and strategic & pedagogical model) shared the same proposition of placing the collegial capacity of ICT implementation strategies as a central focus (Figure 5.10), which affirmed the central role of the collegial capacity of ICT implementation strategies in the social ecology of ICT implementation. Using method of nested model comparison, the fully mediating effect of the collegial capacity of ICT implementation strategies in the strategic model was tested, and results supported the hypothesis that the collegial capacity of ICT implementation strategies fully mediated the effect of organisational interventions (school climate and government ICT policy) on changes in classroom practices (pedagogy and learning).

Discussions

The rejection of the non-mediating model revealed that, in the social ecology of ICT implementation in schools, it was likely that organisational interventions and pedagogical interventions interacted with each other to affect changes in student learning. Specifically, changes in pedagogy as perceived by teacher were found to be a mediating variable transferring the effect of organisational interventions to the outcome variable changes in student learning. Of its own accord, perceived changes in pedagogy were found to be a significant predictor of perceived changes in learning. The findings supported the important role of instructional strategy underlying the use of ICT in determining learning effectiveness.

The rejection of the external model also gave us insights into the dynamics of organisational interventions, which highlighted the central role of internal organisational interventions, in comparison with external policy, in affecting changes in student learning. Notably, the collegial capacity of school's ICT implementation strategies as perceived by teacher was identified as one of the possible focal point of linking up the organisational interventions to the outcome variable (changes in student learning). While the organisational interventions were shown to have no direct impact on the outcome variable, they were capable of exerting indirect influences by shaping the collegial capacity of school's ICT implementation strategies. A transformational leadership had no direct effect on classroom practices (changes in pedagogy and student learning), yet it exerted an indirect effect through shaping the school climate.

The findings of SEM analysis had significant implications on understanding the social ecology of ICT implementation in an education setting. In adopting a sociocultural approach, the study was able to measure the abstract concept of ICT implementation strategies by conceptualising it as the capacity of implementation strategies in enhancing collegial ICT exchange and in mobilising resources. Built on this, the study further adopted an ecological approach to sketch out the social dynamics of ICT implementation by studying the interactions among the relevant contextual factors. This way, the study was successful in demonstrating that the collegial capacity of ICT implementation strategies played a central and mediating role in affecting perceived changes in learning and pedagogy from a teacher-centred approach to a student-centred approach. The results, together with the reliability and validity of the scales and the adequacy of the structural model, guided researchers to believe that the collegial capacity of the ICT implementation strategies brought the effect of organisational interventions to changes in classroom practices.

From the empirical results, schools are advised to give more emphasis on enhancing the collegial exchange and sharing of ICT experiences in order to harness the power of ICT in shifting learning from a teacher-centred to a student-centred approach. The significant path coefficient from changes in pedagogy to changes in learning has clear implication on the crucial role of teachers' pedagogical approach in students' learning outcome. It supported the postulation that ICT implementation strategies will exhibit significant effect on changes in learning outcome if there is a change effected by ICT on teachers' pedagogy in classroom practices. In addition, empirical evidence of the present study supported that a transformational leadership was highly influential in establishing a cohesive and proactive school climate that enhanced the collegial capacity of ICT implementation strategies. It has clear implication on the important role of leadership in schools. School leaders may consider setting higher priority to building a cohesive and proactive school climate, which actually facilitates an effective ICT implementation that can bring about positive changes in learning. Apart from that, an effective government ICT policy was also very important in facilitating effective ICT implementation in schools, though the impact was not as significant as that exerted by the school climate. In the process of implementing ICT in education, an effective government ICT policy as perceived by teachers really made a difference in transforming classroom practices.

Chapter 6: Results (Part 3) – Multilevel Model Building

Introduction

In pooling teacher data across all schools with structural equation modelling
(SEM) analysis (Chapter 5), we have already discussed how the pedagogical and
organisational factors interacted with each other to affect changes in student learning.
In this way, the SEM model helped us to understand the social ecology of ICT
implementation among teachers pooled from the entire sample, irrespective to which
school the teachers belonged. However, Hallinger & Heck (1998) argued that ignoring
the multilevel nature of the data collected might incur significant discrepancies in the
results. In other words, if we regard school as a single unit and do a regression
analysis for each individual school, we may obtain a very different set of regression
coefficients as compared to those obtained from the regression over the data pooled
from the entire sample. To deepen our understanding on the SEM model developed in
Chapter 5, it is therefore necessary to examine the multilevel effects on the outcome
variable, to see (1) whether the regression results obtained from each individual
school are consistent with those in the SEM analysis and; (2) how much variation
found in the outcome variable can be attributed to school-level variables. These
school-level effects can significantly affect the results as it has been reported that
inconsistencies found in principal leadership research results was due to the neglect of
multilevel nature of schooling (Hallinger & Heck, 1998).

Built on top of the SEM model, this chapter is going to use hierarchical linear
modelling (HLM) technique to investigate whether school-level variables will have
any effect on changes in student learning. In the context of the present study, these
school-level variables were mean school leadership, mean school climate, mean ICT

implementation strategies, mean government ICT policy, and mean changes in pedagogy, etc. These school-level variables were basically associated with those (teacher-level) variables identified previously in Chapter 5, but they were derived from teacher aggregate of the same school instead. Here, teacher-level or level-1 variables were referred to the variables measuring the scores from each individual teacher. However, to widen the context and source of school-level variables, some relevant structural characteristics of the school that were not derived from teacher aggregate but from sources like heads of ICT team and school principals were also selected. These school-level variables included ICT infrastructure, ICT competency, and school type (primary or secondary schools). The section on variables will give more details (Table 6.3 and 6.4).

Methods and Results

Underlying Principles

To examine the multilevel effects on the outcome variable, we employed the HLM technique (Raudenbush et al., 2001) to build a 2-level model, with particular emphasis on the school-level effect. To address the school-level effect, the HLM analysis had the advantage of providing a means of partitioning the outcome variable's variance into within units (teacher level: variance among teachers within the same school) and between units (school level: variance among schools); and within the analysis, a means of assigning variables to different levels (Heck & Thomas, 2000). In the 2-level model of the present study, "changes in student learning" was viewed as a function of both teacher-level (level-1) variables and school-level (level-2) variables. In other words, variation in "changes in student learning" was explained in terms of teacher-level variables (perception of teachers) as well as school-level variables (teacher aggregate of the same school and structural characteristics from other sources). At level-1, the units were teachers within the same school, and each teacher's outcome was represented as function of a set of teacher-level variables. At level-2, the units were schools. The regression coefficients (slopes and intercepts) in level-1 model for each school were conceived as outcome variables that were hypothesised to depend on specific school-level variables (Raudenbush & Bryk, 2002).

SEM and HLM are statistical techniques with different approach and emphasis. While the SEM model captures the linear inter-relationships among variables in one model, the HLM uses multilevel models (a 2-level model in our study) to delineate a linear relationship between an outcome variable and the explanatory variables at different levels. The multilevel models allow us to construct separate level-1 linear models for each school, in which each school has the same set of (level-1) variables and the same outcome variable, but with a different set of regression coefficients. It is anticipated that these variations in regression coefficients found among schools can be explained by the school-level (level-2) variables. In other words, the regression coefficients of the level-1 models are regressed on the school-level variables (Kreft & De Leeuw, 1998). In the HLM analysis, it will give each random regression coefficient an expected value ('mean') and a variance. In this way, we are able to estimate the expected value of each random coefficient among all teachers irrespective of the school to which they belong, on the other hand, we may also know from the variance the deviation of each school from the expected value (Kreft & De Leeuw, 1998).

To use HLM to extend our understanding on the SEM model, we focused only on part of the multiple linear relationships from the strategic/basic model in Chapter 5 as a basis to develop the level-1 model (teacher-level model) (Figure 6.1). The strategic/basic model was selected as it excelled the other plausible models in terms of parsimony indexes, indicating that it was the most plausible model among the competing models (Chapter 5).

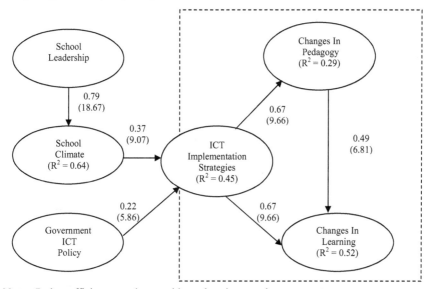

Notes: Path coefficients are shown with t-values in parentheses

Figure 6.1. HLM analysis focusing on part of the linear relationships of the basic model (strategic model).

At the teacher-level, we had the outcome variable "changes in student learning", which was measured by a factor composite of 6 teacher responses (Chapter 3), and we also had two variables "changes in pedagogy" and "ICT implementation strategies". The HLM allowed us to set up separate regression equations in each school to predict the outcome variable by the teacher-level variables:

$$(\text{learning})_{ij} = \beta_{0j} + \beta_{1j}(\text{changes in pedagogy})_{ij}$$

$$+ \beta_{2j}(\text{ICT implementation strategies})_{ij} + r_{ij}$$

Here, the subscripts i, j were the respective running indices for schools (j = 1 to J), and individual teachers clustered within a school (i = 1 to n_j). In a particular school j with n_j teachers, "changes in student learning" was modelled as a function of the level-1 intercept (β_{0j} = school mean for changes in student learning), teacher-level variables, and residual error (r_{ij}). Unlike typical regression that assumed fixed intercept and slope, HLM allowed the level-1 intercept (β_{0j}) and slope coefficients (β_{1j}, β_{2j}, ...) of the level-1 model to vary from school to school. In other words, we might assume that each school had a different level-1 intercept (β_{0j}) and a different set of slope coefficients (β_{1j}, β_{2j}, ...), which were indicated by a subscript j to the regression coefficients. The residual errors (r_{ij}) were assumed to have a mean of zero, and a variance to be estimated (Hox, 2002). Without pooling data across all schools, the HLM level-1 model enabled us to obtain an estimation of effects within teachers of the same school (Heck & Thomas, 2000; Kreft & De Leeuw, 1998; Raudenbush & Bryk, 2002).

As we allowed the level-1 intercept (β_{0j}) and slope coefficients (β_{1j}, β_{2j}, ...) of the level-1 model to vary from school to school, each coefficient would have a distribution with a mean and variance. The next step in the HLM was to explain the variation of the regression coefficients by introducing variables at the school level:

$$\beta_{0j} = \gamma_{00} + \gamma_{01}Z_j + \mu_{0j}$$
$$\beta_{1j} = \gamma_{10} + \gamma_{11}Z_j + \mu_{1j}$$
$$\beta_{2j} = \gamma_{20} + \gamma_{21}Z_j + \mu_{2j}$$
.....
.....

At level-2 (school-level model), the intercept of the level-1 model (β_{0j} = school mean for changes in student learning for school j) was allowed to vary from school to school as a function of the level-2 intercept (γ_{00} = average school mean for changes in student learning), school-level factors (e.g. Z_j = mean school climate), and residual error (μ_{0j}). In this example, mean school climate was proposed to predict the average school mean for changes in student learning. Similarly, slope coefficients (β_{1j}, β_{2j}, ...) would also be allowed to vary from school to school as a function of school-level factors (e.g. Z_j = mean school climate). The residual error terms at school level (μ_{0j}, μ_{1j}, μ_{2j}, ...) were assumed to have a mean of zero, and to be independent from the residual errors at the teacher-level (r_{ij}), and a variance to be estimated. In this way, the level-2 analysis enabled us to obtain an estimation of effect across schools (Heck & Thomas, 2000; Kreft & De Leeuw, 1998; Raudenbush & Bryk, 2002).

While the level-1 model attempted to capture the within-school variation of perceived changes in student learning among teachers clustered in the same school, the level-2 model examined among schools how much variation in the school mean for changes in student learning (level-1 intercept) and variation in the slopes (β_{1j}, β_{2j}, ...) of teacher-level variable(s) could be explained by school-level variables. In this way, the school-level effect behind the scene could be examined. The later sections will give more details.

Strategies

To examine the school effect on the outcome variable, we adopted an exploratory procedure to select a hierarchical model in a systemic manner as suggested by Hox (2000). Exploratory in a sense that we found a plausible model through the means of constructing alternative level-2 models based on different propositions for comparison. A "step-up" strategy (Table 6.2) was adopted in the exploratory process of model building, with theoretical guidance to define a small set of variables as listed in Table 6.3, 6.4 and 6.5 (the next section will give details on the kinds of variables selected) (Raudenbush & Bryk, 2002).

Defining variables. The SEM model derived in Chapter 5 was used as the cornerstone for formulating the level-1 (teacher-level) model for the HLM analysis. In addition, teacher-level variables should also include some demographic data of teachers to adjust for the differences among schools. According to Raudenbush & Bryk (2002), such kind of statistical adjustments for individual background are very important for two reasons. First, because teachers are not usually assigned at random to schools, controlling for background may reduce its bias on estimation of school effects. Second, if variables of individual background are strongly related to the outcome variable (changes in student learning), controlling for them will increase the precision of any estimates of school effect and the power of hypothesis tests by reducing unexplained level-1 error variance. In selecting school-level variables, in addition to those aggregate as derived from variables used in the SEM model building (Chapter 5), we decided to include some relevant structural characteristics of the school that were not derived from teacher aggregate but from sources like heads of ICT team and school principals. The rationale behind was to widen the context and source of school-level variables. Ultimately, we had two types of school-level variables coming from a variety of sources, those measuring the sociocultural factors relating to ICT implementation as identified in the SEM model of Chapter 5, and those measuring structural characteristics of schools (Table 6.3, 6.4, and 6.5).

Step-up strategy. The "step-up" strategy referred to starting with the simplest possible model, and then to add various types of parameters step by step (Hox, 2002) (Table 6.2). As elaborated in later sections, we started with building a fully unconditional model with no variables, which was used to identify the source of variations within the outcome variable "changes in learning" by partitioning the total variance in the outcome variable into its within-school (level-1) and between-school (level-2) components.

Following this, we added demographic variables and teacher-level variables to the fully unconditional model (level-1 model). The demographic variables were supposed to adjust for the differences among schools. This was to generate more accurate estimates of school effect on changes in student learning (Raudenbush & Bryk, 2002; Raudenbush & Willms, 1991). Selection of teacher-level variables was based on part of the linear relationship from the SEM model (basic/strategic model in Chapter 5), which included changes in pedagogy and ICT implementation strategies. Significance of the level-1 variables was judged with reference to the t values. A t-value was obtained from dividing the estimated regression coefficient by its standard error, and a t-value of less than the minimum value of 1.96 indicated that the corresponding parameter was not statistically significantly at the 5% (p = .05) level.

After generating the level-1 model, the next step was to decide whether the slope coefficient of teacher-level variables was conceived as fixed or random, the decision of which was based on the reliability of the random teacher-level coefficient as derived from ordinary least squares (OLS) regression estimator as well as the p-value of its estimated level-2 variances (Raudenbush & Bryk, 2002). The reliability of the estimated level-1 coefficient measured the ratio of the true score relative to the observed score, which was obtained from dividing the parameter variance by the total of parameter variance and error variance. Reliability value of less than .05 indicated that the variances we wished to estimate were likely to be close to zero (Raudenbush & Bryk, 2002). In other words, this implied no variation of the slope among groups. The p-value of estimated level-2 variances with respect to specific level-1 coefficient was associated with the hypothesis of slope homogeneity (chi-square and likelihood-ratio tests). A value of less than .05 indicated that the probability of the estimated variability in the random level-1 coefficient, under a homogeneity hypothesis, was less than .05 (Raudenbush & Bryk, 2002). In other words, p-value of less than .05 implied rejecting the hypothesis of slope homogeneity, which on the other hand implied variation of the slope among groups.

The final step was to assign school-level variables into the level-2 models. Alternative level-2 models were built on adding distinct types of school-level variables, those measuring the sociocultural factors relating to ICT implementation as identified in the SEM model of Chapter 5, and those measuring structural characteristics of schools. In this process, the proportion of reduction in variance as accounted for by the model served as a basis for making judgement about the relative importance of the types of variables (Raudenbush & Bryk, 2002). As suggested by Hox (2000), a variable was considered to have a small effect if it explained 1% variance, and a medium effect if it explained 10% variance, and a large effect it explained 25% variance.

Table 6.2

Step-Up Strategy Summarising Major Steps in HLM Model Building

---

1. Constructing a fully unconditional model

Level 1: $(learning)_{ij} = \beta_{0j} + r_{ij}$

Level 2: $\beta_{0j} = \gamma_{00} + \mu_{0j}$

2. Level-1 model: adding demographic and teacher-level variables

Level 1: $(learning)_{ij} = \beta_{0j} + \beta_{1j}(demographic)_{ij} + \beta_{2j}(teacher\text{-}level)_{ij} + r_{ij}$

Level 2: $\beta_{0j} = \gamma_{00} + \mu_{0j}$

3. Random coefficient model: deciding which slope coefficients be random

Level 1: $(learning)_{ij} = \beta_{0j} + \beta_{1j}(demographic)_{ij} + \beta_{2j}(teacher\text{-}level)_{ij} + r_{ij}$

Level 2: $\beta_{0j} = \gamma_{00} + \mu_{0j}$

                 $B_{2j} = \gamma_{20} + \mu_{2j}$

4. Level-2 model: adding school-level variables

Level 1: $(learning)_{ij} = \beta_{0j} + \beta_{1j}(demographic)_{ij} + \beta_{2j}(teacher\text{-}level)_{ij} + r_{ij}$

Level 2: $\beta_{0j} = \gamma_{00} + \gamma_{01}Z_j + \mu_{0j}$

                 $B_{2j} = \gamma_{20} + \gamma2_1Z_j + \mu_{2j}$

Sample

As stated in the introduction of the chapter, the focus of the HLM analysis was to investigate whether school-level variables would have any effect on changes in student learning, which was of strong interest in the random part (the variance and their standard errors) instead of the fixed parameters ('mean', i.e. γ_{00}, γ_{10} etc.). It was suggested that for such case, the number of groups should be considerably larger, which led to about 100 groups with 10 individuals per group (Hox, 2002). Heading in this direction, the present study recruited 9 or 10 teachers from 130 schools (Chapter 3). Before proceeding to data analysis, listwise deletion was used to handle missing data (Raudenbush et al., 2001), and ultimately 963 teachers nested within 122 schools were obtained, as shown in the descriptive statistics of Table 6.3. The average school sample size was 8 teachers.

Table 6.3

Descriptive Statistics for Teacher-Level and School-Level Variables

Variable	No.	Mean	SD	Min.	Max.
Teacher-level					
Sex	963	---	---	1.00	2.00
Age	963	2.77	1.09	1.00	5.00
Teaching experience	963	2.63	1.28	1.00	5.00
ICT implementation strategies	963	2.23	0.40	1.00	4.00
Changes in pedagogy	963	2.83	0.41	1.33	4.00
Changes in learning	963	2.56	0.46	1.00	4.00
School-level					
ICT infrastructure	122	3.40	0.84	1.20	5.00
ICT competency	122	2.32	0.48	1.00	3.00
School type	122	---	---	1.00	2.00
Mean school leadership	122	2.94	0.19	2.43	3.44
Mean school climate	122	2.91	0.26	2.22	3.50
Mean ICT implementation strategies	122	2.23	0.17	1.80	2.83
Mean government ICT policy	122	2.66	0.16	2.26	3.16
Mean changes in pedagogy	122	2.83	0.15	2.47	3.20

Variables

Teacher-level (Level-1) variables. Teacher-level variables were made up of demographic measures and teachers' perception on the relevant sociocultural factors comprising the social ecology of ICT implementation. Variables measuring teachers' background were introduced to adjust for the differences among schools, which was to generate more accurate estimates of school effect on the outcome variable as teachers were not usually assigned at random to schools (Raudenbush & Bryk, 2002; Raudenbush & Willms, 1991). The measures included age, sex, and teaching experience (Table 6.4 and 6.5). The other teacher-level variables were composite measures made up of several items measured on a 4-point Likert scale (Chapter 3). They were perceived school's ICT implementation strategies and changes in pedagogy (Appendix 1).

School-level (Level-2) variables. School-level variables were of two types, those measuring school's sociocultural factors and those measuring school's structural characteristics. School-level variables measuring sociocultural factors were aggregate of teacher-level variables as defined in Chapter 3, which included mean leadership, mean school climate, mean school's ICT implementation strategies, mean government ICT policy, mean changes in pedagogy. On the other hand, those measuring structural characteristics were generated from responses of heads of ICT team and/ or school principals, which included ICT competency, ICT infrastructure, and school type (Table 6.4 and 6.5).

Table 6.4
Part of the Questionnaire Collecting Supplementary Data

Data	Questions and categorical options
Sex	1. Male 2. Female
Age	1. 18 – 25 2. 26 – 30 3. 31 – 40 4. 41 – 50 5. Above 50
Teaching experience	1. 0 – 3 years 2. 4 – 10 years 3. 11 – 15 years 4. 16 – 20 years 5. Above 20 years
ICT infrastructure	How many LCD/DLP projectors (excluding those in special rooms and laboratories) are available for classroom teaching? 1. 0-5 projectors 2. 6-10 projectors 3. 11-15 projectors 4. 16-20 projectors 5. Over 20 projectors Which one of the following situations best describes the Internet access for classrooms in your school? 1. Less than 1% classrooms are with Internet access 2. 1% to 20% classrooms are with Internet access 3. 21% to 50% classrooms are with Internet access 4. 51% to 80% classrooms are with Internet access 5. 81% to 100% classrooms are with Internet access How many computers/ notebooks are set aside for teachers' use? How many computers/ notebooks are accessible to students after school? How many computers/ notebooks in your school library/ libraries are accessible to students? 1. None 2. 1-20 computers 3. 21-40 computers 4. 41-60 computers 5. Over 60 computers
ICT competency	To date, which one of the following situations best describes the ICT competency of your school? 1. Has not yet reached the *target requirements 2. Has already reached the *target requirements 3. Has reached beyond *target requirements
School type	1. Primary school 2. Secondary school

*Target requirements were ICT requirements for teachers put forward by the Hong Kong government as stated in the document "Information technology for learning in a new era five-year strategy 1998/99 to 2002/03"(Education and Manpower Bureau, 1998).

Table 6.5
Description of Variables Used in the HLM Model

Types	Variable Name	Description
Teacher-level variables		
Demographic	Sex	1 = Male, 2 = Female
	Age	1= 18 – 25, 2 = 26 – 30, 3 = 31 – 40 4 = 41 – 50, 5 = Above 50
	Teaching experience	1 = 0 – 3 years, 2 = 4 – 10 years 3 = 11 – 15 years, 4 = 16 – 20 years 5 = Above 20 years
Non-demographic	ICT implementation strategies	A factor composite of 6 teacher responses
	Changes in pedagogy	A factor composite of 3 teacher responses
School-level variables		
Structural characteristics	ICT infrastructure	A factor composite of 5 responses from head of ICT
	ICT competency	Mean score from the school principal and head of ICT: 1 = Has not yet reached the *target requirements 2 = Has already reached the *target requirements 3 = Has reached beyond *target requirements
	School type	1 = primary school 2 = secondary school
Sociocultural factors relating to ICT implementation	Mean school leadership	Teacher aggregate
	Mean school climate	Teacher aggregate
	Mean government ICT policy	Teacher aggregate
	Mean ICT implementation strategies	Teacher aggregate
	Mean changes in pedagogy	Teacher aggregate

* Target requirements were ICT requirements for teachers put forward by the Hong Kong government as stated in the document "Information technology for learning in a new era five-year strategy 1998/99 to 2002/03" (Education and Manpower Bureau, 1998).

Models

Models considered here were two-level models that consisted of teacher-level

and school-level models. The teacher-level model depicted the relationships between the outcome variable (perceived changes in student learning) and teachers' perception on the relevant sociocultural factors within a given school. Another major issue was to examine at the school-level, how the school-average sociocultural factors and structural characteristics of the school might explain the variation of changes in student learning as perceived by teachers. Unlike typical regression that assuming fixed intercept and slope, HLM allowed the intercept and slopes to vary from school to school. With a step-up strategy, it was finally decided to build a teacher-level model with random intercept and random slope for changes in pedagogy. At level-2, the intercept and random level-1 coefficients (slope for changes in pedagogy) were then modelled as a function of school-level variables. As mentioned in Chapter 2, the theoretical framework of the present study adopted a sociocultural perspective, as a result, it was postulated that sociocultural factors would have relatively greater influence than that of the structural characteristics. To test for the hypothesis, competing level-2 models were constructed, those with structural characteristics as school-level variables, and those with sociocultural factors as school-level variables. The following sections will give more details on why and how the models were constructed.

Fully unconditional model. The exploration started with the simplest possible model that had only a random intercept but no variables at all, the fully unconditional model. As a baseline for reference, the model helped to identify the source of variations within the outcome variable "changes in learning" by partitioning the total variance in the outcome variable into its within-school (level-1) and between-school (level-2) components:

Level 1: (learning)ij = $\beta_{0j} + r_{ij}$

Level 2: $\beta_{0j} = \gamma_{00} + \mu_{0j}$

β_{0j} was the school mean for changes in learning for school j, γ_{00} was the average

school mean (grand mean) for changes in student learning, and r_{ij} and μ_{0j} were the

residual error at level-1 and level-2.

Table 6.6
Results from the Fully Unconditional Model

Fixed Effect	Coefficient	Standard Error	T Ratio
Average school mean for changes in student learning (γ_{00})	2.553	0.021	123.028**

Random Effect	Variance Component	df	Chi-square	p-value
School mean for changes in student learning, μ_{0j}	0.029	121	273.465	.000
Level-1 effect, r_{ij}	0.181			

The results (Table 6.6) indicated that the estimated γ_{00} was equal to 2.55.

Within the total variance of 0.210 (0.181 + 0.029 = 0.210) in the outcome variable,

the estimate for within-school (level-1) variance (i.e. variance of r_{ij}) was 0.181,

whereas the between-school variance (variance of μ_{0j}) was 0.029. In other words, the

teacher-level variance accounted for 86% (0.181/ 0.210 = 86%) of the total variance

in the outcome variable, while 14% (0.029/ 0.210 = 14%) of the total variance was at

the school level.

Level-1 model. The level-1 model examined the effects of selected

sociocultural factors, after controlling for teachers' background, on perceived changes

in learning. At the same time, the level-1 model also enabled us to check whether the

regression results obtained from each individual school were consistent with those in the SEM analysis. Initially, school-level variables were held aside, and data analysis focused on constructing level-1 model by adding teacher-level variables to the fully unconditional model. For the teacher-level variables added to the models, coefficients of which were all fixed. In other words, this means that the corresponding variance components of the slopes were fixed at zero. As it was expected the mean value of variables would vary across schools, each variable was therefore centred around its group mean, which enabled the detection and estimation of slope heterogeneity (Raudenbush & Bryk, 2002).

Demographic variables in addition to teacher-level variables were added to the fully unconditional model (level-1 model). The demographic variables were supposed to adjust for the differences attributed to background of teachers among schools (age, sex and teaching experience). This was to generate more accurate estimates of school effect on changes in student learning (Raudenbush & Bryk, 2002; Raudenbush & Willms, 1991). Selection of teacher-level variables was based on part of the linear relationship from the SEM model (Figure 6.1), which included changes in pedagogy and ICT implementation strategies. The level-1 model was as follows:

Level 1: $(learning)_{ij} = \beta_{0j} + \beta_{1j}(sex) + \beta_{2j}(age) + \beta_{3j}(teaching\ experience)$
$$+ \beta_{4j}(changes\ in\ pedagogy) + \beta_{5j}(ICT\ strategies) + r_{ij}$$
Level 2: $\beta_{0j} = \gamma_{00} + \mu_{0j}$

Table 6.7
Results of the Level-1 Model Showing the Effects of Teacher-Level Variables on Perceived Changes in Student Learning

Fixed Effect	Coefficient	Standard Error	T Ratio
Average school mean for changes in student learning (γ_{00})	2.552	0.021	123.260**
Sex (β_{1j})	-0.039	0.031	-1.270
Age (β_{2j})	-0.061	0.024	-2.524*
Teaching experience (β_{3j})	0.021	0.021	1.014
Changes in pedagogy (β_{4j})	0.439	0.036	12.168**
ICT implementation strategies (β_{5j})	0.216	0.035	6.164**

Random Effect	Variance Component	df	Chi-square	p-value
School mean for changes in student learning, μ_{0j}	0.035	121	375.893	.000
Level-1 effect, r_{ij}	0.132			

Note: * statistically significant at p = .05, ** statistically significant at p = .01.

Results (Table 6.7) showed that the two sociocultural factors (changes in pedagogy, ICT implementation strategies) in the level-1 model were positively related to perceived changes in learning, after controlling for teachers' background. The regression results obtained from each individual school as shown in the level-1 model was consistent with those in the SEM analysis. "Changes in pedagogy" was a strong predictor (coefficient = 0.439, t = 12.168), while "ICT implementation strategies" was a relatively weaker predictor (coefficient = 0.216, t = 6.614). This meant that perceived changes in student learning was substantially higher for teachers who thought that they had made greater changes in their pedagogical practices together with a better perception on the collegial capacity of ICT implementation strategies. Judging from the t-values of the demographic variables (t >1.96 was regarded as significant), only teacher's age was negatively related to changes in student learning with a small negative coefficient of -0.061 (t = -2.524). This suggested that older teachers had slightly lower perceived changes in student learning. As the level-1 model featuring by changes in pedagogy and ICT implementation strategies accounted for 27% of the unexplained variance at level-1 (Table 6.8), the variables

would be considered as having large effect.

Table 6.8

Level-1 Model Showing Reduction in Variance after Adding Teacher-Level Variables

Steps	Model	Variance		
		r_{ij}	μ_{0j}	μ_{2j}
1. Constructing a fully unconditional model				
Level 1:	(learning)ij = β_{0j} + r_{ij}	0.181		
Level 2:	β_{0j} = γ_{00} + μ_{0j}		0.029	---
Variance	0.181/ (0.181 + 0.029) = 86%	86%		
	0.029/ (0.181 + 0.029) = 14%		14%	
2. Level-1 model: adding demographic and teacher-level variables				
Level 1:	(learning)ij = β_{0j} + β_{1j}(sex) + β_{2j}(age)	0.132		
	+ β_{3j}(teaching experience)			
Level 2:	+ β_{4j}(changes in pedagogy)			
	+ β_{5j}(ICT strategies) + r_{ij}			
	β_{0j} = γ_{00} + μ_{0j}		0.035	---
Reduction in variance:	(0.181 – 0.132)/ 0.181 = 27%	27%		

Random coefficient model. After generating the level-1 model, the next step

was to decide whether the slope coefficient of teacher-level variables were conceived

as random, the decision of which was based on the reliability of the random teacher-

level coefficient as derived from ordinary least squares (OLS) regression estimator as

well as the p-value of its estimated level-2 variances (Raudenbush & Bryk, 2002).

Before proceeding to constructing a model with random slope coefficient(s), a

restricted model was obtained first by removing the insignificant level-1 variables. In

removing a level-1 variable, decision was based on whether there was evidence of

fixed effect. A small magnitude of the estimated coefficient and its associated t-ratio

indicated no evidence of a fixed effect. T-ratios near or less than 1 were obvious

candidates for exclusion from the model (Raudenbush & Bryk, 2002). Two indicators

having small magnitude of coefficient and t-ratio were removed from the level-1 model (Table 6.7), which were sex (coefficient = -0.039, t = -1.270) and teaching experience (coefficient = 0.021, t = 1.014). After removing the insignificant teacher-level variables from the level-1 model, a restricted model was then constructed. Thereafter, all the teacher-level coefficients of the restricted model were set random to build the full random-coefficient regression model:

Level 1: $(learning)ij = \beta_{0j} + \beta_{1j}(age) + \beta_{2j}(changes\ in\ pedagogy)$
$+ \beta_{3j}(ICT\ implementation\ strategies) + r_{ij}$

Level 2: $\beta_{0j} = \gamma_{00} + \mu_{0j}$
$\beta_{1j} = \gamma_{10} + \mu_{1j}$
$\beta_{2j} = \gamma_{20} + \mu_{2j}$
$\beta_{3j} = \gamma_{30} + \mu_{3j}$

Table 6.9
Results of the Full Random-Coefficient Regression Model

Fixed Effect	Coefficient	Standard Error	T Ratio	
Average school mean for changes in student learning (γ_{00})	2.552	0.021	123.217**	
Age (γ_{10})	-0.035	0.013	-2.584*	
Changes in pedagogy (γ_{20})	0.449	0.035	12.656**	
ICT implementation strategies (γ_{30})	0.227	0.034	6.725*	

Random Effect	Variance Component	df	Chi-square	p-value
School mean for changes in student learning, μ_{0j}	0.036	113	381.337	.000
Age slope, μ_{1j}	0.004	113	133.650	.090
Changes in pedagogy μ_{2j}	0.033	113	145.305	.022
ICT implementation strategies, μ_{3j}	0.016	113	118.088	.353
Level-1 effect, r_{ij}	0.120			

Reliability of OLS Regression-Coefficient Estimates	
School mean for changes in learning (intercept) (β_{0j})	0.708
Age (β_{1j})	0.147
Changes in pedagogy (β_{2j})	0.165
ICT implementation strategies (β_{3j})	0.081

Note: * statistically significant at $p = .05$, ** statistically significant at $p = .01$.

Table 6.9 presents the results of the full random coefficient regression model. In terms of univariate chi-square tests for the level-1 coefficients, all the p values associated with the hypothesis of slope homogeneity were larger than .05, except that for changes in pedagogy. A p-value of less than .05 for the coefficient slope of changes in pedagogy implied rejecting the hypothesis of slope homogeneity, which on the other hand implied variation of the slope among groups. In addition, the reliability of estimate of its coefficient (reliability = 0.165) was much higher than the suggested value of 0.05. This implied that only the effect of changes in pedagogy was likely to differ across schools. In allowing the slope coefficient for changes in pedagogy to vary from school to school, we constructed the following random-coefficient model:

Level 1: $(\text{learning})_{ij} = \beta_{0j} + \beta_{1j}(\text{age}) + \beta_{2j}(\text{changes in pedagogy})$
$$+ \beta_{3j}(\text{ICT implementation strategies}) + r_{ij}$$

Level 2: $\beta_{0j} = \gamma_{00} + \mu_{0j}$
$$\beta_{2j} = \gamma_{20} + \mu_{2j}$$

For the random coefficient model, the slope coefficient of changes in pedagogy was allowed to vary from school to school. As can be seen from Table 6.10, there were three sources of random effect in the random coefficient model: level-1 variance (variance of $r_{ij} = 0.126$), level-2 variance of school mean for changes in student learning (variance of $\mu_{0j} = 0.021$), and level-2 variance of slope coefficient for changes in pedagogy (variance of $\mu_{2j} = 0.036$). At this stage, the unexplained variance at level-2 was 0.057 (0.021 + 0.036 = 0.057), in which 37% (0.021/ 0.057 = 37%) was from variation in the school mean for changes in student learning, and 63% (0.036/ 0.057 = 63%) was from variation in the effect of changes in pedagogy across schools. In the next step, we added school-level variables. It was expected that adding school-level variables to the random-coefficient model would reduce the level-2 variance. As the proportion of reduction in variance statistics at level-2 were interpretable only for the same level-1 model (Raudenbush & Bryk, 2002), the random-coefficient model would serve as a baseline model in assessing the effect of the school-level variables in explaining the variation of changes in student learning.

Table 6.10
Random Coefficient Model Showing the Change in Variance after Setting Free the Slope Coefficient

Steps	Model	Variance		
		r_{ij}	μ_{0j}	μ_{2j}
1. Constructing a fully unconditional model				
Level 1:	$(learning)ij = \beta_{0j} + r_{ij}$	0.181		
Level 2:	$\beta_{0j} = \gamma_{00} + \mu_{0j}$		0.029	---
Variance	$0.181/ (0.181 + 0.029) = 86\%$	86%		
	$0.029/ (0.181 + 0.029) = 14\%$		14%	
2. Level-1 Model: adding demographic and teacher-level variables				
Level 1:	$(learning)ij = \beta_{0j} + \beta_{1j}(sex) + \beta_{2j}(age)$	0.132		
	$+ \beta_{3j}(teaching\ experience)$			
	$+ \beta_{4j}(changes\ in\ pedagogy)$			
	$+ \beta_{5j}(ICT\ strategies) + r_{ij}$			
Level 2:	$\beta_{0j} = \gamma_{00} + \mu_{0j}$		0.035	---
Reduction in variance:	$(0.181 - 0.132)/ 0.181 = 27\%$	27%		
3. Restricted Model: removing insignificant teacher-level variables				
Level 1:	$(learning)ij = \beta_{0j} + \beta_{1j}(age)$	0.132		
	$+ \beta_{2j}(changes\ in\ pedagogy)$			
	$+ \beta_{3j}(ICT\ strategies) + r_{ij}$			
Level 2:	$\beta_{0j} = \gamma_{00} + \mu_{0j}$		0.035	---
Reduction in variance:	$(0.181 - 0.132)/ 0.181 = 27\%$	27%		
4. Random coefficient model: deciding which slope coefficients be random				
Level 1:	$(learning)ij = \beta_{0j} + \beta_{1j}(age)$	0.126		
	$+ \beta_{2j}(changes\ in\ pedagogy)$			
	$+ \beta_{3j}(ICT\ strategies) + r_{ij}$			
Level 2:	$\beta_{0j} = \gamma_{00} + \mu_{0j}$		0.021	
	$B_{2j} = \gamma_{20} + \mu_{2j}$			0.036
Reduction in variance:	$(0.181 - 0.126)/ 0.181 = 30\%$	30%		

Level-2 models. After obtaining the random coefficient model, school-level variables were introduced to build level-2 models. Alternative level-2 models were built on theoretical consideration, in which school-level variables were added to develop models to illuminate how differences among schools, in terms of structural characteristics and sociocultrual factors, might affect the distribution of perceived changes in learning within schools. The proposed level-2 models had two level-2 equations ($\beta_{0j} = \gamma_{00} + \mu_{0j}$, $\beta_{2j} = \gamma_{20} + \mu_{2j}$), and it was possible that school-level variables entered in one equation might affect variance estimates in another equation. Taking this into account, a common set of school-level variables would be added in both level-2 equations (Raudenbush & Bryk, 2002). In the process of model comparison, with the random coefficient model as a baseline reference, the proportion of reduction in level-2 variance as accounted for by the model served as the basis for making judgement about the importance of variables (Raudenbush & Bryk, 2002).

Structural model. The first proposed model was a structural model, in which school variables relating to structural characteristics (Table 6.4 and 6.5) were added to level-2, which were ICT infrastructure, school type (primary or secondary school), and ICT competency. In the level-2 model (structural model), the intercept (β_{0j}) and random level-1 coefficient (β_{2j}) for changes in pedagogy were regressed on the school-level variables:

Level 1: $(\text{learning})ij = \beta_{0j} + \beta_{1j}(\text{age}) + \beta_{2j}(\text{changes in pedagogy})$
$$+ \beta_{3j}(\text{ICT implementation strategies}) + r_{ij}$$
Level 2: $\beta_{0j} = \gamma_{00} + \gamma_{01}(\text{ICT infrastructure}) + \gamma_{02}(\text{school type})$
$$+ \gamma_{03}(\text{ICT competency}) + \mu_{0j}$$
$$\beta_{2j} = \gamma_{20} + \gamma_{21}(\text{ICT infrastructure}) + \gamma_{22}(\text{school type})$$
$$+ \gamma_{23}(\text{ICT competency}) + \mu_{2j}$$

Sociocultural model. The second one was a sociocultural model, in which the effects of the sociocultural factors as measured by teacher aggregate (Table 6.4 and 6.5, and Chapter 3) were modelled on the school mean for changes in student learning (β_{0j}) and the effect of changes in pedagogy (β_{2j}):

Level 1: (learning)ij = $\beta_{0j} + \beta_{1j}$(age) + β_{2j}(changes in pedagogy)
$\qquad\qquad\qquad\quad + \beta3_j$(ICT implementation strategies) + r_{ij}

Level 2: $\beta_{0j} = \gamma_{00} + \gamma_{01}$(mean leadership) + γ_{02}(mean climate)
$\qquad\qquad + \gamma_{03}$(mean govt ICT policy)
$\qquad\qquad + \gamma_{04}$(mean ICT implementation strategies)
$\qquad\qquad + \gamma_{05}$(mean changes in pedagogy) + μ_{0j}

$\qquad \beta_{2j} = \gamma_{20} + \gamma_{21}$(mean leadership) + γ_{22}(mean climate)
$\qquad\qquad + \gamma_{23}$(mean govt ICT policy)
$\qquad\qquad + \gamma_{24}$(mean ICT implementation strategies)
$\qquad\qquad + \gamma_{25}$(mean changes in pedagogy)+ μ_{2j}

Table 6.11

HLM Model (Level-2 Model) Showing the Effects of School-Level Variables on School Mean for Changes in Student Learning and Changes in Pedagogy

Fixed Effect	Structural Model			Sociocultural Model		
	Coeff.	SE	t	Coeff.	SE	t
School mean for changes in learning (β_{0j})						
Intercept, γ_{00}	2.550	0.017	146.294**	2.550	0.015	168.878**
School type	-0.276	0.045	-6.096**	---	---	---
ICT infrastructure	0.044	0.026	1.671	---	---	---
ICT competency	-0.029	0.037	-0.781	---	---	---
Mean leadership	---	---	---	-0.155	0.107	-1.452
Mean climate	---	---	---	0.323	0.083	3.908**
Mean govt ICT policy	---	---	---	0.440	0.115	3.839**
Mean ICT implementation strategies	---	---	---	0.135	0.094	1.445
Mean changes in pedagogy	---	---	---	0.549	0.099	5.534**
Slope of Changes in Pedagogy (β_{2j})						
Intercept, γ_{20}	0.450	0.036	12.632**	0.452	0.036	12.390**
School type	-0.146	0.079	-1.856	---	---	---
ICT infrastructure	-0.021	0.049	-0.437	---	---	---
ICT competency	-0.061	0.075	-0.815	---	---	---
Mean leadership	---	---	---	-0.174	0.255	-0.682
Mean climate	---	---	---	0.279	0.190	1.470
Mean govt ICT policy	---	---	---	-0.082	0.306	-0.269
Mean ICT implementation strategies	---	---	---	0.166	0.280	0.595
Mean changes in pedagogy	---	---	---	-0.106	0.260	-0.409

Random effect	Variance	df	χ^2	p	Variance	df	χ^2	p
Intercept, μ_{0j}	0.020	117	266.580	.000	0.011	115	201.544	.000
Pedagogy slope, μ_{2j}	0.031	117	154.148	.012	0.033	115	158.666	.005
Level-1, r_{ij}	0.126				0.126			

Proportion of level-2 variance explained		
Intercept	5%	48%
Pedagogy slope	14%	8%

Note: 1. * statistically significant at p = .05, ** statistically significant at p = .01.

2. For the random coefficient model, the variance for level-1 was 0.126, for intercept was 0.021, for pedagogy slope was 0.036.

Table 6.11 presents the results of the two models. Comparatively speaking, the sociocultural model was stronger than the structural model in explaining school-level variance. It accounted for a relatively larger percentage (48.0%) of the unexplained variance in school mean for changes in learning (intercept). This revealed that the sociocultural factors in a school was more important than that of the structural characteristics as a predictor of its school mean for perceived changes in learning. Judging from the coefficients and t-values of the structural characteristics, only school type (coefficient = -0.276. t = -6.096) was found to be a significant school-level variable. On the other hand, most of the sociocultural factors were significantly related to the school mean for perceived changes in learning, which were mean school climate, mean government ICT policy, and mean changes in pedagogy. Among them, mean changes in pedagogy was the strongest predictor (coefficient = 0.549, t = 5.534), while mean school climate was the weakest predictor (coefficient = 0.323, t = 3.908). Surprisingly, both the structural and sociocultural models were unable to account for the unexplained variance of changes in pedagogy. As shown in Table 6.11, none of the school-level variables was significant in predicting differences in the effects of changes in pedagogy.

The unified HLM model. As can be seen from the results of structural model and sociocultural model (Table 6.11), only some of the school-level variables were significant in explaining the variation of the school mean for changes in student learning (intercept). They were school type, mean school climate, mean government ICT policy, and mean changes in pedagogy. In selecting a combination of these significant school-level variables from the structural model and sociocultural model, a unified model was constructed to depict the social ecology of ICT implementation:

Level 1: (learning)ij = β_{0j} + β_{1j}(age) + β_{2j}(changes in pedagogy)

$$+ \beta_{3j}(\text{ICT strategies}) + r_{ij}$$

Level 2: β_{0j} = γ_{00} + γ_{01}(school type) + γ_{02}(mean climate)

$$+ \gamma_{03}(\text{mean govt ICT policy})$$

$$+ \gamma_{04}(\text{mean changes in pedagogy})$$

$$+ \mu_{0j}$$

$$\beta_{2j} = \gamma_{20} + \gamma_{21}(\text{school type}) + \gamma_{22}(\text{mean climate})$$

$$+ \gamma_{23}(\text{mean govt ICT policy})$$

$$+ \gamma_{24}(\text{mean changes in pedagogy}) + \mu_{2j}$$

Table 6.12
Unified HLM Model Showing the Effects of School-Level Variables on School Mean for Changes in Student Learning and Changes in Pedagogy

Fixed Effect	Unified Model		
	Coeff.	SE	t
Mean changes in learning (intercept)			
Intercept	2.547	0.014	183.698**
School type	-0.149	0.027	-5.414**
Mean climate	0.219	0.061	3.606**
Mean govt ICT policy	0.427	0.101	4.229**
Mean changes in pedagogy	0.480	0.077	6.231**
Changes in Pedagogy			
Intercept	0.450	0.036	12.602**
School type	-0.152	0.072	-2.104*
Mean climate	0.178	0.162	1.098
Mean govt ICT policy	-0.079	0.288	-0.275
Mean changes in pedagogy	-0.169	0.237	-0.711

Random effect	Variance	df	χ^2	p
Intercept	0.007	116	168.087	.001
Pedagogy slope	0.028	116	152.914	.012
Level-1	0.126			

Proportion of variance explained	
Intercept	67%
Pedagogy slope	22%

Note: 1. * statistically significant at p = .05, ** statistically significant at p = .01.
2. For the random coefficient model, the variance for level-1 was 0.126, for intercept was 0.021, for pedagogy slope was 0.036.

Table 6.12 shows the specific results of the unified model, and Table 6.13 shows the reduction in variance in the "step-up" process of model construction. As indicated from Table 6.12, in terms of explaining the total variance among schools in mean levels of perceived changes in learning, the unified model accounted for quite a substantial reduction in variation (67%) once the sociocultural and structural characteristics were taken into account. Of the significant school-level variables, the majority were socio-cultural factors shaping the social ecology of ICT implementation in a school. Among the socio-cultural variables affecting the school mean for perceived changes in learning, the mean self-perception of changes in pedagogy was the strongest predictor (coefficient = 0.480, t = 6.231), whereas the mean climate was the weakest one (coefficient = 0.219, t = 3.606). This implied that the school mean for changes in student learning was substantially higher in schools with teachers on average experiencing greater changes in pedagogical practices. In addition, a more cohesive school climate and a better average evaluation of the government ICT policy would also raise the school mean for changes in student learning. In terms of structural characteristics, mean perceived changes in student learning was also slightly lower in secondary schools than in primary schools (coefficient = -0.149, t = -5.414; primary school = 1, secondary school = 2).

On the other hand, in terms of the total variance in the effect of changes in pedagogy among schools, the unified model accounted for quite a small percentage of variance (22%). School type was the only significant school-level predictor on the school effect of changes in pedagogy on changes in student learning. Changes in pedagogy exerted a more positive effect on changes in student learning in primary schools (coefficient = -0.152, t = -2.104; primary school = 1, secondary school = 2).

Anyway, variation among schools in the effect of changes in pedagogy remained largely unexplained. Upon examining the relatively high reliability of the effects of changes in pedagogy (0.191), we would suggest that the variations found in changes in pedagogy among schools were related to other factors not identified here. Still, it was also possible that the unexplained variance was due to the large error inherited in the estimation of variability in slope coefficients. Raudenbush & Bryk (2002) have pointed out that regression coefficients will usually have considerably greater sampling variability than do sample means. The greater sampling variability together with a small sample within a unit will lead to a larger error in the estimation of regression coefficients. The resultant unreliability in slopes weakened our power to detect relationships in the level-2 model. It was because if much of the observed variance in the slope for changes in pedagogy was due to error variance, a level-2 model that explained only a small percentage of the observed variance in the slope coefficient might be discounted, when in fact it was explaining a very large portion of what could, in principle, be explained (Raudenbush & Bryk, 2002).

With a deliberate removal of the non-significant variables from the unified HLM model, the following equation sums-up the results of HLM model building:

Level 1: $(\text{learning})_{ij} = \beta_{0j} + \beta_{1j}(\text{age}) + \beta_{2j}(\text{changes in pedagogy})$
$$+ \beta_{3j}(\text{ICT strategies}) + r_{ij}$$
Level 2: $\beta_{0j} = \gamma_{00} + \gamma_{01}(\text{school type}) + \gamma_{02}(\text{mean climate})$
$$+ \gamma_{03}(\text{mean govt ICT policy})$$
$$+ \gamma_{04}(\text{mean changes in pedagogy})$$
$$+ \mu_{0j}$$
$$\beta_{2j} = \gamma_{20} + \gamma_{21}(\text{school type}) + \mu_{2j}$$

Table 6.13
Reduction in Variance in the Step-up Process of Model Construction

Steps	Model	Variance		
		r_{ij}	μ_{0j}	μ_{2j}
1. Constructing a fully unconditional model				
Level 1:	(learning)ij = $\beta_{0j} + r_{ij}$	0.181		
Level 2:	$\beta_{0j} = \gamma_{00} + \mu_{0j}$		0.029	---
Variance	0.181/ (0.181 + 0.029) = 86%	86%		
	0.029/ (0.181 + 0.029) = 14%		14%	
2. Level-1 model: adding demographic and teacher-level variables				
Level 1:	(learning)ij = $\beta_{0j} + \beta_{1j}$(sex) + β_{2j}(age)	0.132		
	+ β_{3j}(teaching experience)			
	+ β_{4j}(changes in pedagogy)			
	+ β_{5j}(ICT strategies) + r_{ij}			
Level 2:	$\beta_{0j} = \gamma_{00} + \mu_{0j}$		0.035	---
Reduction in variance:	(0.181 – 0.132)/ 0.181 = 27%	27%		
3. Restricted model: removing insignificant teacher-level variables				
Level 1:	(learning)ij = $\beta_{0j} + \beta_{1j}$(age)	0.132		
	+ β_{2j}(changes in pedagogy)			
	+ β_{3j}(ICT strategies) + r_{ij}			
Level 2:	$\beta_{0j} = \gamma_{00} + \mu_{0j}$		0.035	---
Reduction in variance:	(0.181 – 0.132)/ 0.181 = 27%	27%		
4. Random coefficient model: deciding which slope coefficients be random				
Level 1:	(learning)ij = $\beta_{0j} + \beta_{1j}$(age)	0.126		
	+ β_{2j}(changes in pedagogy)			
	+ β_{3j}(ICT strategies) + r_{ij}			
Level 2:	$\beta_{0j} = \gamma_{00} + \mu_{0j}$		0.021	
	$B_{2j} = \gamma_{20} + \mu_{2j}$			0.036
Reduction in variance:	(0.181 – 0.126)/ 0.181 = 30%	30%		
5. Level-2 model: adding school-level variables				
Level 1:	(learning)ij = $\beta_{0j} + \beta_{1j}$(age)	0.126		
	+ β_{2j}(changes in pedagogy)			
	+ β_{3j}(ICT strategies) + r_{ij}			
Level 2:	$\beta_{0j} = \gamma_{00} + \gamma_{01}$(school type)		0.007	
	+ γ_{02}(mean climate)			
	+ γ_{03}(mean government policy)			
	+ γ_{03}(mean changes in pedagogy)			
	+ μ_{0j}			
	$\beta_{2j} = \gamma_{20} + \gamma_{21}$(school type)			0.028
	+ γ_{22}(mean climate)			
	+ γ_{23}(mean government policy)			
	+ γ_{24}(mean changes in pedagogy)			
	+ μ_{2j}			
Reduction in variance:	(0.181 – 0.126)/ 0.181 = 30%	30%		
	(0.021 – 0.007)/ 0.021 = 69%		67%	
	(0.036 – 0.028)/ 0.036 = 22%			22%

Discussions

The multilevel analysis using HLM provided researchers with a plausible 2-level model that portrayed the impact of school-level variables on changes in student learning among schools. In assigning variables at the school-level, we attempted to explain the variation of level-1 intercept (school mean for changes in student learning) and the variation of effect of changes in pedagogy (slope coefficient) on changes in student learning among schools. In this way, the school-level effect behind the scene was examined.

As indicated from the fully unconditional model of the present study, the teacher-level variance accounted for 86% of the total variance in the outcome variable, while 14% of the total variance was at the school level (Table 6.13). This suggested that variation of changes in student learning was relatively smaller at the school-level. Besides, variability in perceived changes in student learning was largely not attributable to the background characteristics of teachers. At the teacher-level, changes in pedagogy and the school's ICT implementation strategies explained 27% of variance in changes in student learning, which indicated that the variables had a large effect (Hox, 2002). The results were also consistent with those of the SEM model in Chapter 5, which indicated that perceived changes in student learning was substantially higher for teachers who believed that they had made greater changes in their pedagogical practices and they worked in schools with a higher collegial capacity of ICT implementation strategies.

Among schools, and beyond teachers within the same school, the school mean for changes in student learning and the effect of changes in pedagogy on student learning did show variations. Results of the study supported that the socio-cultural factors shaping the social ecology of ICT implementation rather than the structural characteristics of schools accounted for the variations. Specifically, the school mean for changes in student learning was substantially higher in schools with teachers on average experiencing greater changes in pedagogical practices. In addition, a more cohesive school climate and a better average evaluation of the government ICT policy would also raise the school mean for changes in student learning. The study further revealed that, in comparison with primary schools, the socio-cultural context effects were less efficient in secondary schools in bringing about a paradigm shift in learning. While variations among schools in the effect of changes in pedagogy remained largely unexplained in our study, pedagogical change was found to exert a more positive effect on changes in student learning in primary schools than in secondary schools.

The findings of this study have significant implications on understanding how school-level variables bring about changes in learning from a teacher-centred approach to a student-centred approach. Results of the study extended our understanding on the postulation that changes in student learning was brought about by pedagogical and organisational interventions (Honey et al., 1999). Honey et al. (1999) have pointed out that effective ICT use is embedded in a larger process of school change, specifically being mediated by factors such as the pedagogical methods of teachers, and the socio-cultural setting of the school. The study revealed that pedagogical methods of teachers, in comparison with the organisational interventions, had a more significant direct impact on enhancing changes in student learning within and among schools. In addition, it was supported by empirical evidence that the socio-cultural setting rather than structural characteristics of the school accounted for the variation of school mean for changes in student learning.

In comparison with primary schools, it was interesting to find out that the socio-cultural context effects were less efficient in bringing about a paradigm shift in learning in secondary schools, and that the pedagogical change exerted a less positive effect on changes in student learning in secondary schools. It remains to be an important issue for future study to find out what differences in primary and secondary school alter the sociocultural context effects and pedagogical effect on changes in student learning, which possibly also requires the exploration of other possible school-level variables to account for the unexplained variations among schools in the effect of changes in pedagogy on perceived changes in learning.

Chapter 7: Conclusions and Discussions

Introduction

The purpose of the present study is to develop models that conceptually integrate ICT implementation with changes in student learning within a context of managing change in schools, with a view to understanding how ICT acts as a lever to bring about a paradigm shift in student learning from a teacher-centred approach to a student-centred approach. In this study, we examined ICT implementation from the perspective of pedagogical and organisational features of schools. Attempting to select variables that were socio-cultural in nature to build models, we wished to capture the prevailing pattern of interactions in schools in which ICT implementation was perceived as a successful initiative. Apart from identifying the relationships among the features as perceived by teachers, this study also examined the effect of the features on student learning as measured at school level. From the research findings of the present study, this chapter focuses on insights gained in understanding ICT implementation in terms of teacher (level-1) and school (level-2) effects on changes in student learning.

The study was primarily a cross-sectional survey with statistical analysis of structural equation modelling (SEM) and hierarchical linear modelling (HLM) to investigate the perception of 1076 teachers from 130 Hong Kong schools on the social ecology of ICT implementation. In adopting a situative perspective (Greeno, 1998; Spillane et al., 2002), the study viewed ICT implementation within the context of schooling in which teachers participate. Along this direction, the survey instrument for studying the social ecology basically consisted of six scales: school leadership, school climate, effectiveness of government ICT policy, collegial capacity of school's

ICT implementation strategies, changes in teacher pedagogy, and changes in student learning. In addition, the questionnaire also collected background information of teachers and structural characteristics of schools: sex, age, teaching experience, school type (primary school or secondary school), school's ICT infrastructure, school's ICT competency.

Conclusions

While the previous chapters have presented individual findings with respect to construct validation, SEM analysis, and HLM analysis, attempting to get a more holistic picture, the individual findings in the previous chapters can be summarized and concluded as follows:

Sociocultural factors were important in bringing about changes in student learning from a teacher-centred approach to a student-centred approach. To answer research question 1 asking what factors affecting changes in student learning from a teacher-centred approach to a student-centred approach, results of HLM analysis echoed that of SEM analysis in supporting that in a context of managing change in schools with ICT implementation, the sociocultural factors shaping the social ecology of ICT implementation rather than the structural characteristics of schools accounted for the variations with respect to changes in student learning among teachers and across schools. The sociocultural factors included school leadership, school climate, collegial capacity of ICT implementation strategies of schools, effectiveness of government ICT policy, and pedagogical changes of teachers.

The collegial capacity of ICT implementation strategies had a significant impact in bringing about changes in pedagogy from a teacher-centred approach to a student centred approach. To answer research question 2 asking how the contextual factors interacted with one another to bring about the paradigm shift in learning, SEM analysis affirmed the central role of collegial capacity of ICT implementation in the social ecology of ICT implementation. Specifically, the collegial capacity of ICT implementation strategies mediated the effect of organisational interventions on changes in pedagogy. In this way, ICT could act as a lever in bringing about changes in pedagogy in the forms of providing a supportive ICT context in schools that might offer collegial exchange of professional-development experiences that enhanced changes in pedagogical practices.

Pedagogical changes were positively related to changes in learning at both teacher level and school level. To answer research question 2 asking how the contextual factors interacted with one another to bring about the paradigm shift in learning, SEM analysis supported the proposition that changes in pedagogy mediated the effect of organisational interventions on changes in student learning. In the social ecology of ICT implementation, teachers were assets of schools and were capable of acting in a more direct way of pedagogical interventions in bringing about changes in student learning. Further, results of HLM informed us that among schools, the school mean for changes in student learning did show variations. Specifically, the school mean for changes in student learning was substantially higher in schools with teachers on average experiencing greater changes in pedagogical practices. In other words, changes in pedagogy modulated changes in student learning among schools. Empirically, the positive relationships between changes in pedagogy and changes in learning demonstrated the paramount role of teachers, which has been however often overlooked (Barber & Phillips, 2000).

Transformational school leadership was relatively less direct in affecting changes in student learning. To answer research question 2 asking how the contextual factors interacted with one another to bring about the paradigm shift in learning, SEM analysis revealed that transformational leadership was important in assuming a leading role in making schools as a workplace that fostered collegiality and a climate for pedagogical innovations, which in turn facilitated improvement in classroom practices. The results have been consistent with findings of a substantial number of empirical studies in supporting the belief that principals exercise a measurable though indirect effect on school effectiveness and student achievement (Hallinger & Heck, 1998).

ICT was able to act as a lever to bring about changes in student learning in the context of establishing collegiality in fostering pedagogical innovations in schools. To answer research question 3 asking how ICT acted as a lever to bring about the paradigm shift in learning, findings of this study has shown that ICT was able to act as a lever to bring about changes in student learning in the context of establishing collegiality in fostering pedagogical innovations in schools. In this study, variables like "school climate" and "ICT implementation strategies" captured the concept of collegiality in fostering pedagogical innovations. While climate referred to the more general feeling of closeness and working together, the collegial capacity of ICT implementation strategies was comparatively more specific in probing into the dissemination, exchange, and sharing of experiences in using ICT. In this study, school climate and collegial capacity of ICT implementation strategies were shown to have significant impact on changes in student learning, which was consistent with the concept of effecting a paradigm shift in learning in the way of making school as a workplace in the form of collegiality that fostered pedagogical innovations (Little, 1982). As shown from the SEM results, the better-fit group of models (Figure 5.1, the basic model; Figure 5.6, the strategic & pedagogical model) indicated that the collegial capacity of ICT implementation strategies might act through changes in pedagogy in effecting a paradigm shift in student learning. At the school level, the HLM results indicated that among schools, the mean school climate did have positive effect on the school mean for changes in student learning. These further have supported the proposition that in and of itself, ICT contains neither pedagogical philosophy nor content basis (Means, 1994), and the benefits of ICT cannot be adequately separated from other variables that impact learning in the larger instructional context (Honey et al., 2000; McCombs, 2000).

School-level factors were less important than within-school factors in bringing about changes in student learning from a teacher-centred approach to a student-centred approach. Results from the HLM analysis showed that the teacher-level (level-1) variance accounted for 86% of the total variance in the outcome variable changes in student learning, while 14% of the total variance was at the school level (level-2). This indicated that organisational features across schools (school-level factors) had relatively smaller effects on perceived changes in student learning in comparison with within-school factors.

Of the organisational features across schools (school-level factors), sociocultural factors played a more important role than structural factors in enhancing a paradigm shift in student learning. To answer research question 4 asking what organisational features of schooling faciltated the paradigm shift in student learning, HLM analysis revealed that the sociocultural factors in a school were more important than the structural factors as predictors of its school mean for perceived changes in learning. The sociocultural factors that were significantly related to the school mean for perceived changes in learning were mean school climate, mean government ICT policy, and mean changes in pedagogy.

School level (primary or secondary schools) had significant though smaller effect on the school mean for changes in student learning among schools. To answer research question 4 asking what organisational features of schooling facilitated the paradigm shift in student learning, among schools, while sociocultural factors played a more important role in enhancing a paradigm shift in student learning from a teacher-centred to a student-centred approach, school level (primary or secondary schools) did have significant though smaller effect on the school mean for changes in student learning. In comparison with secondary schools, primary schools had a little higher school mean for changes in student learning. In addition, among schools, the effect of changes in pedagogy on student learning did show variations. Results of HLM analysis showed that, while variations among schools in the effect of changes in pedagogy remained largely unexplained, pedagogical changes was found to exert a more positive effect on changes in student learning in primary schools than in secondary schools.

Conceptual Model

Based on the above conclusions, a 2-level-3-factor conceptual model (Figure 7.1) was constructed to unfold the social ecology of ICT implementation as interactions of contextual variables in an activity system that comprised two levels and 3 factors. At the teacher level, the pattern of interactions between classroom factors and organisational factors helped us to understand how pedagogical and organisational interventions interacted with one another to enhance changes in student learning. The school level analysis further identified school-level factors that raised the school mean for changes in student learning among schools.

At teacher level, changes in pedagogy was a classroom factor that played a paramount role in advancing changes in student learning. In acquiring such changes, organisational factors in the social ecology needed to be changed towards establishing collegiality that encouraged pedagogical innovations, collegial exchange of ICT knowledge, and mobilisation of resources. Such prevailing pattern of interactions in schools illustrated the notion of schools as a workplace for learning so as to facilitate knowledge creation (Little, 1982; Nonaka & Takeuchi, 1995). Specifically, it could be argued that ICT acted as a lever in bringing about pedagogical innovations in the forms of providing a supportive ICT context in schools that might offer collegial exchange of professional-development experiences that enhanced changes in pedagogical practices. In the change process, transformational school leadership and an effective government ICT policy had substantial effect in making schools as a workplace for fostering collegiality. At school level, schools with higher pedagogical changes raised its school mean for changes in student learning. In addition, other school-level factors also enhanced the average paradigm shift in learning among schools, one of which again was collegiality in the form of mean school climate.

In the social ecology of ICT implementation, notice might also be drawn on the pattern of interactions among variables, which could be a within-factor type of interactions or of cross-factor type of interactions. Collegiality as a school-level factor (mean school climate) and as a within-school organisational factor (school climate and collegial capacity of ICT implementation strategies) played a very important role in enhancing changes in student learning.

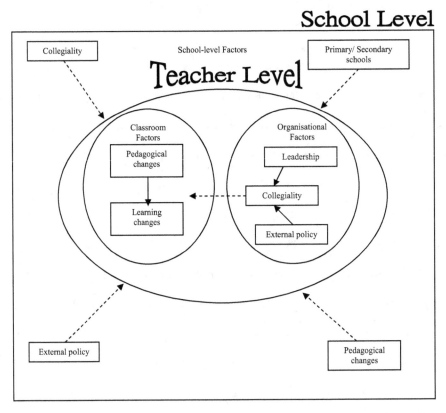

Figure 7.1. The 2-level-3-factor conceptual model depicting the social ecology of ICT implementation in schools.

⟶ Within-factor influence   ----▸ Cross-factor influence

Discussions

Despite the complex dynamics in a real school setting, with the identification of some key contextual factors, the study has been able to use a 2-level-3-factor conceptual model to sketch out the social ecology of ICT implementation. It is hoped that the model may help educators and policymakers to gain insights into the social ecology of ICT implementation, thereby enabling them to unleash the power of organisational interventions and pedagogical interventions to advance changes in student learning.

Theoretically, the study has expanded the concept of ICT implementation from mere ICT use to ICT use in the context of pedagogical and organisational interventions. With the introduction of the concept of social ecology of ICT implementation, ICT use has been conceptualised in the context of establishing collegiality to foster pedagogical innovations in which changes in student learning were attempted.

Methodologically, using the promising SEM technique, the present study has been able to model adequately the complex relationships in the social ecology of ICT implementation in schools given that SEM has the advantage of allowing the researcher to accommodate multiple interrelated dependence relationships in a single model that are not possible with other popular multivariate techniques (Hair et al., 1998). With the use of HLM, this study has also addressed the unit of analysis problem and the multilevel nature of schooling (Hallinger & Heck, 1998). It has extended our understanding on the school effect that was masked in the SEM analysis. Apart from that, the methodological advances have also enabled us to study within-factor type of interactions as well as cross-factor type of interactions. Collegiality as a school-level factor (mean school climate) and as a within-school organisational factor (school climate and collegial capacity of ICT implementation strategies) has been demonstrated to play a very important role in enhancing changes in student learning.

Theoretically, the study has offered insights into the controversy over the relative importance of changes related to classroom practices and changes related to organisational factors. Demonstrating the existence of cross-factor interactions (interactions from organisational factors to classroom factors) and cross-level interactions (interactions from school-level factors to classroom factors), the study has given empirical support to the proposition that attention to organisational changes is essential to changes in classroom practices (Leithwood, 1994).

Findings in this study have affirmed that sociocultural factors, rather than background characteristics of teachers or structural factors of schools, have accounted for the variations with respect to changes in student learning. Practically, an implication to school administrators and policymakers is that in using ICT to advance changes in student learning, more attention should be given to both the social contexts and institutional culture in which teachers situated.

The SEM analysis has shown that at teacher level, pedagogical changes as a classroom factor has played a paramount role in advancing changes in student learning. In acquiring such changes, organisational factors in the social ecology have to be changed towards establishing collegiality that encourages pedagogical innovations, collegial exchange of ICT knowledge, and mobilisation of resources. The findings have supported the proposition that ICT can act as a lever in bringing about pedagogical innovations in the form of providing a supportive ICT context in schools that may offer collegial exchange of professional-development experiences that enhances changes in pedagogical practices. Practically, it alerts school administrators and policymakers of the positive effect of cultivating collegiality in schools in enhancing changes in student learning, which at the same time also offers insights into the appropriate direction of education reform. It is worth pausing to consider whether to adopt a control strategy that involves quality control relying heavily on language proficiency assessment, quality assurance inspection, school self evaluation etc., or whether to adopt a commitment strategy that seeks to develop innovative working arrangements supporting teachers' decision-making and increasing teachers' engagement in the tasks of teaching (Rowen, 1990). Further, policymakers are urged to review the impact of adopting private sector practices to tighten control of educational process and the work of professionals, and to consider whether the dominant ideas of economic rationalism and managerialism will really increase or worsen educational outcomes (Chan, 2002; Mok & Welch, 2002; Tse, 2002).

Given the same level of score in the factors comprising the social ecology of ICT implementation, it is not clear why the changes in student learning as well as the effect of changes in pedagogy were greater in primary schools than in secondary schools. This suggests other important terrain left unexplored. One possible terrain is the clash between new instructional practices and traditional modes of assessment (Valli, Cooper, & Frankes, 1997). In comparison with primary school, assessments in secondary school are more subject-bounded, academically driven, and examination oriented. This possibly renders secondary school teachers less room for adopting new modes of assessment that place greater emphasis in measuring learning attitude and learning ability in terms of active construction of knowledge and collaborative work. This also tends to make them less aware of the actual changes in student learning towards a more constructivist approach. Such happenings in secondary school teachers may have accounted for a relatively lower evaluation of changes in student learning when compared to primary school teachers.

Limitations and Suggestions

In studying the social ecology of ICT implementation from 1076 teachers in 130 schools, we have understood that effective use of ICT in teaching and learning has to be conceptualised in a larger context of pedagogical and organisational changes. Greater changes in student learning have been related directly to schools having greater changes in pedagogy and an atmosphere of higher collegiality, of which a transformational leadership has been found exerting significant effect in fostering the collegiality. Acknowledging that there were limitations in this study, the findings have also led us to think about related issues that may lend themselves to future quantitative and qualitative studies, which are helpful in unleashing irreversible changes in schools. The limitations and suggestions are as follows:

While the recruited sample population (of 1076 teachers from 130 schools) came from all 18 school districts in Hong Kong, it did not truly represent the ratio of district distribution. In terms of distribution according to finance types (government, direct-subsidy scheme, and aided), the sample population closely resembled the true population. With the present sample population, it means that the research was limited to addressing the variations across the 18 school districts, but not the variations within

210

the district. Results of the present study were therefore limited to producing a general pattern, but they were not supposed to produce generalisations to the whole Hong Kong school population. The convenient sample would suffice to serve the primary aim of the present study, which was to deepen our understanding of the sociocultural factors affecting ICT implementation in and across schools rather than to make generalisation over the entire Hong Kong school population.

The outcome variable "changes in student learning" was measured by a cross-sectional survey administered to teachers. As the survey instrument of the present study solicited perceptions of teachers, therefore the data represented teachers' perceived changes in student learning rather than the actual changes in student learning. The gist of the study is not on measuring improvements in traditional processes and knowledge as measured by attainment tests, but rather on discerning whether there are changes in teaching and learning towards a constructivist approach as a result of ICT implementation. In fact, such opinion-based approach has been regarded as informative and was used by a number of Nordic studies in investigating perceived impacts of ICT on student learning such as team work, independent learning and critical thinking (Balanskat, Blamire, & Kefala, 2006). Nevertheless, future studies may consider conducting longitudinal studies tracing actual changes in student learning by assessing actual performance of students at different junctures of time.

Attempting to capture the wider picture of social ecology of ICT implementation in schools that is characterised by interactions of organisational interventions and pedagogical interventions, this study has resorted to variables of macro-nature rather than micro-nature. At the expense of measuring fewer domains of interest in each variable, the study has been able to keep six variables of sociocultural in nature that may adequately represent the dynamics of ICT implementation in a context of managing change in schools. However, the variables can be further developed. For instance, while the study has been able to measure the abstract concept of ICT implementation strategies by conceptualising it as the capacity of implementation strategies in enhancing collegial ICT exchange and in mobilising resources, the concept of collegial capacity of ICT implementation strategies can be further extended to include additional constructs that capture other different domains of interests. It is because different lines of work may focus on different forms and content of collegiality, yet in all cases the conception of collegiality examines the degree to which colleagues constitute a relatively weak or strong source of influence on teachers' practice (Little, 1990). Expanding the domains of interests for the construct, future studies may focus on exploring the collegial capacity of ICT implementation strategies in schools.

The study may have been limited by the common method variance since the measurement scales of the survey instrument were based on a single method of questionnaire survey (Bagozzi et al., 1991). To avoid the common method variance, future study may consider adopting multiple measures obtained with multiple methods. Hence, construct validation can be done with multitrait-multimethod (MTMM) matrix (Bagozzi et al., 1991). Results of the study can also be more convincing if the sample can be split into two halves for validation (Anderson & Gerbing, 1988). Apart from providing a sound theoretical justification to support the causation assumed in the structural model, future studies may consider recruiting a larger sample size so as to allow the split half sample to validate the model developed in the present study.

Issues of leadership and implementation deserve further attention. Acknowledging the important role of leadership in fostering collegiality that facilitates a paradigm shift in classroom practices, we begin to wonder how to bring about a conceptual change of leadership from instructional leadership to transformational leadership. The present study was a cross-sectional one that only measured the level of perceived transformational leadership, but it will be interesting for future studies to probe into changes in transformational leadership. SEM results of the present study have indicated that a transformational leadership has been important in increasing the collegial capacity of ICT implementation strategies. In this process of building collegiality to increase the effectiveness of schools, school principals may also practice continuous learning in the way of being open to new ideas, making decisions, monitoring progress and refining instruction in place (Fleming, 2004). We think it is prospective for future studies to explore whether ICT implementation strategies is a possible source to bring about changes in leadership. In the social ecology of ICT implementation, will ICT implementation strategies bring about changes in leadership from an instructional one to a transformational one?

While the study has identified the important role of transformational leadership in fostering collegiality that facilitates a paradigm shift in classroom practices, and transformational leadership in this study has also measured a principal's capability in trusting and empowering teachers, it will be useful to conduct further studies that focus on investigating the role of trust building in fostering teacher collegiality for advancing changes in student learning, in particular extending the trust building from policymakers and the general public. It has been pointed out that beliefs that most of the general public know something about how schools operate are considered as threatening the ability of schools to successfully perform their own ideology about teaching and learning (Johnson, 1998). Queries have also been raised with regard to whether the widespread development of professional learning communities that foster collegiality can occur if policymakers and the general public do not trust that teaching is a complex and non-routine process, which relies on teachers' expertise and problem solving that are enhanced through collegiality instead of bureaucratic controls (Hord, 2004; Rowen, 1990). Similarly, some scholars believe that teachers but not policymakers are the ones who are nearest to see what needs to be done in each non-standard situation (Bottery, 2000). We suspect that the more private sector practices being adopted to increase control of educational processes and the work of teachers, the less trust building between policymakers and teachers, eventually leaving less room for schools to develop professional learning communities. It is interesting to find out whether the re-positioning of policymakers and the general public from controllers of change to facilitators of change will foster teacher collegiality for advancing changes in student learning.

The study has examined what and how organisational and pedagogical interventions enhance changes in student learning, yet to examine change we may also need to consider sustainability. It has been pointed out that in sustaining change effort in an environment where there is a range of policy developments working their way through, policymakers and principals should be in a position to ensure that each policy is rooted and that the policies are also aligned (Barber & Phillips, 2000). This has enormous consequences for the role of policymakers and leaders during the implementation of innovations. Future studies can go deeper to find out how professional learning communities may enable leaders to appropriately evaluate schools' capacities and readiness for implementing the innovations, and to screen out the less important issues or issues that are not aligned, thereby regulating the pace of implementation and embedding the changes.

This study has explored managing education change along the direction of enhancing a paradigm shift in learning from a teacher-centred approach to a student centred approach, yet to provide what type and quality of education for citizens in a wider context of education policy merits further deliberations. Under global economic competition, education systems are receiving reducing support from governments (A. Hargreaves, 1999; Mok & Welch, 2002). Under such a policy context, schools are increasingly governed by market ideologies and shaped by corporate discourse of efficiency and effectiveness. This discourse of education is largely dominated by the ideas of economic rationalism and managerialism (Mok & Welch, 2002; Tse, 2002). Significant concerns have been raised on whether placing economic reasons above all other objectives in education may marginalise other issues like equity, social justice, tolerance and trust, which will actually run opposite to the notion of public good and civil society (Bottery, 2000; Chan, 2002; Mok & Welch, 2002; Tse, 2002). For that reason, Bottery (2000) has suggested that educators should, as part of their professional development, have a sufficiently broad overview of education. It is also meaningful for future studies to examine quality education in a wider context of societal issues, so as to explore what kind of education policy serves the imperatives of economic development and social justice.

This study has demonstrated that in enhancing student learning, the organisational factors in the social ecology need to be changed towards establishing collegiality that encourages pedagogical innovations, collegial exchange of ICT knowledge, and mobilisation of resources. As mentioned by some scholars, it is noteworthy to notice that tensions always accompany change (Barber & Phillips, 2000; Osguthorpe & Patterson, 1998). To release tensions during implementing changes, school principal and teachers may need to find a balance between the two extremes, the unrestrained spontaneity (subjectivity) and the controlled planning (objectivity) (Osguthorpe & Patterson, 1998). Similarly, Barber & Phillips (2000) have highlighted the apparent conflicting themes in managing changes, including pressure with support, and central direction with empowering the frontline. It will be interesting to conduct in-depth case studies to investigate how the professional learning communities balance the tensions during change in schools.

Appendix 1: Items of the Questionnaire for the Various Scales

Scale	Item	Questions
Leadership	Ld1	My educational beliefs are reflected in the school goals.
	Ld2*	The school goals are reflected in school plans.
	Ld3	The principal trusts me with school matters.
	Ld4	Teachers have autonomy to make decisions relevant to their teaching.
	Ld5	The principal encourages me to experiment with new ideas in classroom practice.
	Ld6	My school provides adequate resources to support staff professional development.
	Ld7	The principal values staff professional development.
Climate	Cm1#	I feel comfortable to talk to the principal about school matters.
	Cm2*	My opinions can be conveyed to policy-making units (e.g. school management board/committees/departments) effectively within school.
	Cm3	I have a sense of belonging to my school.
	Cm4	Teachers in our school work in a collegial manner.
	Cm5	Teachers in our school are willing to experiment with new ideas in classroom practice.
	Cm6*	I share with colleagues my experiences of improving classroom practice.
Government ICT policy	Gp1	The initiative provides adequate professional development for teachers.
	Gp2	The initiative provides adequate technical support for schools.
	Gp3	The initiative enables schools to have autonomy in allocating or recruiting resources relevant to ICT.
	Gp4	The initiative provides adequate hardware facilities for schools.
	Gp5	The initiative provides adequate support on curriculum resources for teachers.
	Gp6	The initiative provides sound network infrastructure for schools.
	Gp7	The initiative enables students to learn more effectively.
	Gp8	The initiative helps to strengthen teachers' quality of teaching.
ICT implementation strategies	Is1	My school has a mechanism to disseminate the experiences of using ICT resources for teaching and learning.
	Is2*	The principal encourages teachers to experiment with new ICT practices.
	Is3	My school has mobilised resources from external parties (e.g. parents/ alumni/ other schools/ organisations) to help to implement ICT in teaching and learning.
	Is4*	The principal is willing to release teachers to attend training/workshops/conferences on using ICT in teaching and learning.
	Is5	Teachers' opinions can be conveyed to ICT policy-making bodies effectively within school.
	Is6	Colleagues in my school exchange experiences of using ICT to enhance teaching and learning.

	Is7	I participate in sharing sessions for exchanging experiences (of using ICT to enhance teaching and learning) with teachers from other schools.
	Is8	I participate in sharing sessions for exchanging experiences (of using ICT to enhance teaching and learning) with educators from tertiary institutions.
Changes in pedagogy	Pd1*	Compared to the past two academic years, our school curriculum is putting more emphasis on infusing generic skills (e.g. critical thinking/creativity/ communication skills) into teaching and learning of various subjects.
	Pd2*	Compared to the past two academic years, I am more willing to have collaborative lesson preparation with my colleagues.
	Pd3*	Compared to the past two academic years, I am more willing to participate in peer classroom observation.
	Pd4	Compared to the past two academic years, I am creating more opportunities for discussions to develop students' expressive and analytical abilities.
	Pd5	Compared to the past two academic years, I am encouraging students to explore and to inquire in learning more.
	Pd6	Compared to the past two academic years, I am providing more opportunities for students to determine their learning activities
Changes in learning	Ln1*	Compared to students of the same level in the past two academic years, my students are more competent in mastering various generic skills (e.g. collaboration skills, communication skills, creativity, critical thinking skills, information technology skills, numeracy skills, problem-solving skills, self-management skills, study skills).
	Ln2	Compared to students of the same level in the past two academic years, my students are more independent in their learning.
	Ln3	Compared to students of the same level in the past two academic years, my students are more active in constructing knowledge.
	Ln4	Compared to students of the same level in the past two academic years, my students are able to make better use of collaborative work to facilitate learning.
	Ln5	Compared to students of the same level in the past two academic years, my students have more courage to express ideas in class.
	Ln6	Compared to students of the same level in the past two academic years, my students are more motivated in their learning.
	Ln7	Compared to students of the same level in the past two academic years, my students enjoy learning more.

Note: Items marked with * were eventually discarded, with # was moved to leadership.

Appendix 2: Part of the Questionnaire Collecting Supplementary Data

Data	Questions and categorical options
Sex	1. Male 2. Female
Age	1. 18 – 25 2. 26 – 30 3. 31 – 40 4. 41 – 50 5. Above 50
Teaching experience	1. 0 – 3 years 2. 4 – 10 years 3. 11 – 15 years 4. 16 – 20 years 5. Above 20 years
ICT infrastructure	How many LCD/DLP projectors (excluding those in special rooms and laboratories) are available for classroom teaching? 1. 0-5 projectors 2. 6-10 projectors 3. 11-15 projectors 4. 16-20 projectors 5. Over 20 projectors Which one of the following situations best describes the Internet access for classrooms in your school? 1. Less than 1% classrooms are with Internet access 2. 1% to 20% classrooms are with Internet access 3. 21% to 50% classrooms are with Internet access 4. 51% to 80% classrooms are with Internet access 5. 81% to 100% classrooms are with Internet access How many computers/ notebooks are set aside for teachers' use? How many computers/ notebooks are accessible to students after school? How many computers/ notebooks in your school library/ libraries are accessible to students? 1. None 2. 1-20 computers 3. 21-40 computers 4. 41-60 computers 5. Over 60 computers
ICT competency	To date, which one of the following situations best describes the ICT competency of your school? 1. Has not yet reached the *target requirements 2. Has already reached the *target requirements 3. Has reached beyond *target requirements
School type	1. Primary school 2. Secondary school

*Target requirements were ICT requirements for teachers put forward by the Hong Kong government as stated in the document "Information technology for learning in a new era five-year strategy 1998/99 to 2002/03" (Education and Manpower Bureau, 1998).

References

Anderson, J. C., & Gerbing, D. W. (1988). Structural equation modeling in practice: A review and recommended two-step approach. *Pyschological Bulletin, 103*(3), 411-423.

Angus, L. (1996). Cultural dynamics and organizational analysis: Leadership, administration and the management of meaning in schools. In K. Leithwood, J. Chapman, D. Corson, P. Hallinger & A. Hart (Eds.), *International handbook of educational leadership and administration*. Dordrecht, Boston, London: Kluwer Academic Publishers.

Babkus, E., Ferguson, C. E., & Joreskog, K. G. (1987). The sensitivity of confirmatory maximum likelihood factor analysis to violations of measurement scale and distributional assumptions. *Journal of Marketing Research, 24*(2), 222-228.

Bagozzi, R. P. (1981). Evaluating structural equation models with unobservable variables and measurement error: A comment. *Journal of Marketing Research, 18*(3), 375-381.

Bagozzi, R. P., & Phillips, L. W. (1982). Representing and testing organizational theories: A holistic construal. *Administrative Science Quarterly, 27*(3), 459-489.

Bagozzi, R. P., Yi, Y., & Phillips, L. W. (1991). Assessing construct validity in organisational research. *Administrative Science Quarterly, 36*(3), 421-458.

Balanskat, A., Blamire, R., & Kefala, S. (2006). *The ICT Impact Report: A review of studies of ICT impact on schools in Europe*. European Commission: Europe Schoolnet.

Barber, M., & Phillips, V. (2000). *Fusion: How to unleash irreversible change*. Hong Kong: Faculty of Education, Hong Kong Institute of Educational Research, The Chinese University of Hong Kong.

Bass, B. M. (1985). *Leadership and performance beyond expectations*. New York: The Free Press.

Bass, B. M. (1990). *Bass & Stogdill's handbook of leadership: Theory, research, and managerial applications* (3rd ed.). New York: The Free Press.

Bass, B. M., & Avolio, B. J. (1994). *Improving organizational effectiveness through transformational leadership*. Thousand Oaks: Sage Publications.

Bober, M. J. (2002). Technology integration: The difficulties inherent in measuring pedagogical change. *TechTrends, 46*(1), 21-23.

Bollen, K. A. (1989). *Structural equations with latent variables*. New York, Chichester: Wiley.

Borko, H., Romagnano, L., Knuth, E., & Willis-Yorker, C. (2000). Teacher education does matter: A situative view of learning to teach secondary Mathematics. *Educational Psychologist, 35*(3), 193-206.

Bottery, M. (2000). *Education, policy and ethics*. London, New York: Continuum.

Brown, J. S., Collins, A., & Duguid, P. (1989). Situated cognition and the culture of learning. *Educational Researcher, 18*(1), 32-42.

Browne, M. W., & Cudeck, R. (1989). Single sample cross-validation indices for covariance structures. *Multivariate Behavioral Research, 24*(4), 445-455.

Browne, M. W., & Cudeck, R. (1993). Alternative ways of assessing model fit. In J. S. Long (Ed.), *Testing structural equation models*. Newbury Park, London: Sage.

Burns, J. M. (1978). *Leadership*. New York: Harper & Row.

Caldwell, B. J. (1999). Education for the public good: Strategic intentions for the 21st century. In D. D. Marsh (Ed.), *Preparing our schools for the 21st century*. Alexandria: Association for Supervision and Curriculum Development.

Centre for Information Technology in School and Teacher Education, U. o. H. K. (2002). *Preliminary study on reviewing the progress and evaluating the Information Technology in education projects, December 2000 to August 2001*. Hong Kong.

Chan, D. K. K. (2002). Policy implications of adopting a managerial approach in education. In J. K. H. Mok & D. K. K. Chan (Eds.), *Globalization and Education: The quest for quality education in Hong Kong*. Hong Kong: Hong Kong University Press.

Cheng, Y. C. (2002a). The changing context of school leadership: Implications for paradigm shift. In K. Leithwood & P. Hallinger (Eds.), *Second international handbook of educational leadership and administration*. Dordrecht, Boston, London: Kluwer Academic Publishers.

Cheng, Y. C. (2002b). Multi-models of education quality and principal leadership. In J. K. H. Mok & D. K. K. Chan (Eds.), *Gloabalization and education: The quest for quality education in Hong Kong*. Hong Kong: Hong Kong University Press.

Cheng, Y. C., & Cheung, W. M. (1995). A framework for the analysis of educational policies. *International Journal of Educational Management, 9*(6), 10-21.

Cheng, Y. C., & Townsend, T. (2000). Educational change and development in the Asia-Pacific region: trends and issues. In T. Townsend & Y. C. Cheng (Eds.), *Educational change and development in the Asia-Pacific region: challenges for the future*. Lisse: Swets & Zeitlinger.

Cheung, W. M., & Cheng, Y. C. (1997). Self-management: implications for teacher training. *Training for Quality, 5*(4), 160-168.

Chou, C. P., & Bentler, P. M. (1995). Estimates and tests in structural equation modeling. In R. H. Hoyle (Ed.), *Structural equation modeling: Concepts, issues, and applications*. Thousand Oaks, London: Sage.

Churchill, G. A. (1979). A paradigm for developing better measures of marketing constructs. *Journal of Marketing Research, 16*(1), 64-73.

Cizek, G. J., & Ramaswamy, V. (1999). American educational policy: constructing crises and crafting solutions. In G. J. Cizek (Ed.), *Handbook of educational policy*. San Diego, London: Academic Press.

Cognition and Technology Group at Vanderbilt. (1993). Designing learning environments that support thinking: The Jasper series as a case study. In D. H. Jonassen (Ed.), *Designing environments for constructive learning*. Springer-Verlag: NATO Scientific Affairs Division.

Cohen, D. K., & Ball, D. L. (1999). *Instruction, capacity, and improvement*. Philadelphia: Consortium for Policy Research in Education.

Collis, B., & Carleer, G. (1992). *Technology enriched schools: Nine case studies with reflections*. Eugene, Oregon: International Society for Technology in Education.

Crow, G. M., & Grogan, M. (2005). The development of leadership thought and practice in the United States. In F. W. English (Ed.), *The SAGE handbook of educational leadership: Advances in theory, research, and practice*. Thousand Oaks, London: Sage.

Curriculum Development Council. (2001). *The way forward in curriculum development: Learning to learn, life-long learning and whole-person development.* Hong Kong: HK Government Press.

Deal, T. E., & Kennedy, A. A. (1982). *Corporate cultures : the rites and rituals of corporate life.* Reading, Mass: Addison-Wesley.

Dexter, S. L., Anderson, R. E., & Becker, H. J. (2000). Teachers' views of computers as catalysts for changes in their teaching practice. *Journal of Research on Computing in Education, 31*(3), 221-239.

Diamantopoulos, A., & Siguaw, J. A. (2000). *Introducing Lisrel.* London, Thousand Oaks: Sage.

Duke, D. L. (2004). *The challenges of educational change.* Boston, New York: Pearson Education.

Education and Manpower Bureau. (1998). *Information technology for learning in a new era five-year strategy 1998/99 to 2002/03.* Hong Kong: Hong Kong Government.

Education and Manpower Bureau. (2000). *Transforming schools into dynamic and accountable professional learning communities.* Hong Kong: Hong Kong Government.

Eib, B. J., & Mehlinger, H. D. (1998). Technology in education: From segregation to integration. *The High School Magazine, 1*, 10-15.

Elmore, R. F., Peterson, P. L., & McCarthey, S. J. (1996). *Restructuring in the classroom: Teaching, learning, and school organization.* San Francisco: Jossey-Bass Publishers.

Ferneding, K. A. (2003). *Questioning technology: Electronic technologies and educational reform.* New York, Washington, D.C.: Peter Lang.

Firestone, W. A., & Corbett, H. D. (1988). Planned organizational change. In N. J.

Boyan (Ed.), *Handbook of research on educational administration: A project
of the American Educational Research Association*. New York: Longman.

Firestone, W. A., & Louis, K. S. (1999). Schools as cultures. In J. Murphy & K. S.

Louis (Eds.), *Handbook of research on educational administration: A project
of the American Educational Research Association*. San Francisco: Jossey-
Bass Publishers.

Fleming, G. L. (2004). Principals and teachers as continuous learners. In S. M. Hord

(Ed.), *Learning together, leading together: Changing schools through
professional learning communities*. New York: Teachers College Press.

Flora, D. B., & Curran, P. J. (2004). An empirical evaluation of alternative methods of

estimation for confirmatory factor analysis with ordinal data. *Psychological
Methods, 9*(4), 446-491.

Fornell, G., & Larcker, D. F. (1981). Evaluating structural equation models with

unobservable variables and measurement error. *Journal of Marketing
Research, 18*(1), 39-50.

Foster, W. (1989). Toward a critical practice of leadership. In J. Smyth (Ed.), *Critical

perspectives on educational leadership*. London, New York: Falmer Press.

Frank, K. A., Zhao, Y., & Borman, K. (2004). Social capital and the diffusion of

innovations within organizations: The case of computer technology in schools.
Sociology of Education, 77(2), 148-171.

Freiberg, H. J., & Stein, T. A. (1999). Measuring, improving and sustaining healthy

learning environments. In H. J. Freiberg (Ed.), *School climate: Measuring,
improving and sustaining healthy learning environments*. London,
Philadelphia: Falmer Press.

Fullan, M. (1992). *Successful school improvement: The implementation perspective and beyond*. Buckingham, Philadelphia: Open University Press.

Fullan, M. (1993). *Change forces: Probing the depths of educational reform*. London, New York: Falmer Press.

Fullan, M. (1995a). The evolution of change and the new work of educational leader. In K. C. Wong & K. M. Cheng (Eds.), *Educational leadership and change: An international perspective*. Hong Kong: Hong Kong University Press.

Fullan, M. (1995b). The school as a learning organisation: Distant dreams. *Theory into Practice, 34*(4), 230-235.

Fullan, M. (1996). Leadership for change. In K. Leithwood, J. Chapman, D. Corson, P. Hallinger & A. Hart (Eds.), *International handbook of educational leadership and administration*. Dordrecht, Boston, London: Kluwer Academic Publishers.

Fullan, M. (1999). *Change forces: the sequel*. Philadelphia: Falmer Press.

Fullan, M. (2001). *The new meaning of educational change* (third ed.). New York, London: Teacher College.

Fullan, M., & Hargreaves, A. (1992). *Teacher development and educational change*. London, New York: Falmer Press.

Fung, A. (1995). Management of educational innovations: The 'Six-A' process model. In K. C. Wong & K. M. Cheng (Eds.), *Educational leadership and change: An international perspective*. Hong Kong: Hong Kong University Press.

Gardner, J. (2000). The nature of leadership. In *The Jossey-Bass reader on educational leadership*. San Franciso: Jossey-Bass.

Gerbing, D. W., & Anderson, J. C. (1988). An updated paradigm for scale development incorporating unidimensionality and its assessment. *Journal of Marketing Research, 25*(2), 186-192.

Goffman, E. (1959). *The presentation of self in everyday life*. New York: Doubleday.

Gonder, P. O. (1994). *Improving school climate & culture*. Arlington: American Association of School Administrators.

Greeno, J. G. (1998). The situativity of knowing, learning, and research. *American Psychologist, 53*(1), 5-26.

Hair, J. F., Anderson, R. E., Tatham, R. L., & Black, W. C. (1998). *Multivariate data analysis*. New Jersey: Prentice Hall, Upper Saddle River.

Hall, G. E. (1995). The local educational change process and policy implementation. In D. S. G. Carter & M. H. O'Neill (Eds.), *International perspectives on educational reform and policy implementation*. London, Washington: Falmer Press.

Hall, G. E., & Carter, D. S. G. (1995). Implementing change in the 1990s: Paradigms, practices and possibilities. In D. S. G. Carter & M. H. O'Neill (Eds.), *International perspectives on educational reform and policy implementation*. London, Washington: Falmer Press.

Hall, G. E., & George, A. A. (1999). The impact of principal change facilitator style on school and classroom culture. In II. J. Freiberg (Ed.), *School climate: Measuring, improving and sustaining healthy learning environments*. London, Philadelphia: Falmer Press.

Hallinger, P., & Heck, R. H. (1998). Exploring the principal's contribution to school effectiveness: 1980-1995. *School Effectiveness and School Improvement, 9*(2), 157-179.

Hargreaves, A. (1999). Schooling in the new millennium: Educational research for the postmodern age. *Discourse: Studies in the Cultural Politics of Education, 20*(3), 333-355.

Hargreaves, D. (2000, 13 -14 March). *Knowledge management in the learning society.* Paper presented at the Forum of OECD Education Ministers on Developing New Tools for Education Policy-Making, Copenhagen, Denmark.

Heck, R. H., & Hallinger, P. (1999). Next generation methods for the study of leadership and school improvement. In J. Murphy & K. S. Louis (Eds.), *Handbook of research on educational administration: A project of the American Educational Research Association.* San Francisco: Jossey-Bass Publishers.

Heck, R. H., & Thomas, S. L. (2000). *An introduction to multilevel modeling techniques.* Mahwah, New Jersey: Lawrence Erlbaum Associates.

Heifetz, R. A. (1994). *Leadership without easy answers.* Cambridge, Massachusetts: Belknap Press.

Henri, J. (1999). The information literate school community: Not just a pretty face. In J. Henri & K. Bonanno (Eds.), *The information literate school community: Best practice.* Wagga, N.S.W.: Centre for Information Studies, Charles Stuart University.

Heppell, S. (1993). Teacher education, learning and the information generation: The progression and evolution of educational computing against a background of change. *Journal of Information Technology for Teacher Education, 2*(2), 229-237.

Hinkin, T. R. (1998). A brief tutorial on the development of measures for use in survey questionnaires. *Organizational Research Methods, 1*(1), 104-121.

Honebein, P. C., Duffy, T. M., & Fishman, B. J. (1993). Constructivism and the design of learning environments: context and authentic activities for learning.

In D. H. Jonassen (Ed.), *Designing environments for constructive learning.*
Springer-Verlag: NATO Scientific Affairs Division.

Honey, M., Culp, K. M., & Carrigg, F. (1999). *Perspectives on technology and education research: Lessons from the past and present.* Paper presented at the Secretary's Conference on Education Technology, Washington, D.C.

Honey, M., Culp, K. M., & Carrigg, F. (2000). Perspectives on technology and education research: Lessons from the past and present. *Journal of Educational Computing Research, 23*(1), 5-14.

Hord, S. M. (1995). From policy to classroom practice: Beyond the mandates. In D. S. G. Carter & M. H. O'Neill (Eds.), *International perspectives on educational reform and policy implementation.* London, Washington: Falmer Press.

Hord, S. M. (2004). Professional learning communities: An overview. In S. M. Hord (Ed.), *Learning together, leading together: Changing schools through professional learning communities.* New York: Teachers College Press.

House, E. R. (1981). Three perspectives on educational innovation: Technological, political and cultural. In R. Lehming & M. Kane (Eds.), *Improving schools: Using what we know.* Beverly Hills, CA: Sage.

Hox, J. (2002). *Multilevel analysis: Techniques and applications.* Mahwah, New Jersey: Lawrence Erlbaum Associates.

Hoyle, R. H. (1995). The structural equation modeling approach: Basic concepts and fundamental issues. In R. H. Hoyle (Ed.), *Structural equation modeling: Concepts, issues, and applications.* Thousand Oaks, London: Sage.

Hu, L. T., & Bentler, P. M. (1995). Evaluating model fit. In R. H. Hoyle (Ed.), *Structural equation modeling: Concepts, issues, and applications.* Thousand Oaks, London: Sage.

Hu, L. T., & Bentler, P. M. (1999). Cutoff criteria for fit indexes in covariance

   structure analysis: Conventional criteria versus new alternatives. *Structural*

   Equation Modeling, 6(1), 1-55.

Itzkan, S. J. (1994). Assessing the future of telecomputing environments: implications

   for instruction and administration. *The Computing Teacher, 22*(4), 60-64.

Jaccard, J., & Wan, C. K. (1996). *LISREL approaches to interaction effects in*

   multiple regression. Thousand Oaks, London.

Jensen, H. S., & Frederiksen, L. (2001). *Knowledge management and education.*

   Paper presented at the OECD Seminar on Knowledge Management: "Learning

   - by comparing" experiences from private firms and public organisations.

Johnson, B. L. (1998). Organizing for collaboration: A reconsideration of some basic

   organizing principles. In D. G. Pounder (Ed.), *Restructuring schools for*

   collaboration: Promises and pitfalls. Albany: State University of New York

   Press.

Jonassen, D. H. (1996). *Computers in the classroom: Mindtools for critical thinking.*

   Englewood Cliffs, New Jersey: Merrill/ Prentice Hall.

Jonassen, D. H., Peck, K. L., & Wilson, B. G. (1999). *Learning with technology: A*

   constructivist perspective. Upper Saddle Rover, New Jersey: Merrill/ Prentice

   Hall.

Joreskog, K. G. (1971). Statistical analysis of congeneric tests. *Psychometrika, 36*,

   109-133.

Joreskog, K. G. (1993). Testing structural equation models. In J. S. Long (Ed.),

   Testing structural equation models. Newbury Park, London, New Delhi: Sage.

Joreskog, K. G., & Sorbom, D. (1993). *LISREL 8: Structural equation modeling with the SIMPLIS command language*. Lincolnwood: Scientific Software International.

Joreskog, K. G., & Sorbom, D. (2001). *LISREL 8: User's reference guide*. Lincolnwood: Scientific Software International.

Kelloway, E. K. (1998). *Using LISREL for structural equation modeling: A researcher's guide*. Thousand Oaks, London: Sage.

Kline, R. B. (1998). *Principles and practice of structural equation modeling*. New York, London: Guilford Press.

Knapp, L. R., & Glenn, A. D. (1996). *Restructuring schools with technology*. Boston, London: Allyn and Bacon.

Kochan, F. K. (2000). Models for enhancing the professional development of teachers. In D. M. Byrd (Ed.), *Research on effective models for teacher education*. Thousand Oaks, California: Corwin.

Kreft, I., & De Leeuw, J. (1998). *Introducing multilevel modeling*. London, Thousand Oaks: Sage.

Kumar, A., & Sharma, S. (1999). A metric measure for direct comparison of competing models in covariance structure analysis. *Structural Equation Modeling, 6*(2), 169-197.

Lankshear, C., Snyder, I., & Green, B. (2000). *Teachers and technoliteracy: Managing literacy, technology and learning in schools*. St Leonards, NSW: Allen & Unwin.

Lave, J., & Wenger, E. (1991). *Situated learning : legitimate peripheral participation*. Cambridge, New York: Cambridge University Press.

Leithwood, K. (1994). Leadership for school restructuring. *Educational Administration Quarterly, 30*(4), 498-518.

Leithwood, K., & Duke, D. L. (1999). A century's quest to understand school leadership. In J. Murphy & K. S. Louis (Eds.), *Handbook of research on educational administration: A project of the American Educational Research Association* (2nd ed.). San Francisco: Jossey-Bass Publishers.

Leithwood, K., Tomlinson, D., & Genge, M. (1996). Transformational school leadership. In K. Leithwood, J. Chapman, D. Corson, P. Hallinger & A. Hart (Eds.), *International handbook of educational leadership and administration.* Dordrecht, Boston, London: Kluwer Academic Publishers.

Levin, B. (2001). *Reforming education: From origins to outcomes.* New York: RoutledgeFalmer.

Lim, C. P. (2002). A theoretical framework for the study of ICT in schools: A proposal. *British Journal of Educational Technology, 33*(4), 411-421.

Lim, C. P., & Hang, D. (2003). An activity theory approach to research of ICT integration in Singapore schools. *Computers and Education, 41*(1), 49-63.

Little, J. W. (1982). Norms of collegiality and experimentation: Workplace conditions of school success. *American Educational Research Journal, 19*(3), 325-340.

Little, J. W. (1990). The persistence of privacy: Autonomy and initiative in teachers' professional relations. *Teachers College Record, 91*(4), 509-536.

Louis, K. S., & Miles, M. B. (1990). *Improving the urban high school : what works and why.* New York: Teachers College Press.

MacCallum, R. C. (1995). Model specification: procedures, strategies, and related issues. In R. H. Hoyle (Ed.), *Structural equation modeling: Concepts, issues, and applications.* Thousand Oaks, London: Sage.

Marks, H. M., & Printy, S. M. (2003). Principal leadership and school performance: An integration of transformational and instructional leadership. *Educational Administration Quarterly, 39*(3), 370-397.

Marsh, H. W., Balla, J. R., & Hau, K. T. (1996). An evaluation of incremental fit indices: A clarification of mathematical and empirical properties. In R. E. Schumacker (Ed.), *Advanced structural equation modeling*. Mahwah, New Jersey: Lawrence Erlbaum Associates.

Maruyama, G. M. (1998). *Basics of Structural Equation Modeling*. Thousand Oaks, London: Sage Publications.

Maydeu-Olivares. (2001). Limited information estimation and testing of thurstonian models for paired comparison data under multiple judgment sampling. *Psychometrika, 66*(2), 209-288.

McAdams, R. P. (1997). A systems approach to school reform. *Phi Delta Kappan, 79*(2), 138-142.

McCombs, B. L. (2000, 11-12 September). *Assessing the role of educational technology in the teaching and learning process: A learner-centred perspective* Paper presented at the The Secretary's Conference on Educational Technology 2000: Measuring Impacts and Shaping the Future, Alexandria, VA.

McDonnell, L. M., & Elmore, R. F. (1991). Getting the job done: alternative policy instruments. In A. R. Odden (Ed.), *Education policy implementation*. Albany: State University of New York Press.

McLaughlin, M. W. (1991). Learning from experience: Lessons from policy implementation. In A. R. Odden (Ed.), *Education policy implementation*. Albany: State University of New York Press.

McLaughlin, M. W. (1993). What matters most in teachers' workplace context? In J. W. Little & M. W. McLaughlin (Eds.), *Teachers' work: Individual, colleagues, and contexts*. New York: Teachers College Press.

Means, B. (1994). Introduction: Using technology to advance educational goals. In B. Means (Ed.), *Technology and education reform*. San Francisco: Jossey-Bass Publishers.

Mehlinger, H. D. (1995). *School reform in the information age*. Bloomington: Center for Excellence in Education, Indiana University.

Mok, J. K. H., & Welch, A. R. (2002). Economic rationalism, managerialism and structural reform in education. In J. K. H. Mok & D. K. K. Chan (Eds.), *Globalization and education: The quest for quality education in Hong Kong*. Hong Kong: Hong Kong University Press.

MuCulloch, G. (1997). Marketing the millennium: Education for the twenty-first century. In A. Hargreaves & R. Evans (Eds.), *Beyond educational reform*. Buckingham: Open University Press.

Murphy, J. (1988). Methodological, measurement, and conceptual problems in the study of instructional leadership. *Educational Evaluation and Policy Analysis, 10*(2), 117-139.

Muthen, B. O. (1993). Goodness of fit with categorical and other nonnormal variables. In J. S. Long (Ed.), *Testing structural equation models*. Newbury Park, London, New Delhi: Sage.

Muthen, B. O., & Muthen, L. (2005). *Exploratory factor analysis*. Retrieved November 8, 2005, from http://www.statmodel.com/discussion/messages/8/8.html?1131281799

Newby, T. J., Stepich, D. A., Lehman, J. D., & Russell, J. D. (2000). *Instructional technology for teaching and learning: Designing instruction, integrating computers, and using media* (2nd ed.). Upper Saddle River, New Jersey: Merrill/ Prentice Hall.

Nonaka, I., & Takeuchi, H. (1995). *The knowledge-creating company - How Japanese companies create the dynamics of innovation*. New York: Oxford University Press.

Nunnally, J. C., & Bernstein, I. H. (1994). *Psychometric theory* (3rd ed.). New York: McGraw-Hill.

Odden, A. R. (1991a). The evolution of education policy implementation. In A. R. Odden (Ed.), *Education policy implementation*. Albany: State University of New York Press.

Odden, A. R. (1991b). New patterns of education policy implementation and challenges for the 1990s. In A. R. Odden (Ed.), *Education policy implementation*. Albany: State University of New York Press.

Ogawa, R. T. (2005). Leadership as social construct: The expression of human agency within organizational constraint. In F. W. English (Ed.), *The SAGE handbook of educational leadership: Advances in theory, research, and practice*. Thousand Oaks, London: Sage.

Ogawa, R. T., & Bossert, S. T. (1995). Leadership as an organisational quality. *Educational Administration Quarterly, 31*(2), 224-243.

Oppenheimer, T. (2003). *The flickering mind: The false promise of technology in the classroom and how learning can be saved*. New York: Random House.

Osguthorpe, R. T., & Patterson, R. S. (1998). *Balancing the tensions of change: Eight keys to collaborative educational renewal*. Thousand Oaks, California: Corwin Press.

Owens, R. G. (2001). *Organizational behavior in education* (7th ed.). Boston: Allyn and Bacon.

Pallant, J. (2001). *SPSS survival manual: A step by step guide to data analysis using SPSS for Windows (version 10)*. Buckingham, Philadelphia: Open University.

Panel On Education Technology. (1997). *Report to the president on the use of technology to strengthen K-12 education in the United States*. Washington, DC: U.S. Government Printing Office.

Papert, S. (1997). Why school reform is impossible. *The Journal of the Learning Sciences, 6*(4), 417-427.

Peressini, D., Borko, H., Romagnano, L., Knuth, E., & Willis, C. (2004). A conceptual framework for learning to teach secondary Mathematics: A situative perspective. *Educational Studies in Mathematics, 56*(1), 67-96.

Peter, J. P. (1981). Construct validity: A review of basic issues and marketing practices. *Journal of Marketing Research, 18*(2), 133-145.

Phillips, D. C. (2000). An opinionated account of the constructivist landscape. In D. C. Phillips (Ed.), *Constructivism in education: opinions and second opinions on controversial issues*. Chicago, Illinois: National Society for the Study of Education.

Piaget, J. (1970). *Genetic epistemology*. New York: Columbia University Press.

Pipho, C. (2000). Governing the American dream of universal public education. In R. S. Brandt (Ed.), *Education in a new era*. Alexandria: Association for Supervision and Curriculum Development.

Plomp, T., & Brummelhuis, A. t. (1999, 12-14 May 1999). *Introducing Information and Communication Technology in teacher education in the Netherlands: A strategy of vision, courage and care.* Paper presented at the First International Conference on Information Technologies in Education Braga, Portugal.

Prawat, R. S., & Peterson, P. L. (1999). Social constructivist views of learning. In K. S. Louis (Ed.), *Handbook of research on educational administration* (2nd ed.). San Francisco: Jossey-Bass Publishers.

Preacher, K. J., & MacCallum, R. C. (2003). Repairing Tom Swift's electric factor analysis machine. *Understanding Statistics, 2*(1), 13-43.

Putnam, R. T., & Borko, H. (2000). What do new views of knowledge and thinking have to say about research on teacher learning. *Educational Researcher, 29*(1), 4-15.

Raudenbush, S. W., & Bryk, A. S. (2002). *Hierarchical linear models: Applications and data analysis methods.* Thousand Oaks, London, New Delhi: Sage.

Raudenbush, S. W., Bryk, A. S., Cheong, Y. F., & Congdon, R. T. (2001). *HLM 5: Hierarchical linear and nonlinear modeling.* Lincolnwood: Scientific Software International.

Raudenbush, S. W., & Willms, J. D. (1991). The organisation of schooling and its methodological implications. In J. D. Willms (Ed.), *Schools, classrooms, and pupils: International studies of schooling from a multilevel perspective.* San Diego, New York: Academic Press.

Raykov, T., & Marcoulides, G. A. (2000). *A first course in structural equation modeling.* Mahwah, New Jersey, London: Lawrence Erlbaum Associates.

Reinhartz, J., & Beach, D. M. (2004). *Educational leadership: Changing schools, changing roles.* Boston: Pearson.

Retallick, J. (1999). Transforming schools into learning communities: Beginning the journey. In J. Retallick, B. Cocklin & K. Coomber (Eds.), *Learning communities in education: Issues, strategies and contexts*. London: Routledge.

Richardson, V. (2003). Constructivist pedagogy. *Teachers College Record, 105*(9), 1623-1640.

Riel, M. (1994). Educational change in a technology-rich environment. *Journal of Research on Computing in Education, 26*(4), 452-474.

Rowen, B. (1990). Commitment and control: Alternative strategies for the organizational design of schools. *Review of Research in Education, 16*, 353-389.

Satorra, A., & Bentler, P. M. (1994). Corrections to test statistics and standard errors on covariance structure analysis. In C. C. Clogg (Ed.), *Latent variables analysis: Applications for developmental research*. Thousand Oaks: Sage.

Scardamalia, M., & Bereiter, C. (1993). Technologies for knowledge-building discourse. *Communications of the ACM, 36*(5), 37-41.

Schein, E. H. (1968). Organizational socialization and the profession of management. *Industrial Management Review, 9*, 1-15.

Schein, E. H. (1985). *Organizational culture and leadership* (1st ed.). San Francisco: Jossey-Bass.

Schein, E. H. (2004). *Organizational culture and leadership* (3rd ed.). San Francisco: Jossey-Bass.

Senge, P. M. (1990). *The fifth discipline: The Art and practice of the learning organization*. New York: Doubleday/Currency.

Senge, P. M., Cambron-McCabe, N., Lucas, T., Smith, B., Dutton, J., & Kleiner, A. (2000). *Schools that learn: A fifth discipline fieldbook for educators, parents, and everyone who cares about education.* New York: Doubleday.

Sergiovanni, T. J. (1995). *The Principalship: A reflective practice perspective* (Third ed.). Boston, London: Allyn and Bacon.

Sergiovanni, T. J. (2001). *Leadership: What's in it for schools?* London, New York: Routledge/Falmer.

Silins, H. C., & Rosalind, M. H. (1999). What makes a good senior secondary school? *Journal of Educational Administration, 37*(4), 329-344.

Sleegers, P., Geijsel, F., & van den Berg, R. (2002). Conditions fostering educational change. In K. Leithwood & P. Hallinger (Eds.), *Second international handbook of educational leadership and administration.* Dordrecht, Boston, London: Kluwer Academic Publishers.

Smylie, M. A. (1997). From bureaucratic control to building human capital: The importance of teacher learning in education reform. *Arts Education Policy Review, 99*(2), 35-38.

Smylie, M. A., & Hart, A. W. (1999). School leadership for teacher learning and change: A human and social capital development perspective. In K. S. Louis (Ed.), *Handbook of research on educational administration: A project of the American Educational Research Association.* San Francisco: Jossey-Bass Publishers.

Smylie, M. A., Miretzky, D., & Konkol, P. (2004). Rethinking teacher workforce development: A strategic human resource management perspective. In D. Miretzky (Ed.), *Developing the teacher workforce.* Chicago: NSSE.

Spillane, J. P., Reiser, B. J., & Todd, R. (2002). Policy implementation and cognition: Reframing and refocusing implementation research. *Review of Educational Research, 72*(3), 387-431.

Squires, D., & McDougall, A. (1994). *Choosing and using educational software : A teachers' guide*. London ; Washington, D.C.: Falmer Press.

Stein, M. K., & Brown, C. A. (1997). Teacher learning in a social context: Integrating collaborative and institutional processes with the study of teacher change. In B. S. Nelson (Ed.), *Mathematics teachers in transition*. Mahwah, NJ: Lawrence Erlbanum Associates.

Sveiby, K. E. (1997). *The New Organisational Wealth - Managing and measuring knowledge-based assets*. San Francisco: Berret-Koehler Publishers.

Tam, W. M., & Cheng, Y. C. (1996). Staff development for school education quality. *Training for Quality, 4*(4), 16-24.

Taylor, R. P. (1980). In *The computer in the school: Tutor, tool, tutee*. New York: Teachers College Press.

Thomas, O., Carswell, L., Price, B., & Petre, M. (1998). A holistic approach to supporting distance learning using the Internet: transformation, not translation. *British Journal of Educational Technology, 29*(2), 149-161.

Tiene, T., & Ingram, A. (2001). *Exploring current issues in educational technology*. Boston: McGraw-Hill.

Tse, T. K. C. (2002). A critical review of the quality education movement in Hong Kong. In J. K. H. Mok & D. K. K. Chan (Eds.), *Globalization and education: The quest for quality education in Hong Kong*. Hong Kong: Hong Kong University Press.

Twining, P. (2002a). Conceptualising computer use in education: Introducing the computer practice framework (CPF). *British Educational Research Journal, 28*(1), 95-110.

Twining, P. (2002b). Enhancing the impact of investments in 'educational' ICT.

Valli, L., Cooper, D., & Frankes, L. (1997). Professional development schools and equity: A critical analysis of rhetoric and research. *Review of Research in Education, 22,* 251-304.

van den Berg, R., & Sleegers, P. (1996). Building innovative capacity and leadership. In K. Leithwood, J. Chapman, D. Corson, P. Hallinger & A. Hart (Eds.), *International handbook of educational leadership and administration.* Dordrecht, Boston, London: Kluwer Academic Publishers.

Van der Vegt, R., Smyth, L. F., & Vandenberghe, R. (2001). Implementing educational policy at the school level - Organization dynamics and teacher concerns. *Journal of Educational Administration, 39*(1), 8 - 23.

Vygotsky, L. S. (1978). *Mind and society: The development of higher psychological processes.* Cambridge, MA: Harvard University Press.

Wagner, T. (1998). Change as collaborative inquiry: A constructivist' methodology for reinventing schools. *Phi Delta Kappan, 79*(7), 512-517.

Wang, W. C. (2005). *A comparison of alternative estimation methods in confirmatory factor analyses of the general health questionnaire across four groups of Australian immigrants.* Unpublished Master of Science thesis, Swinburn University of Technology, Lilydale.

Wang, W. C., & Cunningham, E. G. (2005). Comparison of alternative estimation methods in confirmatory factor analyses of the general health questionnaire. *Psychological Reports, 97*(1), 3-10.

West, S. G., Finch, J. F., & Curran, P. J. (1995). Structural equation models with nonnormal variables: Problems and remedies. In R. H. Hoyle (Ed.), *Structural equation modeling: Concepts, issues, and applications*. Thousand Oaks, London: Sage.

Whitaker, P. (1993). *Managing change in schools*. Buckingham, Philadelphia: Open University Press.

Windschitl, M. (2002). Framing constructivism in practice as the negotiation of dilemmas: An analysis of the conceptual, pedagogical, cultural, and political challenges facing teachers. *Review of Educational Research, 72*(2), 131-175.

Windschitl, M., & Sahl, K. (2002). Tracing teachers' use of technology in a laptop computer school: The interplay of teacher beliefs, social dynamics, and institutional culture. *American Educational Research Journal, 39*(1), 165-205.

Yukl, G. (1998). *Leadership in organizations* (4th ed.). Prentice Hall: Upper Saddle River, New Jersey.

Zhao, Y., & Frank, K. A. (2003). Factors affecting technology use in schools: An ecological perspective. *American Educational Research Journal, 40*(4), 807-840.

Volker Diekert, Manfred Kufleitner, Gerhard Rosenberger
Elemente der diskreten Mathematik
De Gruyter Studium

Volker Diekert, Manfred Kufleitner,
Gerhard Rosenberger

Elemente der diskreten Mathematik

Zahlen und Zählen, Graphen und Verbände

DE GRUYTER

Mathematics Subject Classification 2010
05-01, 05A15, 05A19, 05C10, 05C21, 05C45, 06-01, 11-01, 60-01, 68R10, 94-01

Autoren
Volker Diekert
Universität Stuttgart
Institut für Formale Methoden der Informatik (FMI)
Abteilung Theoretische Informatik
Universitätsstraße 38
70569 Stuttgart
volker.diekert@fmi.uni-stuttgart.de

Manfred Kufleitner
Universität Stuttgart
Institut für Formale Methoden der Informatik (FMI)
Abteilung Theoretische Informatik
Universitätsstraße 38
70569 Stuttgart
manfred.kufleitner@fmi.uni-stuttgart.de

Gerhard Rosenberger
Universität Hamburg
Fachbereich Mathematik
Bereich AZ
Bundesstraße 55 (Geomatikum)
20146 Hamburg
gerhard.rosenberger@math.uni-hamburg.de

Gerhard Rosenberger
Universität Passau
Fakultät für Informatik und Mathematik
Innstraße 33
94032 Passau
rosenber@fim.uni-passau.de

ISBN 978-3-11-027767-8
e-ISBN 978-3-11-027816-3

Library of Congress Cataloging-in-Publication Data
A CIP catalog record for this book has been applied for at the Library of Congress.

Bibliografische Information der Deutschen Nationalbibliothek
Die Deutsche Nationalbibliothek verzeichnet diese Publikation in der Deutschen Nationalbibliografie;
detaillierte bibliografische Daten sind im Internet über http://dnb.dnb.de abrufbar.

© 2013 Walter de Gruyter GmbH, Berlin/Boston
Umschlaggestaltung: Unter Verwendung des magischen Quadrats aus dem Bild „Melencolia I" von
Albrecht Dürer
Satz: le-tex publishing services GmbH, Leipzig
Druck und Bindung: Hubert & Co.GmbH & Co. KG, Göttingen
♾ Gedruckt auf säurefreiem Papier
Printed in Germany

www.degruyter.com

Vorwort

Über den Inhalt. Dieses Buch basiert auf der Vorlesung *Diskrete Mathematik* im Diplomstudiengang Informatik an der Universität Stuttgart sowie auf Teilen der Vorlesung *Lineare Algebra und algebraische Strukturen für Informatiker* im gleichen Studiengang an der Universität Dortmund. Beide Lehrveranstaltungen wurden über viele Jahre hinweg erfolgreich gehalten. Die Erfahrungen daraus haben die Stoffauswahl und deren Darstellung geprägt. Im Laufe der Vorbereitung des Manuskriptes nahm der Stoffumfang immer weiter zu und begann, den vorgegebenen Rahmen eines Lehrbuchs zu sprengen. Es wurde auch klar, dass wir in einem umfassenden Werk überhaupt nur einige wenige *Elemente der diskreten Mathematik* unterbringen können. So haben wir den Titel und Inhalt den Gegebenheiten angepasst und werden im vorliegenden Band nur einen Teil unseres ursprünglichen Manuskripts präsentieren. Die mehr algebraischen Aspekte sind für einen zweiten Band vorgesehen. Der Titel des vorliegenden Lehrbuchs drückt die Beschränkung auf elementaren Inhalt und wenige Elemente aus und gleichzeitig auch unsere Bewunderung für unerreichbare mathematische Vorbilder, die angefangen von Euklid bis zu Dieudonné und Grothendieck mit ihren *Elementen* Mathematikgeschichte geschrieben haben.

Die Grundidee dieses Buches ist, wesentliche Elemente der diskreten Mathematik zu vermitteln, um die modernen Entwicklungen im Informationszeitalter kompetent mathematisch beurteilen zu können. Hierzu gehört das Verständnis von Graphen, das Rechnen mit großen Zahlen und das Rechnen modulo n. Viele Menschen benutzen regelmäßig Onlinebanking oder bezahlen bargeldlos im Internet. Es ist daher wohltuend zu begreifen, warum diese Transaktionen nicht nur Sicherheit vorspielen, sondern unter realistischen Annahmen sogar garantieren, sofern man die wichtigen Spielregeln einhält. Man sollte auch wissen, was passieren kann, wenn man diese Spielregeln verletzt. Hierzu benötigen wir Primzahlen und wichtige Aussagen zu ihrer Dichte. Wir beginnen daher mit einer Darstellung der elementaren Zahlentheorie. Insbesondere wird die Verschlüsselung mit dem RSA-Verfahren erläutert. Danach behandeln wir Abschätzungen, die unerlässlich sind, wenn man Objekte zählen oder Laufzeiten wichtiger Algorithmen verstehen möchte. Diverse in der Praxis vollkommen zuverlässige Algorithmen nehmen den Zufall zu Hilfe, um überhaupt zu einem Ergebnis zu kommen. Daher durfte ein Kaptitel zur diskreten Wahrscheinlichkeit nicht fehlen.

Danach begeben wir uns ins Zentrum der diskreten Mathematik. Wir behandeln Kombinatorik, erzeugende Funktionen und Graphentheorie. Zum Abschluss widmen wir uns Ordnungsstrukturen und Verbänden sowie booleschen Funktionen und Schaltkreisen. Aufgaben nehmen in dem Buch einen hohen Stellenwert ein. Es gibt Musterlösungen, aber natürlich ist es für das Üben nicht zweckmäßig, allzu schnell auf die Lösungen zurückzugreifen.

Für alle wichtigen Aussagen geben wir vollständige Beweise an und verzichten auf Ausreden wie „Dies würde den Rahmen dieses Buches sprengen". Das benötigte Vorwissen ist gering; dadurch sind Teile des Buches bedingt auch für Schüler geeignet.

Die behandelten Grundlagen sind keine bloßen Aneinanderreihungen von Definitionen und elementaren Zusammenhängen. Statt stur zu befolgende Kochrezepte darzustellen, versucht das Buch ein tieferes Verständnis für die behandelten mathematischen Zusammenhänge zu vermitteln. Das Ziel ist es, Wissen, Techniken und Denkweisen vorzustellen, welche den Leser in die Lage versetzen, selbstständig mathematische Probleme zu lösen. Im Zentrum zahlreicher Beweise steht daher eine kombinatorische Interpretation, die einen *bijektiven* Beweis ermöglicht. Traditionell findet man in Lehrbüchern Induktionsbeweise, die es dem Leser relativ leicht ermöglichen sollen, die Korrektheit bekannter Resultate nachzuvollziehen. Das Wesen und der Ursprung dieser Resultate bleibt dann vielfach im Dunkeln. Eine kombinatorische Interpretation vermag den Sachverhalt nachhaltig zu erhellen, sie ist daher vorzuziehen. Allerdings ist diese Denkweise gerade für Anfänger mit einer erkennbaren Hürde verbunden, aber die Erfahrungen mit dem Stoff zeigen, dass diese Hürde genommen wird. Viele stufen am Ende genau das Überwinden dieser Schwierigkeit als gewinnbringend ein und sehen, machmal erst im Nachhinein, hierin den besonderen Reiz bei der Beschäftigung mit dem Thema. Die mathematischen Herleitungen haben wir durch zahlreiche Bilder illustriert.

Wir werden zeigen, dass es sich bei der diskreten Mathematik um ein modernes und spannendes Gebiet mit vielen Anwendungen handelt. Bei den vorgestellten Konzepten haben wir deshalb Wert auf mathematische Ästhetik gelegt, auch wenn dies manchmal zu Lasten der bestmöglichen Ergebnisse ging. Die Lektüre dieses Buches soll Spaß machen und unterhaltsam sein. Dies hat die Auswahl der fortgeschrittenen Themen maßgeblich beeinflusst.

Das Buch ergänzt und vertieft Grundlagen und zeigt mögliche Anwendungen auf. Es werden auch Themen behandelt, die über den Standardstoff hinaus gehen. Wir hoffen, dass ein Leser in jedem Kapitel mindestens ein *Highlight* findet. Wir favorisieren flüssige gegenüber allzu langatmigen Erklärungen, so soll Freiraum für eigene Überlegungen bleiben. Am Ende eines jeden Kapitels haben wir kurze Kapitelzusammenfassungen als Lern- und Merkhilfe hinzugefügt.

Bei der Erstellung des Textes haben wir uns von anderen Mathematikern inspirieren lassen. Hervorheben möchten wir die Lehrbücher [2, 22, 28]. Daneben wurde an vielen Stellen Originalliteratur verwendet, die noch nicht in Lehrbüchern präsentiert wurde. Manchmal ist dann etwas Eigenes entstanden, teilweise sind wir nahe an den Quellen geblieben. In diesen Fällen finden sich häufig explizite Hinweise. Erwähnen wir Mathematiker namentlich, so finden sich biographische Angaben, sofern es uns sinnvoll erschien und die Daten öffentlich zugänglich waren oder wir das Einverständnis zur Nennung der Geburtsjahre erhielten. Wir hoffen, dass eine zeitliche

Einordnung hilft, mathematische Entwicklungen im Kontext zu verstehen. Bei der Umschrift russischer Namen, haben wir die international übliche englische Umschrift bevorzugt, auch wenn sie von der deutschen abweicht. So schreiben wir *Markov-Ungleichung* und nicht *Markow-* oder *Markoff-Ungleichung*. Bei lebenden Mathematikern haben wir, falls uns bekannt, auf ihre selbst verwendete Umschrift zurückgegriffen. Schließlich möchten wir darauf hinweisen, dass wir Satzzeichen am Ende von abgesetzten Formeln unterdrückt haben.

Über die Autoren lässt sich berichten, dass sie sowohl in der Mathematik als auch in der Informatik zu Hause sind:

Der erste Autor hatte das große Glück, dass er bei Alexander Grothendieck in Montpellier (Frankreich) eine Abschlussarbeit anfertigen und bei Ernst Witt in Hamburg regelmäßig Seminare besuchen konnte. Diese beeindruckenden Persönlichkeiten haben nachhaltigen Einfluss auf seine Entwicklung gehabt.

Der zweite hat beim ersten Autor in Stuttgart in Informatik promoviert und dann ebenfalls in Frankreich (Bordeaux) ein Auslandsjahr verbracht. Mathematik und Schachspielen begeistern ihn seit frühester Jugend. Die genaue Vorausschau, durch welche Züge ein Ziel erreicht werden kann, findet sich durchgehend beim Planen der Beweise im Text.

Der im Alphabet letztgenannte Autor verfügt über die größte Lebenserfahrung, die Erfahrungen Mathematik zu unterrichten und Lehrbücher zu verfassen. Geprägt in seiner Lehre und Forschung sowie bei seiner Präsentation von Vorträgen wurde er insbesondere durch längere Aufenthalte in Russland und den USA. In seinen Forschungsarbeiten kann er auf Koautoren aus mehr als 25 verschiedenen Ländern verweisen.

Über Anagramme. Der Arbeitstitel des Buches war schlicht *Diskrete Mathematik*, ein Vorlesungstitel, den Studierende durch Buchstabenvertauschung in *Diekerts Mathematik* umbenannten. Dieses Anagramm ist nun im Titel nicht mehr unmittelbar vorhanden, wurde jedoch durch ein kunstfertigeres ersetzt. Unser Umschlag zeigt einen Ausschnitt aus dem berühmten Kupferstich der Staatlichen Kunsthalle Karlsruhe von Albrecht Dürer mit dem Titel *Melencolia§I*, einem Anagramm von *Cameleon § LI I*. Wir sehen auf dem Umschlag die wandlungsfähigen Elemente aus einem magischen Quadrat davoneilen. Möge uns die Kunst der Wandlungsfähigkeit auf der Reise durch die Mathematik begleiten.

Danksagung. Das Buch wäre ohne Unterstützung und Hilfe nicht zustande gekommen. Namentlich nennen möchten wir Ulrich Hertrampf, Jonathan Kausch, Jürn Laun, Alexander Lauser, Heike Photien, Horst Prote, Aila Rosenberger, Tobias Walter und Armin Weiß. Sie haben diverse Übungsaufgaben erstellt oder gelöst, Text Korrektur gelesen und beim Schreiben in LaTeX sowie Zeichnen mit Till Tantaus „TikZ" (*Ti*lls TikZ

ist kein Zeichenprogramm) geholfen. Alle verbliebenen Fehler gehen zu Lasten der Autoren. Unser Dank gilt auch dem Verlag Walter de Gruyter, der das Buch in seine Lehrbuchreihe aufnahm.

Der größte Dank gilt unseren Partnern für die geduldige und fördernde Begleitung und unseren Kindern und Kindeskindern für die Freude auf die Zukunft.

Stuttgart und Hamburg, Dezember 2012
Volker Diekert
Manfred Kufleitner
Gerhard Rosenberger

Inhalt

1 Elementare Zahlentheorie

1.1 Einführung

> Ja, was ich hier geschrieben habe macht im Einzelnen überhaupt nicht den Anspruch auf Neuheit; und darum gebe ich auch keine Quellen an, weil es mir gleichgültig ist, ob das was ich gedacht habe, vor mir schon ein anderer gedacht hat.[1]

Wir weichen von Wittgensteins Meinung geringfügig ab, denn auf einige wenige Quellen wird verwiesen.

1.1.1 Von natürlichen zu komplexen Zahlen

Die natürlichen Zahlen $\mathbb{N}$ können nach Giuseppe Peano (1858–1932) durch die folgenden Axiome definiert werden:
- Es gibt eine natürliche Zahl *Null*.
- Der Nachfolger einer natürlichen Zahl ist eine natürliche Zahl.
- Null ist kein Nachfolger einer natürlichen Zahl.
- Die Nachfolger zweier verschiedener natürlichen Zahlen sind verschieden.
- Enthält eine Teilmenge natürlicher Zahlen die Null und zusammen mit jeder Zahl auch deren Nachfolger, so ist diese Teilmenge bereits die Menge aller natürlichen Zahlen.

Die letzte Forderung nennt man auch das *Axiom der vollständigen Induktion*. Die Null schreiben wir als 0 und den Nachfolger einer Zahl n bezeichnen wir mit $s(n)$. Die Standardrealisierung definiert jede natürliche Zahl als eine Menge von anderen natürlichen Zahlen. Die Zahl 0 ist die leere Menge $\varnothing$, die Zahl $1 = s(0)$ wird zur einelementigen Menge $\{\varnothing\}$, die Zahl $2 = s(1)$ ist die Menge $\{\varnothing, \{\varnothing\}\}$ mit zwei Elementen. Ausgehend von $0 = \varnothing$ wird $s(n) = \{0, \ldots, n\}$ für alle natürlichen Zahlen n gesetzt. Die Zahl n ist also selbst eine Menge mit „n" Elementen. Die Ordnungsrelation $m \leq n$ wird zur Teilmengenbeziehung $m \subseteq n$. Die Existenz der natürlichen Zahlen wird damit auf die Mengenlehre reduziert. Dies reflektiert unseren Standpunkt, dass alle mathematischen Objekte in diesem Buch als Mengen interpretiert werden können.

Primzahlen spielen die zentrale Rolle in der Zahlentheorie. Eine *Primzahl* ist eine natürliche Zahl n, die genau zwei Teiler in $\mathbb{N}$ hat, nämlich die 1 und sich selbst. Damit ist weder 0 eine Primzahl, da 0 unendlich viele Teiler hat, noch ist 1 eine Primzahl, da nur ein einziger Teiler, nämlich 1, existiert. Die kleinste Primzahl ist 2. Alle weiteren Primzahlen sind ungerade.

---

[1] Aus dem Vorwort der 1918 verfassten Schrift „Logisch-Philosophische Abhandlung" von Ludwig Wittgenstein (1889–1951), die auch als „Tractatus logico-philosophicus" bekannt ist.

Eine Besonderheit der natürlichen Zahlen ist ihre *Wohlordnung*: Jede nichtleere Teilmenge hat ein kleinstes Element. Dies führt direkt zu einem zweiten Induktionsprinzip. Um zu zeigen, dass $P(n)$ für alle $n \in \mathbb{N}$ wahr ist, reicht es, $P(n)$ unter der *Induktionsannahme* zu beweisen, dass $P(m)$ für alle $m < n$ gilt. Denn gilt eine Eigenschaft P nicht für alle natürlichen Zahlen, so muss es eine kleinste Zahl geben, für die P falsch ist und für alle kleineren Zahlen ist P wahr.

Im Aufbau des Zahlensystems erhält man aus $\mathbb{N}$ die ganzen Zahlen $\mathbb{Z}$, dann die rationalen Zahlen $\mathbb{Q}$. Hieraus werden die reellen Zahlen $\mathbb{R}$ konstruiert. Schließlich nimmt man noch eine *imaginäre Zahl* $i = \sqrt{-1}$ hinzu, die die wundersame Eigenschaft hat, dass ihr Quadrat die negative Zahl -1 ist. Hieraus ergeben sich dann die komplexen Zahlen $\mathbb{C}$. Wir gehen davon aus, dass der Leser mit reellen Zahlen vertraut ist. Komplexe Zahlen kommen kaum vor.

1.1.2 Von Halbgruppen zu Körpern

Alle soeben genannten Zahlenbereiche sind in natürlicher Weise mit den beiden Verknüpfungen *„plus"* und *„mal"* ausgestattet. Ähnlich wie man die natürlichen Zahlen durch die Peano-Axiome beschreibt, kann man auch bei Verknüpfungen gewisse Axiome betrachten. Sei $\circ$ eine beliebige Abbildung $M \times M \to M$ mit $(x, y) \mapsto x \circ y$; wir nennen $\circ$ eine *Verknüpfung* (oder *Operation*), und wir sagen $\circ$ ist *assoziativ*, wenn für alle $x, y, z \in M$ die Rechenregel $(x \circ y) \circ z = x \circ (y \circ z)$ gilt. Sie ist *kommutativ* (oder auch *abelsch*), wenn $x \circ y = y \circ x$ für alle $x, y \in M$ gilt. Bei assoziativen Verknüpfungen ist das Ergebnis unabhängig von der Reihenfolge der Auswertung, so dass der Term $x \circ y \circ z$ ein eindeutig bestimmtes Element aus M beschreibt. Ein Element $e \in M$ ist *neutral*, falls $x \circ e = e \circ x = x$ für alle $x \in M$ gilt. Insbesondere gibt es höchstens ein neutrales Element, denn sind e und e' neutral, so folgt $e' = e \circ e' = e$. Wenn e ein neutrales Element ist, dann ist $x \in M$ *invertierbar*, falls ein Element $y \in M$ mit $x \circ y = y \circ x = e$ existiert; man nennt dann y das *Inverse* von x. Wenn die zugrunde liegende Verknüpfung klar ist, dann schreiben wir auch oft xy anstelle von $x \circ y$. Eine *Halbgruppe* ist eine Menge mit einer assoziativen Verknüpfung. Für jedes $n \in \mathbb{N}$ bilden die Zahlen $\{m \in \mathbb{N} \mid m \geq n\}$ mit der Addition als Verknüpfung eine kommutative Halbgruppe. Ein *Monoid* ist eine Halbgruppe mit einem neutralen Element. Ein typisches Beispiel für ein kommutatives Monoid sind die natürlichen Zahlen $\mathbb{N}$ mit der Addition, und das neutrale Element ist 0. Eine *Gruppe* ist ein Monoid, bei dem jedes Element invertierbar ist. Die ganzen Zahlen $\mathbb{Z}$ mit der Addition bilden eine kommutative Gruppe. Hierbei ist $-x$ das Inverse von x. Vor allem in der Gruppentheorie spricht man anstatt von kommutativen Gruppen häufig von abelschen Gruppen nach Niels Henrik Abel (1802–1829). Bei kommutativen Operationen benutzt man häufig $+$ als Zeichen für die Verknüpfung und 0 für das neutrale Element.

Wenn nun M mit zwei Verknüpfungen $+$ und $\cdot$ ausgestattet ist, dann kann man durch Rechengesetze das Verhältnis der beiden Operationen zueinander beschreiben. Die wichtigsten Axiome sind hier die *Distributivgesetze*; das heißt, es gilt $x \cdot (y+z) = x \cdot y + x \cdot z$ und $(x + y) \cdot z = x \cdot z + y \cdot z$ für alle $x, y, z \in M$. Wir sagen, dass M ein *Ring* ist, wenn die Distributivgesetze gelten, wenn M mit der Verknüpfung $+$ eine kommutative Gruppe ist, und wenn M mit $\cdot$ ein Monoid bildet. Ein Ring ist *kommutativ*, wenn $\cdot$ eine kommutative Verknüpfung ist. Die ganzen Zahlen $\mathbb{Z}$ mit Addition und Multiplikation bilden einen kommutativen Ring. Sei M ein kommutativer Ring mit 0 als neutralem Element für die Verknüpfung $+$; dann ist M ein *Körper*, wenn $M \setminus \{0\}$ bezüglich der Multiplikation eine Gruppe ist. Die Zahlenbereiche $\mathbb{Q}$, $\mathbb{R}$ und $\mathbb{C}$ sind Körper.

Für eine algebraische Struktur X sagen wir, $Y \subseteq X$ ist eine *Unterstruktur*, wenn Y selbst auch wieder dieselben Struktureigenschaften erfüllt, wie sie bei X gefordert werden. Beispielsweise bilden die geraden Zahlen mit der Addition eine Untergruppe der ganzen Zahlen. Ein anderes Beispiel ist die Halbgruppe $M = \{1,0\}$ mit der Multiplikation; hier sind $\{1\}$ und $\{0\}$ Unterhalbgruppen. Die Halbgruppe M ist auch ein Monoid, aber nur $\{1\}$ ist ein Untermonoid, da $\{0\}$ zwar ein Monoid bildet, aber nicht die 1 von M enthält. Die von $Y \subseteq X$ *erzeugte* Unterstruktur von X ist die kleinste Unterstruktur, welche die Menge Y enthält.

Eine Abbildung zwischen algebraischen Strukturen, die mit den jeweiligen Operationen verträglich ist (wie $+$ und $\cdot$) sowie neutrale Elemente aufeinander abbildet, heißt *Homomorphismus*. Ein Gruppenhomomorphismus ist eine Abbildung $\varphi : G \to G'$ zwischen Gruppen mit $\varphi(x \cdot y) = \varphi(x) \cdot \varphi(y)$ für alle $x, y \in G$. Denn durch die Gruppeneigenschaft wird das neutrale Element 1_G von G automatisch auf das neutrale Element $1_{G'}$ von G' abgebildet. Für Monoide ist dies eine zusätzliche Forderung. Daher ist ein Ringhomomorphismus eine Abbildung $\varphi : R \to R'$ zwischen Ringen mit $\varphi(x + y) = \varphi(x) + \varphi(y)$, $\varphi(x \cdot y) = \varphi(x) \cdot \varphi(y)$ für alle $x, y \in R$ und $\varphi(1_R) = 1_{R'}$. Eine Bijektion φ besitzt stets eine Umkehrabbildung φ^{-1}. Sind beide Abbildungen φ und φ^{-1} Homomorphismen, so nennt man φ einen *Isomorphismus*. In vielen Fällen ist ein bijektiver Homomorphismus bereits ein Isomorphismus.

1.2 Der euklidische Algorithmus

Eine ganze Zahl k *teilt* ℓ, geschrieben $k \mid \ell$, falls $m \in \mathbb{Z}$ existiert mit $km = \ell$. Den *größten gemeinsamen Teiler* von zwei ganzen Zahlen k und ℓ bezeichnen wir mit $\mathrm{ggT}(k, \ell)$; es ist die größte natürliche Zahl, die sowohl k als auch ℓ teilt. Den größten gemeinsamen Teiler von k und 0 definieren wir als die Zahl $|k|$. Zwei Zahlen heißen *teilerfremd*, wenn ihr größter gemeinsamer Teiler 1 ist. Der *euklidische Algorithmus* (Euklid von Alexandria, Wirken um 300 v. Chr.) ist ein effizientes Verfahren zur Berechnung des größten gemeinsamen Teilers. Da $\mathrm{ggT}(k, \ell) = \mathrm{ggT}(-k, \ell) = \mathrm{ggT}(\ell, k)$ gilt, genügt es $k, \ell \in \mathbb{N}$ zu betrachten. Sei $0 < k \leq \ell$ und schreibe $\ell = qk + r$,

wobei $0 \le r < k$ der *Rest* ist. Für diesen Rest r schreiben wir auch „ℓ mod k". Hierauf gehen wir im nächsten Abschnitt näher ein. Jede Zahl, die k und den Rest r teilt, teilt auch die Summe $\ell = qk + r$. Jede Zahl, die k und ℓ teilt, teilt auch die Differenz $r = \ell - qk$. Dies liefert uns die folgende rekursive Version des euklidischen Algorithmus:

```
/* Voraussetzung ist k ≥ 0, ℓ ≥ 0 */
function ggT(k, ℓ)
begin
    if k = 0 then return ℓ
    else return ggT(ℓ mod k, k) fi
end
```

Beispiel 1.1. Wir wollen $\mathrm{ggT}(21, 59)$ bestimmen. Links geben wir im Folgenden den Rechenweg für das obige Programm an; rechts befindet sich eine kürzere Rechnung unter der Zuhilfenahme von negativen Zahlen.

$$59 = 2 \cdot 21 + 17 \qquad 59 = 3 \cdot 21 - 4$$
$$21 = 1 \cdot 17 + 4 \qquad 21 = 5 \cdot 4 + 1$$
$$17 = 4 \cdot 4 + 1$$

Damit ist $\mathrm{ggT}(21, 59) = 1$. ◇

Satz 1.2 wird häufig Étienne Bézout (1730–1783) zugeschrieben, der eine entsprechende Aussage für Polynome gezeigt hat. Die Aussage des Satzes war schon früher durch Arbeiten von Claude Gaspard Bachet de Méziriac (1581–1638) bekannt.

Satz 1.2 (Lemma von Bézout). *Seien $k, \ell \in \mathbb{Z}$. Dann existieren $a, b \in \mathbb{Z}$ mit:*

$$\mathrm{ggT}(k, \ell) = ak + b\ell$$

Beweis. Wir können $\ell > k > 0$ annehmen, die anderen Fälle sind offensichtlich oder können auf diesen Fall reduziert werden. Setze $r_0 = \ell$ und $r_1 = k$. Der euklidische Algorithmus berechnet nacheinander Reste $r_0 > r_1 > r_2 \ldots > r_n \ge r_{n+1} = 0$, die die Beziehungen

$$r_{i-1} = q_i r_i + r_{i+1}$$

für geeignete $q_i \in \mathbb{N}$ erfüllen. Es folgt $\mathrm{ggT}(k, \ell) = \mathrm{ggT}(r_{i+1}, r_i) = \mathrm{ggT}(0, r_n) = r_n$. Wir zeigen nun, dass für alle $i \in \{0, \ldots, n\}$ ganze Zahlen a_i und b_i derart existieren, dass $a_i r_{i+1} + b_i r_i = r_n$ gilt. Für $i = n$ ist $a_n = 0$ und $b_n = 1$. Sei nun $i < n$ und seien a_{i+1} und b_{i+1} bereits definiert, d.h. $a_{i+1} r_{i+2} + b_{i+1} r_{i+1} = r_n$. Mit $r_{i+2} = r_i - q_{i+1} r_{i+1}$ folgt $(b_{i+1} - a_{i+1} q_{i+1}) r_{i+1} + a_{i+1} r_i = r_n$. Damit haben $a_i = b_{i+1} - a_{i+1} q_{i+1}$ und $b_i = a_{i+1}$ die gewünschte Eigenschaft. □

Der obige Beweis liefert das folgende Verfahren. Der *erweiterte euklidische Algorithmus* berechnet zusätzlich zu $\text{ggT}(k, \ell)$ auch Zahlen a und b mit der Eigenschaft $ak + b\ell = \text{ggT}(k, \ell)$.

```
/* Voraussetzung ist k ≥ 0, ℓ ≥ 0 */
/* Berechnet wird (a, b, t) mit ak + bℓ = t = ggT(k, ℓ) */
function erw-ggT(k, ℓ)
begin
   if k = 0 then return (0, 1, ℓ)
   else
      (a, b, t) := erw-ggT(ℓ mod k, k);
      return (b − a · ⌊ℓ/k⌋, a, t)
   fi
end
```

Beispiel 1.3. Wir führen Beispiel 1.1 fort. Rückwärts Einsetzen bei der ersten Rechnung ergibt

$$
\begin{aligned}
1 &= 17 - 4 \cdot 4 \\
&= 17 - 4 \cdot (21 - 1 \cdot 17) \\
&= -4 \cdot 21 + 5 \cdot 17 \\
&= -4 \cdot 21 + 5 \cdot (59 - 2 \cdot 21) \\
&= -14 \cdot 21 + 5 \cdot 59
\end{aligned}
$$

Die Darstellung der 1 als Linearkombination von 21 und 59 ist nicht eindeutig. Beispielsweise gilt $1 = -14 \cdot 21 + (59 \cdot 21 - 21 \cdot 59) + 5 \cdot 59 = 45 \cdot 21 - 16 \cdot 59$.   ◊

Das Korollar 1.4 von Satz 1.2 sagt aus, dass die zu n teilerfremden Zahlen bezüglich der Multiplikation ein Untermonoid von $\mathbb{Z}$ bilden.

Korollar 1.4. *Sei $n \in \mathbb{Z}$ und seien $k, \ell \in \mathbb{Z}$ zu n teilerfremde Zahlen, also $\text{ggT}(n, k) = \text{ggT}(n, \ell) = 1$. Dann ist n auch teilerfremd zum Produkt $k\ell$.*

Beweis. Schreibe $1 = an + bk = cn + d\ell$ für gewisse $a, b, c, d \in \mathbb{Z}$. Dann gilt $1 = (an + bk)(cn + d\ell) = (anc + bkc + ad\ell)n + bd(k\ell)$. Ein gemeinsamer Teiler von n und $k\ell$ teilt damit auch 1. Hieraus folgt $\text{ggT}(n, k\ell) = 1$.   □

1.3 Der Fundamentalsatz der Arithmetik

Mit dem Fundamentalsatz der Arithmetik bezeichnet man die Aussage, dass jede positive natürliche Zahl eine eindeutige *Primfaktorzerlegung* hat.

Satz 1.5. *Sei* $\mathbb{P} = \{2, 3, 5, 7, 11, \ldots\}$ *die Menge der Primzahlen und* $n \in \mathbb{N}$ *mit* $n \geq 1$. *Dann gibt es eine eindeutig bestimmte Produktdarstellung*

$$n = \prod_{p \in \mathbb{P}} p^{n_p}$$

Die Zahlen n_p *sind dabei genau dann von Null verschieden, wenn die Primzahl* p *ein Teiler von* n *ist.*

Beweis. Die Zahl 1 ermöglicht die Primfaktorzerlegung mit $n_p = 0$ für alle $p \in \mathbb{P}$ und dies ist auch die einzige Zerlegung. Gilt $n > 1$, so teilt eine Primzahl p die Zahl n und wir erhalten eine Darstellung der gewünschten Form induktiv aus der Primfaktorzerlegung von n/p. Darüber hinaus gibt es für jede Primzahl p, die n teilt, eine Primfaktorzerlegung mit $n_p \geq 1$. Klar ist auch, dass nur diejenigen n_p ungleich Null sein können, für die p ein Teiler von n ist.

Es bleibt daher nur, die Eindeutigkeit der Primfaktorzerlegung zu zeigen. Angenommen, es gäbe für eine Zahl zwei verschiedene Zerlegungen. Dann würden wir durch Kürzen maximaler Primzahlpotenzen zwei disjunkte endliche Teilmengen $R \subseteq \mathbb{P}$ und $Q \subseteq \mathbb{P}$ finden sowie eine Gleichung

$$m = \prod_{p \in R} p^{n_p} = \prod_{q \in Q} q^{n_q}$$

Wir können $n_q > 0$ für ein $q \in Q$ annehmen. Also teilt q die Zahl m. Andererseits kommt q in R nicht vor und q ist teilfremd zu allen Zahlen in R, da verschiedene Primzahlen teilerfremd sind. Nach Korollar 1.4 ist dann q teilerfremd zum Produkt m. Dies ist ein Widerspruch, da $q > 1$. □

1.4 Modulare Arithmetik

Es sei n eine ganze Zahl, aber auch die Annahme $n \in \mathbb{N}$ ist keine wirkliche Einschränkung. Die Zahl n teilt die Menge der ganzen Zahlen in Restklassen ein. Die *Restklasse* zu $k \in \mathbb{Z}$ ist die Teilmenge $k + n\mathbb{Z} \subseteq \mathbb{Z}$. Es gilt $\ell \in k + n\mathbb{Z}$ genau dann, wenn $k + n\mathbb{Z} = \ell + n\mathbb{Z}$; und Restklassen sind entweder identisch oder disjunkt. Die Menge $\{k + n\mathbb{Z} \mid k \in \mathbb{Z}\}$ wird als Ring der Restklassen $\mathbb{Z}/n\mathbb{Z}$ bezeichnet. Es ist ein Ring, da man Klassen wie ganze Zahlen addieren und multiplizieren kann. Wir setzen:

$$(k + n\mathbb{Z}) + (\ell + n\mathbb{Z}) = k + \ell + n\mathbb{Z}$$
$$(k + n\mathbb{Z}) \cdot (\ell + n\mathbb{Z}) = k\ell + n\mathbb{Z}$$

Man beachte, dass die Operationen wohldefiniert sind, denn das Ergebnis der Operationen hängt nicht von den Repräsentanten ab. Wir können also k durch ein $k' \in k + n\mathbb{Z}$ und ℓ durch ein $\ell' \in \ell + n\mathbb{Z}$ ersetzen; dann gelten $k + \ell + n\mathbb{Z} = k' + \ell' + n\mathbb{Z}$ sowie $k\ell + n\mathbb{Z} = k'\ell' + n\mathbb{Z}$. Für $k \in \ell + n\mathbb{Z}$ schreiben wir

$$k \equiv \ell \mod n$$

und sagen, k und ℓ sind *kongruent modulo* n. Für $n = 0$ ist $\mathbb{Z} = \mathbb{Z}/n\mathbb{Z}$, und $k \equiv \ell \bmod 0$ bedeutet $k = \ell$. Sei deshalb ab jetzt $n \neq 0$. Mit $k \bmod n$ meinen wir die eindeutig bestimmte Zahl $r \in \{0, \ldots, |n| - 1\}$ mit $k \equiv r \bmod n$. Die Zahl r ist der Rest beim Teilen von k durch n. Jede Klasse $k + n\mathbb{Z}$ wird durch den Rest $k \bmod n$ eindeutig repräsentiert. Daher ist es manchmal nützlich, die Menge $\mathbb{Z}/n\mathbb{Z}$ mit der Menge $\{0, \ldots, n - 1\}$ zu identifizieren. Wir erhalten:

$$(k \bmod n) + (\ell \bmod n) \equiv k + \ell \quad \bmod n$$
$$(k \bmod n) \cdot (\ell \bmod n) \equiv k \cdot \ell \quad \bmod n$$

Beispiel 1.6. Wir wollen $11^{561} \bmod 12$ ausrechnen. Zuerst den Wert 11^{561} auszurechnen ist zu mühsam. Für die Basis gilt $11 \equiv -1 \bmod 12$. Daraus folgt $11^{561} \equiv (-1)^{561} \equiv -1 \equiv 11 \bmod 12$ und wir erhalten $11^{561} \bmod 12 = 11$. ◇

Beim Modulo-Rechnen mit Addition und Multiplikation kann man jederzeit zwischen beliebigen konkreten Zahlen und ihren Restklassen hin und her wechseln. Das Ergebnis stimmt dann „modulo n". Nur bei der Division mit Teilern von n müssen wir aufpassen.

Mit $(\mathbb{Z}/n\mathbb{Z})^*$ bezeichnen wir die Gruppe der *Einheiten* des Ringes $\mathbb{Z}/n\mathbb{Z}$. Dies sind die Restklassen, die ein multiplikatives Inverses besitzen. Die Anzahl $|(\mathbb{Z}/n\mathbb{Z})^*|$ wird mit $\varphi(n)$ bezeichnet und im Abschnitt 1.10 genauer untersucht. Der Wert $\varphi(n)$ ist die Anzahl der zu n teilerfremden natürlichen Zahlen im Bereich von 1 bis n. Für eine Primzahl p ist $\varphi(p) = p - 1$.

Satz 1.7. *Sei $n \in \mathbb{N}$. Dann gelten die folgenden Aussagen.*

(a) $(\mathbb{Z}/n\mathbb{Z})^* = \{k + n\mathbb{Z} \mid \mathrm{ggT}(k, n) = 1\}$.

(b) Die Zahl n ist genau dann eine Primzahl, wenn $\mathbb{Z}/n\mathbb{Z}$ ein Körper ist.

(c) Sei $k \in \mathbb{Z}$. Die Multiplikation $\mathbb{Z}/n\mathbb{Z} \to \mathbb{Z}/n\mathbb{Z}$, $x \mapsto kx$ mit k ist genau dann bijektiv, wenn $\mathrm{ggT}(k, n) = 1$ gilt.

Beweis. (a) Es gilt $k \in (\mathbb{Z}/n\mathbb{Z})^*$ genau dann, wenn wir $1 = k\ell + mn$ schreiben können. Nach Satz 1.2 ist dies genau dann der Fall, wenn $\mathrm{ggT}(k, n) = 1$.

(b) Der Ring $\mathbb{Z}/n\mathbb{Z}$ ist genau dann ein Körper, wenn alle von Null verschiedenen Elemente invertierbar sind. Dies bedeutet nach (a), dass alle natürlichen Zahlen von 1 bis $n - 1$ zu n teilerfremd sind. Dies wiederum ist gleichwertig zur Primzahleigenschaft.

(c) Für $\mathrm{ggT}(k, n) = 1$ ist k in $(\mathbb{Z}/n\mathbb{Z})^*$ invertierbar, und die Multiplikation mit k hat eine inverse Abbildung; sie ist damit bijektiv. Haben k und n einen gemeinsamen Teiler $m \neq 1$, so ist $k \cdot (n/m) \in n\mathbb{Z}$ und damit $k \cdot (n/m) \equiv 0 \equiv k \cdot 0 \bmod n$, aber $n/m \not\equiv 0 \bmod n$. □

1.5 Anwendungen der modularen Arithmetik

1.5.1 Bits und Bytes

Eine der vielen Anwendungen der modularen Arithmetik (oder auch Restklassen-arithmetik) findet sich in der internen Darstellung ganzer Zahlen im Rechner. Wir nehmen an, dass k Bits für die Darstellung zur Verfügung stehen und dass die Multiplikation mit Minus-Eins weitgehend möglich sein soll. Da bereits ein Bit für das Vorzeichen verloren geht und da aufgrund von $-0 = +0$ nicht genau gleich viele positive wie negative Zahlen dargestellt werden können, ist ein maximaler Zahlenbereich:

$$\underbrace{-2^{k-1},\ldots,-1}_{2^{k-1}\ \text{Zahlen}},\underbrace{0,1,\ldots,2^{k-1}-1}_{2^{k-1}\ \text{Zahlen}}$$

Für viele Fälle erweist es sich als günstiger, statt explizit mit Vorzeichen zu rechnen, zur Restklassenarithmetik modulo 2^k überzugehen. Wir rechnen im Ring $\mathbb{Z}/2^k\mathbb{Z}$. Der obige Zahlenbereich ist dann ein Repräsentantensystem. Bei einem Stellenwert-system zur Basis 2 liefert das erste (höchstwertige) der k Bits die Information über das Vorzeichen. Es ist genau dann Eins, wenn die dargestellte Zahl negativ ist. Natürlich muss ein Überschreiten des gültigen Zahlenbereichs (ein „Overflow") gesondert verwaltet werden. Angenommen, 8 Bits stehen zur Verfügung. Dann können wir den Bereich von -128 bis $+127$ darstellen. Mit $k = 8$ gilt nämlich:

$$
\begin{aligned}
\underbrace{
\begin{aligned}
01111111 &= 2^{k-1}-1 &= 127 &\equiv 127 &&\mod 2^8\\
10000000 &= 2^{k-1} &= 128 &\equiv -128 &&\mod 2^8\\
11111111 &= 2^k-1 &= 255 &\equiv -1 &&\mod 2^8
\end{aligned}
}_{k\ \text{Bits}}
\end{aligned}
$$

Ein Vorteil dieser Arithmetik ist, dass die Subtraktion nicht schwieriger als die Addition ist, da die Multiplikation mit Minus-Eins nur die Bildung des *Zweierkomplements* erfordert. Sei $x = x_1 \cdots x_k$ mit $x_i \in \{0, 1\}$ und definiere $\overline{x} = \overline{x_1} \cdots \overline{x_k}$ mit der Bezeichnung $\overline{0} = 1$ und $\overline{1} = 0$. Dann gilt

$$x + \overline{x} = \underbrace{1 \cdots 1}_{k\text{-mal}} = 2^k - 1$$

Unsere Rechnungen gelten stets nur modulo 2^k, liegt das Endergebnis jedoch im gültigen Zahlenbereich, so ist es korrekt. Insbesondere gilt dort $-x = \overline{x} + 1$. Sei beispielsweise $k = 8$ und $x = 01101011 = +107$. Wir berechnen $-x = -107$ durch:

$$
\begin{aligned}
\overline{x} &= 10010100\\
\overline{x}+1 &= 10010101 = 149 \equiv -107 \mod 256
\end{aligned}
$$

Im Fall einer negativen Zahl $x = 10010000 = -112 \equiv 144 \mod 256$ ergibt sich $-x = +112$ wie folgt:

$$
\begin{aligned}
\overline{x} &= 01101111\\
\overline{x}+1 &= 01110000 = 112
\end{aligned}
$$

1.5.2 Fehlererkennung bei Artikelnummern

Weitere Anwendungen der modularen Arithmetik findet man bei der EAN (European Article Number) und der ISBN (International Standard Book Number). Die EAN dient der Kennzeichnung von Handelsartikeln, wohingegen die ISBN diese Aufgabe bei Büchern erfüllt. Bei beiden Systemen ist die letzte Stelle eine Prüfziffer, die aus der gewichteten Quersumme der übrigen Stellen entsteht.

Eine korrekte EAN ist eine 13-stellige Dezimalzahl $x_{13}x_{12}\cdots x_1$ mit Prüfziffer x_1 und der Eigenschaft

$$x_{13} + 3 \cdot x_{12} + x_{11} + 3 \cdot x_{10} + \cdots + x_1 \equiv 0 \mod 10$$

Als Gewichte treten abwechselnd 1 und 3 auf. Falls nun beim Bestimmen der gewichteten Quersumme ein Wert ungleich 0 herauskommt, so entspricht dies einer Fehlermeldung. Welche Arten von Fehlern können wir mit diesem Verfahren erkennen? Die Abweichung einer Stelle um a führt je nach Gewicht dieser Stelle zu einer Abweichung a bzw. $3a$ in der Prüfsumme. Wegen $\gcd(3, 10) = 1$ folgt, dass wir bei $a \not\equiv 0 \mod 10$ eine Fehlermeldung erhalten. Diese sogenannten Einfachfehler hätten wir schon erkennen können, wenn wir alle Stellen mit 1 gewichtet hätten. Die unterschiedlichen Gewichte benachbarter Stellen ermöglichen es, die Vertauschung zweier Stellen x_{i+1} und x_i einer EAN zu erkennen, falls $x_{i+1} \not\equiv x_i \mod 5$ gilt. Wir können aber zum Beispiel keine Vertauschung von 7 und 2 erkennen. Bei der 10-stelligen ISBN rechnet man modulo 11, so dass hier die 10 möglichen Gewichte $10, 9, 8, \ldots, 1$ zur Verfügung stehen, die alle teilerfremd zu 11 sind. Eine korrekte ISBN $x_{10}x_9 \cdots x_1$ mit Prüfziffer x_1 erfüllt

$$10 \cdot x_{10} + 9 \cdot x_9 + 8 \cdot x_8 + \cdots + 1 \cdot x_1 \equiv 0 \mod 11$$

Da die Differenz zweier benachbarter Gewichte stets 1 ist, lassen sich zusätzlich zu einzelnen Tippfehlern alle Vertauschungen benachbarter Stellen erkennen. Allerdings benötigt man bei der Darstellung der Prüfziffer von ISBNs wegen des Rechnens modulo 11 ein weiteres Symbol X für den Wert 10.

1.6 Der chinesische Restsatz

Die Grundlage für diesen Abschnitt ist das Lemma 1.8.

Lemma 1.8. *Für teilerfremde Zahlen $k, \ell \in \mathbb{Z}$ ist die folgende Abbildung surjektiv:*

$$\pi : \mathbb{Z} \to \mathbb{Z}/k\mathbb{Z} \times \mathbb{Z}/\ell\mathbb{Z}$$
$$x \mapsto (x + k\mathbb{Z}, x + \ell\mathbb{Z})$$

Die Abbildung π induziert durch $(x \mod k\ell) \mapsto (x \mod k, x \mod \ell)$ eine Bijektion zwischen $\mathbb{Z}/k\ell\mathbb{Z}$ und $\mathbb{Z}/k\mathbb{Z} \times \mathbb{Z}/\ell\mathbb{Z}$.

Beweis. Betrachte $(x + k\mathbb{Z}, y + \ell\mathbb{Z})$. Aufgrund der Teilerfremdheit von k und ℓ gibt es Zahlen $a, b \in \mathbb{Z}$ mit $ak + b\ell = 1$. Damit gilt $b\ell \equiv 1 \bmod k$ und $ak \equiv 1 \bmod \ell$. Für $x, y \in \mathbb{Z}$ hat $yak + xb\ell$ die folgenden Eigenschaften:

$$yak + xb\ell \equiv 0 + x \cdot 1 \equiv x \quad \bmod k$$
$$yak + xb\ell \equiv y \cdot 1 + 0 \equiv y \quad \bmod \ell$$

Es folgt $\pi(yak + xb\ell) = (x + k\mathbb{Z}, y + \ell\mathbb{Z})$ und π ist surjektiv. Es gilt $\pi(x') = \pi(x)$ für alle $x' \in x + k\ell\mathbb{Z}$, daher ist $(x \bmod k\ell) \mapsto (x \bmod k, x \bmod \ell)$ wohldefiniert. Daher induziert π eine Surjektion von $\mathbb{Z}/k\ell\mathbb{Z}$ auf $\mathbb{Z}/k\mathbb{Z} \times \mathbb{Z}/\ell\mathbb{Z}$. Schließlich erkennen wir, dass es jeweils genau $k\ell$ Elemente in $\mathbb{Z}/k\ell\mathbb{Z}$ und in $\mathbb{Z}/k\mathbb{Z} \times \mathbb{Z}/\ell\mathbb{Z}$ gibt. Also ist die induzierte Abbildung bijektiv. □

Der erweiterte euklidische Algorithmus aus Abschnitt 1.2 liefert ein effektives Verfahren, für ein Paar $(y, z) \in \mathbb{Z}/k\mathbb{Z} \times \mathbb{Z}/\ell\mathbb{Z}$ eine Zahl $x \in \mathbb{Z}$ mit $\pi(x) = (y, z)$ zu berechnen.

Beispiel 1.9. Sei $k = 5$ und $\ell = 7$. Wir wollen eine Zahl z ausrechnen mit $\pi(z) = (3 + 5\mathbb{Z}, 4 + 7\mathbb{Z})$, d. h., wir suchen ein z mit $z \equiv 3 \bmod 5$ und $z \equiv 4 \bmod 7$. Der erweiterte euklidische Algorithmus liefert uns $-4 \cdot 5 + 3 \cdot 7 = 1$. Wie im Beweis von Lemma 1.8 setzen wir $z = 4 \cdot (-4) \cdot 5 + 3 \cdot 3 \cdot 7 = -17$ und es gilt $-17 \equiv 3 \bmod 5$ und $-17 \equiv 4 \bmod 7$. Eine andere Lösung ist $-17 + 5 \cdot 7 = 18$. Auch hier gilt $18 \equiv 3 \bmod 5$ und $18 \equiv 4 \bmod 7$. ◇

Lemma 1.8 ergibt den sogenannten chinesischen Restsatz, der eine algebraische Interpretation liefert. Sind R_1 und R_2 Ringe, so können wir auf dem kartesischen Produkt $R_1 \times R_2$ durch komponentenweise Addition und Multiplikation eine Ringstruktur erklären. Konkret ist die Addition definiert durch $(x_1, y_1) + (x_2, y_2) = (x_1 + x_2, y_1 + y_2)$. Die Null ist das Paar $(0, 0) \in R_1 \times R_2$. Die Multiplikation ist analog definiert durch $(x_1, y_1) \cdot (x_2, y_2) = (x_1 \cdot x_2, y_1 \cdot y_2)$. Das Einselement ist $(1, 1)$. Wir nennen diesen Ring das *direkte Produkt* von R_1 und R_2.

Satz 1.10 (Chinesischer Restsatz). *Seien $k, \ell \in \mathbb{Z}$ teilerfremd. Dann definiert folgende Zuordnung einen Ringisomorphismus:*

$$\mathbb{Z}/k\ell\mathbb{Z} \to \mathbb{Z}/k\mathbb{Z} \times \mathbb{Z}/\ell\mathbb{Z}$$
$$x + k\ell\mathbb{Z} \mapsto (x + k\mathbb{Z}, x + \ell\mathbb{Z})$$

Beweis. Die Zuordnung ist mit der Addition und Multiplikation verträglich und entspricht der bijektiven Abbildung aus Lemma 1.8. □

Beispiel 1.11. Die Ringe $\mathbb{Z}/35\mathbb{Z}$ und $\mathbb{Z}/5\mathbb{Z} \times \mathbb{Z}/7\mathbb{Z}$ sind isomorph. In der folgenden Tabelle geben wir die Entsprechungen der Elemente an.

0	1	2	3	4	5	6
(0,0)	(1,1)	(2,2)	(3,3)	(4,4)	(0,5)	(1,6)
7	8	9	10	11	12	13
(2,0)	(3,1)	(4,2)	(0,3)	(1,4)	(2,5)	(3,6)
14	15	16	17	18	19	20
(4,0)	(0,1)	(1,2)	(2,3)	(3,4)	(4,5)	(0,6)
21	22	23	24	25	26	27
(1,0)	(2,1)	(3,2)	(4,3)	(0,4)	(1,5)	(2,6)
28	29	30	31	32	33	34
(3,0)	(4,1)	(0,2)	(1,3)	(2,4)	(3,5)	(4,6)

$\Diamond$

Der chinesische Restsatz 1.10 hat viele wichtige Konsequenzen. Die invertierbaren Elemente von $\mathbb{Z}/k\mathbb{Z} \times \mathbb{Z}/\ell\mathbb{Z}$ sind genau diejenigen, die sowohl in der ersten als auch in der zweiten Komponente invertierbar sind.

Korollar 1.12. *Seien $k, \ell \in \mathbb{Z}$ teilerfremd. Dann ist*

$$(\mathbb{Z}/k\ell\mathbb{Z})^* \rightarrow (\mathbb{Z}/k\mathbb{Z})^* \times (\mathbb{Z}/\ell\mathbb{Z})^*$$
$$x + k\ell\mathbb{Z} \mapsto (x + k\mathbb{Z}, x + \ell\mathbb{Z})$$

ein Gruppenisomorphismus bezüglich der Multiplikation der invertierbaren Elemente.

Eine typische Anwendung des chinesischen Restsatzes 1.10 ist, dass für teilerfremde Zahlen k und ℓ die Kongruenz $x \equiv y \bmod k\ell$ genau dann gilt, wenn die beiden Kongruenzen $x \equiv y \bmod k$ und $x \equiv y \bmod \ell$ erfüllt sind. Mit Korollar 1.12 gilt weiter, dass x modulo $k\ell$ genau dann invertierbar ist, wenn x sowohl modulo k invertierbar ist als auch modulo ℓ.

Bei einem Produkt von mehr als zwei teilerfremden Zahlen lässt sich der chinesische Restsatz 1.10 auch mehrfach anwenden.

Korollar 1.13. *Seien $m_1, \ldots, m_n \in \mathbb{Z}$ paarweise teilerfremd und sei $m = m_1 \cdots m_n$. Folgende Zuordnung definiert einen Ringisomorphismus:*

$$\mathbb{Z}/m\mathbb{Z} \rightarrow \mathbb{Z}/m_1\mathbb{Z} \times \cdots \times \mathbb{Z}/m_n\mathbb{Z}$$
$$x + m\mathbb{Z} \mapsto (x + m_1\mathbb{Z}, \ldots, x + m_n\mathbb{Z})$$

Wenn wir Korollar 1.13 nach „kongruent modulo n" übersetzen, erhalten wir die Form des chinesischen Restsatzes von Sun Zi in Korollar 1.14 aus dem 3. Jhd. Allerdings wurde sein Ergebnis erst später im Jahre 1247 durch Qin Jiushao veröffentlicht.

Korollar 1.14. *Seien $m_1, \ldots, m_n \in \mathbb{Z}$ paarweise teilerfremd und sei $m = m_1 \cdots m_n$. Für alle $x_1, \ldots, x_n \in \mathbb{Z}$ existiert genau ein $x \in \{0, \ldots, m-1\}$, das simultan die folgenden n Kongruenzen erfüllt:*

$$x \equiv x_1 \quad \bmod m_1$$
$$\vdots$$
$$x \equiv x_n \quad \bmod m_n$$

Die Lösungsmenge dieses Systems von Kongruenzen ist $x + m\mathbb{Z}$.

Es gibt diverse einfache Beweise für die Tatsache, dass es unendlich viele Primzahlen gibt. Ein solcher Beweis kann etwa aus Korollar 1.14 abgeleitet werden.

Korollar 1.15. *Es gibt unendlich viele Primzahlen.*

Beweis. Angenommen, es gäbe nur endlich viele Primzahlen. Dann könnten wir eine Zahl n finden, die die Kongruenz $n \equiv p - 1 \bmod p$ für alle Primzahlen erfüllt. Diese Zahl ist größer als 1 und wird von keiner Primzahl geteilt. Dies ist nach Satz 1.5 unmöglich. $\qquad\square$

1.7 Ein erster Primzahltest nach Fermat

Ein sich wiederholendes Thema ist der *kleine Satz von Fermat* (benannt nach Pierre de Fermat, ca. 1607–1665). Hauptberuflich war Fermat Jurist und später Richter in Toulouse. Mathematischen Einfluss gewann er vor allem durch Korrespondenzen mit bedeutenden Mathematikern seiner Zeit. Legendär ist seine Notiz, dass er einen „wahrhaft wunderbaren Beweis" für die Unlösbarkeit der diophantischen Gleichungen $a^n + b^n = c^n$ mit ganzen Zahlen $a, b, c \neq 0$ und $n > 2$ gefunden hätte. Aber der Rand sei zu klein, den Beweis zu fassen. Diese Behauptung ging als *großer Satz von Fermat* in die Mathematikgeschichte ein und war bis Anfang der 1990er Jahre eine der berühmtesten zahlentheoretischen Vermutungen. Aufgrund der einfachen Formulierung des Problems versuchten sich diverse Hobby-Mathematiker an der Lösung und ließen nicht nach, immer wieder falsche Lösungen vorzulegen. Der große Satz von Fermat wurde erst 1993 von Wiles (Sir Andrew John Wiles, geb. 1953) bewiesen. Allerdings enthielt sein erster Beweis noch eine Lücke, die er dann 1995 in einer gemeinsamen Arbeit mit Taylor (Richard Lawrence Taylor, geb. 1962) schließen konnte.

Wir behandeln hier nur den *kleinen Satz von Fermat*. Auch hier gibt es keinen erhaltenen Beweis, der aus der Feder von Fermat stammt. Der Beweis ist aber genügend einfach, dass kein Zweifel daran besteht, dass Fermat einen Beweis kannte.

Satz 1.16 (Kleiner Satz von Fermat). *Sei p eine Primzahl und $a \in \mathbb{Z}$ eine ganze Zahl. Dann gelten:*

$$a^p \equiv a \mod p$$
$$a^{p-1} \equiv 1 \mod p \quad \text{für } \mathrm{ggT}(a, p) = 1$$

Beweis. Sei zunächst $\mathrm{ggT}(a, p) = 1$. Nach Satz 1.7 ist die Multiplikation mit a auf $(\mathbb{Z}/p\mathbb{Z})^*$ bijektiv. Also gilt

$$(p-1)! = \prod_{i \in \{1, \dots, p-1\}} i \equiv \prod_{i \in \{a, \dots, a(p-1)\}} i \equiv (p-1)! \cdot a^{p-1} \mod p$$

Die Restklasse $(p-1)!$ besitzt in $(\mathbb{Z}/p\mathbb{Z})^*$ ein multiplikatives Inverses b, da $(p-1)!$ und p teilerfremd sind. Wenn wir in der obigen Gleichung beide Seiten mit b (bzw. mit ba) multiplizieren, erhalten wir die Aussagen des Satzes für $\mathrm{ggT}(a, p) = 1$. Der

verbleibende Fall ist, dass p ein Teiler von a ist. Dann ist sowohl $a \bmod p$ als auch $a^p \bmod p$ Null. □

Die Idee für einen einfachen Primzahltest, die aus Satz 1.16 abgeleitet werden kann, führt auf den *Fermat-Test*. Der Test geht wie folgt vor:

(1) Wähle $a \in \{1, \ldots, n-1\}$ zufällig.

(2) Berechne $a^{n-1} \bmod n$.

(3) Falls $a^{n-1} \not\equiv 1 \bmod n$, so ist n sicher keine Primzahl, ansonsten *möglicherweise*.

Dieser Test liegt explizit oder implizit fast allen verwendeten Primzahltests zugrunde. In technischen Anwendungen sind a und n häufig Binärzahlen mit mehreren hundert oder tausend Stellen. Man kann unmöglich 2^{1000} Rechenoperationen in der Zeitspanne dieses Universums durchführen. Für realistische Anwendungen des Fermat-Tests benötigen wir eine schnelle Exponentiation.

Beispiel 1.17. Wir wollen den Wert $z = 14^{2222} \bmod 77$ ohne Rechner bestimmen. Wir können das Ergebnis $z = 1$ ausschließen, da $14^{2222} \equiv 0 \bmod 7$ nicht invertierbar ist. Aus dem kleinen Satz von Fermat 1.16 wissen wir $14^{10} \equiv 1 \bmod 11$. Daraus folgt $14^{2222} = (14^{10})^{222} \cdot 14^2 \equiv 1^{222} \cdot 14^2 \equiv 14^2 \bmod 11$. Mit dem chinesischen Restsatz erhalten wir $14^{2222} \equiv 14^2 \bmod 77$ und damit $z = 14^2 \bmod 77 = 196 \bmod 77 = 42$. ◇

1.8 Die schnelle Exponentiation

Der erste überlieferte Beweis für den kleinen Satz von Fermat stammt aus der Feder von Leonhard Euler (1707–1783). Euler war extrem produktiv und hat im Laufe seines Lebens die Mathematik um viele fundamentale Erkenntnisse bereichert. Im Jahr 1732 hat Euler festgestellt, dass $2^{2^5} + 1 = 4294967297$ keine Primzahl ist und damit eine Vermutung von Fermat widerlegt, dass alle Zahlen der Form $2^{2^n} + 1$ Primzahlen sind. Die ersten fünf Zahlen dieser Folge sind die Primzahlen $3, 5, 17, 257$ und 65537 und eine Primzahl der Form $2^{2^n} + 1$ nennt man *Fermat-Primzahl*. Nur, außer den fünf genannten ist keine weitere Fermat-Primzahl bekannt; und aus heutiger Sicht vermutet man eher, dass auch keine weiteren existieren. Außerdem hat Euler den Teiler 641 der Zahl $2^{2^5} + 1$ finden können. Tatsächlich gilt

$$3^{4294967296} \bmod 4294967297 = 3029026160$$

Die sechste Fermat-Zahl besteht also schon für $a = 3$ nicht den Fermat-Test. Hätte Euler damals (ohne moderne Hilfsmittel) den Wert $3^{4294967296} \bmod 4294967297$ überhaupt bestimmen können? Sicher hätte er nicht 4294967296 Mal die Zahl 3 multiplizieren können, um danach die entstandene Zahl durch 4294967297 zu teilen. Dies ginge aus zwei Gründen nicht. Zum einen hätte das Ergebnis der Multipli-

kationen viel mehr Stellen, als eine Person jemals aufschreiben könnte, und zum anderen hätte Euler 4294967296 Rechenschritte aus Zeitgründen nicht ausführen können. Die sogenannte *schnelle Exponentiation* löst beide Probleme. Dass die Zahl zu groß wird, verhindert man, indem das Ergebnis jeder Multiplikation modulo der Zahl 4294967297 gerechnet wird. Damit sind alle Zwischenergebnisse kleiner als 4294967297. Das zweite Problem löst man dadurch, dass man $3^{4294967296}$ bestimmt, indem man 3 lediglich $2^5 = 32$ Mal sukzessive hintereinander quadriert. Dieser Rechenaufwand wäre für Euler bereits 1732 möglich gewesen. Ob Euler tatsächlich so vorgegangen ist, wissen wir nicht. Es bleibt eine Spekulation.

Untersuchen wir nun das Problem etwas allgemeiner. Wir wollen $a^b \bmod n$ mit $a, b, n \in \mathbb{N}$ berechnen und stellen uns vor, dass diese Zahlen viele hundert Stellen haben. Betrachten wir das folgende Programm:

```
/* Voraussetzung ist a, b, n ∈ ℕ */
/* Berechnet wird a^b mod n */
function modexp(a, b, n)
begin
    e := 1;
    while b > 0 do
        if b ungerade then e := e · a mod n fi;
        a := a² mod n; b := ⌊b/2⌋
    od;
    return e
end
```

Da b in jedem Schleifendurchlauf halbiert wird, wird die **while**-Schleife höchstens so oft durchlaufen, wie b Binärstellen hat; dies sind $\lfloor \log_2 b \rfloor + 1$ viele. Des Weiteren werden in jedem Schleifendurchlauf höchstens 2 (modulare) Multiplikationen ausgeführt. Einen Extremfall stellen hier Zweierpotenzen im Exponenten dar. Bei diesen Zahlen wird bis auf den letzten Schleifendurchlauf immer nur eine Multiplikation ausgeführt. Der andere Extremfall sind Zahlen von der Form $2^q - 1$. Die Binärdarstellung von solchen Zahlen besteht aus lauter Einsen. Bei diesen Zahlen werden in jedem Schleifendurchlauf 2 Multiplikationen ausgeführt. Allgemein gilt, dass bei k Einsen und m Nullen in der Binärdarstellung von $b > 0$ genau $2k + m$ Multiplikationen durchgeführt werden. Genau genommen wird $(k + m)$-mal quadriert und k-mal multipliziert. Quadrieren ist eine spezielle Form des Multiplizierens. Wegen

$$ab = \frac{(a + b)^2 - (a - b)^2}{4}$$

kann umgekehrt Quadrieren nicht wesentlich schneller als Multiplizieren möglich sein. Wir halten fest: Sind $a, b, n, k \in \mathbb{N}$ natürliche Zahlen mit $a, b, n \le 2^k$, so kann der Wert $a^b \bmod n$ in einer in k polynomiellen Zeit bestimmt werden.

1.9 Verschlüsselung mit dem RSA-Verfahren

Das bekannteste Verschlüsselungsverfahren mit öffentlichen Schlüsseln ist das RSA-Verfahren von Ronald Linn *Rivest* (geb. 1947), Adi *Shamir* (geb. 1952) und Leonard *Adleman* (geb. 1945). Mit nur wenigen Kenntnissen der modularen Arithmetik lässt sich dieses Verfahren einfach beschreiben und als korrekt nachweisen. Wir benötigen nur den kleinen Satz von Fermat 1.16 sowie den chinesischen Restsatz 1.10. Für die algorithmische Umsetzung benötigen wir zuverlässige Primzahltests, schnelle Exponentation und den erweiterten euklidischen Algorithmus.

Im folgenden Protokoll möchte eine Person A, genannt „Alice", von einer Person B, genannt „Bob", Informationen erhalten, die über einen öffentlichen Kanal gesendet werden und dennoch geheim bleiben müssen. Das Verfahren ist asymmetrisch, es werden nur Nachrichten von Bob an Alice verschlüsselt, es ist auch asymmetrisch in dem Sinne, dass die Ressourcen von Bob möglicherweise beschränkter als die von Alice sind. Dies ist durchaus realistisch, wenn Nachrichten von einer mobilen Station aus gesendet werden sollen und schon die Energieressourcen beschränkt sein können.

Das RSA-Verfahren

(1) Alice wählt Primzahlen p, q mit $3 < p < q$.

(2) Sie berechnet $n = pq$ und setzt $\varphi(n) = (p-1)(q-1)$.

(3) Sie wählt einen Exponenten $e > 1$ mit $\mathrm{ggT}(e, \varphi(n)) = 1$.

(4) Sie berechnet s mit $es \equiv 1 \bmod \varphi(n)$.

(5) Sie veröffentlicht (n, e). Alle anderen Parameter bleiben ihr Geheimnis, insbesondere darf sie das „secret" s nicht weitergeben.

(6) Bob verschlüsselt eine Nachricht $0 \le x \le n-1$ durch $y = x^e \bmod n$ und sendet y and Alice.

(7) Alice entschlüsselt y durch $y^s \bmod n$.

Betrachten wir das folgende Beispiel. Alice wählt $p = 5$ und $q = 11$. Daraus ergibt sich $n = 55$ und $\varphi(n) = 4 \cdot 10 = 40$. Als Exponenten wählen wir $e = 3$. Dann gilt $\mathrm{ggT}(3, 40) = 1$. Um s zu berechnen, wendet Alice den erweiterten euklidischen Algorithmus auf die Zahlen $e = 3$ und $\varphi(n) = 40$ an, so dass sie $s = 27$ erhält mit $3 \cdot s \equiv 1 \bmod 40$. In diesem Beispiel veröffentlicht Alice das Paar $(55, 3)$. Wenn Bob die Nachricht $x = 23$ an Alice übermitteln will, dann berechnet er $y = 23^3 \bmod 55 = 12$ und verschickt y. Alice erhält die Nachricht $y = 12$ und berechnet zum Entschlüsseln $12^{27} \bmod 55 = 23$. Mit Hilfe des chinesischen Restsatzes und des kleinen Satzes von Fermat ist dies sogar von Hand möglich. Es gilt

$$12^{27} \equiv 2^3 = 8 \equiv 3 \quad \bmod 5$$
$$12^{27} \equiv 1^{27} = 1 \quad \bmod 11$$

Wir wissen $1 = -2 \cdot 5 + 11$, also erhalten wir den Wert $12^{27} \bmod 55$ durch:

$$(1 \cdot (-2) \cdot 5 + 3 \cdot 11) \bmod 55 = 23$$

Der nächste Satz sagt, dass es kein Zufall ist, dass man nach dem Entschlüsseln einer verschlüsselten Nachricht wieder den ursprünglichen Text bekommt.

Satz 1.18. *Das RSA-Verfahren ist korrekt: Verschlüsselt Bob eine Zahl x in dem Bereich $0 \leq x \leq n-1$ durch $y = x^e \bmod n$, so gilt $x = y^s \bmod n$.*

Beweis. Mit dem chinesischen Restsatz reicht es, $x \equiv y^s \bmod r$ für $r = p$ und $r = q$ zu zeigen. Ohne Einschränkung sei $r = p$. Wegen $y = x^e \bmod p$, zeigen wir $x \equiv x^{es} \bmod p$ für alle $x \in \mathbb{Z}/p\mathbb{Z}$. Dies stimmt für $x \equiv 0 \bmod p$. Sei nun $\mathrm{ggT}(x, p) = 1$. Es gilt $es = 1 + k(p-1)(q-1)$ für ein $k \in \mathbb{N}$ und $x^{p-1} \equiv 1 \bmod p$ nach dem kleinen Satz von Fermat. Damit erhalten wir

$$x^{es} = x^{1+k(p-1)(q-1)} = x \cdot \left(x^{(p-1)} \right)^{k(q-1)} \equiv x \quad \bmod p \qquad \square$$

Die Sicherheit von RSA beruht darauf, dass kein effizientes Verfahren bekannt ist, welches bei einer zufälligen Wahl der Primzahlen p und q die Zahl $n = pq$ faktorisiert. Nach gegenwärtigem Stand der Forschungen 2012 gelten 1000 Bits für n noch als sicher, und es liegen keine belastbaren Ideen vor, Zahlen mit 2000 Bits oder mehr zu faktorisieren. Man kann zwar nicht beweisen, dass man n faktorisieren muss, um RSA zu brechen, denn es würde beispielsweise reichen, $\varphi(n)$ oder den geheimen Exponenten s zu kennen. Allerdings sind die folgenden drei Probleme etwa gleich schwierig: (1) Faktorisiere n. (2) Berechne $\varphi(n)$. (3) Berechne ein s mit $es \equiv 1 \bmod \varphi(n)$. Wir gehen an dieser Stelle nicht genauer auf diese Tatsache ein. Die Aussage findet sich in der Literatur und wird auch ausführlich in unserem Band *Diskrete algebraische Methoden* behandelt [12].

Viele Implementierungen verwenden tatsächlich kleine Exponenten wie $e = 3$ oder $e = 17$, damit die Verschlüsselung möglichst schnell ist. Allerdings muss Bob bei kleinen öffentlichen Exponenten gewisse Vorsichtsmaßnahmen beachten, ansonsten kann ein möglicher Angreifer die Nachrichten entschlüsseln; siehe beispielsweise Aufgabe 1.17. für die einfachste Form von *Håstad's Broadcast Attack* [24] oder den Übersichtsartikel von Boneh [7] aus dem Jahr 1999. Wirklich problematisch ist es, den geheimen Entschlüsselungsexponenten klein zu wählen. Ist er sehr klein, sagen wir kleiner als 2^{30}, so kann man s mit einer vollständigen Suche finden, da ja e bekannt ist. Nach Boneh und Durfee [8] gilt sogar $s \leq n^{0,292}$ für den privaten Schlüssel s als unsicher. Die Situation für e und s ist also asymmetrisch.

In dem Projekt *The RSA Challenge Numbers* wurden RSA-ℓ Zahlen wachsender Binärlängen ℓ mit der Aufforderung veröffentlicht, diese zu faktorisieren. Wir geben eine kleine Übersicht, die den Stand von Ende 2012 reflektiert.

- RSA-640 wurde am 2.11.2005 faktorisiert.
- RSA-768 wurde am 12.12.2009 faktorisiert.
- Die Faktoren von RSA-704 und von RSA-1024 sind nicht öffentlich bekannt.

$$RSA\text{-}1024 = 135066410865995223349603216278805969938881 4756056670$$
$$2752448514385152651060485953383394028715057190944179$$
$$8207282164471551373680419703964191743046496589274256$$
$$2393410208643832021103729587257623585096431105640735$$
$$0150818751067659462920556368552947521350085287941637$$
$$7328533906109750544334999811150056977236890927563$$

Die Stromkosten, die bei der Faktorisierung der Zahl RSA-768 mit Hilfe diverser welt-
weit vernetzter Rechner anfielen, wurden nicht berechnet, aber ein Betrag zwischen
50 000 und 200 000 Euro mag realistisch sein. Wenn man den Aufwand zur Faktorisierung von RSA-1024 als 1000 Mal höher einschätzt, fallen aus heutiger Sicht hierfür
(mindestens) 50 Millionen Euro Stromkosten an. Für die allermeisten Anwendungen
bleiben Schlüssellängen von 1024 Bits daher auf absehbare Zeit vollkommen sicher.
Mit einem Kapitaleinsatz von 50 Millionen Euro sollte es nämlich wesentlich einfachere Wege geben, an fast beliebige „Geheimnisse" zu gelangen.

1.10 Die Euler'sche phi-Funktion

Wir erinnern uns, dass mit *Einheiten* die (multiplikativ) invertierbaren Elemente eines Rings R gemeint sind. Diese bilden eine Untergruppe von $(R, \cdot, 1)$, welche die
Einheitengruppe R^* genannt wird. In diesem Abschnitt wollen wir die Einheitengruppe $(\mathbb{Z}/n\mathbb{Z})^*$ des Rings $\mathbb{Z}/n\mathbb{Z}$ untersuchen.

Hier stellen sich uns zwei Fragen. Die erste ist, wie erkennt man, ob ein Element
von $\mathbb{Z}/n\mathbb{Z}$ eine Einheit ist. Die zweite Frage, die uns interessiert, ist, wie viele Einheiten es in $\mathbb{Z}/n\mathbb{Z}$ gibt. Die erste Frage wurde in Satz 1.7 beantwortet. Ein Element
$k \in \mathbb{Z}/n\mathbb{Z}$ ist genau dann invertierbar, wenn $\mathrm{ggT}(k, n) = 1$ gilt, d. h., wenn k und n
teilerfremd sind. In diesem Fall kann man mit dem erweiterten euklidischen Algo-
rithmus effizient eine Zahl ℓ berechnen mit $k\ell \equiv 1 \bmod n$. Wir widmen uns nun der
zweiten Frage und betrachten die Euler'sche φ-Funktion:

$$\varphi(n) = |(\mathbb{Z}/n\mathbb{Z})^*|$$

Dies liefert uns $\varphi(1) = 1$ und die Abschätzung $1 \leq \varphi(n) \leq n-1$. Da n genau
dann eine Primzahl ist, wenn alle Zahlen zwischen 1 und $n-1$ teilerfremd zu n sind,
gilt

$$\varphi(n) = n-1 \iff n \text{ ist eine Primzahl}$$

Dies deckt sich auch mit unserer Beobachtung, dass $\mathbb{Z}/n\mathbb{Z}$ genau dann ein Körper ist,
wenn n eine Primzahl ist. Mit Korollar 1.12 erhalten wir

$$\mathrm{ggT}(m, n) = 1 \implies \varphi(mn) = \varphi(m)\varphi(n)$$

Um den Wert der φ-Funktion für beliebige Zahlen zu bestimmen, müssen wir nun nur
noch klären, was der Wert bei Primzahlpotenzen ist. Sei p eine Primzahl und $k \geq 1$.

Von den Zahlen $0, 1, \ldots, p^k - 1$ sind genau die p^{k-1} Zahlen $0, p, 2p, \ldots, (p^{k-1} - 1)p$ durch p teilbar und damit nicht teilerfremd zu p^k. Die übrigen $p^k - p^{k-1}$ Zahlen sind alle teilerfremd zu p^k. Dies liefert uns

$$p \text{ ist Primzahl} \Rightarrow \varphi(p^k) = (p - 1)p^{k-1}$$

Wir sind nun in der Lage, den Wert $\varphi(n)$ für beliebige Zahlen n auszurechnen, falls uns die Primfaktorzerlegung von n bekannt ist. Sei $n = \prod_i p_i^{e_i}$ die Primfaktorzerlegung von n. Dann gilt $\varphi(n) = \prod_i \varphi(p_i^{e_i}) = \prod_i (p_i - 1)p_i^{e_i - 1}$. Durch Umformung ergibt sich daraus die *Euler'sche Formel*:

$$\varphi(n) = n \cdot \prod_{\substack{p \text{ Primzahl} \\ p \mid n}} \left(1 - \frac{1}{p}\right) \tag{1.1}$$

Eine wichtige Eigenschaft der Euler'schen φ-Funktion ergibt sich aus der folgenden Verallgemeinerung des kleinen Satzes von Fermat. Wir können den Satz 1.19 von Euler vollkommen analog und elementar herleiten.

Satz 1.19 (Euler). *Aus* $\mathrm{ggT}(a, n) = 1$ *folgt* $a^{\varphi(n)} \equiv 1 \bmod n$.

Beweis. Schreibe $(\mathbb{Z}/n\mathbb{Z})^* = \{g_1 \bmod n, \ldots, g_{\varphi(n)} \bmod n\}$ für gewisse $g_i \in \mathbb{Z}$. Wegen $\mathrm{ggT}(a, n) = 1$ ist die Multiplikation mit a in $(\mathbb{Z}/n\mathbb{Z})^*$ bijektiv und wir erhalten $(\mathbb{Z}/n\mathbb{Z})^* = \{ag_1 \bmod n, \ldots, ag_{\varphi(n)} \bmod n\}$. Sei $g = \prod_{i=1}^{\varphi(n)} g_i$. Dann ist

$$g \equiv \prod_{i=1}^{\varphi(n)} ag_i \equiv a^{\varphi(n)}g \bmod n$$

Wegen $y \in (\mathbb{Z}/n\mathbb{Z})^*$ hat g in $(\mathbb{Z}/n\mathbb{Z})^*$ ein multiplikatives Inverses $g^{-1} \in (\mathbb{Z}/n\mathbb{Z})^*$. Damit folgt:

$$1 \equiv gg^{-1} \equiv a^{\varphi(n)}gg^{-1} \equiv a^{\varphi(n)} \bmod n \qquad \square$$

Beispiel 1.20. Wir wollen die letzten zwei Dezimalstellen von 3^{4444} bestimmen. Hierzu berechnen wir $3^{4444} \bmod 100$. Es gilt $\varphi(100) = \varphi(2^2)\varphi(5^2) = (2 - 1) \cdot 2 \cdot (5 - 1) \cdot 5 = 40$. Da 3 und 100 teilerfremd sind, folgt aus dem Satz von Euler 1.19, dass $3^{40} \equiv 1 \bmod 100$ gilt. Daraus ergibt sich folgende Rechnung: $3^{4444} = (3^{40})^{111} \cdot 3^4 \equiv 1 \cdot 3^4 \equiv 81 \bmod 100$. Dies zeigt, dass die Dezimaldarstellung von 3^{4444} mit 81 endet. $\diamond$

In Satz 1.21 stellen wir eine weitere Eigenschaft der Euler'schen φ-Funktion dar. Mit $\sum_{t \mid n} \varphi(t)$ meinen wir hierbei, dass wir für alle positiven Teiler $t > 0$ von n die Werte $\varphi(t)$ aufsummieren.

Satz 1.21.

$$\sum_{t \mid n} \varphi(t) = n$$

Beweis. Betrachten wir folgende Menge

$$N = \left\{ \frac{0}{n}, \frac{1}{n}, \ldots, \frac{n-1}{n} \right\}$$

von n verschiedenen Brüchen. Durch Kürzen lassen sich alle Brüche $\frac{m}{n}$ darstellen durch $\frac{m}{n} = \frac{k}{t}$ mit teilerfremden Zahlen k und t. Außerdem ist nach dem Kürzen t immer noch ein Teiler von n. Deshalb gilt

$$N = \left\{ \frac{k}{t} \;\middle|\; t \mid n,\, 0 \le k < t,\, \mathrm{ggT}(k,t) = 1 \right\}$$

Gruppieren nach den verschiedenen Nennern t liefert uns eine Einteilung in disjunkte Teilmengen von N:

$$N = \bigcup_{t \mid n} \left\{ \frac{k}{t} \;\middle|\; 0 \le k < t,\, \mathrm{ggT}(k,t) = 1 \right\}$$

Aus $|\{k/t \mid 0 \le k < t,\, \mathrm{ggT}(k,t) = 1\}| = |\{k \mid 0 \le k < t,\, \mathrm{ggT}(k,t) = 1\}| = \varphi(t)$ folgt die Behauptung. $\qquad\square$

Die Ähnlichkeit der Beweise von den Sätzen von Euler und vom kleinen Satz von Fermat ist kein Zufall und führt in die elementare Gruppentheorie. Sei G eine endliche Gruppe und $a \in G$. Die Anzahl der Elemente in G, also $|G|$, nennt man die *Ordnung der Gruppe* G und die kleinste positive ganze Zahl m mit $a^m = 1$ nennt man die *Ordnung von a*. Wir formulieren den folgenden Satz von Lagrange (1736–1813) nur für den Spezialfall abelscher Gruppen (obwohl er auch für nichtkommutative Gruppen gilt) und geben dafür einen direkten und einfachen Beweis an.

Satz 1.22. *Sei G eine endliche abelsche Gruppe und $a \in G$ ein Element der Ordnung m. Dann gilt $a^{|G|} = 1$. Ist $a^k = 1$ mit $k \in \mathbb{Z}$, so gilt $m \mid k$. Insbesondere teilt die Ordnung m die Gruppenordnung $|G|$.*

Beweis. Es sei $|G| = n$. Wir schreiben $G = \{g_1, \ldots, g_n\}$. Die Multiplikation mit a in G ist bijektiv und wir erhalten $G = \{ag_1, \ldots, ag_n\}$. Sei $g = \prod_{i=1}^{n} g_i$. Dann ist $g = \prod_{i=1}^{n} ag_i = a^n g$, da G kommutativ ist. Hieraus folgt $a^n = 1$ in G. Sei jetzt $a^k = 1$. Zu zeigen ist $m \mid k$. Wir dürfen $k \ge 1$ annehmen (für negative Zahlen k folgt die Behauptung dann aus $m \mid qm + k$). Insbesondere ist $1 \le m \le k$. Zusammen mit $a^m = 1$ erhalten wir $a^r = 1$ für $r = k \bmod m$. Da m die kleinste positive Zahl mit $a^m = 1$ ist, folgt $r = 0$ und $m \mid k$. $\qquad\square$

Nach dieser Vorbereitung kann man den Satz von Euler in Verbindung mit dem kleinen Satz von Fermat für ein exaktes Primzahlzertifikat verwenden.

Satz 1.23. *Sei $n \geq 2$ eine natürliche Zahl. Gibt es für jeden Primfaktor p von $n-1$ eine ganze Zahl a mit*

    1. $a^{n-1} \equiv 1 \bmod n$ und

    2. $a^{\frac{n-1}{p}} \not\equiv 1 \bmod n$,

so ist n eine Primzahl.

Beweis. Sei $n \geq 2$ eine natürliche Zahl, die die Bedingungen des Satzes erfüllt. Um zu zeigen, dass n eine Primzahl ist, genügt es zu zeigen, dass $\varphi(n) = n-1$ ist. Hierfür reicht es zu zeigen, dass $(n-1) \mid \varphi(n)$ gilt, da $\varphi(n) \leq n-1$. Betrachte einen Primfaktor p von $n-1$ und einen (maximalen) Exponenten $r \geq 1$ mit $p^r \mid (n-1)$. Zu zeigen ist $p^r \mid \varphi(n)$.

Für die Primzahl p gibt es nach Voraussetzung nun eine ganze Zahl a, die die beiden obigen Bedingungen erfüllt. Sei nun m die Ordnung von a in der abelschen Gruppe $(\mathbb{Z}/n\mathbb{Z})^*$. Dann muss gelten $m \mid (n-1)$ nach der ersten Bedingung und Satz 1.22. Nach der zweiten Bedingung ist aber m kein Teiler von $\frac{n-1}{p}$. Deshalb gilt $p^r \mid m$. Nach dem Satz von Euler und Satz 1.22 erhalten wir $m \mid \varphi(n)$ und damit $p^r \mid \varphi(n)$. Also ist $n-1 = \varphi(n)$ und damit n eine Primzahl. □

Wie wir später sehen werden, sind die Voraussetzungen in Satz 1.23 für alle Primzahlen n erfüllt. Vaughan Pratt (geb. 1944) zeigte 1975 mit diesem Zertifikat, dass die Primzahleigenschaft effizient verifiziert werden kann. Dies bedeutet, es gibt kurze Beweise, die in polynomieller Zeit überprüft werden können, die entweder belegen, dass eine Binärzahl eine Primzahl ist oder dass sie keine Primzahl ist.

Die Grundidee zum Beweis dieses Resultats ist wie folgt: Ist n keine Primzahl, so ist ein kurzer Beweis hierfür die Angabe eines nichttrivialen Teilers. Ist n eine Primzahl, so gibt man die Primfaktorzerlegung von $n-1$ an, und für jeden Primfaktor p von $n-1$ gibt man eine Zahl a wie in Satz 1.23 an. Weiter zertifiziert man die Primfaktoren von $n-1$ nach demselben Verfahren. Erst seit 2002 ist bekannt, dass die Primzahleigenschaft von Binärzahlen in polynomieller Zeit entschieden werden kann [1].

Satz 1.24 über endliche zyklische Gruppen ist ein Bindeglied zwischen der Gruppentheorie und der Euler'schen φ-Funktion. Eine Gruppe heißt dabei *zyklisch*, wenn sie von einem einzigen Element erzeugt wird. Jedes Element der Gruppe kann dann als ganzzahlige Potenz des Erzeugers geschrieben werden.

Satz 1.24. *Sei $n \in \mathbb{N}$ und t ein positiver Teiler von n. Eine zyklische Gruppe der Ordnung n hat genau $\varphi(t)$ Elemente der Ordnung t.*

Beweis. Sei G eine zyklische Gruppe der Ordnung n und $g \in G$ ein erzeugendes Element. Mit $\psi(t)$ bezeichnen wir die Anzahl der Elemente der Ordnung t. Nach dem Satz 1.22 ist die Ordnung von jedem Element aus G ein Teiler von n. Deshalb gilt $\sum_{t \mid n} \psi(t) = n$. Wir zeigen $\psi(t) \geq \varphi(t)$ falls t ein Teiler von n ist. Mit Satz 1.21 folgt dann die Behauptung. Sei $k \in \{1, \dots, t-1\}$ mit $\mathrm{ggT}(k, t) = 1$. Betrachten wir das

Element
$$h = g^{\frac{kn}{t}}$$

Dann gilt $h^t = 1$. Für die Ordnung d von h gilt also $d \mid t$. Da g die Ordnung n hat, erhalten wir zusammen mit $h^d = 1$, dass

$$n \ \Big| \ \frac{dkn}{t}$$

Wir schließen daraus $t \mid dk$. Aus $\mathrm{ggT}(k, t) = 1$ folgt nun $t \mid d$ und insgesamt $t = d$. Damit hat h die Ordnung t. Insgesamt können wir jedem $k \in (\mathbb{Z}/t\mathbb{Z})^*$ ein Gruppenelement h mit Ordnung t zuordnen. Außerdem ist diese Zuordnung wegen $\frac{kn}{t} < n$ injektiv. Dies zeigt $\psi(t) \geq \varphi(t)$. $\qquad\qquad\qquad\qquad\qquad\qquad\qquad\qquad\quad$ □

1.11 Fibonacci-Zahlen

Die Fibonacci-Zahlen F_n (Leonardo Pisano Bigollo, genannt Leonardo da Pisa, genannt Fibonacci „Filius Bonacci", ca. 1175–1240) sind durch die folgenden Bedingungen definiert:

$$F_0 = 0, \quad F_1 = 1, \quad F_{n+1} = F_n + F_{n-1} \qquad\qquad (1.2)$$

Die ersten Werte der Folge sind damit:

$$0, 1, 1, 2, 3, 5, 8, 13, 21, 34, 55, 89, 144, 233, 377, 610, 987, 1597, 2584, \ldots$$

Diese Zahlenfolge gehört zu den beliebtesten Zahlenfolgen überhaupt. So darf man nicht davon ausgehen, als einzige Person 6 Richtige im Lotto zu haben, wenn man etwa die Zahlen $3, 5, 8, 13, 21, 34$ ankreuzt. Falls man es doch tut, sollte man sich wenigstens nicht zu früh über einen hohen Auszahlungsbetrag freuen.

In Fibonaccis Buch „Liber Abaci" von 1202 wird das berühmte *Kaninchenproblem* eher beiläufig erwähnt, welches wie folgt lautet: Wie viele Kaninchenpaare existieren nach einem Jahr, wenn man mit einem erwachsenen Paar startet, Kaninchenpaare nach einem Monat erwachsen sind, sich danach monatlich reproduzieren und alle Kaninchen länger als ein Jahr leben? Die Lösung findet sich in der folgenden Tabelle, in dieser steht A für ein erwachsenes Paar und B für ein Kinderpaar.

Die Wörter in der zweiten Spalte ergeben sich von Zeile zu Zeile indem man ein A durch AB ersetzt und ein B durch ein A. Ein erwachsenes Paar lebt fort und bekommt ein Kinderpaar. Ein Kinderpaar wird erwachsen. Jedes Wort ist Anfangsstück des Wortes in der Nachfolgerzeile, auf diese Weise erhalten wir ein unendliches Wort, das *Fibonacci-Wort*.

Neben dem Wachstumsverhalten von Kaninchen lassen sich weitere kombinatorische Interpretationen der Fibonacci-Zahlen finden. Wir geben einige weitere Beispiele.

	Paare	Zahl der A's	B's	Gesamt
1. Jan.	A	1	0	1
1. Feb.	A B	1	1	2
1. März	A B A	2	1	3
1. April	$ABAAB$	3	2	5
1. Mai	$ABAABABA$	5	3	8
1. Juni	$ABAABABAABAAB$	8	5	13
1. Juli		13	8	21
1. Aug.		21	13	34
1. Sept.		34	21	55
1. Okt.		55	34	89
1. Nov.		89	55	144
1. Dez.		144	89	233
1. Jan.		233	144	377

Beispiel 1.25. Angenommen, wir haben beliebig viele Dominosteine der Längen 1 und 2. Dann ist F_{n+1} die Anzahl der Möglichkeiten, Dominosteine zu einer Kette der Länge n hintereinander zu legen. Die Zahl F_n ist die Anzahl der Wörter über zwei Buchstaben a und b, die die Länge n haben, mit einem a beginnen, aber in denen keine zwei a's benachbart sind. Die Zahl F_{n+2} ist die Anzahl der Wörter über zwei Buchstaben a und b, die die Länge n haben und in denen keine zwei a's benachbart sind.

Erklären wir es für die Zahl der Wörter, die mit einem a beginnen und keine zwei benachbarten a's besitzen. Die Startwerte $F_0 = 0$ und $F_1 = 1$ geben die richtigen Zahlen für die Längen 0 und 1, denn es gibt nur das leere Wort mit der Länge Null und das Wort a als einziges Wort der Länge 1, welches mit a beginnt. Betrachte Wörter der Länge $n \geq 2$. Die Zahl derjenigen Wörter der Länge n, die mit einem b aufhören, mit einem a beginnen, aber in denen keine zwei a's benachbart sind, ist F_{n-1}. Diejenigen Wörter der Länge n, die mit einem a aufhören, mit einem a beginnen, aber in denen keine zwei a's benachbart sind, haben als vorletzten Buchstaben ein b. Ihre Zahl ist F_{n-2}. Verzichten wir auf die Forderung, dass Wörter mit a anfangen, so argumentieren wir genauso. Der Unterschied ist nur, dass wir ein Wort der Länge Null zählen (das leere Wort) und zwei Wörter der Länge 1 (die Wörter a und b), was die Startwerte $F_2 = 1$ und $F_3 = 2$ festlegt. ◇

Die Fibonacci-Zahlen wachsen schnell. Aufgrund des Bildungsgesetzes erhalten wir für $n \geq 3$ die Abschätzungen:

$$F_n \leq 2^n \leq F_{2n}$$

Das Wachstum ist also *irgendwie* exponentiell. Dies lässt sich sehr genau fassen. Wir machen den folgenden Ansatz. Angenommen, es wäre $F_n = x^n$ für ein $x \in \mathbb{R}$, dann würde für alle $n \geq 1$ gelten:

$$x^{n+1} = x^n + x^{n-1}$$

Aus $F_n \geq 1$ für $n \geq 1$ folgt $x \neq 0$. Deshalb können wir durch x^{n-1} dividieren, und die obige Gleichung ist äquivalent zu $x^2 = x + 1$. Diese quadratische Gleichung hat zwei Lösungen:

$$\Phi = \frac{1 + \sqrt{5}}{2} \quad \text{und} \quad \hat{\Phi} = \frac{1 - \sqrt{5}}{2}$$

Hierbei ist $\Phi = \frac{1+\sqrt{5}}{2}$ der *goldene Schnitt*. Dies ist das Seitenverhältnis b/a eines Rechtecks mit den Seitenlängen a und b, wenn $a/b = b/(a + b)$ gilt. Eine Annäherung dieses Verhältnisses findet sich in einer berühmten Zeichnung des vitruvianischen Menschen von Leonardo da Vinci (1452–1519), die auf italienischen 1-Euro-Münzen abgebildet ist.

Eine bessere Annäherung ist:

$$\Phi \quad = \quad 1{,}61803\,39887\,49894\,84820\,45868\,34365\,63811\,77203\,09179 \cdots$$

Aus $x^2 = x+1$ folgt $x(x-1) = 1$. Deshalb ist $\Phi^{-1} = \Phi - 1 = -\hat{\Phi}$. Die beiden Zahlen Φ und $\hat{\Phi}$ genügen den Bildungsgesetzen $\Phi^{n+1} = \Phi^n + \Phi^{n-1}$ und $\hat{\Phi}^{n+1} = \hat{\Phi}^n + \hat{\Phi}^{n-1}$, denn so wurden sie ja gerade bestimmt. Also gehorcht auch jede Linearkombination $F_n(a, b) = a\Phi^n + b\hat{\Phi}^n$ dem Bildungsgesetz

$$F_{n+1}(a, b) = F_n(a, b) + F_{n-1}(a, b)$$

Um $F_n(a, b) = F_n$ zu finden, reicht es, das folgende Gleichungssystem mit 2 Unbekannten zu lösen

$$a\Phi^0 + b\hat{\Phi}^0 = F_0 = 0$$

$$a\Phi^1 + b\hat{\Phi}^1 = F_1 = 1$$

Wir erhalten $b = -a$ und $a = \frac{1}{\sqrt{5}}$. Insgesamt ergibt sich

$$F_n = \frac{\Phi^n - \hat{\Phi}^n}{\Phi - \hat{\Phi}} = \frac{1}{\sqrt{5}} \left(\left(\frac{1 + \sqrt{5}}{2} \right)^n - \left(\frac{1 - \sqrt{5}}{2} \right)^n \right) \tag{1.3}$$

Nun ist $-0{,}7 < \frac{1-\sqrt{5}}{2} < -0{,}6$. Die Folge $(\frac{1-\sqrt{5}}{2})^n$ fällt (alternierend) exponentiell schnell gegen Null. Wir erhalten

$$F_n = \left[\frac{1}{\sqrt{5}} \left(\frac{1 + \sqrt{5}}{2} \right)^n \right]$$

wobei $[x]$ die nächste ganze Zahl für $x \in \mathbb{R}$ bedeutet (auf- oder abgerundet). Die Approximation wird mit wachsendem n immer besser. Damit können wir auch große Fibonacci-Zahlen wie

$$F_{234} = 3\,577\,855\,662\,560\,905\,981\,638\,959\,513\,147\,239\,988\,861\,837\,901\,112$$

sehr schnell berechnen, wenn wir eine genügend genaue Arithmetik zur Verfügung haben. Wie steht es bei einer Rechnung mit exakter Arithmetik? Auch dies ist keine Hürde. Wir benutzen die folgenden 2×2 Matrizen

$$M_1 = \begin{pmatrix} 0 & 1 \\ 1 & 1 \end{pmatrix} = \begin{pmatrix} F_0 & F_1 \\ F_1 & F_2 \end{pmatrix} \quad \text{und} \quad M_n = \begin{pmatrix} F_{n-1} & F_n \\ F_n & F_{n+1} \end{pmatrix}$$

Eine elementare Matrix-Multiplikation zeigt $M_{n+1} = M_n \cdot M_1 = M_1 \cdot M_n$. Dies bedeutet $M_n = (M_1)^n$ für alle $n \in \mathbb{Z}$. Also:

$$\begin{pmatrix} F_{n-1} & F_n \\ F_n & F_{n+1} \end{pmatrix} = \begin{pmatrix} 0 & 1 \\ 1 & 1 \end{pmatrix}^n \tag{1.4}$$

Mit schneller Exponentation lässt sich damit F_n mit $\mathcal{O}(\log|n|)$ Multiplikationen von 2×2 Matrizen über ganzen Zahlen berechnen. Die maximale Bitlänge der Einträge bei der Rechnung ist dabei linear in n.

Der Zusammenhang zwischen dem goldenen Schnitt Φ und den Fibonacci-Zahlen ist gut verstanden: So ist Φ eine irrationale Zahl, die sich in rationalen Zahlen am besten durch die Quotienten zweier aufeinander folgender Fibonacci-Zahlen approximieren lässt. Dies hat wiederum zur Folge, dass sich der goldene Schnitt nur schlecht durch ein Verhältnis zweier ganzer Zahlen annähern lässt. Dies sieht man daran, dass in der *Kettenbruchentwicklung* von Φ nur Einsen vorkommen. Durch wiederholte Anwendung der Identität $\Phi = 1 + \frac{1}{\Phi}$ erhalten wir:

$$\Phi = 1 + \frac{1}{\Phi} = 1 + \frac{1}{1 + \frac{1}{\Phi}} = 1 + \frac{1}{1 + \frac{1}{1 + \frac{1}{\Phi}}} = 1 + \frac{1}{1 + \frac{1}{1 + \frac{1}{1 + \cdots}}}$$

Erstaunlicher Weise scheint gerade diese ausgeprägte *Irrationalität* von Φ die Bedeutung in Kunst und Natur zu untermauern. In der Kunst werden Bilderrahmen im Verhältnis des goldenen Schnittes angefertigt, in der Natur weisen viele Pflanzen in ihrem Bauplan Spiralen auf, deren Anzahl durch Fibonacci-Zahlen gegeben sind. So gibt es in Sonnenblumen Spiralen aus 34 und 55 Blättern.

Es gibt zahlreiche Identitäten für Fibonacci-Zahlen. Eine besonders hübsche bezieht sich auf den größten gemeinsamen Teiler von F_m und F_n. Es gilt:

Satz 1.26.

$$\text{ggT}(F_m, F_n) = F_{\text{ggT}(m,n)}$$

Beweis. Wir gliedern die Behauptung des Satzes in die beiden Teilaussagen $F_{\text{ggT}(m,n)} \mid$ ggT(F_m, F_n) und ggT$(F_m, F_n) \mid F_{\text{ggT}(m,n)}$. Sei $n = kp$. Wir zeigen $F_k \mid F_n$. Ohne Einschränkung können wir annehmen, dass $n \geq 1$ gilt. Wir betrachten den Zusammenhang zwischen den Matrizen M_n und M_k:

$$\begin{pmatrix} F_{n-1} & F_n \\ F_n & F_{n+1} \end{pmatrix} = M_1^n = M_1^{kp} = \begin{pmatrix} F_{k-1} & F_k \\ F_k & F_{k+1} \end{pmatrix}^p$$

Rechnen wir modulo F_k, so erhalten wir

$$\begin{pmatrix} F_{n-1} & F_n \\ F_n & F_{n+1} \end{pmatrix} \equiv \begin{pmatrix} F_{k-1}^p & 0 \\ 0 & F_{k+1}^p \end{pmatrix} \quad \mod F_k$$

Insbesondere gilt $F_n \equiv 0 \mod F_k$ und damit $F_k \mid F_n$. Dies zeigt, dass $F_{\text{ggT}(m,n)}$ sowohl ein Teiler von F_m als auch von F_n ist. Also gilt $F_{\text{ggT}(m,n)} \mid$ ggT(F_m, F_n).

Für die andere Richtung bemerken wir zunächst, dass F_n und F_{n+1} teilerfremd sind: Dies folgt unmittelbar mit Induktion aus der Rekursionsgleichung $F_{n+1} = F_n + F_{n-1}$ und wegen ggT$(F_1, F_0) = 1$. Sei $m > n$. Zu zeigen ist ggT$(F_m, F_n) \mid F_{\text{ggT}(m,n)}$. Dies ist trivial für $n = 0$ und für $n = 1$. Wir setzen $g = $ ggT(F_m, F_n) und schreiben $m = np + r$ mit $0 \leq r < n$. Es gilt $M_m = M_{np} M_r$. Rechnen wir modulo g so ergibt sich

$$M_m \equiv \begin{pmatrix} F_{m-1} & 0 \\ 0 & F_{m+1} \end{pmatrix} \equiv \begin{pmatrix} F_{n-1}^p & 0 \\ 0 & F_{n+1}^p \end{pmatrix} \begin{pmatrix} F_{r-1} & F_r \\ F_r & F_{r+1} \end{pmatrix} \quad \mod g$$

Hieraus folgt $0 \equiv F_{n+1}^p F_r \mod g$, und g ist ein Teiler von $F_{n+1}^p F_r$. Zusammen mit ggT$(F_n, F_{n+1}) = 1$ und $g \mid F_n$ sehen wir, dass die Zahlen g und F_{n+1}^p teilerfremd sind. Deshalb teilt g die Zahl F_r. Damit teilt g auch ggT(F_n, F_r) und mit Induktion folgt $g \mid F_{\text{ggT}(n,r)}$, denn es ist $r < n$. Nun ist ggT$(m, n) = $ ggT(n, r), und wir erhalten schließlich $g \mid F_{\text{ggT}(m,n)}$. $\qquad\square$

Ein weiterer Zusammenhang zwischen Fibonacci-Zahlen und dem größten gemeinsamen Teiler lässt die algorithmische Bedeutung der Fibonacci-Zahlen erkennen. Damit beschäftigen wir uns im nächsten Abschnitt.

1.12 Laufzeitanalyse des euklidischen Algorithmus

Wir beschäftigen uns hier mit der Frage, wie viele rekursive Aufrufe es beim euklidischen Algorithmus maximal geben kann. Der euklidische Algorithmus berechnet den größten gemeinsamen Teiler ggT(k, ℓ). Wir wollen hier seine Funktionsweise kurz wiederholen. Wegen ggT$(k, \ell) = $ ggT$(\ell, k) = $ ggT$(|k|, |\ell|)$ können wir dabei stets von $0 \leq k \leq \ell$ ausgehen. Für $k = 0$ gilt ggT$(0, \ell) = \ell$, und wir sind fertig. Sei nun $0 < k \leq \ell$, dann berechnen wir zunächst $\ell = qk + r$ mit $0 \leq r < k$. Jede Zahl, die k und ℓ teilt, teilt dann auch r; und umgekehrt, jede Zahl, die k und r teilt, teilt auch ℓ. Daher gilt ggT$(k, \ell) = $ ggT(r, k), und wir können die Berechnung rekursiv mit $0 \leq r < k$ fortsetzen.

Beginnen wir mit einem Beispiel. Angenommen, wir setzen den euklidischen Algorithmus auf die benachbarten Fibonacci-Zahlen F_{n-1} und F_n an. Wegen $F_n = F_{n-1} + F_{n-2}$ erhalten wir rekursiv die Aufrufe:

$$\mathrm{ggT}(F_{n-1}, F_n) = \mathrm{ggT}(F_{n-2}, F_{n-1}) = \cdots = \mathrm{ggT}(F_0, F_1) = 1$$

Dies lehrt uns, dass die Zahl der Aufrufe bei der Berechnung von $\mathrm{ggT}(k, \ell)$ logarithmisch in k werden kann. Benachbarte Fibonacci-Zahlen sind jedoch schon der schlechteste Fall. Seien $0 \le k \le \ell$ Zahlen mit $\mathrm{ggT}(k, \ell) = g$ und nehmen wir an, dass zur Berechnung von g insgesamt n rekursive Aufrufe im euklidischen Algorithmus fällig wurden. Dann gibt es eine Folge von Zahlen

$$f_0 = 0, f_1 = g, \ldots, f_{n-1} = k, f_n = \ell$$

mit $f_{i+1} = q_i f_i + f_{i-1}$ wobei gilt $0 \le f_{i-1} < f_i$ und $q_i \ge 1$. Hieraus folgt $f_i \ge F_i$ für alle $0 \le i \le n$. Insbesondere ist $k \ge F_{n-1}$. Wir halten dies in einem Satz fest.

Satz 1.27. *Sei* $\Phi = \frac{1+\sqrt{5}}{2}$ *der goldene Schnitt und seien* $k, \ell \in \mathbb{N} \setminus \{0\}$. *Dann erfordert die Berechnung des größten gemeinsamen Teilers* $\mathrm{ggT}(k, \ell)$ *mit dem euklidischen Algorithmus höchstens* $\lceil \log_\Phi k \rceil$ *rekursive Aufrufe.*

Wir bemerken noch, dass $\log_\Phi k < \frac{3}{2} \log_2 k$ für $k > 1$ gilt. Hat man es zum Beispiel mit 100-stelligen Binärzahlen zu tun, so werden höchstens 150 rekursive Aufrufe benötigt.

Aufgaben

1.1. Sei p eine Primzahl. Zeigen Sie, dass $\log_{10}(p)$ nicht rational ist.

1.2. Anwendung des euklidischen Algorithmus:

(a) Bestimmen Sie zwei Zahlen $x, y \in \mathbb{Z}$ mit $x \cdot 35 - y \cdot 56 = \mathrm{ggT}(35, 56)$.

(b) Bestimmen Sie $x, y \in \mathbb{N}$ mit obiger Eigenschaft.

1.3. Bestimmen Sie alle Lösungen der linearen Kongruenz

$$3x - 7y \equiv 11 \mod 13$$

1.4. Sei für die natürlichen Zahlen a, b mit $a \ge b$ der größte gemeinsame Teiler $\mathrm{ggT}(a, b) = 1$. Zeigen Sie, dass $\mathrm{ggT}(a + b, a - b) \in \{1, 2\}$ gilt.

1.5. Teilbarkeitsregeln:

(a) Zeigen Sie, dass eine Zahl im Zehnersystem genau dann durch 3 teilbar ist, wenn ihre Quersumme durch 3 teilbar ist.

(b) Leiten Sie eine analoge Regel für die Teilbarkeit durch 11 im Zehnersystem her.

1.6. (Satz von Wilson) Für $n \geq 2$ gilt $(n - 1)! \equiv -1 \bmod n$ genau dann, wenn n eine Primzahl ist. Diese Charakterisierung ist nach John Wilson (1741–1793) benannt.

1.7. Zeigen Sie, dass für $n \geq 2$ die Zahl $n^4 + 4^n$ keine Primzahl ist.

Hinweis: Betrachten Sie das Polynom $(x^2 + 2y^2)^2 - 4x^2y^2$.

1.8. Sei $n \in \mathbb{N}$. Zeigen Sie:

(a) Wenn $2^n - 1$ eine Primzahl ist, dann ist n eine Primzahl.

(b) Wenn $2^n + 1$ eine Primzahl ist, dann ist n eine Zweierpotenz.

(c) Sei $f_n = 2^{2^n} + 1$ die n-te Fermat-Zahl. Zeigen Sie $\mathrm{ggT}(f_m, f_n) = 1$ für $m \neq n$. Folgern Sie daraus, dass es unendlich viele Primzahlen gibt.

Hinweis: Betrachten Sie $(f_n - 2)/f_m$ für $n > m$.

1.9. Bestimmen Sie das kleinste $x \in \mathbb{N}$ mit:

$$x \equiv 1 \bmod 2, \quad x \equiv 0 \bmod 3, \quad x \equiv 1 \bmod 5, \quad x \equiv 6 \bmod 7$$

1.10. Zeigen Sie, dass das Kongruenz-System $x \equiv a \bmod n$, $x \equiv b \bmod m$ genau dann eine Lösung besitzt, wenn $\mathrm{ggT}(n, m) \mid (a - b)$ gilt. Bestätigen Sie, dass eine Lösung, vorausgesetzt dass sie existiert, eindeutig modulo $\mathrm{kgV}(n, m)$ ist.

1.11. Zeigen Sie für alle $n \in \mathbb{N}$:

(a) $n^5 \equiv n \bmod 30$

(b) $3^{n^4 + n^2 + 2n + 4} \equiv 21 \bmod 60$

(c) $7^{n+2} + 8^{2n+1} \equiv 0 \bmod 57$

1.12. Sei p eine ungerade Primzahl, und sei $a \in \mathbb{N}$ ungerade und nicht durch p teilbar. Zeigen Sie:

$$a^{p-1} \equiv 1 \bmod 4p$$

1.13. Das RSA-Verfahren:

(a) Wieviele Elemente enthält die multiplikative Gruppe $(\mathbb{Z}/51\mathbb{Z})^*$?

(b) Bestimmen Sie den geheimen Entschlüsselungsexponenten, welcher zu dem öffentlichen RSA-Schlüssel $(n, e) = (51, 11)$ gehört.

(c) Der Geheimtext 7 wurde nach dem RSA-Verfahren mit dem öffentlichen Schlüssel $(n, e) = (51, 11)$ verschlüsselt, d. h. $7 = x^{11} \bmod 51$. Wie lautet der Klartext x?

(d) Wieviele Elemente der Ordnung 10 gibt es in $(\mathbb{Z}/51\mathbb{Z})^*$?

(e) Ist die multiplikative Gruppe $(\mathbb{Z}/51\mathbb{Z})^*$ zyklisch?

1.14. Seien p und q Primzahlen, $n = pq$ und $e \in \mathbb{N}$ mit $\mathrm{ggT}(e, \varphi(n)) = 1$. Um Nachrichten zu entschlüsseln, welche mit dem RSA-Verfahren und dem öffentlichen

Schlüssel (n, e) verschlüsselt sind, empfiehlt Ihnen der Haushaltsausschuss den privaten Schlüssel s kostengünstiger durch die Vorschrift

$$es \equiv 1 \mod \mathrm{kgV}(p - 1, q - 1)$$

zu bestimmen (anstatt $es \equiv 1 \mod \varphi(n)$) zu verwenden). Dies sei eine Einsparung, da $\mathrm{kgV}(p - 1, q - 1)$ definitiv kleiner ist als das Produkt $\varphi(n) = (p - 1)(q - 1)$. Bleibt das Verfahren korrekt?

1.15. Wir erweitern das RSA-Verfahren auf drei Primzahlen. Seien p, q, r drei verschiedene Primzahlen, sei $n = pqr$ und sei $s \cdot e \equiv 1 \mod \varphi(n)$. Nachrichten $x, y \in \{0, \ldots, n-1\}$ werden mit Hilfe der Vorschrift $c(x) = x^e \mod n$ verschlüsselt und durch $d(y) = y^s \mod n$ entschlüsselt.

(a) Zeigen Sie, dass das Verfahren korrekt ist, d. h., dass $d(c(x)) = x$ gilt für alle $x \in \{0, \ldots, n - 1\}$.

(b) Der Geheimtext $y = 14$ wurde mit dem öffentlichen Schlüssel $(n, e) = (66, 27)$ verschlüsselt. Bestimmen Sie $y^s \mod k$ für $k = 2, 3, 11$ sowie den Klartext $x = y^s \mod 66$.

1.16. Wir nehmen an, zwei Benutzer A_1 und A_2 des RSA-Systems verwenden die öffentlichen Schlüssel (n, e_1) und (n, e_2). Nun sendet Bob den Text m verschlüsselt an A_1 und A_2. Zeigen Sie, dass Oskar den Klartext m aus den beiden Geheimtexten entschlüsseln kann, sofern e_1 und e_2 teilerfremd sind.

1.17. Wir nehmen an, eine Bank sendet die gleiche Nachricht m an drei verschiedene Kunden. Die Nachricht m wird jeweils mit dem RSA-Verfahren unter Verwendung der öffentlichen Schlüssel $(n_1, 3)$, $(n_2, 3)$ und $(n_3, 3)$ verschlüsselt, wobei die n_i alle verschieden sind. Zeigen Sie, dass der Angreifer Oskar unter diesen Voraussetzungen die Nachricht m entschlüsseln kann.

1.18. Zeigen Sie: $\forall a, b \in \mathbb{Z} : \mathrm{ggT}(a, b) = 1 \Rightarrow a^{\varphi(b)} + b^{\varphi(a)} \equiv 1 \mod ab$.

1.19. Sei F_n mit $n \in \mathbb{N}$ die Folge der Fibonacci-Zahlen. Zeigen Sie:

(a) $F_1 + \cdots + F_n = F_{n+2} - 1$

(b) $\sum_{k=0}^{n} F_k^2 = F_n F_{n+1}$

(c) $\forall n \geq 0 \; \forall k \geq 1 : F_{n+k} = F_k F_{n+1} + F_{k-1} F_n$

(d) $\forall n \geq 1 : F_{n+1} F_{n-1} - F_n^2 = (-1)^n$

1.20. Sei M eine Menge, p eine Primzahl und $f : M \to M$ eine Abbildung mit $f^p(m) = m$ für alle $m \in M$. Hierbei bezeichnet f^p die p-fache Hintereinanderausführung der Abbildung f.

(a) Sei $m \in M$. Zeigen Sie, dass die Werte $f(m), f^2(m), \ldots, f^p(m)$ alle gleich oder alle verschieden sind.

(b) Sei jetzt M endlich und $F = \{m \in M \mid f(m) = m\}$ die Menge der Fixpunkte. Zeigen Sie $|M| \equiv |F| \bmod p$.

1.21. Wir wollen auf elementarem Wege zeigen, dass $F_{p+1} + F_{p-1} \equiv 1 \bmod p$ gilt, falls p eine Primzahl ist. (Für eine algebraische Lösung siehe Aufgabe 1.24.) Hierzu definieren wir $L_1 = 1, L_2 = 3$ und $L_{n+2} = L_{n+1} + L_n$. Zeigen Sie:

(a) $L_n = F_{n+1} + F_{n-1}$.

(b) L_n ist die Mächtigkeit der Menge $\mathcal{L}_n$, wenn $\mathcal{L}_n$ die Menge der Teilmengen $M \subseteq \{1, \ldots, n\}$ bezeichnet, in welchen keine zwei aufeinander folgenden Zahlen auftauchen, wenn wir modulo n rechnen (und somit 1 ein Nachfolger von n ist).

(c) Wenn p eine Primzahl ist, dann gilt $L_p \equiv 1 \bmod p$.

1.22. Sei $\mathbb{F}$ ein Körper, in dem 2 und 5 invertierbar sind und 5 ein Quadrat ist. Definieren Sie in $\mathbb{F}$ die Fibonacci-Zahlen sowie einen *goldenen Schnitt* φ. Zeigen Sie Gleichung 1.3 und bestimmen Sie φ und $F_{12} + F_{10}$ in dem Körper $\mathbb{Z}/11\mathbb{Z}$.

1.23. Sei p eine Primzahl mit $2 \neq p \neq 5$ und $\mathbb{F}_p$ der Körper $\mathbb{Z}/p\mathbb{Z}$.

(a) In $\mathbb{F}_p$ sei 5 kein Quadrat. Beispielsweise $p = 3$ oder $p = 7$. Es soll gezeigt werden, dass ein Körper $\mathbb{F}$ mit $|\mathbb{F}| = p^2$ existiert, der $\mathbb{F}_p$ als Unterkörper hat und in dem 5 ein Quadrat ist. Insbesondere ist ein Element $\sqrt{5}$ definiert, daher wird $\mathbb{F}$ auch als $\mathbb{F}_p(\sqrt{5})$ bezeichnet.

(b) Sei $\mathbb{F} = \mathbb{F}_p$ falls 5 ein Quadrat in $\mathbb{F}_p$ ist und $\mathbb{F} = \mathbb{F}_p(\sqrt{5})$ sonst. Setze $\varphi = \frac{1+\sqrt{5}}{2}$ und $\hat{\varphi} = \frac{1-\sqrt{5}}{2}$. Zeigen Sie $\{\varphi^p, \hat{\varphi}^p\} = \{\varphi, \hat{\varphi}\}$. Man beachte, im Gegensatz zu der positiven reellen Zahl $\sqrt{5}$ ist $q \in \mathbb{F}$ mit $q^2 = 5$ nur bis auf Vorzeichen definiert. Wir müssen $\sqrt{5} \in \{q, -q\}$ wählen und legen damit fest, welcher Wert φ ist.

1.24. Sei p eine Primzahl. Geben Sie einen algebraischen Beweis für die schon aus Aufgabe 1.21. bekannte Kongruenz $F_{p+1} + F_{p-1} \equiv 1 \bmod p$.

Hinweis: Verwenden Sie Aufgabe 1.23. und Aufgabe 1.22. bzw. Gleichung (1.4).

1.25. Für $p, q \in \mathbb{Z}$ mit $p > 1$ sei $q \operatorname{rem} p$ (für *remainder*) die ganze Zahl r mit $-\frac{p}{2} \leq r < \frac{p}{2}$ und $q \equiv r \bmod p$. Der ggT werde mit dem euklidischen Algorithmus berechnet, allerdings mit rem anstelle von mod. Zeigen Sie, dass die Rekursionstiefe dieses Algorithmus höchstens $\lceil \log_\Psi k \rceil$ ist mit $\Psi = \sqrt{2} + 1$.

Hinweis: Betrachten Sie hierzu $G_{n+1} = G_n + 2G_{n-1}$.

Zusammenfassung

Begriffe

- Induktion
- Primzahl
- natürliche Zahlen $\mathbb{N}$
- ganze Zahlen $\mathbb{Z}$
- rationale Zahlen $\mathbb{Q}$
- reelle Zahlen $\mathbb{R}$
- komplexe Zahlen $\mathbb{C}$
- assoziativ
- neutrales Element
- Inverses, invertierbar
- kommutativ
- abelsch
- distributiv
- Halbgruppe
- Monoid
- Gruppe
- Ring
- Körper

- Einheit
- Einheitengruppe
- Unterstruktur
- erzeugen
- Homomorphismus
- Isomorphismus
- teilen
- $\mathrm{ggT}(k, \ell)$
- teilerfremd
- Primfaktorzerlegung
- Restklasse
- kongruent modulo n
- Euler'sche φ-Funktion
- Ordnung einer Gruppe
- Ordnung eines Elements
- zyklische Gruppe
- Fibonacci-Zahlen F_n
- goldener Schnitt

Methoden und Resultate

- (Erweiterter) euklidischer Algorithmus
- Lemma von Bézout:
  Für alle $k, \ell \in \mathbb{Z}$ existieren $a, b \in \mathbb{Z}$ mit $\mathrm{ggT}(k, \ell) = ak + b\ell$.
- Fundamentalsatz der Arithmetik: Alle $n \in \mathbb{N}$ haben eindeutige Primfaktorzerlegung.
- $k \in (\mathbb{Z}/n\mathbb{Z})^* \Leftrightarrow \mathrm{ggT}(k, n) = 1 \Leftrightarrow$ die Abbildung $x \mapsto kx$ auf $\mathbb{Z}/n\mathbb{Z}$ ist bijektiv
- Berechnen von Inversen modulo n
- $\mathbb{Z}/n\mathbb{Z}$ ist Körper $\Leftrightarrow n$ ist Primzahl
- Chinesischer Restsatz: Für $\mathrm{ggT}(k, \ell) = 1$ definiert $\mathbb{Z}/k\ell\mathbb{Z} \to \mathbb{Z}/k\mathbb{Z} \times \mathbb{Z}/\ell\mathbb{Z}$ mit $x \bmod k\ell \mapsto (x \bmod k, x \bmod \ell)$ einen Ringisomorphismus.
- Lösen von simultanen Kongruenzen
- Es gibt unendlich viele Primzahlen.
- Kleiner Satz von Fermat: p Primzahl $\Rightarrow a^p \equiv a \bmod p$. Fermat-Test
- Schnelle (modulare) Exponentiation
- RSA-Verfahren
- Berechnung der Euler'schen φ-Funktion

- Satz von Euler: $\mathrm{ggT}(a, n) = 1 \Rightarrow a^{\varphi(n)} \equiv 1 \bmod n$
- $\sum_{t \mid n} \varphi(t) = n$
- In endlichen kommutativen Gruppen G gilt $a^{|G|} = 1$ für alle $a \in G$.
- In endlichen Gruppen gilt: $a^k = 1 \Leftrightarrow$ die Ordnung von a teilt k
- Primzahlzertifizierung nach Pratt
- Kombinatorische Interpretation der Fibonacci-Zahlen
- Sei $t \mid n$. Eine zyklische Gruppe der Ordnung n hat $\varphi(t)$ Elemente der Ordnung t.
- $F_n = \frac{1}{\sqrt{5}}\left(\left(\frac{1+\sqrt{5}}{2}\right)^n - \left(\frac{1-\sqrt{5}}{2}\right)^n\right)$. Schnelle Berechnung von F_n durch Matrizen.
- $\mathrm{ggT}(F_m, F_n) = F_{\mathrm{ggT}(m,n)}$
- Rekursionstiefe beim euklidischen Algorithmus

2 Einige nützliche Abschätzungen

Wir wissen schon, dass die n-te Fibonacci-Zahl F_n der Rundungswert von $\Phi^n / \sqrt{5}$ ist. Die Zahlen wachsen also exponentiell mit der Basis des goldenen Schnitts Φ. In diesem Abschnitt untersuchen wir das Wachstumsverhalten der folgenden Funktionen:

$$n!, \quad \binom{2n}{n}, \quad \text{kgV}(n) \quad \text{sowie} \quad \pi(x)$$

Hierbei ist weiterhin $n \in \mathbb{N}$ und $n! = n(n-1) \cdots 1$ die *Fakultät* von n. Mit $\binom{2n}{n}$ $= \frac{(2n)!}{n!n!}$ betrachten wir den *mittleren* Binomialkoeffizienten und $\text{kgV}(n) = \text{kgV}\{1,$ $\dots, n\}$ bezeichnet das *kleinste gemeinsame Vielfache* der ersten n positiven ganzen Zahlen. Die Funktion $\pi(x)$ meint für eine positive reelle Zahl x die Anzahl der Primzahlen, die kleiner oder gleich x sind.

Das Ziel ist nicht, in allen Fällen bestmögliche Abschätzungen für das Wachstumsverhalten der obigen Funktionen anzugeben, sondern solche, die sich leicht herleiten und damit auch einprägen lassen. Die Funktion $\pi(x)$ haben wir aufgenommen, da faszinierende Aussagen über Primzahldichten ohne weitere Schwierigkeiten aus dem übrigen Stoff folgen.

2.1 Das Wachstum der Fakultät

Die Folgen $n!$ und 2^n lassen sich induktiv definieren:

$$0! = 2^0 = 1 , \quad (n+1)! = (n+1)n! \quad \text{und} \quad 2^{n+1} = 2 \cdot 2^n$$

Einige Anfangswerte finden sich in der folgenden Tabelle:

n	0	1	2	3	4	5	...	10	...	20
2^n	1	2	4	8	16	32	...	1024	...	1048576
$n!$	1	1	2	6	24	120	...	3628800	...	2432902008176640000

Grobe, aber häufig brauchbare Schätzwerte für 2^{10} und 2^{20} sind also 1000 und 1 Million, wobei der Fehler bei 1 Million bei etwa 5% liegt. Die Zahlen $n!$ für $n \leq 5$ sind leicht zu merken.

Für $n \geq 4$ gilt stets $n! > 2^n$. Aber um wie viel schneller wächst $n!$ als 2^n? Wächst $n!$ schneller als 2^{n^2}? Die Antwort ist nein, und dies ist wie folgt einzusehen. Offensichtlich ist $n! \leq n^n = 2^{n \log_2 n}$ für alle n, und $n \log_2 n$ ist kleiner als n^2 für alle $n \geq 1$. Eine unmittelbare untere Schranke für $n!$ erhalten wir durch die Beobachtung, dass in dem Produkt $n!$ mindestens die Hälfte der Faktoren so groß sind wie $\frac{n}{2}$. Hieraus ergibt sich $(\frac{n}{2})^{\frac{n}{2}} \leq n!$; was zusammen mit der oberen Schranke n^n immerhin die wichtige Regel

$$\log n! \in \Theta(n \log n)$$

liefert. Wir wollen jetzt genauere Schranken für $n!$ herleiten. Es gilt:

$$\ln n! = \ln 2 + \ln 3 + \cdots + \ln n$$

Hieraus ergibt sich für $n \geq 2$:

$$\ln(n-1)! < \int_1^n \ln x \, dx < \ln n!$$

Die Stammfunktion von $\ln x$ ist $x \ln x - x + C$. Damit erhalten wir:

$$\ln(n-1)! < n \ln n - n + 1 < \ln n!$$

Es folgt:

$$(n-1)! < e \cdot \left(\frac{n}{e}\right)^n < n!$$

Wir sind am Ziel unserer Betrachtung; es gilt für $n \geq 1$ (mit Gleichheit nur bei $n = 1$):

$$e \cdot \left(\frac{n}{e}\right)^n \leq n! \leq ne \cdot \left(\frac{n}{e}\right)^n \tag{2.1}$$

Tatsächlich lassen sich durch eine genauere Untersuchung bessere Resultate erzielen. Insbesondere gilt die *Stirling'sche Formel*:

$$n! \sim \sqrt{2\pi n} \left(\frac{n}{e}\right)^n \tag{2.2}$$

Für $n = 20$ liefert die Stirling'sche Formel den Wert $2{,}42 \cdot 10^{18}$, was verglichen mit dem Tabelleneintrag für $20!$ von etwas mehr als $2{,}43 \cdot 10^{18}$ ziemlich gut ist. Die Abschätzung nach Gleichung (2.1) liefert $0{,}58 \cdot 10^{18} \leq 20! \leq 11{,}75 \cdot 10^{18}$.

2.2 Das Wachstum der Binomialkoeffizienten

Viele kennen *Binomialkoeffizienten* schon aus der Schule. Üblicherweise wird dort (nur) für natürliche Zahlen k, n mit $k \leq n$ der Binomialkoeffizient $\binom{n}{k}$ durch die folgende Gleichung definiert

$$\binom{n}{k} = \frac{n!}{k! \, (n-k)!}$$

Damit ist $\binom{n}{k}$ gerade die Anzahl der k-elementigen Teilmengen in $\{1, \ldots, n\}$; dies ist die *kombinatorische Interpretation* der Zahl $\binom{n}{k}$. Später behandeln wir Binomialkoeffizienten ausführlich in Abschnitt 4.2, aber diese einfache Tatsache begründen wir sofort: Eine Folge $(i_1, \ldots, i_k)$ mit k paarweise verschiedenen Zahlen zwischen 1 und n definiert die k-elementige Teilmenge $\{i_1, \ldots, i_k\}$. Die Anzahl dieser Folgen ist

$n(n-1)\cdots(n-k+1)$. Da es auf die Reihenfolge in der Mengendarstellung nicht ankommt, dürfen wir die i_j beliebig permutieren. Also definieren je $k!$ Folgen die gleiche Teilmenge. Dies liefert die Behauptung, da

$$\frac{(n-1)\cdots(n-k+1)}{k!} = \frac{n!}{k!\,(n-k)!}$$

Da es insgesamt 2^n Teilmengen von $\{1,\ldots,n\}$ gibt, erkennen wir auch:

$$2^n = (1+1)^n = \sum_k \binom{n}{k}$$

Die Summendarstellung folgt auch direkt aus dem Binomialsatz 4.3. Hier genügt uns zunächst die unmittelbare Folgerung $\binom{n}{k} \le 2^n$ für alle $0 \le k \le n$. Direkt aus der Definition der Binomialkoeffizienten können wir auch die Gleichung

$$k\binom{n}{k} = (n-k+1)\binom{n}{k-1}$$

ableiten. Als Konsequenz erhalten wir:

$$1 = \binom{n}{0} < \binom{n}{1} < \cdots < \binom{n}{\lfloor \frac{n}{2} \rfloor} = \binom{n}{\lceil \frac{n}{2} \rceil} > \cdots > \binom{n}{n-1} > \binom{n}{n} = 1$$

Die Folge steigt bis zur Mitte hin an und fällt dann wieder (siehe Abbildung 2.1). Für $n \ge 2$ ist $\binom{n}{\lfloor \frac{n}{2} \rfloor}$ also der größte Wert unter den n folgenden Werten $2, \binom{n}{1}, \ldots, \binom{n}{n-1}$. Damit muss $\binom{n}{\lfloor \frac{n}{2} \rfloor}$ mindestens so groß sein wie der Mittelwert. Wir erhalten für $n \ge 2$:

$$\frac{2^n}{n} \le \binom{n}{\lfloor \frac{n}{2} \rfloor} = \binom{n}{\lceil \frac{n}{2} \rceil} \tag{2.3}$$

Als Merkregel notieren wir noch für $n \ge 1$:

$$\frac{4^n}{2n} \le \binom{2n}{n} < \binom{2n+1}{n} < 4^n \tag{2.4}$$

Hierbei ist $\binom{2n}{n} < \binom{2n+1}{n}$ eine triviale Konsequenz der Definition; und $\binom{2n+1}{n} < 4^n$ gilt wegen $\binom{2n+1}{n} = \binom{2n+1}{n+1}$ und $\binom{2n+1}{n} + \binom{2n+1}{n+1} < 2^{2n+1}$. Vergleichen wir die Abschätzung $\frac{4^n}{2n} \le \binom{2n}{n} < 4^n$ mit einer Abschätzung vermöge der Stirling'schen Formel, so liegt der wirkliche Wert in der Nähe des geometrischen Mittels, denn wir haben die folgende Asymptotik.

$$\binom{2n}{n} \sim \frac{\sqrt{4\pi n}\left(\frac{2n}{e}\right)^{2n}}{2\pi n \left(\frac{n}{e}\right)^{2n}} = \frac{4^n}{\sqrt{\pi n}}$$

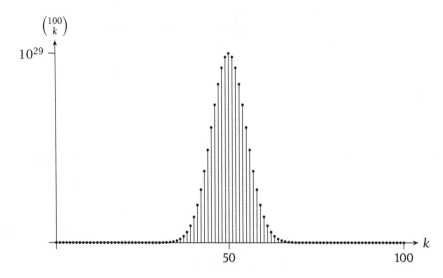

$\binom{100}{k}$

10^{29}

50   100

k

Abb. 2.1. Wachstum der Binomialkoeffizienten $\binom{100}{k}$ für $k = 0, \ldots, 100$.

Der Wert $\binom{20}{10}$ sollte also bei $\frac{2^{20}}{\sqrt{10\pi}}$ sein, was ungefähr 187000 liefert. Der tatsächliche Wert von $\binom{20}{10}$ ist 184756.

Falls k klein und n groß ist, kann die folgende Abschätzung brauchbar sein. Für $0 < k \le n$ gilt:

$$\left(\frac{n}{k}\right)^k \le \binom{n}{k} < \left(\frac{en}{k}\right)^k$$

Die untere und obere Schranke unterscheiden sich also *nur* um einen Faktor e^k, selbst wenn n sehr groß ist. Dies kann man wie folgt sehen: Zunächst ist $\frac{n}{k}$ minimal unter den k Faktoren $\frac{n-i}{k-i}$ von $\binom{n}{k}$; also gilt $(\frac{n}{k})^k \le \binom{n}{k}$. Für die zweite Ungleichung betrachten wir:

$$\binom{n}{k}\left(\frac{k}{n}\right)^k = \frac{n^{\underline{k}}\, k^k}{n^k\, k!} \le \frac{k^k}{k!} < \sum_{i \ge 0} \frac{k^i}{i!} = e^k$$

Hierbei haben wir die Reihendarstellung der Exponentialfunktion $e^x = \sum_{i \ge 0} \frac{x^i}{i!}$ benutzt.

2.3 Das Wachstum des kleinsten gemeinsamen Vielfachen

Dieser Abschnitt basiert auf einem schönen Artikel von Mohan Nair aus dem Jahre 1982 [29]. Alle Rechnungen sind elementar und auf Schulniveau. Es ist die Kunstfertigkeit von Nair gewesen, sie so zusammenzustellen, um die gewünschten exponentiellen Schranken für kgV(n) herzuleiten. Bevor wir anfangen, sollten wir uns

klarmachen, dass wir die Primfaktorzerlegung von kgV(n) unmittelbar hinschreiben können. Beginnen wir mit einigen einfachen Beispielen:

$$
\begin{aligned}
\mathrm{kgV}(6) &= 2^2 \cdot 3 \cdot 5 & &= & 60 \\
\mathrm{kgV}(7) &= 2^2 \cdot 3 \cdot 5 \cdot 7 & &= & 420 \\
\mathrm{kgV}(8) &= 2^3 \cdot 3 \cdot 5 \cdot 7 & &= & 840 \\
\mathrm{kgV}(9) &= 2^3 \cdot 3^2 \cdot 5 \cdot 7 & &= & 2\,520 \\
\mathrm{kgV}(23) &= \mathrm{kgV}(24) = 2^4 \cdot 3^2 \cdot 5 \cdot 7 \cdot 11 \cdot 13 \cdot 17 \cdot 19 \cdot 23 & &= & 140\,900\,760 \\
\mathrm{kgV}(25) &= \mathrm{kgV}(26) = \mathrm{kgV}(23) \cdot 5 & &= & 704\,503\,800 \\
\mathrm{kgV}(27) &= \mathrm{kgV}(28) = \mathrm{kgV}(25) \cdot 3 & &= & 2\,113\,511\,400
\end{aligned}
$$

Das allgemeine Resultat ist:

$$
\mathrm{kgV}(n) = \prod_{p \leq n} p^{\lfloor \log_p n \rfloor}
$$

Hierbei bezeichnet p eine Primzahl. Je größer das kleinste gemeinsame Vielfache der ersten n Zahlen ist, desto mehr Primzahlen bis n muss es also geben. Wir zeigen daher als erstes eine untere Schranke für das kleinste gemeinsame Vielfache der ersten n Zahlen. Zunächst konstruieren wir eine Menge von Teilern von kgV(n). Wir schreiben $m \mid n$ falls m die Zahl n teilt, d. h., falls $k \in \mathbb{Z}$ existiert, so dass $mk = n$ ist. Zum Beispiel gilt $m \mid 0$ für alle $m \in \mathbb{Z}$.

Lemma 2.1. *Für alle $m, n \in \mathbb{N}$ mit $1 \leq m \leq n$ gilt*

$$
m \binom{n}{m} \;\bigg|\; \mathrm{kgV}(n)
$$

Beweis. Wir untersuchen das Integral

$$
I = \int_0^1 x^{m-1}(1-x)^{n-m} \, \mathrm{d}x
$$

Die Auswertung geschieht auf zweifache Weise. Zunächst wenden wir wieder den Binomialsatz an, um $(1-x)^{n-m}$ als $\sum_k (-1)^k \binom{n-m}{k} x^k$ zu schreiben. Damit folgt:

$$
x^{m-1}(1-x)^{n-m} = \sum_k (-1)^k \binom{n-m}{k} x^{m-1+k}
$$

Die Auswertung des Integrals ergibt also:

$$
I = \sum_k (-1)^k \binom{n-m}{k} \int_0^1 x^{m-1+k} \, \mathrm{d}x = \sum_k (-1)^k \binom{n-m}{k} \frac{1}{m+k}
$$

Multiplizieren wir I mit dem kleinsten gemeinsamen Vielfachen aller Zahlen bis n, also mit $\mathrm{kgV}(n)$, so wird $I \cdot \mathrm{kgV}(n)$ eine alternierende Summe über ganze Zahlen, da $\frac{\mathrm{kgV}(n)}{m+k} \in \mathbb{N}$ für $0 \leq k \leq n - m$. Da der Wert von I positiv ist, muss gelten

$$I \cdot \mathrm{kgV}(n) \in \mathbb{N} \tag{2.5}$$

Induktiv nach $n - m$ zeigen wir als Nächstes

$$I = \frac{1}{m\binom{n}{m}}$$

Für $m = n$ gilt:

$$I = \int_0^1 x^{m-1}(1-x)^{n-n}\,dx = \int_0^1 x^{m-1}\,dx = \left[\frac{1}{m}x^m\right]_0^1 = \frac{1}{m} = \frac{1}{m\binom{m}{m}}$$

Sei nun $1 \leq m < n$. Durch Verwendung partieller Integration

$$\int u' \cdot v = u \cdot v - \int u \cdot v'$$

$$\text{mit} \quad u = \frac{1}{m}x^m \quad v = (1-x)^{n-m}$$

$$u' = x^{m-1} \quad v' = -(n-m)(1-x)^{n-m-1}$$

ergibt sich wegen $u(1) \cdot v(1) = u(0) \cdot v(0) = 0$ zunächst

$$I = \int_0^1 x^{m-1}(1-x)^{n-m}\,dx$$

$$= \int_0^1 -u \cdot v' = \frac{n-m}{m}\int_0^1 x^{(m+1)-1}(1-x)^{n-(m+1)}\,dx$$

Mit Induktion erhalten wir

$$I = \frac{n-m}{m} \cdot \frac{1}{(m+1)\binom{n}{m+1}} = \frac{1}{m\binom{n}{m}}$$

Mit Gleichung (2.5) folgt nun $\frac{\mathrm{kgV}(n)}{m\binom{n}{m}} \in \mathbb{N}$ und damit $m\binom{n}{m} \mid \mathrm{kgV}(n)$. $\qquad\square$

Aus Lemma 2.1 folgt zusammen mit Gleichung (2.3) aus dem vorigen Abschnitt die Abschätzung

$$2^{n-1} \leq \frac{n}{2}\binom{n}{\lceil\frac{n}{2}\rceil} \leq \left\lceil\frac{n}{2}\right\rceil\binom{n}{\lceil\frac{n}{2}\rceil} \leq \mathrm{kgV}(n)$$

für alle $n \geq 1$. Der nächste Satz verbessert diese Abschätzung noch etwas für Zahlen $n \geq 7$.

Satz 2.2. *Für alle $n \geq 7$ gilt:*
$$2^n < \mathrm{kgV}(n)$$

Beweis. Mit Lemma 2.1 lassen sich zwei Teiler von $\mathrm{kgV}(2n+1)$ herleiten:

$$(2n+1)\binom{2n}{n} = (n+1)\binom{2n+1}{n+1} \;\Big|\; \mathrm{kgV}(2n+1)$$

$$n\binom{2n}{n} \;\Big|\; \mathrm{kgV}(2n) \;\Big|\; \mathrm{kgV}(2n+1)$$

Da n und $2n+1$ teilerfremd sind, folgt

$$n(2n+1)\binom{2n}{n} \;\Big|\; \mathrm{kgV}(2n+1)$$

Mit der unteren Schranke aus (2.4) für $\binom{2n}{n}$ ergibt sich aus dieser Teilbarkeitseigenschaft folgende Größenabschätzung:

$$n \cdot 4^n < n\,(2n+1)\binom{2n}{n} \leq \mathrm{kgV}(2n+1) \qquad (2.6)$$

Sei $n \geq 4$. Dann gilt

$$2^{2n+2} = 4 \cdot 2^{2n} \leq n \cdot 4^n < \mathrm{kgV}(2n+1) \leq \mathrm{kgV}(2n+2)$$

Damit gilt die Aussage
$$2^n < \mathrm{kgV}(n)$$

für alle $n \geq 9$. Es bleiben noch die Fälle $n = 7$ und $n = 8$ zu untersuchen. Dies weisen wir mit den folgenden Rechnungen direkt nach: $2^7 = 128 < 420 = \mathrm{kgV}(7)$, $2^8 = 256 < 840 = \mathrm{kgV}(8)$. $\qquad\qquad\square$

Es gilt $\mathrm{kgV}(6) = 60 < 2^6 = 64$. Die Schranke $n \geq 7$ ist also optimal. Sei $n \geq 2$, dann gilt $2^{2n+1} \leq n \cdot 4^n \leq \mathrm{kgV}(2n+1)$. Also gilt der Satz für ungerade $n \geq 5$, ohne dass wir 7 hätten extra untersuchen müssen. Aus Gleichung (2.6) folgt $\mathrm{kgV}(n) \in \omega(2^n)$.

Lemma 2.3. *Für alle $m, n \in \mathbb{N}$ mit $n/2 \leq m \leq n$ gilt:*

$$\mathrm{kgV}(n) \;\Big|\; \mathrm{kgV}(m) \cdot \binom{n}{m}$$

Beweis. Es gilt

$$k\binom{n}{k}\binom{n-k}{m-k} = k\binom{m}{k}\binom{n}{m}$$

Mit Lemma 2.1 folgt daraus für alle $1 \le k \le m$, dass $k\binom{n}{k}$ ein Teiler von $\mathrm{kgV}(m)\binom{n}{m}$ ist. Wegen $k\binom{n}{k} = (n - k + 1)\binom{n}{n-k+1}$ und $n/2 \le m$ lässt sich jeder Term $k\binom{n}{k}$ für $m < k \le n$ auch schreiben als $k'\binom{n}{k'}$ mit $k' \le m$. Also gilt

$$k\binom{n}{k} \,\Big|\, \mathrm{kgV}(m) \cdot \binom{n}{m}$$

für alle $1 \le k \le n$. Insbesondere ist jede Zahl k zwischen 1 und n ein Teiler von $\mathrm{kgV}(m)\binom{n}{m}$; daraus folgt die Aussage des Lemmas. ☐

Wir kommen nun zu einer oberen Schranke für das kleinste gemeinsame Vielfache der ersten n Zahlen.

Satz 2.4. *Für alle $n \ge 1$ gilt:*

$$\mathrm{kgV}(n) \le 4^{n-1}$$

Beweis. Für $n = 1$ gilt die Aussage. Wir unterscheiden nun, ob n gerade oder ungerade ist. Für $n = 2m$ gilt:

$$\mathrm{kgV}(2m) \le \mathrm{kgV}(m) \cdot \binom{2m}{m} \le \mathrm{kgV}(m) \cdot 4^m \le 4^{m-1}4^m = 4^{2m-1}$$

Hierbei folgt die erste Ungleichung aus Lemma 2.3, die zweite aus der Abschätzung (2.4) und die dritte mit Induktion. Analog ergibt sich für $n = 2m + 1$:

$$\mathrm{kgV}(2m + 1) \le \mathrm{kgV}(m + 1)\binom{2m + 1}{m + 1} \le \mathrm{kgV}(m + 1)4^m \le 4^m 4^m = 4^{2m}$$

Daraus ergibt sich für alle $n \ge 1$ die obere Schranke $\mathrm{kgV}(n) \le 4^{n-1}$. ☐

2.4 Aussagen zur Primzahldichte

In diesem Abschnitt sei p stets eine Primzahl und n eine positive ganze Zahl. Es sei $x \in \mathbb{R}$ mit $x \ge 1$ und $\pi(x)$ die Anzahl der Primzahlen kleiner oder gleich x. Es ist eine der ältesten mathematischen Beobachtungen, dass $\pi(x)$ nicht beschränkt werden kann. Schon Euklid argumentierte in etwa wie folgt: Angenommen, es gäbe eine größte Primzahl p, dann betrachte das Produkt über alle Primzahlen bis einschließlich p, addiere 1 hinzu und nenne diese Zahl q. Dann ist q eine Primzahl oder es muss zwischen p und q eine weitere Primzahl geben. Hieraus lässt sich eine untere Schranke für das Wachstum von $\pi(x)$ herleiten, allerdings ist diese Schranke sehr schlecht, denn, wie wir gleich erkennen werden, wissen wir schon viel mehr.

Der Beweis aus dem vorigen Abschnitt zur Aussage $2^n < \mathrm{kgV}(n)$ für alle $n \ge 7$ war einigermaßen mühsam. Hier kommt die Belohnung. Wir haben schon weiter oben festgestellt, dass wenn $a \in \mathbb{N}$ maximal ist mit $p^a \mid \mathrm{kgV}(n)$, dann muss $a = \lfloor \log_p n \rfloor$

gelten; oder anders ausgedrückt:

$$\mathrm{kgV}(n) = \prod_{p \le n} p^{\lfloor \log_p n \rfloor} \le \prod_{p \le n} n = n^{\pi(n)}$$

Aus dem Satz 2.2 erhalten wir also für $n \ge 7$:

$$2^n < n^{\pi(n)}$$

Dies liefert uns eine untere Schranke für $\pi(n)$, die für fast alle Anwendungen vollkommen ausreicht: Für alle $n \ge 4$ gilt $n \le \pi(n) \log_2 n$ und damit

$$\frac{n}{\log_2 n} \le \pi(n) \tag{2.7}$$

Suchen wir also eine Primzahl mit bis zu 100 Binärstellen, so ist die grobe Idee, dass etwa jede 100-ste Zahl eine Primzahl ist. Da wir nicht unter den geraden Zahlen suchen werden und mit Probedivisionen auch alle Zahlen ausschließen können, die durch $3, 5, 7$ oder 11 teilbar sind, stehen die Chancen gut, schnell auf eine tatsächliche Primzahl zu stoßen.

Wir wollen nun eine obere Schranke für $\pi(n)$ herleiten. Aus Satz 2.4 erhalten wir für $n \ge 1$ die Abschätzung:

$$\prod_{p \le n} p \le \mathrm{kgV}(n) \le 4^{n-1} \tag{2.8}$$

Ganz analog sehen wir

$$t^{\pi(n) - \pi(t)} \le \prod_{t < p \le n} p < 4^n$$

für jedes $1 < t \le n$. Hieraus folgt $(\pi(n) - \pi(t)) \log t \le 2n$. Der Logarithmus ist hier zur Basis 2 gemeint. Die Angabe der Basis fehlt, um die Lesbarkeit zu erleichtern. Zusammen mit $\pi(t) \le t$ ergibt sich

$$\pi(n) \le \frac{2n}{\log t} + t$$

Wenn wir $t = \frac{n}{(\log n)^2}$ setzen, liefert dies

$$\pi(n) \le \frac{2n}{\log n} \cdot \frac{\log n}{\log n - 2 \log \log n} + \frac{n}{(\log n)^2}$$

Der Faktor $\frac{\log n}{\log n - 2 \log \log n}$ geht für großes n gegen 1 und der Summand $\frac{n}{(\log n)^2}$ wird von $\frac{2n}{\log n}$ dominiert. Hieraus folgt für jedes $\varepsilon > 0$ die Abschätzung

$$\pi(n) \le \frac{(2 + \varepsilon)n}{\log n}$$

falls n genügend groß ist. Zusammen mit der unteren Schranke für $\pi(n)$ lässt sich aus der obigen Rechnung für jedes $\varepsilon > 0$ leicht ein n_ε bestimmen mit der Eigenschaft, dass für alle $n > n_\varepsilon$ gilt:

$$\frac{n}{\log_2 n} \leq \pi(n) \leq \frac{(2+\varepsilon)n}{\log_2 n} \tag{2.9}$$

Für die verwendeten rein elementaren Methoden ist dies ein erstaunlich gutes Ergebnis. Abschätzungen dieser Form wurden erstmals mit anderen Methoden von Tschebyschev 1851 gezeigt [30] (Pafnuty Lvovich Tschebyschev, 1821–1894). Sie waren ein Vorläufer für den berühmten Primzahlsatz, der im Jahre 1896 unabhängig von Jacques Salomon Hadamard (1865–1963) und Charles-Jean Étienne Gustave Nicolas de la Vallée Poussin (1866–1962) bewiesen [11, 23] und bereits von Carl Friedrich Gauß (1777–1855) um 1800 vermutet wurde. Er gibt die genaue Asymptotik der Primzahldichte an:

$$\pi(x) \sim \frac{x}{\ln x}$$

Nach einer Anekdote wurde für diesen Nachweis den Mathematikern *Ewigkeit* versprochen, und in der Tat, beide wurden mehr als 95 Jahre alt.

2.5 Das Bertrand'sche Postulat

Moderne kryptographische Verfahren wie RSA basieren auf dem Finden großer Primzahlen, z. B. mit 200 Dezimalstellen. Dies stellt uns vor das Problem, solch große Primzahlen schnell zu finden. Außerdem sollten genügend viele 200-stellige Primzahlen vorhanden sein, damit unsere gewählte Zahl einmalig bleibt. Die Chance, dass jemand anderes auf dieselbe Zahl verfällt, muss verschwindend gering sein. Es reicht hier nicht, dass es bis 10^{200} schon mindestens 10^{197} Primzahlen gibt; wir benötigen viele Primzahlen für eine genau festgelegte moderate Stellenzahl. Zum Glück gibt es wirklich eine gewaltige Zahl von Primzahlen mit genau 200 Dezimalstellen. Unsere Herleitung liefert weit mehr als 10^{196} Stück. Genauer zeigen wir für $n \geq 2^{12}$, dass zwischen n und $2n$ mindestens $\frac{n}{3\log_2 n}$ Primzahlen vorkommen. Das primäre Ziel in diesem Abschnitt ist jedoch die Aussage in Satz 2.5. Der Beweis benutzt in unserer Darstellung an einer entscheidenden Stelle eine Idee von Paul Erdős (ungarisch: Erdős Pál, 1913–1996), die 1932 veröffentlicht wurde [18]. Erdős war damals 19 Jahre alt.

Satz 2.5 (Bertrand'sches Postulat). *Für alle $n \geq 1$ gibt es mindestens eine Primzahl p mit $n < p \leq 2n$.*

Beweis. Um das Bertrand'sche Postulat (benannt nach Joseph Louis François Bertrand, 1822–1900) bis $n = 4048$ zu überprüfen, reicht es, die folgende Liste von Primzahlen anzugeben:

$$2, 3, 5, 7, 13, 23, 43, 83, 163, 317, 631, 1259, 2503, 4049$$

Interessant für uns sind daher nur die n ab 4048. Für eine Primzahl p definieren wir $e_p(n)$ als die größte natürliche Zahl a, so dass p^a die Zahl n teilt. Diese Definition ist aufgrund der Existenz und Eindeutigkeit der Primfaktorzerlegung äquivalent zur Forderung:

$$n = \prod_p p^{e_p(n)}$$

Aus Lemma 2.1 folgt:

$$n \binom{2n}{n} \ \Big|\ \mathrm{kgV}(2n)$$

Da für alle p die Beziehung $e_p(\mathrm{kgV}(2n)) = \lfloor \log_p(2n) \rfloor$ gilt, muss also auch $e_p\left(\binom{2n}{n}\right) \leq \log_p(2n)$ gelten, und wir erhalten

$$p^{e_p\left(\binom{2n}{n}\right)} \leq 2n$$

Für Primzahlen $p > \sqrt{2n}$ kann damit $e_p\left(\binom{2n}{n}\right)$ nur 0 oder 1 sein. Hier kommt die für den Beweis entscheidende Beobachtung von Erdős:

$$\text{Für } \tfrac{2}{3}n < p \leq n \text{ und } n \geq 3 \text{ gilt } e_p\left(\binom{2n}{n}\right) = 0$$

In diesem Bereich teilt p den Wert $n!$ genau einmal. Die Zahl p erscheint im Nenner von $\binom{2n}{n} = \frac{(2n)!}{n!n!}$ also zweimal. Im Zähler $(2n)!$ gibt es genau zwei Faktoren, die p teilen, nämlich p und $2p$. Fügen wir dies zusammen, so sehen wir:

$$\frac{4^n}{2n} \leq \binom{2n}{n} \leq \left(\prod_{p \leq \sqrt{2n}} 2n\right)\left(\prod_{\sqrt{2n} < p \leq \frac{2}{3}n} p\right)\left(\prod_{n < p \leq 2n} p\right)$$

Das erste Produkt wird durch $(2n)^{\sqrt{2n}-1}$ abgeschätzt (da 1 keine Primzahl ist), und Gleichung (2.8) liefert eine obere Schranke für das zweite Produkt. Es folgt:

$$4^n \leq (2n)^{\sqrt{2n}} 4^{\frac{2}{3}n} \prod_{n < p \leq 2n} p$$

Schließlich erhalten wir mit $m = 2n > 853$ die Aussage:

$$1 < 2^{\frac{1}{3}m - \sqrt{m}\log_2 m} \leq \prod_{n < p \leq 2n} p \tag{2.10}$$

Also existiert eine Primzahl p mit $n < p \leq 2n$. □

Tatsächlich lässt sich aus der letzten Ungleichung eine bessere Abschätzung herleiten:

Satz 2.6. *Für alle $n \geq 2^{12}$ gilt:*

$$\left| \{ p \mid n < p \leq 2n \} \right| \geq \frac{n}{3 \log_2 n} > 113$$

Beweis. Aus Gleichung (2.10) folgt mit $m = 2n$ die Beziehung:

$$(2n)^{|\{p \mid n < p \leq 2n\}|} \geq \prod_{n < p \leq 2n} p \geq 2^{\frac{1}{3}m - \sqrt{m}\log_2 m}$$

Dies liefert die Abschätzung

$$|\{p \mid n < p \leq 2n\}| \geq \frac{m}{3\log_2 m} - \sqrt{m} \geq \frac{n}{3\log_2 n} > 113$$

was die Behauptung zeigt. □

Übrigens gibt es schon 461 Primzahlen zwischen 4048 und 8096.

Aufgaben

2.1. Zeigen Sie:

(a) Bernoulli-Ungleichung: $(1 + x)^n \geq 1 + nx$ für $x \geq -1$ und $n \geq 0$.

(b) $e^x \geq 1 + x$ für $x \in \mathbb{R}$ sowie $\ln x \leq x - 1$ für $x > 0$.

(c) $e^x \geq (1 + \frac{x}{n})^n$ für $x \geq -1$ und $n \geq 1$ oder $n > |x|$.

(d) $\ln(x + 1) \geq \frac{x}{x+1}$ für $x > -1$.

2.2. Seien $a_1 \leq \cdots \leq a_n$ und $b_1 \leq \cdots \leq b_n$ zwei Folgen reeller Zahlen, und sei $\pi : \{1,\ldots,n\} \to \{1,\ldots,n\}$ eine Permutation. Zeigen Sie, dass die Summe $S(\pi) = \sum_{i=1}^{n} a_i b_{\pi(i)}$ maximal ist, wenn π die Identität ist. Die Summe $S(\pi)$ ist minimal, wenn $\pi(i) = n + 1 - i$ gilt (d. h., wenn π die Reihenfolge umkehrt).

2.3. Seien $a_1,\ldots,a_n$ positive reelle Zahlen und sei

-   $H = n/(\frac{1}{a_1} + \cdots + \frac{1}{a_n})$ das harmonische Mittel,
-   $G = \sqrt[n]{a_1 \cdots a_n}$ das geometrische Mittel,
-   $A = (a_1 + \cdots + a_n)/n$ das arithmetische Mittel und
-   $Q = \sqrt{(a_1^2 + \cdots + a_n^2)/n}$ das quadratische Mittel.

Zeigen Sie: $\min(a_1,\ldots,a_n) \leq H \leq G \leq A \leq Q \leq \max(a_1,\ldots,a_n)$.

2.4. Sei s eine reelle Zahl. Zeigen Sie, dass die Reihe $\sum_{i \geq 1} \frac{1}{i^s}$ genau dann konvergiert, wenn $s > 1$ gilt.

2.5. Für $n \geq 1$ sei $t(n)$ die Anzahl der positiven Teiler von n. Wir definieren die durchschnittliche Teilerzahl durch $\bar{t}(n) = \frac{1}{n}\sum_{i=1}^{n} t(i)$. Zeigen Sie $|\bar{t}(n) - \ln n| \leq 1$.

2.6. Zeigen Sie, dass es beliebig große Lücken zwischen zwei aufeinander folgenden Primzahlen gibt. Das heißt, für jedes $n \in \mathbb{N}$ existiert ein Index i mit $p_{i+1} - p_i \geq n$. Hierbei ist $p_1, p_2, \ldots$ die aufsteigende Folge der Primzahlen.

2.7. Sei $p_1, p_2, \ldots$ die aufsteigende Folge der Primzahlen. Zeigen Sie:

(a) $p_n \leq 2n \log n$

(b) Für jede genügend große Zahl n gilt $p_n \geq \frac{1}{3} n \log n$.

(c) Zeigen Sie, dass die Reihe $\sum_{i \geq 1} \frac{1}{p_i}$ divergiert.

Zusammenfassung

Begriffe

- Fakultät $n!$
- Binomialkoeffizient $\binom{n}{k}$
- kleinstes gemeinsames Vielfaches

- $\text{kgV}(n) = \text{kgV}(\{1, \ldots, n\})$
- Primzahlfunktion $\pi(x)$
- k teilt ℓ, $k \mid \ell$

Methoden und Resultate

- $e \cdot \left(\frac{n}{e}\right)^n \leq n! \leq ne \cdot \left(\frac{n}{e}\right)^n$ für $n \geq 1$

- $\binom{n}{\lfloor n/2 \rfloor}$ ist der größte Binomialkoeffizient aus $\binom{n}{1}, \ldots, \binom{n}{n}$

- $\frac{4^n}{2n} \leq \binom{2n}{n} < \binom{2n+1}{n} < 4^n$ für $n \geq 1$

- $\left(\frac{n}{k}\right)^k \leq \binom{n}{k} < \left(\frac{en}{k}\right)^k$ für $0 < k \leq n$

- $\text{kgV}(n) = \prod_{p \leq n} p^{\lfloor \log_p n \rfloor}$ für $n \geq 1$ (p Primzahl)

- $m\binom{n}{m} \mid \text{kgV}(n)$ für $1 \leq m \leq n$

- $2^n < \text{kgV}(n) \leq 4^{n-1}$ für $n \geq 7$

- $\frac{n}{\log_2 n} \leq \pi(n)$ für $n \geq 4$

- $\prod_{p \leq n} p \leq 4^{n-1}$ für $n \geq 1$ (p Primzahl)

- Für jedes $\varepsilon > 0$ existiert n_0, so dass $\pi(n) \leq \frac{(2+\varepsilon)n}{\log n}$ für alle $n \geq n_0$ gilt.

- Bertrand'sches Postulat: $\forall n \geq 1$ existiert Primzahl p mit $n < p \leq 2n$.

- Für alle $n \geq 2^{12}$ gibt es mindestens $\frac{n}{3 \log_2 n}$ Primzahlen p mit $n < p \leq 2n$.

3 Diskrete Wahrscheinlichkeitsrechnung

Viele Abschätzungen deuten zunächst einmal auf das Verhalten im schlechtesten Fall hin. Häufig interessiert man sich jedoch mehr für ein Verhalten im „Normalfall". Im schlechtesten Fall gewinnt man beim Roulette niemals. Im Mittel gewinnt man wenigstens ab und zu, aber viel zu selten, um den Bestand der Spielbank zu gefährden. Um solches Verhalten präziser beschreiben zu können, entwickeln wir hier einige elementare Begriffe aus der diskreten Wahrscheinlichkeitstheorie, wie wir sie für die Anwendungen später brauchen werden.

3.1 Wahrscheinlichkeitsräume und Erwartungswerte

Ein diskreter *Wahrscheinlichkeitsraum* ist eine endliche oder abzählbare Menge Ω zusammen mit einer Abbildung $\Pr : \Omega \to [0, 1]$ in das reelle 0-1-Intervall, welche die folgende Bedingung erfüllt:

$$\sum_{\omega \in \Omega} \Pr[\omega] = 1$$

Ist Ω endlich und $\Pr[\omega]$ ein konstanter Wert, also $\Pr[\omega] = \frac{1}{|\Omega|}$ für alle $\omega \in \Omega$, so sprechen wir von einer *Gleichverteilung*. Ein *Ereignis* ist eine Teilmenge $A \subseteq \Omega$. Die *Wahrscheinlichkeit* von A ist

$$\Pr[A] = \sum_{\omega \in A} \Pr[\omega]$$

Wenn Ω endlich ist, dann gilt im Falle eine Gleichverteilung:

$$\Pr[A] = \frac{|A|}{|\Omega|} = \frac{\text{„Anzahl der günstigen Fälle"}}{\text{„Anzahl der möglichen Fälle"}}$$

Dies ist eine der Motivationen für das nächste Kapitel, wo wir Techniken lernen wollen, die jeweiligen Anzahlen zu bestimmen. Bei einer Runde des Roulettespiels ist der Wahrscheinlichkeitsraum die Menge $\{0, \ldots, 36\}$ und die Ereignisse *rot* und *schwarz* haben die gleiche Wahrscheinlichkeit, nämlich $18/37$. Im Prinzip ist es diese Differenz $1 - 36/37 = 1/37$, die gegen die Spieler spricht.

Eine *Zufallsvariable* X ist hier stets eine reellwertige Funktion

$$X : \Omega \to \mathbb{R}$$

Der *Erwartungswert* von X wird wie folgt definiert:

$$E[X] = \sum_{\omega \in \Omega} X(\omega) \Pr[\omega]$$

Falls die Menge Ω unendlich viele Elemente hat, muss die Reihe absolut konvergieren, ansonsten ist der Erwartungswert nicht definiert. In den meisten betrachteten

Fällen ist der Wahrscheinlichkeitsraum endlich, und es kann keine Probleme mit der Konvergenz geben. In den anderen Fällen machen wir implizite Konvergenzvoraussetzungen, die wir häufig gar nicht extra erwähnen. Bei einer Gleichverteilung ist der Erwartungswert der Mittelwert über die Funktionswerte der Zufallsvariablen. Es gilt dann:

$$E[X] = \frac{1}{|\Omega|} \sum_{\omega} X(\omega)$$

Der Erwartungswert einer gewürfelten Augenzahl mit einem Würfel ist zum Beispiel $3,5$. Man beachte, dass diese Zahl keiner beim Würfeln auftretenden Augenzahl entspricht. Jedes Ereignis $A \subseteq \Omega$ kann über die charakteristische Funktion $\chi_A : \Omega \to \{0,1\}$ (mit $\chi_A(a) = 1$ für $a \in A$ und $\chi_A(a) = 0$ sonst) direkt als eine Zufallsvariable gelesen werden. Die Wahrscheinlichkeit des Ereignisses A ist dann der Erwartungswert der charakteristischen Funktion: $\Pr[A] = E[\chi_A]$. Ist $x \in \mathbb{R}$, so bezeichnet $\Pr[X = x]$ die Wahrscheinlichkeit des Ereignisses:

$$\{\omega \in \Omega \mid X(\omega) = x\}$$

Damit gilt $\Pr[X = x] = \Pr[X^{-1}(x)]$. Direkt aus der Definition ergibt sich auch die folgende Aussage:

$$E[X] = \sum_{\omega} X(\omega) \Pr[\omega] = \sum_{x} x \Pr[X = x]$$

Nimmt X keine negativen Werte an und ist $X(\omega) > 0$ für ein ω mit $\Pr[\omega] > 0$, so gilt offenbar $E[X] > 0$. Außerdem erhalten wir den nach Andrei Andrejewitsch Markov (1856–1922) benannten Zusammenhang 3.1 zwischen Wahrscheinlichkeit und Erwartungswert.

Satz 3.1 (Markov-Ungleichung). *Sei X eine Zufallsvariable mit $X(\omega) \geq 0$ für alle ω und $E[X] > 0$. Dann gilt für alle $\lambda > 0$:*

$$\Pr\left[X \geq \lambda E[X]\right] \leq \frac{1}{\lambda}$$

Beweis. Es gilt:

$$E[X] = \sum_{\omega} X(\omega) \Pr[\omega] \geq \sum_{\substack{\omega \in \Omega \\ X(\omega) \geq \lambda E[X]}} X(\omega) \Pr[\omega] \geq \lambda E[X] \Pr\left[X \geq \lambda E[X]\right]$$

Dies zeigt die Behauptung. □

Eine wichtige Eigenschaft ist die *Linearität des Erwartungswertes*:

$$E[aX + bY] = aE[X] + bE[Y]$$

Hierbei sind $a, b \in \mathbb{R}$ und $X, Y : \Omega \to \mathbb{R}$ Zufallsvariablen. Die Zufallsvariable $aX + bY : \Omega \to \mathbb{R}$ ist definiert durch $(aX + bY)(\omega) = aX(\omega) + bY(\omega)$. Ist $X : \Omega \to \mathbb{R}$

eine Zufallsvariable, so assoziiert man mit X ihre *diskrete Dichte* $f_X : \mathbb{R} \to [0,1]$ und ihre *Verteilung* $F_X : \mathbb{R} \to [0,1]$. Diese sind wie folgt definiert:

$$f_X : \mathbb{R} \to [0,1], \; f_X(x) = \Pr[X = x]$$
$$F_X : \mathbb{R} \to [0,1], \; F_X(x) = \Pr[X \leq x]$$

Aus der Dichte lässt sich die Verteilung berechnen, und die Verteilung bestimmt die Dichte. Sehr verschiedene Zufallsvariablen können auf die gleiche Verteilung (Dichte) führen. Viele interessante Eigenschaften ergeben sich schon allein aus der Verteilung (oder der Dichte), ohne die konkrete Zufallsvariable genau zu kennen. Daher spielt der konkrete Wahrscheinlichkeitsraum häufig gar keine Rolle. Insbesondere ist:

$$E[X] = \sum_{x \in \mathbb{R}} x \, f_X(x)$$

Um möglichst nahe an einer konkreten Vorstellung zu bleiben, arbeiten wir weiterhin meistens mit diskreten Zufallsvariablen. Wir bemerken jedoch, dass es dieser Ansatz ist, der den Übergang zu kontinuierlichen Zufallsvariablen ermöglicht. Im Wesentlichen ersetzt man Summen durch ein Integral, wobei $f_X(x)$ zu einem dx wird. Dabei muss man jedoch gewährleisten, dass Ausdrücke sinnvoll und wohldefiniert bleiben, was einen erheblichen theoretischen Unterbau erfordern würde.

Zwei Zufallsvariablen X und Y heißen *unabhängig*, wenn für alle $x, y \in \mathbb{R}$

$$\Pr[X = x \wedge Y = y] \; = \; \Pr[X = x] \cdot \Pr[Y = y]$$

gilt. Hierbei steht $X = x \wedge Y = y$ für den Durchschnitt der Ereignisse $X = x$ und $Y = y$. Die Intuition ist, dass sich unabhängige Zufallsvariablen nicht gegenseitig beeinflussen. Beispielsweise ist die Wahrscheinlichkeit bei zwei Würfeln für einen Wurf mit zwei Sechsen $1/36$, da das Ergebnis von einem Würfel nicht das Ergebnis des anderen Wurfs beeinflusst. Analog gilt, dass die Wahrscheinlichkeit für einen Pasch $1/6$ ist, und dass die Wahrscheinlichkeit für einen *Kniffel* (5 gleiche Augenzahlen) in einem einzigen Wurf mit fünf Würfeln $1/6^4 = 1/1296$ ist. Falls X und Y unabhängig sind, so gilt:

$$E[XY] = E[X]E[Y]$$

Dies folgt aus der folgenden Betrachtung:

$$E[XY] = \sum_z z \Pr[XY = z]$$
$$= \sum_z \sum_{xy = z} xy \Pr[X = x \wedge Y = y]$$
$$= \left(\sum_x x \Pr[X = x] \right) \cdot \left(\sum_y y \Pr[Y = y] \right)$$
$$= E[X]E[Y]$$

Betrachtet man die Zufallsvariable $X - E[X]$, so ist deren Erwartungswert 0. Interessanter ist das Quadrat dieser Zufallsvariablen $(X - E[X])^2$. Der Erwartungswert kann nicht negativ sein. Er ist positiv, sowie er definiert ist und $\Pr[X \neq E[X]] > 0$ gilt. Der Erwartungswert von $(X - E[X])^2$ heißt die *Varianz* $\mathrm{Var}[X]$ von X und misst, wie stark X von $E[X]$ abweicht. Es gilt:

$$
\begin{aligned}
\mathrm{Var}[X] &= E[(X - E[X])^2] \\
&= E[X^2 - 2E[X]X + E[X]^2] \\
&= E[X^2] - 2E[X]E[X] + E[X]^2 \\
&= E[X^2] - E[X]^2
\end{aligned}
$$

Die erste Gleichung gilt nach Definition. Die dritte folgt aus der Linearität des Erwartungswertes. Der Erwartungswert der Zufallsvariablen X^2 ist also mindestens so groß wie $E[X]^2$. Die Differenz misst die Varianz.

Beispiel 3.2. Bei einem *Bernoulli-Experiment* (Jacob Bernoulli, 1654–1705) misst man Erfolg oder Misserfolg durch ein 0-1-Ereignis. Typischerweise setzt man $\Pr[X = 1] = p$ und $\Pr[X = 0] = q = 1 - p$. Damit ist $E[X] = p$ und $\mathrm{Var}[X] = p - p^2 = pq$. ◇

Mit σ_X wird die *Standardabweichung* von X bezeichnet, sie ist definiert durch $\sigma_X = \sqrt{\mathrm{Var}[X]}$. Der Name ergibt sich aus der Beziehung 3.3.

Satz 3.3 (Tschebyschev-Ungleichung). *Sei $\lambda > 0$. Dann gilt:*

$$
\Pr\left[|X - E[X]| \geq \lambda\,\sigma_X\right] \leq \frac{1}{\lambda^2}
$$

Beweis. Nach der Markov-Ungleichung und der Definition von σ_X und $\mathrm{Var}[X]$ gilt:

$$
\Pr\left[|X - E[X]| \geq \lambda\,\sigma_X\right] = \Pr[(X - E[X])^2 \geq \lambda^2\,\mathrm{Var}[X]] \leq \frac{1}{\lambda^2} \qquad \square
$$

Die Abschätzung aus Satz 3.3 liefert erst für Abweichungen oberhalb der Standardabweichung (also für $\lambda > 1$) eine sinnvolle Aussage.

Satz 3.4. *Für unabhängige Zufallsvariablen X und Y gilt:*

$$
\mathrm{Var}[X + Y] = \mathrm{Var}[X] + \mathrm{Var}[Y]
$$

Beweis. Mit $E[XY] = E[X]E[Y]$ erhalten wir:

$$
\begin{aligned}
\mathrm{Var}[X + Y] &= E[(X + Y)^2] - E[X + Y]^2 \\
&= E[X^2] + 2E[XY] + E[Y^2] - E[X]^2 - 2E[X]E[Y] - E[Y]^2 \\
&= E[X^2] - E[X]^2 + E[Y^2] - E[Y]^2 \\
&= \mathrm{Var}[X] + \mathrm{Var}[Y] \qquad \square
\end{aligned}
$$

3.2 Die Jensen'sche Ungleichung

Eine Funktion $f : \mathbb{R} \to \mathbb{R}$ heißt *konvex*, wenn für alle $\lambda \in [0,1]$ und $x, y \in \mathbb{R}$ folgende Ungleichung gilt:

$$f((1 - \lambda)x + \lambda y) \leq (1 - \lambda)f(x) + \lambda f(y)$$

Konvexität bedeutet, dass, wenn man in der Ebene $\mathbb{R}^2$ die Strecke von dem Punkt $(x, f(x))$ zum Punkt $(y, f(y))$ zieht, diese oberhalb des Graphen von f liegt. Das Schaubild einer konvexen Funktion f sieht etwa wie folgt aus:

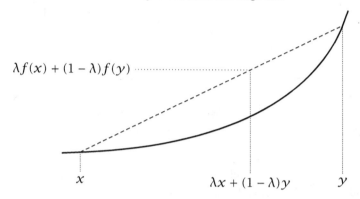

Eine zweimal differenzierbare Funktion f ist genau dann konvex, wenn die zweite Ableitung f'' nirgends negativ ist. Die Funktionen $f(x) = x^2$ und $g(x) = 2^x$ sind jeweils konvex. Die zweiten Ableitungen sind $f''(x) = 2$ und $g''(x) = (\ln 2)^2 \cdot 2^x$ und damit nirgends negativ.

Die Beziehung 3.5 ist nach Johan Ludwig William Valdemar Jensen (1859–1925) benannt.

Satz 3.5 (Jensen'sche Ungleichung). *Sei $f: \mathbb{R} \to \mathbb{R}$ eine konvexe Funktion und $k \geq 1$. Seien $\lambda_1, \ldots, \lambda_k \in [0,1] \subseteq \mathbb{R}$ mit $\sum_{i=1}^{k} \lambda_i = 1$. Dann gilt:*

$$f\left(\sum_{i=1}^{k} \lambda_i x_i\right) \leq \sum_{i=1}^{k} \lambda_i f(x_i)$$

Beweis. Ohne Einschränkung gilt $\lambda_i > 0$ für alle $1 \leq i \leq k$. Wir führen eine Induktion nach k. Für $k = 1$ ist $\lambda_1 = 1$, und die Aussage ist erfüllt. Sei also $k > 1$ und $\lambda_1 < 1$. Damit gilt jetzt:

$$f\left(\sum_{i=1}^{k} \lambda_i x_i\right) = f\left(\lambda_1 x_1 + (1 - \lambda_1) \sum_{i=2}^{k} \frac{\lambda_i}{1 - \lambda_1} x_i\right)$$

$$\leq \lambda_1 f(x_1) + (1 - \lambda_1) f\left(\sum_{i=2}^{k} \frac{\lambda_i}{1 - \lambda_1} x_i\right) \qquad \text{da } f \text{ konvex}$$

$$\overset{IV}{\le} \lambda_1 f(x_1) + (1 - \lambda_1) \sum_{i=2}^{k} \frac{\lambda_i}{1 - \lambda_1} f(x_i)$$

$$= \sum_{i=1}^{k} \lambda_i f(x_i) \qquad \qquad \square$$

Ist $X : \Omega \to \mathbb{R}$ eine Zufallsvariable und $f : \mathbb{R} \to \mathbb{R}$ eine Funktion, so bezeichnet $f(X) : \Omega \to \mathbb{R}$ die Zufallsvariable mit $f(X)(\omega) = f(X(\omega))$. Es gilt:

$$E(f(X)) = \sum_{y} y \, \Pr[f(X) = y]$$

$$= \sum_{x} y \sum_{y = f(x)} \Pr[X = x]$$

$$= \sum_{x} f(x) \Pr[X = x]$$

Dies ermöglicht die Bestimmung des Erwartungswertes von $f(X)$, ohne die Dichte von $f(X)$ explizit zu bestimmen. Wir wenden das Korollar 3.6 in Abschnitt 4.10 mit der konvexen Funktion 2^x an, um die mittlere Höhe binärer Suchbäume zu berechnen.

Korollar 3.6. *Sei $f : \mathbb{R} \to \mathbb{R}$ eine konvexe Funktion und X eine Zufallsvariable auf einem endlichen Wahrscheinlichkeitsraum. Dann gilt:*

$$f(E[X]) \le E[f(X)]$$

Beweis. Es sei $X : \Omega \to \mathbb{R}$ die Zufallsvariable. Wir können annehmen, dass $X(\Omega) = \{x_1, \dots, x_k\}$ mit $\Pr[X = x_i] = \lambda_i$ gilt. Nach der Jensen'schen Ungleichung gilt:

$$f(E[X]) = f\left(\sum_{i=1}^{k} \lambda_i x_i \right) \le \sum_{i=1}^{k} \lambda_i f(x_i) = E[f(X)] \qquad \square$$

Bemerkung 3.7. Die Erfahrung lehrt, dass man sich zwar gut merken kann, dass für konvexe Funktionen f eine Ungleichung zwischen den Werten $f(E[X])$ und $E[f(X)]$ besteht, aber dass man sich weniger gut die Richtung der Ungleichung merken kann. Gilt $f(E[X]) \le E[f(X)]$ oder $f(E[X]) \ge E[f(X)]$? Hier hilft die Erinnerung an die Varianz; diese ist durch $E[X^2] - E[X]^2$ definiert, sie ist positiv und $x \mapsto x^2$ ist eine konvexe Funktion. Also gilt $f(E[X]) \le E[f(X)]$. ◇

3.3 Das Geburtstagsparadoxon

Eine Kurvendiskussion der Funktion $(1 + x) - e^x$ ergibt, dass $(1 + x) \le e^x$ für alle x mit Gleichheit nur bei $x = 0$ gilt (siehe Übungsaufgabe 2.1.(b)). Falls x nahe bei

Null ist, erhalten wir eine durchaus brauchbare Abschätzung. Diese wichtige Technik erklärt das *Geburtstagsparadoxon*:

> Sind mehr als 23 Personen auf einer Party, so ist die Wahrscheinlichkeit größer als $1/2$, dass zwei Gäste am gleichen Tag Geburtstag haben.

Das Beiwort *Paradoxon* kommt daher, dass die Zahl 23 bei maximal 366 möglichen Geburtstagen pro Jahr auf den ersten Blick viel zu klein erscheint, um diese Wahrscheinlichkeit vorherzusagen. Aber schauen wir es uns genauer an. Angenommen, wir haben n mögliche Geburtstage und m Gäste. Stellen wir die Gäste in eine Reihe und jeder nennt seinen Geburtstag, so erhalten wir eine Zufallsfolge (na ja, wenigstens so halbwegs). Die Wahrscheinlichkeit, dass die ersten $i + 1$ Folgenglieder alle verschieden sind, ist dann:

$$\frac{n}{n} \cdot \frac{n-1}{n} \cdots \frac{n-i}{n} = 1 \cdot \left(1 - \frac{1}{n}\right) \cdots \left(1 - \frac{i}{n}\right)$$

Die Wahrscheinlichkeit, dass alle m Geburtstage verschieden sind, ist daher:

$$\prod_{i=0}^{m-1} \left(1 - \frac{i}{n}\right)$$

Haben wir bisher einen Fehler gemacht? Nun, die Annahme einer Zufallsfolge bedeutet eine Gleichverteilung, von der die Realität womöglich abweicht. Es ist jedoch intuitiv klar, dass wir auf der sicheren Seite sind (wenn sich die Wahrscheinlichkeit bei gewissen Tagen häuft, dann wird es leichter, eine Übereinstimmung zu erreichen). Außerdem werden wir den Ausdruck jetzt noch vergrößern. Im nächsten Schritt verwenden wir die oben erwähnte Ungleichung $(1 + x) \le e^x$. Damit ergibt sich für die Wahrscheinlichkeit, dass alle Geburtstage verschieden sind, folgende Abschätzung

$$\prod_{i=0}^{m-1} \left(1 - \frac{i}{n}\right) \le \prod_{i=0}^{m-1} e^{-\frac{i}{n}} = e^{-\sum_{i=0}^{m-1} \frac{i}{n}} = e^{-\frac{m(m-1)}{2n}}$$

Der Grenzwert $1/2$ wird also spätestens im Bereich von $m = \sqrt{2n\ln 2}$ unterschritten. Für $n = 365$ (oder 366) ist dies 23. Experimente auf Geburtstagsfeiern und in Vorlesungen bestätigen diesen Wert sehr gut.

Aufgaben

3.1. Ein Jäger hat die Treffsicherheit $1/2$. Wie groß ist die Wahrscheinlichkeit, dass er bei 10 Schüssen mindestens 3 Treffer landet?

3.2. Eine Familie hat vier Kinder. Gehen Sie davon aus, dass die Wahrscheinlichkeit, ein Mädchen zu bekommen bei $0,5$ liegt, und berechnen Sie die Wahrscheinlichkeit, dass

(a) die Familie genau ein Mädchen hat,

(b) das erste und zweite Kind ein Junge ist,

(c) mindestens zwei Kinder männlich sind,

(d) alle Kinder weiblich sind.

3.3. Seien $m, n \in \mathbb{N}$ mit $n < m$. Alice und Bob denken sich jeweils unabhängig voneinander eine Zahl aus der Menge $M = \{1, 2, \ldots, m\}$ aus. Wie groß ist die Wahrscheinlichkeit, dass sich die beiden Zahlen höchstens um n unterscheiden? Bestimmen Sie hierzu die Mächtigkeit der Menge

$$\{ (a, b) \mid a, b \in M \quad \text{und} \quad |a - b| \leq n \}$$

3.4. Wir wollen eine Folge von unterschiedlichen Zahlen $a = (a_1, \ldots, a_n)$ mittels *Quicksort* sortieren. Hierfür wählen wir ein zufälliges Pivotelement a_i und bilden die Teilsequenzen $a' = (a_{i_1}, \ldots, a_{i_k})$ und $a'' = (a_{j_1}, \ldots, a_{j_\ell})$ mit

- $a_{i_s} < a_i < a_{j_t}$ für alle $1 \leq s \leq k$ und alle $1 \leq t \leq \ell$,

- $i_1 < \cdots < i_k$ und $j_1 < \cdots < j_\ell$ und $k + \ell + 1 = n$.

Dies ist mit $n-1$ Vergleichen möglich („pivotieren"). Danach werden a' und a'' rekursiv sortiert zu b' und b''. Hieraus ergibt sich durch (b', a_i, b'') die Sortierung von a. Die Rekursion bricht ab, wenn $n = 0$ gilt. Wieviele Vergleiche benötigt Quicksort im Durchschnitt?

3.5. Sei wieder $a = (a_1, \ldots, a_n)$ eine Folge unterschiedlicher Zahlen. Wir wollen das k-t größte Element bestimmen, ohne vorher die Folge zu sortieren. Wir gehen dafür ähnlich wie bei Quicksort aus Aufgabe 3.4. vor. Wir wählen zufällig ein Pivotelement p und bilden damit erneut die beiden Teilsequenzen der Elemente, die kleiner bzw. größer als p sind. Wir können gleichzeitig die Anzahl der Elemente in der vorderen Teilsequenz festhalten und dann entscheiden, ob wir das gesuchte Element bereits mit p gefunden haben oder in welcher der beiden Listen das gesuchte Element zu bestimmen ist. Die Prozedur nennt man *Quickselect*. Zeigen Sie, dass die mittlere Zahl der Vergleiche $Q(n)$ bei Quickselect durch $2(1 + \ln 2)n$ begrenzt werden kann. *Hinweis*: Nehmen Sie an, dass die Folge a aus den Zahlen $1, \ldots, n$ besteht und dass die Position des Elements k bestimmt werden soll. Bezeichnet π eine Reihenfolge der Pivotelemente, so benutzen Sie die 0-1-wertigen Zufallsvariablen $X_{ij}(\pi) = $ „i wird mit j verglichen". Unterscheiden Sie drei Fälle, je nachdem wie k zu i und j steht.

3.6. Sei $n \geq 1$ und $H_n = \sum_{k=1}^{n} \frac{1}{k}$. Gegeben sei eine Zufallsvariable $X : \Omega \to \{1, \ldots, n\}$ mit der Zipf-Verteilung $\Pr[X = k] = (H_n \cdot k)^{-1}$. Sie ist nach *George Kingsley Zipf* (1902–1950) benannt, der empirisch feststellte, dass in natürlichsprachlichen

Texten das k-t häufigste Wort mit einer Wahrscheinlichkeit proportional zu $1/k$ auftritt. Berechnen Sie die Asymptotik des Erwartungswerts und der Standardabweichung von X.

Zusammenfassung

Begriffe

– (diskreter) Wahrscheinlichkeitsraum	– Verteilung F_X
– Gleichverteilung	– unabhängige Zufallsvariablen
– Wahrscheinlichkeit $\Pr[A]$	– Varianz $\mathrm{Var}[X]$
– Zufallsvariable X	– Bernoulli-Experiment
– Erwartungswert $E[X]$	– Standardabweichung σ_X
– diskrete Dichte f_X	– konvexe Funktion

Methoden und Resultate

- Ω endlich, gleichverteilt $\Rightarrow \Pr[A] = \frac{|A|}{|\Omega|}$
- $E[X] = \sum_{\omega \in \Omega} X(\omega)\Pr[\omega]$
- Ω endlich, gleichverteilt $\Rightarrow E[X] = \left(\sum_\omega X(\omega)\right)/|\Omega|$
- Markov-Ungleichung: $X \geq 0, E[X] > 0, \lambda > 0 \Rightarrow \Pr[X \geq \lambda E[X]] \leq \frac{1}{\lambda}$
- Linearität des Erwartungswertes: $E[aX + bY] = aE[X] + bE[Y]$
- $E[X] = \sum_x x \Pr[X = x] = \sum_x x f_X(x)$
- X, Y unabhängig $\Rightarrow E[XY] = E[X]E[Y]$
- $\mathrm{Var}[X] = E[(X - E[X])^2] = E[X^2] - E[X]^2 \geq 0$
- $\sigma_X = \sqrt{\mathrm{Var}[X]}$
- Tschebyschev-Ungleichung: Für $\lambda > 0$ gilt $\Pr[|X - E[X]| \geq \lambda\,\sigma_X] \leq \frac{1}{\lambda^2}$
- X, Y unabhängig $\Rightarrow \mathrm{Var}[X + Y] = \mathrm{Var}[X] + \mathrm{Var}[Y]$
- Jensen'sche Ungleichung: $f : \mathbb{R} \to \mathbb{R}$ konvex, $\lambda_i \in [0,1]$, $\sum_{i=1}^{k} \lambda_i = 1$
  $\Rightarrow f\left(\sum_{i=1}^{k} \lambda_i x_i\right) \leq \sum_{i=1}^{k} \lambda_i f(x_i)$
- Ω endlich, f konvex $\Rightarrow f(E[X]) \leq E[f(X)]$
- Geburtstagsparadoxon: Für zufällige Folgen von m Ereignissen aus Ω mit $m \geq \sqrt{2|\Omega|\ln 2}$ ist $\Pr[\text{zwei gleiche Folgenglieder}] > 1/2$.

4 Kombinatorik

Wir beginnen diesen Abschnitt mit einer kurzen Einführung in die *abzählende Kombinatorik*. Danach lernen wir anhand der Binomialkoeffizienten das kombinatorische Prinzip eines *bijektiven Beweises* kennen. Die Idee ist, eine Identität der Form $f(n) = g(n)$ dadurch zu beweisen, dass man einerseits Mengen F und G findet mit $|F| = f(n)$ und $|G| = g(n)$, und andererseits eine Bijektion zwischen F und G nachweist. Den Schritt von einer Funktion $f(n)$ zu einer Menge F mit $|F| = f(n)$ bezeichnet man als *kombinatorische Interpretation*. Zwei Mengen sind disjunkt, falls ihr Durchschnitt leer ist. Der folgende Zusammenhang zwischen den Mengen F und G ist typisch. Man zerlegt die Menge F in Klassen, also schreibt man F als eine Vereinigung paarweise disjunkter Teilmengen G_k mit $|G_k| = g_k(n)$. Die Bijektion zwischen F und $\bigcup_k G_k$ ist die Identität. Aus $F = \bigcup_k G_k$ folgt dann $f(n) = \sum_k g_k(n)$.

Eine weit verbreitete kombinatorische Interpretation ist das *Urnenmodell* von Pólya (George Pólya, 1887–1985). In diesem Modell werden Kugeln aus einem Gefäß (der Urne) gezogen und man zählt, auf wie viele Weisen dies geschehen kann. Bei der Methode des Ziehens von Kugeln kann man unter verschiedenen Modi unterscheiden, was zu verschiedenen Zählfunktionen führt. Ein Vorteil dieses Modells ist, dass sich dadurch verschiedene Zählfunktionen einheitlich interpretieren lassen. Nun lassen sich zum einen nicht alle für uns interessanten Funktionen in diesem Modell darstellen, und zum anderen gibt es häufig bessere und naheliegendere Interpretationen, daher verfolgen wir einen allgemeineren Ansatz.

4.1 Abzählende Kombinatorik

Wir schreiben $|A| = |B|$, falls eine Bijektion zwischen A und B existiert; wir sagen dann, A und B sind *gleichmächtig*. Wir können $|A|$ als Anzahl der Elemente in A interpretieren. Ist A endlich und enthält A genau n Elemente, so schreiben wir $|A| = n$. Natürlich bleibt $|A| = |n|$ richtig, denn der Betrag $|n|$ ist die Anzahl der Elemente von n, wenn man die Zahl $n \in \mathbb{N}$ z. B. durch die Menge $\{0, \dots, n-1\}$ definiert.

Die Menge aller Abbildungen von A nach B bezeichnen wir mit B^A. Dies ist sinnvoll, denn eine Abbildung $f : A \to B$ kann mit einem A-Tupel $(b_a)_{a \in A}$ identifiziert werden, indem man $b_a = f(a)$ für alle $a \in A$ setzt. Sind A und B endliche Mengen mit $|A| = n$ und $|B| = m$, dann gibt es genau m^n Abbildungen von A nach B; für jedes der n Elemente $a \in A$ gibt es m mögliche Bilder $f(a) \in B$. Wir halten fest:

$$\left| \{ f : A \to B \mid f \text{ ist Abbildung} \} \right| = |B^A| = |B|^{|A|} = m^n$$

Was passiert, wenn sowohl A als auch B leer ist? Dann steht links die Zahl 1, da für $A = B = \emptyset$ die Identität die einzige Abbildung von A nach B ist, und rechts der Ausdruck 0^0. Es ergibt sich also auf natürliche Weise, dass $0^0 = 1$ gesetzt wird. Damit gilt $x^0 = 1$ für jede Zahl x. Dies gehört zu einer ganzen Palette sinnvoller Konventio-

nen. So ist ein leeres Produkt stets 1, eine leere Summe stets 0, also $\prod_{k\in\varnothing} a_k = 1$ und $\sum_{k\in\varnothing} a_k = 0$. Das Analogon aus der Prädikatenlogik ist, dass eine *für-alle*-quantifizierte Aussage $\forall x \in \varnothing : \varphi(x)$ über der leeren Menge stets wahr ist, und $\exists x \in \varnothing : \varphi(x)$ ist stets falsch.

Angenommen, es gilt $|A| = |B| = n$. Wie viele Bijektionen zwischen A und B gibt es dann? Die Antwort ist $n!$ (gesprochen: „*n-Fakultät*"), wobei die Fakultät definiert ist durch $n! = n \cdot (n-1) \cdots 1$. Mit unserer Konvention, dass leere Produkte 1 sind, ergibt sich $0! = 1$. Die Behauptung ist also:

$$|\{ f \in B^A \mid f \text{ ist bijektiv} \}| = n!$$

Für $n = 0$ ist dies richtig. Allgemein müssen wir für beliebiges n die Tupel $(b_i)_{1\le i\le n}$ mit paarweise verschiedenen $b_i \in B$ zählen. Für b_1 gibt es n Möglichkeiten, für b_2 gibt es noch $n-1$ Möglichkeiten und so fort. Auf diese Weise erhalten wir $n!$ Möglichkeiten.

Dieser Ansatz lässt sich auf beliebige endliche Mengen mit $|A| = k$ und $|B| = n$ verallgemeinern:

$$|\{ f \in B^A \mid f \text{ ist injektiv} \}| = n(n-1) \cdots (n-k+1) \qquad (4.1)$$

$$= \frac{n!}{(n-k)!} \quad \text{für } k \le n$$

Dies ist erneut leicht einzusehen. Wieder müssen wir die Tupel $(b_i)_{1\le i\le n}$ mit paarweise verschiedenen $b_i \in B$ zählen. Für b_1 gibt es n Möglichkeiten, für b_2 gibt es noch $n-1$ Möglichkeiten und so fort, bis für b_k noch $(n-k+1)$ Möglichkeiten verbleiben.

Der Anteil der Bijektionen unter allen Abbildungen von $\{1,\dots,n\}$ nach $\{1,\dots,n\}$ nimmt exponentiell ab, wenn n wächst: Mit Gleichung (2.1) sehen wir

$$\frac{1}{e^{n-1}} \le \frac{n!}{n^n} \le \frac{n}{e^{n-1}}$$

und mit der Stirling'schen Formel (2.2) erhalten wir sogar $\frac{n!}{n^n} \sim e^{-n}\sqrt{2\pi n}$. Hier und an vielen anderen Stellen ist π die Kreiszahl und e die Euler'sche Zahl:

$$\pi = 3{,}14159\,26535\,89793\,23846\,26433\,83279\,50288\,41971\,69399\,37510\cdots$$

$$e = 2{,}71828\,18284\,59045\,23536\,02874\,71352\,66249\,77572\,47093\,69995\cdots$$

Die Menge aller Teilmengen von A heißt die *Potenzmenge* von A und wird mit 2^A bezeichnet. Die Bezeichnung verdeutlicht, dass eine Teilmenge $B \subseteq A$ mit ihrer *charakteristischen Abbildung* $\chi_B : A \to \{0, 1\} = 2$ identifiziert werden kann. Dabei gilt $\chi_B(a) = 1$, falls $a \in B$ und $\chi_B(a) = 0$ sonst. Wir erkennen

$$|2^A| = 2^{|A|}$$

Enthält die Menge A also n Elemente, so gibt es 2^n Teilmengen von A. Es gilt $n < 2^n$ für alle $n \in \mathbb{N}$. Diese Beobachtung ist ein Spezialfall der mengentheoretischen Aussage, dass keine Surjektion einer Menge auf ihre Potenzmenge existiert (siehe Übungsaufgabe 4.1. (c)); die Potenzmenge ist also immer „größer". Daher kann es auch nicht die *Menge aller Mengen* geben, denn, salopp gesagt, dies wäre die größte aller Mengen, aber ihre Potenzmenge wäre noch größer.

4.2 Binomialkoeffizienten

In Abschnitt 2.2 hatten wir schon erwähnt, dass der Binomialkoeffizient $\binom{n}{k} = \frac{n!}{k!(n-k)!}$ die Anzahl der k-elementigen Teilmengen einer Menge von n Elementen wiedergibt und dies seine kombinatorische Interpretation ist. Als Nächstes erweitern wir diese kombinatorische Interpretation auf $k \in \mathbb{Z}$ und beliebige Mengen A. Wir bezeichnen mit $\binom{A}{k}$ die Menge der k-elementigen Teilmengen von A:

$$\binom{A}{k} = \{B \subseteq A \mid |B| = k\}$$

Offensichtlich ist $\binom{A}{k} = \varnothing$, falls $k < 0$ oder $k > |A|$ gilt. Andererseits gilt stets $\binom{A}{0} = \{\varnothing\}$. Sei jetzt A endlich mit $|A| = n$. Es gilt $|\binom{A}{1}| = n$, da wir die einelementigen Teilmengen von A mit den Elementen von A identifizieren können. Zwischen den Mengen A und $\binom{A}{1}$ gibt es also eine Bijektion. Außerdem gibt es auch für die Mengen $\binom{A}{k}$ und $\binom{A}{n-k}$ eine Bijektion: Wir müssen nur jeder Teilmenge $B \in \binom{A}{k}$ ihr Komplement $A \setminus B \in \binom{A}{n-k}$ zuordnen.

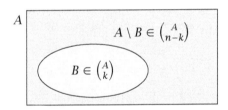

Hieraus folgt: $|\binom{A}{k}| = |\binom{A}{n-k}|$ für alle $k \in \mathbb{Z}$. Die Potenzmenge von A enthält 2^n Elemente, gleichzeitig ist sie die disjunkte Vereinigung aller $\binom{A}{k}$. Daher ergibt sich unmittelbar und ohne eine weitere Rechnung:

$$2^n = \sum_k \left| \binom{A}{k} \right|$$

Da $\binom{A}{k}$ und $\binom{x}{k}$ beides Standardbezeichnungen sind, bleibt eigentlich gar nichts anderes übrig, als dass Satz 4.1 gilt, den wir jetzt formal beweisen.

Satz 4.1. *Sei A eine Menge mit n Elementen. Dann gilt:*

$$\left| \binom{A}{k} \right| = \binom{n}{k}$$

Beweis. Der Satz ist richtig für $k < 0$ oder $k > n$, dann sind nämlich beide Terme 0. Für $0 \leq k \leq n$ gibt es $n(n-1) \cdots (n-k+1)$ Folgen $(a_1, \ldots, a_k)$ mit paarweise verschiedenen a_i. Man beachte die Konvention $n(n-1) \cdots (n-k+1) = 1$, falls $k = 0$ ist. Zwei solcher Folgen repräsentieren genau dann dieselbe Menge, wenn die Folgen bis auf eine Permutation (d. h. Vertauschung) der Indizes übereinstimmen. Es gibt $k!$ solcher Permutationen, also ist der Satz bewiesen. □

Wir erweitern den Definitionsbereich von Binomialkoeffizienten $\binom{x}{k}$ auf komplexe Zahlen x und ganze Zahlen k wie folgt:

$$\binom{x}{k} = \frac{x(x-1) \cdots (x-k+1)}{k!} \qquad \text{für } x \in \mathbb{C} \text{ und } k \in \mathbb{N}$$

Das Produkt im Zähler der oberen Zeile nennt man die *fallende Faktorielle*; es wird mit $x^{\underline{k}}$ bezeichnet. Also gilt stets:

$$\binom{x}{k} = \frac{x^{\underline{k}}}{k!}, \quad \text{wobei } x^{\underline{k}} = x(x-1) \cdots (x-k+1)$$

Wir beachten, dass dies für $x = n \in \mathbb{N}$ die übliche Definition $\binom{n}{k} = \frac{n!}{k!(n-k)!}$ liefert. Für $k > 0$ stehen im Zähler und Nenner von $\binom{x}{k}$ jeweils k Faktoren. Für $k = 0$ steht in $x^{\underline{k}}$ das leere Produkt; dies ist nach der obigen Konvention 1, also $x^{\underline{0}} = \binom{x}{0} = 1$. Sind k, x natürliche Zahlen mit $x < k$, so durchläuft das Produkt $x(x-1) \cdots (x - k+1)$ die Null, also gilt $x^{\underline{k}} = \binom{x}{k} = 0$ für $x \in \mathbb{N}$ mit $0 \leq x < k$. Für alle anderen x und $k \geq 0$ wird die Null nicht getroffen, also gilt $x^{\underline{k}} \neq 0 \neq \binom{x}{k}$ für $x \notin \mathbb{N}$ und $0 \leq k$. So gilt zum Beispiel $\binom{1/10}{4} < 0$ und $\binom{1/10}{5} > 0$; insbesondere sind beide Werte nicht Null.

Als Nächstes erweitern wir den Definitionsbereich für alle $k \in \mathbb{Z}$ und $x \in \mathbb{C}$ durch $\binom{x}{k} = 0$, falls k negativ ist. Insbesondere gilt:

$$\binom{n}{k} = \binom{n}{n-k} \qquad \text{für } n \in \mathbb{N} \text{ und } k \in \mathbb{Z}$$

Für $k \geq 0$ gilt die hübsche Beziehung:

$$\binom{-1}{k} = \frac{(-1) \cdot (-2) \cdots (-k)}{k!} = (-1)^k$$

Wir können den Binomialkoeffizienten $\binom{x}{k}$ als ein Polynom in x vom Grad k auffassen mit den Nullstellen $0, \ldots, k-1$. Dabei ist $\binom{x}{0}$ das konstante Polynom mit Wert 1.

Dies führt direkt zur *Polynommethode*. Wenn wir eine Identität für Binomialkoeffizienten beweisen möchten, bei denen nur $\binom{x}{m}$ mit $m \leq k$ auftreten, so ist dies eine Identität für Polynome in x (mit Koeffizienten in $\mathbb{C}$) vom Grad kleiner oder gleich k. Ein Satz der Algebra besagt, dass zwei verschiedene Polynome (mit Koeffizienten aus $\mathbb{C}$) vom Grad kleiner oder gleich k schon dann gleich sind, wenn sie an $k + 1$ verschiedenen Stellen übereinstimmen. Kann man die Identität dann für mindestens $k + 1$ natürliche Zahlen x zeigen, so gilt sie automatisch für alle $x \in \mathbb{C}$. Dies ist Leitmotiv für dieses Kapitel:

> *Beweise (sofern möglich) Identitäten mit Hilfe einer kombinatorischen Interpretation und versuche dann die Polynommethode.*

Der Ansatz ist extrem hilfreich, denn er vermittelt ein *Verständnis* für Identitäten. Außerdem erspart er einige Induktionsbeweise. Diese eignen sich zwar vielfach sehr gut zum Nachvollziehen von Identitäten, sind aber kaum geeignet, Identitäten zu finden oder im Kopf zu behalten.

Dadurch, dass Binomialkoeffizienten außerdem für alle $k \in \mathbb{Z}$ definiert sind, können wir uns oft Summationsgrenzen ersparen, was die Formeln übersichtlicher macht und Induktionsbeweise vereinfacht. Für den Rest dieses Abschnittes sind x, y stets komplexe Zahlen (oder Unbekannte) und k, ℓ, m, n stets ganze Zahlen.

Binomialkoeffizienten $\binom{n}{k}$ sind für $n, k \in \mathbb{N}$ selbst natürliche Zahlen, aber dies ist der Darstellung $\binom{n}{k} = \frac{n!}{k!(n-k)!}$ nicht sofort anzusehen. Mit Satz 4.1 wird diese Aussage trivial. Die folgende Identität ist die Grundlage für das *Pascal'sche Dreieck* (Blaise Pascal, 1623–1662) und die vielleicht wichtigste Eigenschaft von Binomialkoeffizienten (siehe Abbildung 4.1). Hieraus lässt sich ebenfalls die Ganzzahligkeit der Werte $\binom{n}{k}$ für $n, k \in \mathbb{Z}$ ableiten.

Satz 4.2 (Additionstheorem).

$$\binom{x}{k} = \binom{x-1}{k} + \binom{x-1}{k-1}$$

Beweis. Mit Hilfe der kombinatorischen Interpretation (Satz 4.1) ist die Identität direkt zu verstehen, wenn $x = n$ eine natürliche Zahl ist: Wir können annehmen, dass $A = \{1, \ldots, n\}$ gilt. Die Menge der k-elementigen Teilmengen von A zerlegt sich in zwei Klassen. Diejenigen, die das Element n enthalten und diejenigen, die es nicht tun. Wir wenden jetzt Satz 4.1 an: Von der ersten Sorte gibt es so viele, wie es $(k-1)$-elementige Teilmengen von $A \setminus \{n\}$ gibt, dies sind $\binom{n-1}{k-1}$. Von der zweiten Sorte gibt es so viele, wie es k-elementige Teilmengen von $A \setminus \{n\}$ gibt, dies sind $\binom{n-1}{k}$. Die Summe muss $\binom{n}{k}$ sein. Daraus ergibt sich die Identität zunächst für alle $x \in \mathbb{N}$. Die behauptete Identität gilt also für unendlich viele Werte. Da links und rechts Polynome vom Grad k stehen, gilt sie für alle $x \in \mathbb{C}$ nach der Polynommethode. □

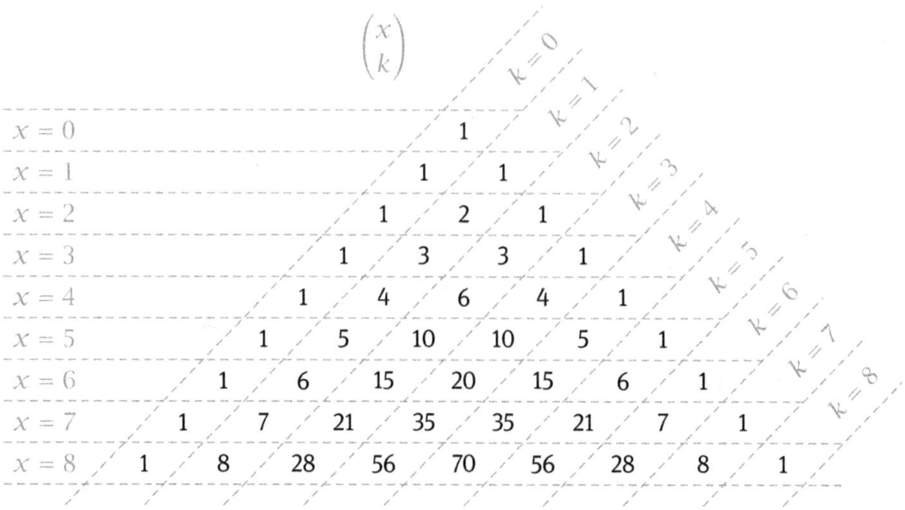

$$\binom{x}{k}$$

$x = 0$					1					
$x = 1$					1	1				
$x = 2$				1	2	1				
$x = 3$			1	3	3	1				
$x = 4$		1	4	6	4	1				
$x = 5$	1	5	10	10	5	1				
$x = 6$	1	6	15	20	15	6	1			
$x = 7$	1	7	21	35	35	21	7	1		
$x = 8$	1	8	28	56	70	56	28	8	1	

$k = 0, k = 1, k = 2, k = 3, k = 4, k = 5, k = 6, k = 7, k = 8$

Abb. 4.1. Pascal'sches Dreieck.

Eine unmittelbare Konsequenz des Addititionstheorems ist die Ganzzahligkeit der Binomialkoeffizienten $\binom{n}{k}$ für $n \in \mathbb{Z}$, die dem Bruch $\frac{n\cdots(n-k+1)}{k!}$ nicht direkt anzusehen ist. Eine weitere einfache Folgerung aus Satz 4.2 ist der Binomialsatz. Hierbei bedeutet $\sum_k$, dass wir über alle $k \in \mathbb{Z}$ summieren. Dies ist erlaubt, da in den betrachteten Summen fast alle Terme 0 sind.

Satz 4.3 (Binomialsatz).

$$(x + y)^n = \sum_k \binom{n}{k} x^k y^{n-k}$$

Beweis. Wir betrachten das Produkt in Unbestimmten x und y

$$(x + y)^n = \underbrace{(x + y)(x + y) \cdots (x + y)}_{n \text{ Faktoren}}$$

Den Term $x^k y^{n-k}$ können wir erzeugen, indem wir in k der n Faktoren den Summanden x wählen und in den übrigen $n - k$ Faktoren den Summanden y. Damit liefert jede k-elementige Teilmenge der n Faktoren den Term $x^k y^{n-k}$. Nach Satz 4.1 gibt es $\binom{n}{k}$ solcher Teilmengen. Also ist $\binom{n}{k}$ der Koeffizient von $x^k y^{n-k}$. □

Eine weitere wichtige Identität ist die *trinomiale Revision*. Die Namensgebung wurde entsprechend in Anlehnung an das Buch von Graham, Knuth und Patashnik gewählt [22]. Sie beruht darauf, dass für $x = k + \ell + n$ und $m = k + \ell$ das Produkt $\binom{x}{m}\binom{m}{k} = \binom{k+\ell+n}{k+\ell}\binom{k+\ell}{k}$ zum *Trinomialkoeffizienten* $\frac{(k+\ell+n)!}{k!\,\ell!\,n!}$ wird. Die trinomiale

Revision erlaubt es, Produkte von Binomialkoeffizienten zu vereinfachen:

Satz 4.4 (Trinomiale Revision).

$$\binom{x}{m}\binom{m}{k} = \binom{x}{k}\binom{x-k}{m-k} \text{ für } x \in \mathbb{C} \text{ und } m,k \in \mathbb{Z}$$

Beweis. Für $m < 0$ oder $m < k$ steht auf beiden Seiten 0. Nach der Polynommethode reicht es daher, die Aussage für $0 \leq k \leq m \leq n = x \in \mathbb{N}$ zu zeigen. Stellen wir uns vor, dass wir n Kugeln haben und hiervon k rot, $m - k$ grün und die restlichen $n - m$ blau färben möchten. Dann können wir uns zunächst für die Teilmenge der rot oder grün gefärbten entscheiden und danach unter diesen die roten wählen. Oder wir entscheiden uns erst nur für die Teilmenge der roten Kugeln und wählen danach unter den restlichen $n - k$ Kugeln die grünen. □

Wenn man eine Identität von der Form $f_n = \sum_k \binom{n}{k} g_k$ nach g_k umformen will, dann kann man sich des folgenden Tricks bedienen. Betrachte die $(n + 1) \times (n + 1)$ Matrizen P und Q mit Einträgen $P_{ij} = \binom{i}{j}$ und $Q_{ij} = (-1)^{i-j}\binom{i}{j}$. Die Indizes sind aus $\{0,\ldots,n\}$. Beide Matrizen P und Q sind untere Dreiecksmatrizen. Die Matrix P ist ein Ausschnitt des Pascal'schen Dreiecks. Für das Produkt $R = PQ$ gilt

$$R_{ij} = \sum_k P_{ik}Q_{kj} = \sum_k \binom{i}{k}(-1)^{k-j}\binom{k}{j}$$

Insbesondere ist $R_{ij} = 0$ für $i < j$. Für $i \geq j$ ergibt sich mit der trinomialen Revision:

$$R_{ij} = \binom{i}{j}\sum_k \binom{i-j}{k-j}(-1)^{k-j} = \binom{i}{j}\sum_k \binom{i-j}{k}(-1)^k$$

$$= \binom{i}{j}(-1+1)^{i-j} = \begin{cases} 1 & \text{für } i - j \\ 0 & \text{für } i > j \end{cases}$$

Also ist R die Einheitsmatrix. Dies liefert Satz 4.5.

Satz 4.5 (Binomialinversion). *Seien $f_0,\ldots,f_n$ und $g_0,\ldots,g_n$ Zahlen, so dass $f_i = \sum_k \binom{i}{k} g_k$ für alle $0 \leq i \leq n$ gilt. Dann ist $g_n = \sum_k (-1)^{n-k}\binom{n}{k}f_k$.*

Beweis. Mit den obigen Matrizen gilt $(f_0,\ldots,f_n) = (g_0,\ldots,g_n) \cdot P^t$. Es folgt $(g_0,\ldots,g_n) = (f_0,\ldots,f_n) \cdot Q^t$ und damit die Behauptung. Hierbei bezeichnet P^t die zu P transponierte Matrix mit dem Eintrag P_{ij} an der Stelle (j,i). □

Im nächsten Beispiel fassen wir einige wichtige kombinatorische Interpretationen zusammen.

Beispiel 4.6. Eine Urne enthalte n Kugeln, die von 1 bis n nummeriert sind. Wir ziehen k Kugeln. Insbesondere gilt $k \leq n$, wenn wir keine Kugel zurücklegen. Dann gibt es

(a) $n^{\underline{k}} = \frac{n!}{(n-k)!}$ Ziehungen ohne Zurücklegen und mit Reihenfolge,

(b) $\binom{n}{k}$ Ziehungen ohne Zurücklegen und ohne Reihenfolge,

(c) n^k Ziehungen mit Zurücklegen und mit Reihenfolge, und

(d) $\binom{n+k-1}{k}$ Ziehungen mit Zurücklegen und ohne Reihenfolge.

Zurücklegen bedeutet hierbei, dass jede Kugel nach dem Ziehen vermerkt und wieder zu den Kugeln in der Urne zurück gelegt wird. Mit Reihenfolge bedeutet, dass man zwischen den verschiedenen Reihenfolgen unterscheidet, in denen die Kugeln gezogen werden. Aus der ersten Formel sehen wir zum Beispiel, dass es 336 mögliche Verteilungen der ersten 3 Plätze bei einem Rennen mit 8 Teilnehmern gibt, denn es ist $8 \cdot 7 \cdot 6 = 336$.

Die ersten drei Formeln haben wir bereits unter den Stichworten Injektionen, k-elementige Teilmengen und beliebige Abbildungen behandelt. Die Formel in Situation (d) kann man wie folgt einsehen. Ein Auswahl von Elementen mit Zurücklegen und ohne Reihenfolge kann man durch Werte $b_1, \ldots, b_n \in \mathbb{N}$ darstellen. Die Zahl b_i gibt an, wie oft die Kugel i gezogen wurde. Wenn wir k Kugeln ziehen, dann ist $\sum_i b_i = k$. Wir setzen $a_0 = 0$ und $a_{i+1} = a_i + b_{i+1} + 1$ für $0 \leq i < n$. Dann gilt $1 \leq a_1 < \cdots < a_{n-1} < a_n = n + k$. Insbesondere ist $\{a_1, \ldots, a_{n-1}\}$ eine Auswahl von $n - 1$ Elementen aus $\{1, \ldots, n + k - 1\}$. Aus den a_i's kann man auch wieder die b_i's berechnen. Mit $\binom{n+k-1}{k} = \binom{n+k-1}{n-1}$ folgt daraus die Behauptung. ◇

Die Herleitung in Beispiel 4.6 (d) enthält eine häufig anzutreffende Methode, und zwar dass man die a_i's nach einer vorgegebenen Ordnung anordnet (hier ist dies die natürliche Ordnung auf Zahlen), obwohl sie in einer beliebigen Reihenfolge gezogen wurden. Wir werden diese Technik im nächsten Teil weiter verfeinern.

Nach einer Anekdote sollten die Schüler in der Klasse von Gauß einige Zeit still beschäftigt werden. Der Lehrer verlangte daher, die Zahlen von 1 bis 100 aufzusummieren. Gauß löste diese Aufgabe sofort, indem er auf einen Zettel schrieb:

$$(1 + 100) + (2 + 99) + \cdots + (49 + 52) + (50 + 51) = 50 \cdot 101 = 5050$$

Wieso gehört dies hierher? Nun, die bekannte Formel, die aufgrund der obigen Anekdote häufig nach Gauß benannt wird, lautet:

$$\sum_{0 \leq k \leq n} k = \frac{n(n+1)}{2} = \binom{n+1}{2}$$

Ein einfacher Induktionsbeweis ist möglich, aber langweilig. Versuchen wir eine kombinatorische Interpretation und betrachten die Menge $A = \{1, \ldots, n+1\}$. Es gilt

$$\left| \binom{A}{2} \right| = \sum_{1 \leq k \leq n+1} \left| \{ \{j, k\} \mid 1 \leq j < k \} \right| = \sum_{1 \leq k \leq n+1} (k-1) = \sum_{0 \leq k \leq n} k$$

Gauß in Reinform! Wie steht es mit der Summe der Quadrate? Kennt man das Ergebnis, so ist es erneut eine leichte Übung, es mit Induktion zu zeigen. Aber was tun, wenn man diese Formel vergessen hat? Besser wir lernen sie herzuleiten. Wie eben überzeugen wir uns von der folgenden Identität:

$$\binom{n+1}{3} = \sum_{1 \leq k \leq n+1} \left| \{ \{\ell, j, k\} \mid 1 \leq \ell < j < k \} \right|$$

$$= \sum_{1 \leq k \leq n+1} \left| \{ \{\ell, j\} \mid 1 \leq \ell < j < k \} \right|$$

$$= \sum_{1 \leq k \leq n+1} \binom{k-1}{2} = \sum_{0 \leq k \leq n} \binom{k}{2}$$

Da $2 \cdot \binom{k}{2} = k^2 - k$ gilt, erhalten wir im nächsten Schritt:

$$2 \cdot \binom{n+1}{3} = \left(\sum_{1 \leq k \leq n} k^2 \right) - \left(\sum_{1 \leq k \leq n} k \right)$$

Zusammen mit dem Wissen über die Gauß-Summe ergibt sich:

$$\sum_{1 \leq k \leq n} k^2 = 2 \cdot \binom{n+1}{3} + \binom{n+1}{2} = \frac{2n^3 + 3n^2 + n}{6}$$

Diese Idee lässt sich verallgemeinern. Die $(m+1)$-elementigen Teilmengen von $A = \{1, \ldots, n+1\}$ lassen sich in Klassen nach ihrem maximalen Element k einteilen. Hieraus folgt:

$$\binom{n+1}{m+1} = \sum_{1 \leq k \leq n+1} \left| \binom{\{1, \ldots, k-1\}}{m} \right| = \sum_{1 \leq k \leq n+1} \binom{k-1}{m} = \sum_{0 \leq k \leq n} \binom{k}{m}$$

Wir können also Satz 4.7 festhalten.

Satz 4.7 (Obere Summation). *Für $m, n \in \mathbb{N}$ gilt:*

$$\binom{n+1}{m+1} = \sum_{0 \leq k \leq n} \binom{k}{m}$$

Ganz ähnlich erhalten wir die Identität 4.8. Sie gilt für $n \in \mathbb{Z}$ und x beliebig.

Satz 4.8 (Parallele Summation).

$$\binom{x+n+1}{n} = \sum_{k \leq n} \binom{x+k}{k}$$

Beweis. Seien zunächst $x \in \mathbb{N}$ und $n \in \mathbb{N}$. Die n-elementigen Teilmengen von $A = \{1, \ldots, x + n + 1\}$ lassen sich in Klassen nach dem größten Element aus A einteilen, welches nicht (!) in der Teilmenge enthalten ist. Es sei $B \in \binom{A}{n}$ und $x + k + 1$ das größte Element von $A \setminus B$. Es gilt also $x + k + 1 \notin B$ und $x + k + 2, \ldots, x + n + 1 \in B$. Hieraus folgt $0 \leq k \leq n$ und $|B \cap \{1, \ldots, x + k\}| = k$. Die Menge B ist also eindeutig durch den Wert k und eine k-elementige Teilmenge von $\{1, \ldots, x + k\}$ bestimmt. Damit erhalten wir:

$$\left| \binom{A}{n} \right| = \sum_{k \leq n} \left| \binom{\{1, \ldots, x + k\}}{k} \right| = \sum_{k \leq n} \binom{x + k}{k}$$

Mit Hilfe der Polynommethode folgt die Identität für alle $x \in \mathbb{C}$ und $n \in \mathbb{N}$. Die Erweiterung auf $n \in \mathbb{Z}$ ist trivial, da alle Terme 0 werden. □

Die Gleichung in Satz 4.9 ist nach dem französischen Mathematiker, Chemiker und Musiker Alexandre-Théophile Vandermonde (1735–1796) benannt.

Satz 4.9 (Vandermonde'sche Identität).

$$\binom{x + y}{n} = \sum_{k} \binom{x}{k} \binom{y}{n - k}$$

Beweis. Wir beweisen die Identität zunächst durch kombinatorische Interpretation für $x, y \in \mathbb{N}$ und wenden dann die Polynommethode an. Seien X und Y disjunkte Mengen mit $|X| = x$ und $|Y| = y$. Auf der linken Seite steht die Anzahl der Möglichkeiten, n Elemente aus $X \cup Y$ auszuwählen. Auf der rechten Seite zählen wir die Möglichkeiten zunächst k Elemente aus X auszuwählen und dann $n - k$ Elemente aus Y auszuwählen. Insgesamt wählen wir auch hier wieder n Elemente aus $X \cup Y$ aus. Die Summation über k hat zur Folge, dass wir alle Aufteilungen der n Elemente in Teilmengen von X und in Teilmengen von Y genau einmal zählen.

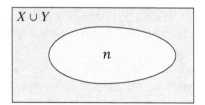 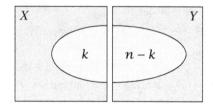

Für $n < 0$ und beliebige $x, y \in \mathbb{C}$ sind die Terme auf beiden Seiten 0. Betrachten wir den Fall $n \in \mathbb{N}$. Als Nächstes folgt die Polynommethode für zwei Variablen. Für ein festes $y \in \mathbb{N}$ steht sowohl links als auch rechts ein Polynom in x vom Grad n. Die kombinatorische Interpretation hat gezeigt, dass diese beiden Polynome für alle $x \in \mathbb{N}$ übereinstimmen, insbesondere stimmen die Polynome vom Grad n an mindestens $n + 1$ Stellen überein. Daraus folgt, dass die Polynome gleich sind und obige Identität für alle $x \in \mathbb{C}$ und alle $y \in \mathbb{N}$ gilt. Sei nun $x \in \mathbb{C}$ fest, dann stehen links

und rechts zwei Polynome in y vom Grad n, die an mindestens $n + 1$ Stellen übereinstimmen. Dies zeigt, dass die Vandermondesche Identität für alle $n \in \mathbb{Z}$ und alle $x, y \in \mathbb{C}$ gilt. $\qquad\qquad\qquad\qquad\qquad\qquad\qquad\qquad\qquad\qquad\qquad\qquad$ $\square$

Als Spezialfall erhalten wir die folgende Aussage. Für $m, n \in \mathbb{N}$ ist:

$$\binom{m + n}{m} = \sum_{k} \binom{m}{k}\binom{n}{m - k}$$

Wir geben noch eine direkte kombinatorische Interpretation der Aussage und betrachten hierfür das Rechteckgitter mit den Endpunkten $(0, 0)$ und (n, m) und die Gesamtheit M aller kürzesten Wege im Gitter von $(0, 0)$ nach (n, m). Jeder dieser Wege besteht aus m senkrechten Einheitswegstrecken und n waagerechten Einheitswegstrecken in beliebiger Reihenfolge. Wenn wir in der Reihenfolge des Weges jede senkrechte Strecke durch „0" und jede waagerechte Strecke durch „1" kennzeichnen, so wird jeder kürzeste Weg eindeutig dargestellt durch eine 0-1-Sequenz der Länge $m + n$ mit genau n Einsen. Durch Auswählen der n Positionen der Einsen sehen wir:

$$|M| = \binom{m + n}{n}$$

Zum anderen benutzt jeder Weg aus M genau einen der Punkte $x_k = (m - k, k)$ mit $0 \leq k \leq m$.

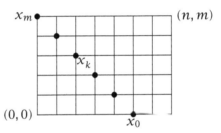

Es gibt genau $\binom{m - k + k}{k} = \binom{m}{k}$ kürzeste Wege von $(0, 0)$ nach x_k und genau $\binom{n - m + k + m - k}{m - k} = \binom{n}{m - k}$ kürzeste Wege von x_k nach (n, m). Das ergibt die Behauptung.

Für $x, y \in \mathbb{N}$ wollen wir die Vandermonde'sche Identität alternativ noch mit Hilfe einer anderen sehr lehrreichen Technik beweisen. Wir interpretieren hierzu den Binomialsatz 4.3 als Gleichheit zwischen zwei Polynomen. Sei Z eine Unbestimmte. Dann gilt

$$\sum_{n} \binom{x + y}{n} Z^n = (Z + 1)^{x + y} = (Z + 1)^x \cdot (Z + 1)^y$$

$$= \left(\sum_{k} \binom{x}{k} Z^k\right) \cdot \left(\sum_{\ell} \binom{y}{\ell} Z^\ell\right)$$

$$= \sum_{n} \left(\sum_{k} \binom{x}{k}\binom{y}{n - k}\right) Z^n$$

Hierbei haben wir bei der letzten Gleichheit alle Koeffizienten von Z^n zusammenge-fasst. Aus obiger Gleichheit von Polynomen folgt nun durch Koeffizientenvergleich bei Z^n die Vandermonde'sche Identität.

Angenommen, wir möchten bis zu t identische Objekte in ℓ Behälter aufteilen. Wie viele Möglichkeiten gibt es hierfür? Die Antwort liefert der nächste Satz, welchen wir in Beispiel 4.6 (d) bereits in anderer Form kennen gelernt haben.

Satz 4.10.

$$\left| \left\{ (e_1, \ldots, e_\ell) \in \mathbb{N}^\ell \ \middle| \ \sum_{1 \leq k \leq \ell} e_k \leq t \right\} \right| = \binom{t + \ell}{\ell}$$

Beweis. Wir stellen uns $t + \ell$ Punkte vor, die waagerecht in einer Reihe liegen. Aus diesen Punkten wählen wir ℓ Punkte aus und ersetzen diese durch Striche. Hierfür gibt es $\binom{t+\ell}{\ell}$ Möglichkeiten. Jede solche Auswahl entspricht genau einem ℓ-Tupel $(e_1, \ldots, e_\ell) \in \mathbb{N}^\ell$ mit $\sum_{k=1}^{\ell} e_k \leq t$.

$$\underbrace{\overbrace{\bullet \bullet \cdots \bullet \bullet}^{e_1 \text{ Punkte}} \mid \overbrace{\bullet \bullet \cdots \bullet \bullet}^{e_2 \text{ Punkte}} \mid \cdots \mid \overbrace{\bullet \bullet \cdots \bullet \bullet}^{e_\ell \text{ Punkte}} \mid \overbrace{\bullet \bullet \cdots \bullet \bullet}^{\text{Überschuss}}}_{t \text{ Punkte und } \ell \text{ Striche}}$$

Zunächst werden e_1 Punkte bis zum ersten Strich abgetragen. Nach dem ersten Strich werden e_2 Punkte abgetragen, so fahren wir fort. Nach dem ℓ-ten Strich kann noch ein Überschuss an Punkten folgen, um insgesamt t Punkte zu erhalten. So lassen sich die Lösungen der Ungleichung und Auswahlen an Punkten und Strichen bijektiv aufeinander abbilden. $\qquad \square$

Kehren wir zum Binomialsatz 4.3 zurück. Wir geben einen leicht modifizierten kombinatorischen Beweis und formulieren den Satz in voller Allgemeinheit.

Satz 4.11 (Allgemeiner Binomialsatz)**.** *Seien* $r, x, y \in \mathbb{C}$. *Dann gilt*

$$(x + y)^r = \sum_k \binom{r}{k} x^k y^{r-k} \quad \text{für } |x| < |y| \text{ oder für } r \in \mathbb{N}$$

Beweis. Seien $r, x, y \in \mathbb{N}$ und seien R, X, Y Mengen mit $|R| = r$, $|X| = x$, $|Y| = y$ und $X \cap Y = \emptyset$. Auf der linken Seite des Binomialsatzes steht die Anzahl der Ab-bildungen von R in die disjunkte Vereinigung von X und Y. Jede Teilmenge K von R definiert eine Klasse von Abbildungen:

$$F_K = \left\{ f : R \to X \cup Y \ \middle| \ f^{-1}(X) = K \right\}$$

Jede Abbildung $f \in F_K$ setzt sich zusammen aus einer Abbildung $f_X : K \to X$ und einer Abbildung $f_Y : R \setminus K \to Y$. Sei $|K| = k$. Für f_X gibt es x^k Möglichkeiten und für f_Y gibt es y^{r-k} Möglichkeiten. Deshalb gilt $|F_K| = x^k y^{r-k}$. Die Anzahl der Teilmengen K mit $|K| = k$ ist $\binom{r}{k}$ und die Behauptung folgt für $r, x, y \in \mathbb{N}$.

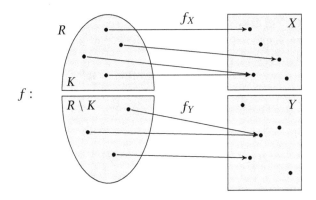

Als nächster Schritt folgt die Polynommethode. Es gibt hier wie in Satz 4.9 zwei Variablen, also machen wir wiederum zwei Schritte. Aus $r, x, y \in \mathbb{N}$ wird $r, x \in \mathbb{N}$, $y \in \mathbb{C}$ mit Hilfe der Polynommethode im ersten Schritt. Dasselbe Argument führt von $r, x \in \mathbb{N}, y \in \mathbb{C}$ zu $r \in \mathbb{N}, x, y \in \mathbb{C}$.

Hier sind wir mit unserem Standardansatz am Ende. Für $r \in \mathbb{C} \setminus \mathbb{N}$ ist $\sum_k \binom{r}{k} x^k \cdot y^{r-k}$ eine unendliche Reihe. Wir betrachten daher den Fall $|x| < |y|$ mit $r \in \mathbb{C}$ gesondert. Wir setzen $z = \frac{x}{y}$ und $f(z) = (1 + z)^r$. Dann gilt $|z| < 1$ und es reicht,

$$f(z) = \sum_k \binom{r}{k} z^k$$

zu zeigen. Betrachte die k-te Ableitung von f:

$$f^{(k)}(z) = r^{\underline{k}} (1 + z)^{r-k}$$

Damit ist $f^{(k)}(0) - r^{\underline{k}}$. Die Taylorreihe von f ist folglich:

$$\sum_{k \geq 0} \frac{f^{(k)}(0)}{k!} z^k = \sum_k \binom{r}{k} z^k$$

Diese Reihe konvergiert für $|z| < 1$ absolut, da $|\binom{r}{k}|$ als Funktion von k durch ein Polynom in k vom Grad $\lfloor |r| \rfloor$ begrenzt ist. Dies kann man direkt aus der Definition der Binomialkoeffizienten ableiten, indem man Faktoren kleiner als 1 herausstreicht. Die absolute Konvergenz impliziert die Gleichheit $f(z) = \sum_k \binom{r}{k} z^k$. $\square$

In dem Spezialfall $r = -1$, $x = -z$ und $y = 1$ liefert Satz 4.11 den Grenzwert der geometrischen Reihe:

$$\sum_{k \geq 0} z^k = \frac{1}{1 - z} \quad \text{für } |z| < 1$$

Eine Integralabschätzung zeigt, dass $\binom{k+z}{k} k^{-z}$ für $k \to \infty$ gegen eine Konstante konvergiert (die übrigens $\frac{1}{z!}$ für alle komplexen Zahlen z definiert). Hieraus folgt, dass $\binom{r}{k}$ in $\mathcal{O}(k^{-1-r})$ liegt. Aufgrund des Leibniz-Kriteriums (Gottfried Wilhelm Leibniz, 1646–1716) über alternierende Reihen, konvergiert damit die Reihe $\sum_k \binom{r}{k}$ für reelle Zahlen $r > -1$. Insbesondere gilt $(1+1)^{\frac{1}{2}} = \sqrt{2} = \sum_k \binom{1/2}{k}$.

Haben wir eine n-te Potenz aus mehr als zwei Summanden zu bilden, so können wir den nächsten Satz verwenden.

Satz 4.12 (Multinomialsatz). *Sei $d \geq 1$. Dann gilt:*

$$(x_1 + \cdots + x_d)^n = \sum_{k_i \geq 0,\, k_1 + \cdots + k_d = n} \frac{n!}{k_1! \cdots k_d!} x_1^{k_1} \cdots x_d^{k_d}$$

Beweis. Die kombinatorische Interpretation ist wie beim Binomialsatz 4.11 möglich, wir müssen nur beachten:

$$\frac{n!}{k_1! \cdots k_d!} = \frac{n^{\underline{k_1}}(n-k_1)^{\underline{k_2}} \cdots k_d!}{k_1! \cdots k_d!} = \binom{n}{k_1}\binom{n-k_1}{k_2} \cdots \binom{k_d}{k_d}$$

Eine Induktion nach n ist auch möglich, aber das kennen wir schon aus dem Beweis von Satz 4.3. Diesmal machen wir eine Induktion nach d. Für $d = 1$ ist der Satz richtig. Für $d > 1$ schreiben wir $y = x_2 + \cdots + x_d$ und erhalten aus dem Binomialsatz 4.3:

$$(x_1 + \cdots + x_d)^n = (x_1 + y)^n = \sum_{k_1} \binom{n}{k_1} x_1^{k_1} y^{n-k_1}$$

Mit Induktion nach d sehen wir:

$$(x_1 + \cdots + x_d)^n = \sum_{k_1} \binom{n}{k_1} x_1^{k_1} \left(\sum_{k_i \geq 0,\, k_2 + \cdots + k_d = n - k_1} \frac{(n-k_1)!}{k_2! \cdots k_d!} x_2^{k_2} \cdots x_d^{k_d} \right)$$

$$= \sum_{0 \leq k_1 \leq n} \frac{n!}{k_1!(n-k_1)!} x_1^{k_1} \left(\sum_{k_i \geq 0,\, k_2 + \cdots + k_d = n - k_1} \frac{(n-k_1)!}{k_2! \cdots k_d!} x_2^{k_2} \cdots x_d^{k_d} \right)$$

$$= \sum_{k_i \geq 0,\, k_1 + \cdots + k_d = n} \frac{n!}{k_1! \cdots k_d!} x_1^{k_1} \cdots x_d^{k_d} \qquad \square$$

Sind $d, k_i, n \in \mathbb{N}$ mit $k_1 + \cdots + k_d = n$ so definiert man den *Multinomialkoeffizienten* durch:

$$\binom{n}{k_1, \ldots, k_d} = \frac{n!}{k_1! \cdots k_d!}$$

Er gibt an, wie viele Möglichkeiten es gibt, eine n-elementige Menge so in d disjunkte Klassen zu zerlegen, dass die i-te Klasse C_i genau k_i Elemente enthält. Um dies einzusehen, sei s die Zahl solcher Zerlegungen. Wir schreiben jede Zerlegung als Sequenz $(C_1, \ldots, C_d)$. Es gibt jeweils $k_i!$ Möglichkeiten, die Elemente aus C_i anzuordnen. Wenn wir alle Klassen angeordnet haben, erhalten wir eine beliebige Permutation der n Elemente. Umgekehrt können wir aus einer Permutation π wieder die Zerlegung zurückgewinnen, indem wir in $(\pi(1), \ldots, \pi(n))$ jeweils nacheinander von

links nach rechts d Blöcke bilden, so dass der i-te Block genau k_i Elemente enthält. Wenn wir nun die Reihenfolge bei den Elementen des Blocks i ignorieren, liefert das die Klasse C_i. Dies zeigt $s \cdot k_1! \cdots k_d! = n!$ und damit $s = \binom{n}{k_1,\ldots,k_d}$.

Beispiel 4.13. Sei $n = 4$, $d = 3$ und $(k_1, k_2, k_3) = (1, 1, 2)$. Wenn wir die n-elementige Menge $\{1, 2, 3, 4\}$ als Positionen eines Worts interpretieren und Positionen der Klasse 1 mit a beschriften, Positionen der Klasse 2 mit b und Positionen der Klasse 3 mit c, dann zählt $\binom{4}{1,1,2} = 12$ die folgenden Wörter

$$
\begin{array}{cccc}
abcc & bacc & cabc & cbca \\
acbc & bcac & cacb & ccab \\
accb & bcca & cbac & ccba
\end{array}
$$

so dass wir auf diese Weise eine weitere Interpretation von Multinomialkoeffizienten erhalten. ◇

4.3 Durchschnittsanalyse von Bubble-Sort

Mit dem Begriff *Bubble-Sort* verbindet sich ein einfaches Sortierverfahren, welches auf lokalen Vertauschungen basiert. Sei $\pi = (\pi_1, \ldots, \pi_n)$ eine Folge von Zahlen. Diese können wir dadurch sortieren, dass wir immer wieder von links nach rechts die Folge durchgehen und die Elemente π_i und π_{i+1} vertauschen, sofern die Bedingung $\pi_i > \pi_{i+1}$ vorliegt. Jede Vertauschung verändert natürlich π. Dieses Verfahren bricht ab, sobald ein Durchgang ohne Vertauschung möglich ist. Dies ist ein Vorteil von Bubble-Sort: Der letzte Durchlauf verifiziert, dass die Folge tatsächlich sortiert wurde. Es ist auch klar, dass Bubble-Sort gut für fast vorsortierte Folgen geeignet ist.

Wir wollen die Zeit betrachten, die dieses Sortierverfahren benötigt. Nach dem ersten Durchlauf steht die größte Zahl ganz hinten, nach dem zweiten stimmen die letzten beiden Positionen und so weiter. Nach n Durchläufen sind wir fertig, und jeder Durchlauf kostet uns höchstens n Schritte. Es ist also ein quadratisches Verfahren, und man muss sich schon anstrengen, es nicht mit einer Laufzeit von $\mathcal{O}(n^2)$ zu implementieren.

Es gibt diverse Vorschläge, Bubble-Sort zu optimieren, aber kommen diese mit einer Zeit in $o(n^2)$ aus? Die Antwort ist ein sehr deutliches *Nein*. Selbst beliebig optimierte Bubble-Sort Varianten sind im Mittel immer noch quadratische Sortierverfahren: Nehmen wir als Maß die Zahl der Fehlstellungen einer Permutation $\pi = (\pi_1, \ldots, \pi_n)$. Dies ist definiert durch die Zahl der Paare (i, j) mit $i < j$ und $\pi_j < \pi_i$.

Satz 4.14. *Die Anzahl der Fehlstellungen einer Permutation $\pi = (\pi_1, \ldots, \pi_n)$ beträgt im Mittel*

$$
\frac{1}{2} \cdot \binom{n}{2}
$$

Beweis. Für $n \leq 1$ ist die Behauptung erfüllt. Sei also $n \geq 2$. Für $\pi = (\pi_1, \ldots, \pi_n)$ definieren wir eine Permutation $\overline{\pi}$ durch $\overline{\pi} = (\pi_n, \ldots, \pi_1)$. Die Zuordnung $\pi \mapsto \overline{\pi}$ definiert eine Involution (es gilt $\overline{\overline{\pi}} = \pi$) ohne Fixpunkte. Dadurch wird die Menge der Permutationen so in Klassen eingeteilt, dass jede Klasse von der Form $\{\pi, \overline{\pi}\}$ ist und aus genau zwei Elementen besteht. Die Summe der Fehlstellungen für eine Klasse $\{\pi, \overline{\pi}\}$ ist genau $\binom{n}{2}$. □

Was immer wir auch Bubble-Sort nennen, gehen wir davon aus, dass pro Zeiteinheit höchstens eine Fehlstellung beseitigt wird. Die mittlere Zahl der Fehlstellungen unterschätzt daher den Zeitverbrauch jeder realistischen Bubble-Sort-Implementierung.

Korollar 4.15. *Bubble-Sort benötigt im Mittel und im schlechtesten Fall $\Theta(n^2)$ Vergleiche, um eine Folge mit n Elementen zu sortieren.*

Die Aussage in Korollar 4.15 gilt für jede Verteilung der Eingaben, sofern die Folgen π und $\overline{\pi}$ jeweils gleich wahrscheinlich sind oder jeweils zufällig entschieden wird, ob die Folge von links nach rechts oder in umgekehrter Richtung sortiert werden soll.

4.4 Das Prinzip von Inklusion und Exklusion

Sind A und B disjunkte Mengen, so gilt $|A \cup B| = |A| + |B|$. Allgemeiner gilt für beliebige, nicht nowendigerweise disjunkte Mengen A und B die Formel $|A \cup B| = |A| + |B| - |A \cap B|$, denn durch Aufzählen der Elemente aus A und der Elemente aus B werden diejenigen Elemente doppelt gezählt, die im Schnitt der beiden Mengen liegen. Diese müssen deswegen einmal abgezogen werden. Für drei endliche Mengen A, B, C gilt

$$|A \cup B \cup C| = |A| + |B| + |C| - |A \cap B| - |B \cap C| - |A \cap C| + |A \cap B \cap C|$$

Dies erkennt man aus dem folgenden Venn-Diagramm (John Venn, 1834–1923) dreier Mengen A, B, C:

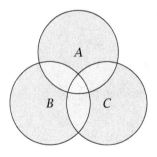

Beispiel 4.16. Wie viele Zahlen zwischen 1 und 1000 sind durch 3 oder 5 oder 8 teilbar? Sei A bzw. B bzw. C die Menge der durch 3 bzw. 5 bzw. 8 teilbaren Zahlen zwischen 1 und 1000. Es gilt:

$$|A| = \left\lfloor \frac{1000}{3} \right\rfloor = 333 \qquad |A \cap B| = \left\lfloor \frac{1000}{15} \right\rfloor = 66$$

$$|B| = \left\lfloor \frac{1000}{5} \right\rfloor = 200 \qquad |B \cap C| = \left\lfloor \frac{1000}{40} \right\rfloor = 25$$

$$|C| = \left\lfloor \frac{1000}{8} \right\rfloor = 125 \qquad |A \cap C| = \left\lfloor \frac{1000}{24} \right\rfloor = 41$$

$$|A \cap B \cap C| = \left\lfloor \frac{1000}{120} \right\rfloor = 8$$

Daraus folgt $|A \cup B \cup C| = 333 + 200 + 125 - 66 - 25 - 41 + 8 = 534.$ ◊

Das Prinzip von Inklusion und Exklusion verallgemeinert diese Formel und erlaubt das genaue Zählen der Elemente einer Vereinigung von Mengen. Die Beziehung 4.17 ist nach James Joseph Sylvester (1814–1897) benannt.

Satz 4.17 (Siebformel von Sylvester). *Für endliche Mengen $A_1, \ldots, A_n$ gilt:*

$$|A_1 \cup \cdots \cup A_n| = \sum_{k \geq 1} (-1)^{k+1} \sum_{1 \leq r_1 < \cdots < r_k \leq n} |A_{r_1} \cap \cdots \cap A_{r_k}|$$

Beweis. Sei $x \in A = A_1 \cup \cdots \cup A_n$ und x komme in genau m Mengen A_i vor, $1 \leq m \leq n$. Das Element x wird auf der linken Seite genau einmal gezählt. Wir zählen jetzt, wie oft x auf der rechten Seite vorkommt. Hierfür können wir $x \in A_i$ für $1 \leq i \leq m$ und $x \notin A_i$ für $m < i \leq n$ voraussetzen. Ein Summand $|A_{r_1} \cap \cdots \cap A_{r_k}|$ liefert also keinen Beitrag für x, wenn $r_k > m$. Die Gleichung lässt sich daher für dieses x wie folgt umschreiben:

$$1 = \sum_{k \geq 1} (-1)^{k+1} \sum_{I \in \binom{\{1, \ldots, m\}}{k}} 1 = \sum_{k \geq 1} (-1)^{k+1} \binom{m}{k}$$

Dies ist äquivalent zur folgenden Identität, welche aus der Anwendung des Binomialsatzes auf $(-1 + 1)^m$ folgt:

$$\sum_k (-1)^k \binom{m}{k} = 0 \quad \text{für } m \geq 1$$

Insgesamt erhalten wir die Gleichheit für jedes $x \in A$ und damit die Siebformel von Sylvester. □

Eine Anwendung der Siebformel von Sylvester wird häufig als (das Prinzip von) Inklusion und Exklusion bezeichnet. Wir betrachten nun zwei Beispiele. Als erstes erhalten wir einen weiteren Beweis für die Euler'sche Formel (1.1). Sei $n = p_1^{n_1} \cdots p_r^{n_r}$

die Primfaktorzerlegung von $n \in \mathbb{N}$ mit $n_i > 0$ und p_i paarweise verschiedenen Primzahlen. Dann gilt für die Anzahl $\varphi(n)$ der zu n teilerfremden Zahlen zwischen 1 und n die Beziehung:

$$\varphi(n) = n \left(1 - \frac{1}{p_1}\right)\left(1 - \frac{1}{p_2}\right) \cdots \left(1 - \frac{1}{p_r}\right)$$

Für $n \leq 1$ ist dies klar. Sei nun $n \geq 2$. Für jeden Teiler $d \in \mathbb{N}$ von n gibt es $\frac{n}{d}$ unter den Zahlen zwischen 1 und n, die durch d teilbar sind. Sei $A = \{1, \ldots, n\}$, und für $i = 1, \ldots, r$ sei $A_i = \{x \in A \mid p_i \text{ teilt } x\}$. Für $\varphi(n) = |A \setminus (A_1 \cup \cdots \cup A_r)|$ gilt:

$$|A \setminus (A_1 \cup \cdots \cup A_r)| = |A| + \sum_{k=1}^{r} (-1)^k \sum_{1 \leq r_1 < \cdots < r_k \leq r} |A_{r_1} \cap \cdots \cap A_{r_k}|$$

$$= n + \sum_{k=1}^{r} (-1)^k \sum_{1 \leq r_1 < \cdots < r_k \leq r} \frac{n}{p_{r_1} \cdots p_{r_k}} = n \left(1 - \frac{1}{p_1}\right) \cdots \left(1 - \frac{1}{p_r}\right)$$

Als Nächstes wollen wir die Beweistechnik aus der Siebformel von Sylvester auf ein ähnliches Problem anwenden. Seien $A_1, \ldots, A_r$ Teilmengen einer endlichen Menge A. Die Anzahl der Elemente aus A, die zu genau m der Teilmengen A_i gehören, ist gegeben durch

$$\sum_{k=m}^{r} (-1)^{k+m} \binom{k}{m} \sum_{1 \leq r_1 < \cdots < r_k \leq r} |A_{r_1} \cap \cdots \cap A_{r_k}|$$

Für den Beweis der Formel sei $x \in A$ ein Element, das zu genau s der Teilmengen gehört. Wir zählen, wieviel x zur obigen Summe beiträgt. Das Element x gehört genau zu den k-fachen Durchschnitten derjenigen s Mengen, in denen x liegt. Deren Anzahl ist $\binom{s}{k}$. Also ist der Beitrag von x in der Summe genau

$$\sum_{k=m}^{s} (-1)^{k+m} \binom{k}{m}\binom{s}{k}$$

Im Falle von $s < m$ ist die Summe 0 und x kommt auch in keinem Durchschnitt von m Mengen vor. Für $s = m$ hat die Summe den Wert 1. Sei jetzt $s > m$. Wir zeigen, dass die Summe den Wert 0 hat, womit die Behauptung bewiesen ist. Mittels der trinomialen Revision in Satz 4.4 ergibt sich dies aus dem Binomialsatz, angewendet auf $(1 - 1)^{s-m} = 0$:

$$\sum_{k=m}^{s} (-1)^{k+m} \binom{s}{k}\binom{k}{m} = \sum_{k} (-1)^{k+m} \binom{s}{m}\binom{s-m}{k-m}$$

$$= \binom{s}{m} \sum_{k} (-1)^{k-m} \binom{s-m}{k-m} = \binom{s}{m} \cdot 0 = 0$$

4.5 Rencontres-Zahlen

Es sei S_n die Gruppe der Permutationen über der Menge $\{1,\dots,n\}$ und $\mathcal{R}_n$ die Teilmenge der fixpunktfreien Permutationen. Also ist $|S_n|$ die Anzahl der Bijektionen einer n-elementigen Menge und $R_n = |\mathcal{R}_n|$ zählt die Bijektionen, die kein Element auf sich selbst abbilden. Die Zahlen R_n heißen *Rencontres-Zahlen* nach dem „Problème des rencontres" (Treffen) nach Pierre Rémond de Montmort (1678–1719). Treffen sich n Ehepaare zu einem Tanzabend, so gibt es R_n Möglichkeiten, n Tanzpaare zu bilden, ohne dass Ehepaare zusammen tanzen.

Satz 4.18. *Für $n \geq 1$ gilt:*

$$R_n = n! \sum_{k=0}^{n} \frac{(-1)^k}{k!}$$

Beweis. Sei P_m die Menge aller Permutationen aus S_n, die das Element m fest lassen. Die Anzahl der Permutationen, die eine ausgewählte Teilmenge $I \in \binom{\{1,\dots,n\}}{k}$ fest lassen, ist $(n-k)!$. Es folgt:

$$R_n = |S_n \setminus (P_1 \cup \cdots \cup P_n)| = n! + \sum_{k=1}^{n} (-1)^k \sum_{1 \leq i_1 < \cdots < i_k \leq n} (n-k)!$$

$$= n! + \sum_{k=1}^{n} (-1)^k \binom{n}{k} (n-k)! = n! \sum_{k=0}^{n} (-1)^k \frac{1}{k!}$$

Hierbei verwendet die zweite Gleichheit die Siebformel von Sylvester. □

Wir können den Rencontres-Zahlen eine weitere Interpretation geben: Wenn jemand n Briefe und die zugehörigen Umschläge schreibt und dann die Briefe willkürlich in die Umschläge steckt, wie groß ist die Wahrscheinlichkeit p_n, dass keiner der Adressaten den ihm zugedachten Brief erhält? Diese Wahrscheinlichkeit ist gegeben durch:

$$p_n = \frac{R_n}{n!} = \sum_{k=0}^{n} \frac{(-1)^k}{k!}$$

Für $n \to \infty$ strebt diese Wahrscheinlichkeit gegen $\frac{1}{e} \approx 0{,}37$. Insbesondere erhalten wir mit der Stirling'schen Formel 2.2 das folgende Wachstumsverhalten:

$$R_n \sim \frac{\sqrt{2\pi n}}{e} \left(\frac{n}{e}\right)^n$$

Als Nächstes betrachten wir eine Verallgemeinerung der Rencontres-Zahlen. Sei $R_{n,m}$ die Anzahl der Permutationen in S_n mit genau m Fixpunkten. Insbesondere ist $R_{n,0} = R_n$. Dann gilt:

$$R_{n,m} = \frac{n!}{m!} \sum_{k=0}^{n-m} (-1)^k \frac{1}{k!}$$

Dies sieht man wie folgt. Für die Auswahl der m Fixpunkte hat man $\binom{n}{m}$ Möglichkeiten. Die restlichen Elemente sind dann fixpunktfrei abzubilden. Dafür gibt es R_{n-m} Möglichkeiten. Also ist $R_{n,m} = \binom{n}{m} R_{n-m}$, und hieraus folgt die Behauptung.

Wir geben noch eine alternative Herleitung für den Wert der Rencontres-Zahlen. Setze $Q_n = \sum_{k=0}^{n} \binom{n}{k} R_k$. Wie eben bemerkt, gilt $R_{n,k} = \binom{n}{k} R_{n-k}$. Daraus folgt

$$Q_n = \sum_{k=0}^{n} \binom{n}{k} R_k = \sum_{k=0}^{n} \binom{n}{k} R_{n-k} = \sum_{k=0}^{n} R_{n,k} = n!$$

Die Binomialinversion in Satz 4.5 liefert nun die gewünschte Beziehung:

$$R_n = \sum_{k=0}^{n} (-1)^{n-k} \binom{n}{k} Q_k = \sum_{k=0}^{n} (-1)^k \binom{n}{k} Q_{n-k}$$
$$= \sum_{k=0}^{n} (-1)^k \binom{n}{k} (n-k)! = n! \sum_{k=0}^{n} (-1)^k \frac{1}{k!}$$

4.6 Stirling-Zahlen

Mit Hinweis auf eine 1730 veröffentlichte Arbeit von James Stirling (1692–1770) führte Niels Nielsen (1865–1931) in seinem *Handbuch der Theorie der Gammafunktion* (Teubner Verlag, Leipzig 1906) die Bezeichnung „Stirling-Zahlen der ersten und zweiten Art" ein. Wir benutzen hier die Karamata-Notation für die Stirling-Zahlen, die auf Jovan Karamata (1902–1967) zurückgeht. Damit bezeichnet $\left[\begin{smallmatrix} n \\ k \end{smallmatrix}\right]$ die Stirling-Zahlen erster Art und $\left\{\begin{smallmatrix} n \\ k \end{smallmatrix}\right\}$ diejenigen der zweiten Art. In der kombinatorischen Interpretation ist $\left[\begin{smallmatrix} n \\ k \end{smallmatrix}\right]$ die Anzahl der Permutationen über $M = \{1, \dots, n\}$, die sich in k Zykel zerlegen lassen, und $\left\{\begin{smallmatrix} n \\ k \end{smallmatrix}\right\}$ ist die Anzahl der Partitionen von M in k Klassen. Die Karamata-Notation wurde insbesondere durch die Arbeiten von Donald Ervin Knuth (geb. 1938) zu einem anerkannten Standard. Sie betont die Analogie der Bildungsgesetze zu den Binomialkoeffizienten.

$$\binom{0}{k} = \left[\begin{matrix} 0 \\ k \end{matrix}\right] = \left\{\begin{matrix} 0 \\ k \end{matrix}\right\} = \begin{cases} 1 & \text{für } k = 0 \\ 0 & \text{sonst} \end{cases}$$

$$\binom{n+1}{k} = \binom{n}{k-1} + \binom{n}{k}$$

$$\left[\begin{matrix} n+1 \\ k \end{matrix}\right] = \left[\begin{matrix} n \\ k-1 \end{matrix}\right] + n\left[\begin{matrix} n \\ k \end{matrix}\right]$$

$$\left\{\begin{matrix} n+1 \\ k \end{matrix}\right\} = \left\{\begin{matrix} n \\ k-1 \end{matrix}\right\} + k\left\{\begin{matrix} n \\ k \end{matrix}\right\}$$

Wir untersuchen zunächst die Stirling-Zahlen der zweiten Art.

4.6.1 Die Stirling-Zahlen zweiter Art

Die Zahlen $\left\{{n \atop k}\right\}$ geben die Anzahl der Partitionen einer Menge von n Elementen in k Klassen an. Eine *Partition* einer Menge A ist eine Menge $P = \{P_1, \ldots, P_k\}$ mit $\bigcup_{1 \le i \le k} P_i = A$, mit $P_i \ne \emptyset$ für alle $1 \le i \le k$ und mit $P_i \cap P_j = \emptyset$ für alle $1 \le i < j \le k$.

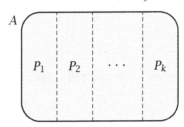

Die Mengen P_i sind die *Klassen* der Partition P. Die Menge P definiert also eine disjunkte Zerlegung von A in k nichtleere Klassen. Formal definieren wir für $n \in \mathbb{N}$ und $k \in \mathbb{Z}$ die Zahlen

$$\left\{{n \atop k}\right\} = |\{ P \mid P \text{ ist eine Partition von } \{1, \ldots, n\} \text{ in } k \text{ Klassen} \}|$$

Beispiel 4.19.

$$\left\{{n \atop n}\right\} = 1 \quad \text{für } n \ge 0 \qquad\qquad \left\{{n \atop 0}\right\} = 0 \quad \text{für } n \ge 1$$

$$\left\{{n \atop k}\right\} = 0 \quad \text{für } n < 0 \le k \qquad\qquad \left\{{n \atop k}\right\} = 0 \quad \text{für } 0 \le n < k$$

$$\left\{{n \atop k}\right\} > 1 \quad \text{für } n > k > 1 \qquad\qquad \left\{{n \atop 1}\right\} = 1 \quad \text{für } n \ge 1$$

$$\left\{{n \atop 2}\right\} = 2^{n-1} - 1 \quad \text{für } n \ge 1 \qquad \left\{{n \atop n-1}\right\} = \binom{n}{2} \quad \text{für } n \ge 1 \qquad \diamond$$

Satz 4.20 (Additionstheorem für Stirling-Zahlen zweiter Art).

$$\left\{{n \atop k}\right\} = \left\{{n-1 \atop k-1}\right\} + k\left\{{n-1 \atop k}\right\}$$

Beweis. Wir beweisen die Identität zunächst nur für den Fall $n \ge 1$ und $k \in \mathbb{Z}$, da sonst rechts noch nicht alle Terme definiert sind. Der allgemeine Fall wird sich später aus Gleichung (4.3) ergeben und wird bis dahin nicht benutzt. Sei also $n \ge 1$. Wir teilen die Menge der Partitionen in zwei Typen ein. Der erste Typ enthält $\{n\}$ als eine Klasse. Dies ergibt den Summanden $\left\{{n-1 \atop k-1}\right\}$, da jede solche Partition mit einer Partition von $\{1, \ldots, n-1\}$ in $k-1$ Klassen identifiziert werden kann. Der zweite Typ von Partitionen entsteht dadurch, dass wir bei einer Partition $\{P_1, \ldots, P_k\}$ von

$\{1, \dots, n - 1\}$ das Element n zu einer der k Klassen hinzufügen. Es gibt $\left\{ {n-1 \atop k} \right\}$ Partitionen von $\{1, \dots, n - 1\}$, und es gibt k mögliche Klassen, zu denen wir n hinzufügen können. Damit gibt es genau $k \left\{ {n-1 \atop k} \right\}$ Partitionen vom zweiten Typ. Jede Partition gehört entweder zu Typ eins oder zu Typ zwei. $\qquad\square$

Satz 4.21. *Seien A und B endliche Mengen mit $|A| = n$ und $|B| = m$. Dann gibt es $m! \left\{ {n \atop m} \right\}$ Surjektionen von A auf B.*

Beweis. Ohne Einschränkung können wir $A = \{1, \dots, n\}$ und $B = \{1, \dots, m\}$ annehmen. Sei $P = \{P_1, \dots, P_m\}$ eine Partition von A in m Klassen und sei π eine Permutation von B. Es gibt $m! \left\{ {n \atop m} \right\}$ Paare (P, π). Jedem dieser Paare können wir eine Surjektion $f : A \to B$ zuordnen, indem wir $f(i) = \pi(j)$ für $i \in P_j$ definieren. Dies zeigt, dass es mindestens $m! \left\{ {n \atop m} \right\}$ Surjektionen von A auf B gibt. Umgekehrt können wir jeder Surjektion $f : A \to B$ eine Partition P zuordnen:

$$P = \left\{ f^{-1}(j) \mid j \in B \right\}$$

Für jede Permutation $\pi : B \to B$ führt $\pi \circ f$ zur selben Partition wie f. Dies zeigt, dass es höchstens $m! \left\{ {n \atop m} \right\}$ Surjektionen von A auf B gibt. $\qquad\square$

Der letzte Satz liefert eine untere Schranke für $\left\{ {2n \atop n} \right\}$. Hierfür betrachten wir nur gewisse Surjektionen von $\{1, \dots, 2n\}$ auf $\{1, \dots, n\}$. Wir wählen eine Permutation π von $\{1, \dots, n\}$ und eine Abbildung f von $\{n + 1, \dots, 2n\}$ nach $\{1, \dots, n\}$. Jedes solche Paar (π, f) liefert eine Surjektion der Menge $\{1, \dots, 2n\}$ auf $\{1, \dots, n\}$ und es gibt $n! \cdot n^n$ solche Paare. Mit Satz 4.21 erhalten wir jetzt:

$$\left\{ {2n \atop n} \right\} \geq n^n \tag{4.2}$$

Mit Satz 4.22 lassen sich Potenzen in Summen von fallenden Faktoriellen umrechnen.

Satz 4.22.

$$x^n = \sum_k \left\{ {n \atop k} \right\} \cdot x^{\underline{k}}$$

Beweis. Sei zunächst $x \in \mathbb{N}$ und seien A und X Mengen mit $|A| = n$ und $|X| = x$. Auf der linken Seite der Gleichung zählen wir alle Abbildungen von A nach X. Als Nächstes untersuchen wir die rechte Seite der Gleichung. Nach der Formel (4.1) gibt es $x^{\underline{k}}$ Injektionen von $\{1, \dots, k\}$ nach X. Nun lässt sich jeder Partition $P = \{P_1, \dots, P_k\}$ von A in k Klassen und jeder injektiven Abbildung $g : \{1, \dots, k\} \to X$ eindeutig eine Abbildung $f : A \to X$ zuordnen:

$$f(i) = g(j) \quad \text{für } i \in P_j$$

Damit bildet f alle Elemente aus der Klasse P_j auf $g(j)$ ab. Jede Abbildung kann eindeutig auf diese Weise erzeugt werden. Mit der Polynommethode folgt die Behauptung für $x \in \mathbb{C}$. $\qquad\square$

Nach der Umformung $\sum_k \left\{{n \atop k}\right\} \cdot x^{\underline{k}} = \sum_k k! \left\{{n \atop k}\right\} \binom{x}{k}$ könnte man Satz 4.22 auch mit Hilfe von Surjektionen beweisen. Die Idee hierbei ist, bei jeder Abbildung zuerst den Wertebereich festzulegen und dann in diesen Wertebereich surjektiv abzubilden.

Als Anwendung des Prinzips von Inklusion und Exklusion beweisen wir noch die Formel 4.23 über die Stirling-Zahlen, die wir später für eine Darstellung der Bell-Zahlen in Satz 4.34 benutzen werden.

Satz 4.23.

$$m! \left\{{n \atop m}\right\} = \sum_k (-1)^{m-k} \binom{m}{k} k^n \quad \text{für } n, m \geq 0$$

Beweis. Sei $A = \{1, \dots, n\}$ und $B = \{1, \dots, m\}$. Für $i \in B$ definieren wir $F_i = \{f : A \to B \mid i \notin f(A)\}$. Mit $|B^A| = m^n$ gilt nach Satz 4.21 die Gleichung $m! \left\{{n \atop m}\right\} = m^n - |\bigcup_{i=1}^{m} F_i|$. Mit der Siebformel 4.17 lässt sich dies schreiben als

$$m! \left\{{n \atop m}\right\} = m^n - \sum_{k=1}^{m} (-1)^{k+1} \sum_{1 \leq r_1 < \cdots < r_k \leq m} |F_{r_1} \cap \dots \cap F_{r_k}|$$

$$= \sum_k (-1)^k \sum_{I \in \binom{\{1,\dots,m\}}{k}} \left| \left\{ f \in B^A \mid f(A) \cap I = \varnothing \right\} \right|$$

$$= \sum_k (-1)^k \sum_{I \in \binom{\{1,\dots,m\}}{k}} (m-k)^n = \sum_k (-1)^k \binom{m}{k} (m-k)^n$$

$$= \sum_k (-1)^{m-k} \binom{m}{k} k^n$$

In der zweiten Zeile ist der Term mit $k = 0$ genau m^n. Für $|I| = k$ enthält die Menge $\{f \in B^A \mid f(A) \cap I = \varnothing\}$ genau $(m - k)^n$ Abbildungen; dies erklärt die dritte Gleichung. Von der vorletzten zur letzten Zeile wurde die Symmetrie $\binom{m}{k} = \binom{m}{m-k}$ benutzt. □

Korollar 4.24. *Für* $0 \leq n \leq m$ *gilt:*

$$\sum_k (-1)^{m-k} \binom{m}{k} k^n = \begin{cases} n! & \text{falls } n = m \\ 0 & \text{falls } n < m \end{cases}$$

Beweis. Für $0 \leq n < m$ gilt $\left\{{n \atop m}\right\} = 0$, und für $0 \leq n = m$ gilt $\left\{{n \atop m}\right\} = \left\{{n \atop n}\right\} = 1$. Die Aussage folgt nun aus dem vorigen Satz. □

Das Korollar 4.24 ist überraschend. Als Übung prüfe man etwa direkt:

$$\binom{5}{1} - \binom{5}{2} 2^3 + \binom{5}{3} 3^3 - \binom{5}{4} 4^3 + \binom{5}{5} 5^3 = 0$$

$$\binom{4}{1} - \binom{4}{2}2^3 + \binom{4}{3}3^3 - \binom{4}{4}4^3 = 0$$

$$\binom{4}{1} - \binom{4}{2}2^4 + \binom{4}{3}3^4 - \binom{4}{4}4^4 = 4! = 24$$

Das Ziel der folgenden Überlegungen ist die Erweiterung der Stirling-Zahlen $\left\{{n \atop k}\right\}$ auf alle $k, n \in \mathbb{Z}$, so dass das Additionstheorem 4.20 allgemein gilt:

$$\left\{{n \atop k}\right\} = \left\{{n-1 \atop k-1}\right\} + k\left\{{n-1 \atop k}\right\} \in \mathbb{N} \tag{4.3}$$

Bisher ist $\left\{{n \atop k}\right\}$ definiert für $n \geq 0$ und $k \in \mathbb{Z}$. Hierfür liegt bereits eine kombinatorische Interpretation vor, und die obige Formel gilt für alle $n \geq 1$. Für $n < 0$ definieren wir zunächst $\left\{{n \atop 0}\right\} = 0$. Damit gilt dann

$$\left\{{n \atop 0}\right\} = \left\{{0 \atop n}\right\} = \begin{cases} 1 & \text{falls } n = 0 \\ 0 & \text{sonst} \end{cases}$$

Dies erweitert die Definition auf alle $n \in \mathbb{Z}$ und $k = 0$. Sei jetzt $k < 0$ und $\left\{{n \atop k+1}\right\}$ für alle $n \in \mathbb{Z}$ schon definiert. Wir setzen für $n \in \mathbb{Z}$ und $k < 0$ induktiv

$$\left\{{n \atop k}\right\} = \left\{{n+1 \atop k+1}\right\} - (k+1)\left\{{n \atop k+1}\right\}$$

Dies definiert die Erweiterung auf $n \in \mathbb{Z}$ und $k \leq 0$. Sei schließlich $n < 0$ und $k > 0$. Wegen

$$\left\{{n \atop k}\right\} = \frac{1}{k}\left(\left\{{n+1 \atop k}\right\} - \left\{{n \atop k-1}\right\}\right) \quad \text{und} \quad \left\{{n \atop 0}\right\} = \left\{{0 \atop k}\right\} = 0$$

ist

$$\left\{{n \atop k}\right\} = 0 \quad \text{falls } nk < 0$$

notwendig und hinreichend, um die Identität auf ganz $\mathbb{Z} \times \mathbb{Z}$ fortzusetzen. Das Additionstheorem 4.20 ist also vollständig bewiesen. Aufgrund dieser Tatsache können wir die Identität aus (4.3) wie folgt umschreiben:

$$\left\{{-k \atop -n}\right\} = \left\{{-(k-1) \atop -(n-1)}\right\} + (n-1)\left\{{-k \atop -(n-1)}\right\} \tag{4.4}$$

Dies sieht zwar etwas seltsam aus, wird aber im nächsten Abschnitt zur Dualität zwischen den Stirling-Zahlen erster Art und zweiter Art führen. Das folgende Schaubild des *Stirling'schen Schmetterlings* gibt ein paar der Werte $\left\{{n \atop k}\right\}$ an.

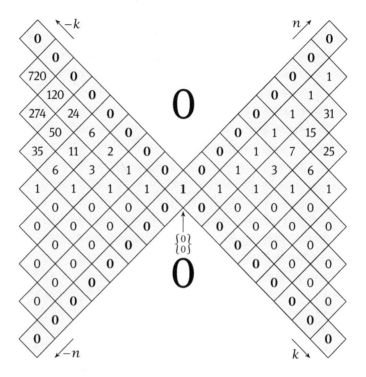

Stirling'scher Schmetterling

4.6.2 Die Stirling-Zahlen erster Art

Die Stirling-Zahlen $\begin{bmatrix} n \\ k \end{bmatrix}$ der ersten Art geben die Anzahl der Möglichkeiten an, n Objekte in k Zykel zu arrangieren. In der Sprache der Permutationen ist dies die Anzahl der Permutationen mit k Zykeln über n Elementen. Sei π eine Permutation der Elemente $A = \{1, \ldots, n\}$ und sei $i \in A$. Dann ist $(i, \pi(i), \pi^2(i), \ldots, \pi^{\ell-1}(i))$ der *Zykel* von i, falls $\pi^\ell(i) = i$ und die Elemente $i, \pi(i), \pi^2(i), \ldots, \pi^{\ell-1}(i)$ paarweise verschieden sind. Hierbei bezeichnet π^j die j-fache Hintereinanderausführung der Permutation π. Jede Permutation lässt sich als Menge von disjunkten Zykeln beschreiben. Die mathematische Sprechweise verwendet hier tatsächlich den Term „Zykel" und nicht „Zyklus" oder „Zyklen". Für $n \in \mathbb{N}$ und $k \in \mathbb{Z}$ definieren wir

$$\begin{bmatrix} n \\ k \end{bmatrix} = |\{\pi \mid \pi \text{ ist Permutation von } \{1, \ldots, n\} \text{ mit } k \text{ Zykeln}\}|$$

Beispiel 4.25. Sei $A = \{1, 2, 3, 4, 5, 6\}$. Dann entspricht $\pi = (1, 2, 4)(3)(5, 6)$ folgender Permutation:

i	1	2	3	4	5	6
$\pi(i)$	2	4	3	1	6	5

Diese Permutation besteht aus drei Zykeln. Weitere Schreibweisen der selben Zykeldarstellungen sind z. B. $(2,4,1)(5,6)(3)$ oder $(3)(6,5)(4,1,2)$. Diese Darstellung lässt sich graphisch folgendermaßen veranschaulichen:

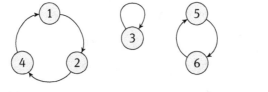

$\Diamond$

Da jede Permutation eine Darstellung durch disjunkte Zykel besitzt, folgt direkt aus der Definition der Stirling-Zahlen erster Art der Satz 4.26.

Satz 4.26.

$$n! = \sum_k \begin{bmatrix} n \\ k \end{bmatrix}$$

Jede Zerlegung in k Zykel definiert eine Partition in k Klassen und jede Partition in k Klassen wird auch durch eine Zerlegung in k Zykel getroffen. Damit können wir festhalten:

$$\begin{Bmatrix} n \\ k \end{Bmatrix} \leq \begin{bmatrix} n \\ k \end{bmatrix} \quad \text{und} \quad \sum_k \begin{Bmatrix} n \\ k \end{Bmatrix} \leq n! \tag{4.5}$$

Beispiel 4.27.

$$\begin{bmatrix} 0 \\ 0 \end{bmatrix} = 1 \qquad\qquad \begin{bmatrix} n \\ 0 \end{bmatrix} = 0 \quad \text{für } n \geq 1$$

$$\begin{bmatrix} n \\ n \end{bmatrix} = 1 \quad \text{für } n \geq 0 \qquad \begin{bmatrix} n \\ k \end{bmatrix} = 0 \quad \text{für } 0 \leq n < k$$

$$\begin{bmatrix} n \\ k \end{bmatrix} = 0 \quad \text{für } n < 0 \leq k \qquad \begin{bmatrix} n \\ 1 \end{bmatrix} = (n-1)! \quad \text{für } n \geq 1$$

$$\begin{bmatrix} n \\ k \end{bmatrix} > 1 \quad \text{für } n > k > 1 \qquad \begin{bmatrix} n \\ n-1 \end{bmatrix} = \begin{Bmatrix} n \\ n-1 \end{Bmatrix} = \binom{n}{2} \quad \text{für } n \geq 1 \qquad \Diamond$$

Satz 4.28 (Additionstheorem für Stirling-Zahlen erster Art).

$$\begin{bmatrix} n \\ k \end{bmatrix} = \begin{bmatrix} n-1 \\ k-1 \end{bmatrix} + (n-1)\begin{bmatrix} n-1 \\ k \end{bmatrix}$$

Beweis. Wir zeigen das Additionstheorem für $n \geq 1$ und $k \in \mathbb{Z}$. Auf der linken Seite zählen wir alle Permutationen von $\{1, \ldots, n\}$ mit k Zykeln. Diese lassen sich in zwei

Typen einteilen: Der erste Typ enthält (n) als Zykel und der zweite Typ nicht. Die Permutationen des ersten Typs entstehen, indem man den Zykel (n) zu einer Permutation von $\{1, \ldots, n-1\}$ mit $k-1$ Zykeln hinzunimmt. Hierfür gibt es $\left[\begin{smallmatrix} n-1 \\ k-1 \end{smallmatrix}\right]$ Möglichkeiten. Als Nächstes überzeugen wir uns davon, dass der zweite Summand den Permutationen des zweiten Typs entspricht. Alle diese Permutationen erhält man, indem man bei einer Permutation von $\{1, \ldots, n-1\}$ mit k Zykeln das Element n einfügt. Es gibt $\left[\begin{smallmatrix} n-1 \\ k \end{smallmatrix}\right]$ solche Permutationen, und bei jeder davon können wir das Element n direkt hinter einem der $n-1$ übrigen Elemente in einen Zykel einzufügen. $\square$

Aus Satz 4.28 und der Gleichung $\left[\begin{smallmatrix} n \\ 1 \end{smallmatrix}\right] = (n-1)!$ in Beispiel 4.27 erhalten wir mit Induktion nach n die Abschätzung

$$\begin{bmatrix} n \\ k \end{bmatrix} \geq (n-k)! \quad \text{für } n \geq k \geq 1 \tag{4.6}$$

Die *steigende Faktorielle* $x^{\overline{n}}$ bezeichnet das Polynom $x^{\overline{n}} = x(x+1) \cdots (x+n-1)$. Das Korollar 4.29 zum Additionstheorem sagt, dass die Werte $\left[\begin{smallmatrix} n \\ k \end{smallmatrix}\right]$ als Koeffizienten vor x^k für $k, n \in \mathbb{N}$ erscheinen, wenn wir dieses Polynom $x^{\overline{n}}$ ausmultiplizieren.

Korollar 4.29.

$$x^{\overline{n}} = \sum_k \begin{bmatrix} n \\ k \end{bmatrix} x^k$$

Beweis. Die Aussage ist richtig für $n = 0$, da $x^{\overline{0}} = 1$ und $\left[\begin{smallmatrix} 0 \\ 0 \end{smallmatrix}\right] = 1$ gilt. Für $k \neq 0$ ist $\left[\begin{smallmatrix} 0 \\ k \end{smallmatrix}\right] = 0$. Wir betrachten jetzt $n \geq 1$:

$$\sum_k \begin{bmatrix} n \\ k \end{bmatrix} x^k = \sum_k \left(\begin{bmatrix} n-1 \\ k-1 \end{bmatrix} + (n-1) \begin{bmatrix} n-1 \\ k \end{bmatrix} \right) x^k$$

$$= \sum_k \begin{bmatrix} n-1 \\ k-1 \end{bmatrix} x^k + \sum_k (n-1) \begin{bmatrix} n-1 \\ k \end{bmatrix} x^k$$

$$= x \cdot x^{\overline{n-1}} + (n-1) x^{\overline{n-1}} = (x+n-1) x^{\overline{n-1}} = x^{\overline{n}}$$

Die letzte Zeile folgt mit Induktion nach n. $\square$

Betrachten wir nun das Polynom der fallenden Faktoriellen:

$$x^{\underline{n}} = \sum_k s(n,k) x^k$$

Die Zahlen $s(n,k)$ erfüllen für $n, k \in \mathbb{N}$ wegen $(-1)^n x^{\overline{n}} = (-x)^{\underline{n}}$ nach Korollar 4.29 die Beziehung:

$$s(n,k) = (-1)^{n-k} \begin{bmatrix} n \\ k \end{bmatrix}$$

Die Werte $s(n,k)$ sind also Stirling-Zahlen der ersten Art „mit Vorzeichen". Das Polynom $x^{\underline{n}}$ hat die Nullstellen $0,\dots,n-1$ und nimmt bei $x = n$ den Wert $n!$ an. Also gilt:

Korollar 4.30. *Für* $0 \le m \le n$ *gilt:*

$$\sum_k (-1)^{n-k} \begin{bmatrix} n \\ k \end{bmatrix} m^k = \begin{cases} n! & \text{falls } n = m \\ 0 & \text{falls } m < n \end{cases}$$

Aus den Additionstheoremen 4.20 und 4.28 und aus den Startwerten in den Beispielen 4.19 und 4.27 folgt zusammen mit der Erweiterung der Stirling-Zahlen zweiter Art auf ganze Zahlen auch Satz 4.31.

Satz 4.31.

$$\begin{bmatrix} n \\ k \end{bmatrix} = \begin{Bmatrix} -k \\ -n \end{Bmatrix}$$

Dies zeigt insbesondere, dass das Additionstheorem 4.28 für die Stirling-Zahlen erster Art für ganze Zahlen gilt. Als Abschluss dieses Abschnitts über die Stirling-Zahlen geben wir noch eine Formel zur Berechnung von $\begin{bmatrix} n \\ 2 \end{bmatrix}$ an. Hierzu benötigen wir die n-te *harmonische Zahl* H_n

$$H_n = \sum_{k=1}^n \frac{1}{k}$$

Die harmonischen Zahlen H_n liegen aufgrund der Integralabschätzung

$$\int_{t=1}^{n+1} \frac{1}{t}\, dt < \sum_{k=1}^n \frac{1}{k} < 1 + \int_{t=1}^n \frac{1}{t}\, dt$$

nahe bei $\ln n$. Bekanntermaßen konvergiert die Differenz $H_n - \ln n$ gegen die Euler'sche Konstante

$$\gamma \quad = \quad 0{,}57772\,15664\,90153\,28606\,06512\,09008\,24024\,31042\,15933\,\cdots$$

Satz 4.32. *Für* $n \ge 1$ *gilt:*

$$\begin{bmatrix} n \\ 2 \end{bmatrix} = (n-1)!\,H_{n-1}$$

Beweis. Wir geben eine „Bauanleitung" für Permutationen von $\{1,\dots,n\}$ mit 2 Zykeln an. Im ersten Schritt schreiben wir die Elemente aus $\{1,\dots,n-1\}$ in einer beliebigen Reihenfolge hintereinander. Hierfür gibt es $(n-1)!$ Möglichkeiten. Am Ende dieser Folge fügen wir n hinzu. Im zweiten Schritt teilen wir die entstandene Folge in zwei nichtleere Folgen auf. Die erste Folge hat die Länge $k \ge 1$, die zweite Folge hat die Länge $n-k \ge 1$ und endet mit dem Element n.

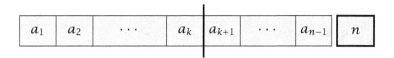

Dies liefert uns $\sum_{k=1}^{n-1}(n-1)!$ verschiedene „Baupläne". Die beiden Folgen repräsentieren zwei Zykel. Die Darstellung des zweiten Zykel ist durch das Element n am Ende eindeutig festgelegt. Bei dem ersten Zykel hingegen gibt es k gleichwertige zyklische Vertauschungen (wir könnten mit jedem der k Elemente beginnen). Dies führt dazu, dass wir für jede Permutation mit 2 Zykeln k verschiedene „Baupläne" angegeben haben, falls der Zykel ohne das Element n die Länge k hat.

$$\begin{bmatrix} n \\ 2 \end{bmatrix} = \sum_{k=1}^{n-1} \frac{(n-1)!}{k} = (n-1)!\,H_{n-1}$$

Umgekehrt findet man für jede Permutation mit 2 Zykeln auch k Baupläne, wenn der Zykel ohne das Element n die Länge k hat. □

4.7 Bell-Zahlen

Die n-te *Bell-Zahl* B_n gibt die Anzahl aller Partitionen einer n-elementigen Menge an. Die Bell-Zahlen sind nach Eric Temple Bell (1883–1960) benannt, der auch unter dem Pseudonym John Taine Science-Fiction Romane schrieb. Da die Stirling-Zahl $\begin{Bmatrix} n \\ k \end{Bmatrix}$ die Anzahl aller Partitionen in k Klassen angibt, erhalten wir die folgende Identität:

$$B_n = \sum_k \begin{Bmatrix} n \\ k \end{Bmatrix}$$

Mit Hilfe der beiden Ungleichungen in (4.2) und (4.5) erhalten wir eine grobe Abschätzung:

$$\left(\frac{n}{2}\right)^{\frac{n}{2}} \leq B_n \leq n! \qquad (4.7)$$

Diese Abschätzung reicht insbesondere für eine untere Schranke $B_n \in 2^{\omega(n)}$.

Satz 4.33. *Es gilt $B_0 = 1$ und*

$$B_{n+1} = \sum_k \binom{n}{k} B_k$$

Beweis. Sei $M = \{1, \ldots, n+1\}$. Für jedes $k \in \{0, \ldots, n\}$ hat man $\binom{n}{k}$ Möglichkeiten für die Auswahl einer Teilmenge A von M mit $|A| = k+1$ und $n+1 \in A$. Weiter gibt es B_{n-k} Möglichkeiten für die Partitionierung der Restmenge $M \setminus A$. Hieraus folgt die Behauptung. □

Mit Hilfe des Additionstheorems 4.20 erhalten wir eine weitere Summendarstellung der Bell-Zahlen durch

$$B_{n+1} = \sum_k (k+1) \begin{Bmatrix} n \\ k \end{Bmatrix}$$

denn es gilt

$$B_{n+1} = \sum_k \begin{Bmatrix} n+1 \\ k \end{Bmatrix} = \sum_k \left(\begin{Bmatrix} n \\ k-1 \end{Bmatrix} + k \begin{Bmatrix} n \\ k \end{Bmatrix} \right)$$

$$= \sum_k \begin{Bmatrix} n \\ k \end{Bmatrix} + \sum_k k \begin{Bmatrix} n \\ k \end{Bmatrix} = \sum_k (k+1) \begin{Bmatrix} n \\ k \end{Bmatrix}$$

Die Formel 4.34 wurde von G. Dobiński im Jahre 1877 gefunden.

Satz 4.34 (Dobiński-Formel).

$$B_n = \frac{1}{e} \sum_{k \geq 0} \frac{k^n}{k!}$$

Beweis. Für alle $N \geq n$ gilt nach Satz 4.23:

$$B_n = \sum_m \begin{Bmatrix} n \\ m \end{Bmatrix} = \sum_{0 \leq m \leq N} \left(\frac{1}{m!} \sum_k (-1)^{m-k} \binom{m}{k} k^n \right)$$

$$= \sum_{0 \leq m \leq N} \left(\frac{1}{m!} \sum_{0 \leq k \leq m} (-1)^{m-k} \frac{m!}{k!(m-k)!} k^n \right)$$

$$= \sum_{0 \leq k \leq N} \left(\frac{k^n}{k!} \sum_{k \leq m \leq N} (-1)^{m-k} \frac{1}{(m-k)!} \right)$$

$$= \sum_{0 \leq k \leq N} \left(\frac{k^n}{k!} \sum_{0 \leq \ell \leq N-k} \frac{(-1)^\ell}{\ell!} \right)$$

Wegen $e^{-1} = \sum_{\ell \geq 0} \frac{(-1)^\ell}{\ell!}$ ergibt sich mit $N \to \infty$ die Behauptung. □

4.8 Partitionszahlen

Eine Partition einer Menge $A = A_1 \cup \cdots \cup A_k$ mit n Elementen in k nichtleere disjunkte Teilmengen bewirkt eine Zerlegung von n in k positive Summanden, nämlich

$$n = |A_1| + \cdots + |A_k|$$

Die Anzahl dieser Partitionen wird durch die Stirling-Zahlen $\begin{Bmatrix} n \\ k \end{Bmatrix}$ der zweiten Art beschrieben, jedoch können verschiedene Partitionen dieselbe Summenzerlegung erzeugen. Da es auf die Reihenfolge nicht ankommt, können wir die Summanden der Größe nach ordnen. Unter einer *Partition* (oder *Zerlegung*) einer Zahl n in k positive

Summanden verstehen wir eine Folge $(n_1, \ldots, n_k) \in \mathbb{N}^k$ mit $n_1 \geq \cdots \geq n_k \geq 1$ sowie

$$n = \sum_{i=1}^{k} n_i$$

Wir bezeichnen mit $\mathcal{P}(n, k)$ die Menge dieser Folgen, und $P(n, k) = |\mathcal{P}(n, k)|$ ist deren Anzahl. Die Zahlen $P(n, k) \in \mathbb{N}$ heißen *(arithmetische) Partitionszahlen.* Sie drücken die Anzahl der Zerlegungen von n in genau k positive Summanden ohne Berücksichtigung der Reihenfolge aus. Die *summatorischen Partitionszahlen* $P(n)$ sind erklärt durch die Anzahl der Zerlegungen von n in positive Summanden, also

$$P(n) = \sum_{k} P(n, k)$$

Beispiel 4.35. Es gilt $(5, 3, 3, 2, 1, 1) \in \mathcal{P}(15, 6)$, denn $15 = 5 + 3 + 3 + 2 + 1 + 1$. Wegen $4 = 3 + 1 = 2 + 2$ ergibt sich $P(4, 2) = 2$. Und es ist $P(7, 3) = 4$, weil $7 = 5 + 1 + 1 = 4 + 2 + 1 = 3 + 3 + 1 = 3 + 2 + 2$. Etwas allgemeiner gilt für $n \geq 0$:

$$P(n, n) = 1 \qquad\qquad P(n, 1) = 1 \ \text{ für } n \geq 1$$

$$P(n, 2) = \left\lfloor \frac{n}{2} \right\rfloor \qquad P(n, n - 1) = 1 \ \text{ für } n \geq 2 \qquad\qquad \Diamond$$

Es gilt $P(0, 0) = 1$, denn es gibt genau eine, nämlich die leere Summe, deren Summanden alle positiv sind und addiert 0 ergeben. Ferner gilt $P(n, 0) = 0$ für alle $n \neq 0$ und $P(n, k) = 0$ für alle $n < 0$.

Einer Partition $(n_1, \ldots, n_k) \in \mathcal{P}(n, k)$ ordnen wir ihr *Ferrers-Diagramm* zu. Dies ist ein Punkteschema, das im kartesischen Koordinatensystem durch die Punkte (i, j) mit $i \in \{1, \ldots, k\}$, $j \in \{1, \ldots, n_i\}$ gegeben wird. Dies bedeutet, dass wir in Spalte i genau n_i viele Punkte zeichnen. Diese Diagramme wurden von Norman Macleod Ferrers (1829–1903) eingeführt, um die Partitionszahlen zu visualisieren. Die Spiegelung an der Hauptdiagonalen führt Ferrers-Diagramme in Ferrers-Diagramme zur selben Zahl n über, siehe Abbildung 4.2. Man erhält so eine Bijektion zwischen der Menge der Zerlegungen von n in genau k positive Summanden und der Menge der Zerlegungen von n in positive Summanden mit k als größtem Summanden. Also ist $P(n, k)$ gleich der Anzahl aller Zerlegungen von n in positive Summanden, in denen k als größter Summand auftritt.

Bisher haben wir die Zahlen $P(n, k)$ für $n \in \mathbb{Z}$ und $k \in \mathbb{N}$ definiert, für die eine kombinatorische Interpretation vorliegt. Wir erweitern den Bereich jetzt auf alle Paare $(n, k) \in \mathbb{Z} \times \mathbb{Z}$, indem wir $P(n, k) = 0$ für $k < 0$ setzen.

Satz 4.36 (Rekursionsformeln für Partitionszahlen).

$$P(n, k) = P(n - 1, k - 1) + P(n - k, k)$$

$$= \sum_{j \leq k} P(n - k, j)$$

$$= \sum_{j \geq 0} P(n - jk, k - 1)$$

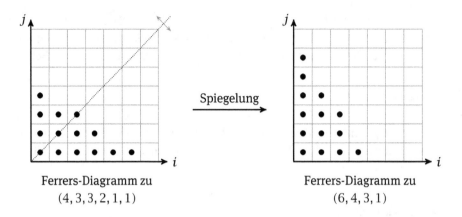

Abb. 4.2. Ferrers-Spiegelung.

Insbesondere gilt $P(n, k) = \sum_{j \leq n-k} P(n - k, j) = P(n - k)$ *für* $k \geq n/2$.

Beweis. Wir zeigen zunächst die erste Gleichung. Diese gilt für $k \leq 0$ oder $n \leq 0$. Seien also $n, k > 0$. Die Zerlegungen von n in genau k positive Summanden zerfallen in zwei Klassen, nämlich in solche, bei denen 1 als Summand auftritt und in solche, bei denen sämtliche Summanden größer als 1 sind. Lassen wir bei den Zerlegungen des ersten Typs einen Summanden 1 weg, so bleibt eine Zerlegung von $n - 1$ in genau $k - 1$ positive Summanden, und dafür gibt es $P(n - 1, k - 1)$ Möglichkeiten. Bei den Zerlegungen des zweiten Typs können wir von jedem Summanden 1 abziehen und erhalten eine Zerlegung von $n - k$ in genau k positive Summanden, wofür es $P(n - k, k)$ Möglichkeiten gibt. Die anderen Gleichungen ergeben sich aus der ersten durch Induktion. □

Beispiel 4.37. Ist $n \geq 4$, so folgt $P(n, n - 2) = \sum_{j=0}^{2} P(2, j) = P(2, 0) + P(2, 1) + P(2, 2) = 0+1+1 = 2$. Ist $n \geq 2k$, so hängt $P(n, n-k) = \sum_{j \leq n-k} P(k, j) = P(2k, k)$ nur von k ab.   ◇

Wir haben gesehen, dass $P(n, k)$ auch die Zahl der Zerlegungen von n in positive Summanden mit k als größtem Summanden ist. Entsprechend kann man die *unteren Partitionszahlen* $p(n, k)$ wie folgt definieren. Es sei $p(n, k)$ die Zahl der Zerlegungen von n in positive Summanden, die alle mindestens den Wert k haben. Also gilt $p(n, n) = 1$ für $n \in \mathbb{N}$ und $p(n, k) = 0$ für $n < 0$ oder $k > n$. Ferner gilt

$$P(n) = p(n, k) \quad \text{für } k \leq 1$$

Satz 4.38 (Rekursionsformel für untere Partitionszahlen). *Für* $k \geq 1$ *gilt*

$$p(n, k) = p(n, n) + \sum_{j \geq k} p(n - j, j)$$

Beweis. Die Gleichung gilt für $n \le 0$. Seien also $n, k \ge 1$. Es gibt eine Zerlegung mit genau einem Summanden. Entfernen wir aus den anderen den kleinsten Summanden, so hat dieser einen Wert j mit $k \le j \le \lfloor n/2 \rfloor$. Dies liefert die Beiträge $p(n - j, j)$. $\qquad\square$

Aus Satz 4.38 folgt für $n \in \mathbb{N}$ und $k \ge 1$ die Gleichung $p(n, k) = 1 + \sum_{j=k}^{\lfloor n/2 \rfloor} p(n - j, j)$, denn es gilt $p(n, n) = 1$ und $p(n - j, j) = 0$ für $j > \lfloor n/2 \rfloor$. Im Kapitel über erzeugende Funktionen zeigen wir $\log P(n) \in \Theta(\sqrt{n})$. Das Wachstum der Partitionszahlen $P(n)$ kann also durch kein Polynom begrenzt werden, es ist aber deutlich langsamer als etwa 2^n.

4.9 Catalan-Zahlen

Die *Catalan-Zahlen* (Eugène Charles Catalan, 1814–1894) tauchen bei einer Vielzahl kombinatorischer Probleme auf. Hiervon werden wir in diesem Abschnitt Klammerausdrücke, Dyck-Wörter und Binärbäume behandeln. Die n-te Catalan-Zahl C_n ist wie folgt definiert:

$$C_n = \frac{1}{n+1}\binom{2n}{n}$$

Durch Umformung sieht man $C_n = \frac{1}{2n+1}\binom{2n+1}{n}$. Das Wachstum des *mittleren* Binomialkoeffizienten $\binom{2n}{n}$ haben wir in Abschnitt 2.2 untersucht. Nach Gleichung (2.4) gilt $\frac{4^n}{2n(n+1)} \le C_n \le 4^n$ und mit Hilfe der Stirling'schen Formel sehen wir

$$C_n \sim \frac{4^n}{n \cdot \sqrt{\pi n}}$$

In der nächsten Tabelle geben wir einige Werte der Catalan-Zahlen an:

n	0	1	2	3	4	5	6	$\cdots$	10	$\cdots$	20
C_n	1	1	2	5	14	42	132	$\cdots$	16796	$\cdots$	6564120420

4.9.1 Dyck-Wörter und Catalan-Zahlen

Dyck-Wörter[1] beschreiben korrekte Klammerungen. So ist ()((()))() korrekt geklammert, während es ())()((() nicht ist. Der besseren Lesbarkeit halber verwenden wir den Buchstaben *a* für „Klammer auf" und *b* für „Klammer zu". Also ist *abaaabbbab* ein Dyck-Wort und *abbabaaabb* ist keins.

---

1 Ritter Walther Franz Anton von Dyck (1856–1934) war der erste Rektor der heutigen Technischen Universität München, als diese 1903 die Rektoratsverfassung erhielt.

Die Menge der Zeichenketten oder Wörter über a, b sei $\{a, b\}^*$. Für ein Wort $w \in \{a, b\}^*$ bezeichne $|w|$ die Länge, $|w|_a$ die Anzahl der a in w und $|w|_b$ die Anzahl der b. Insbesondere gilt $|w| = |w|_a + |w|_b$. Wir schreiben $u \leq w$, falls u ein *Präfix* von w ist, also wenn $uv = w$ für ein Wort $v \in \{a, b\}^*$ gilt. Hierbei ist uv das Wort, das entsteht, wenn man die Wörter u und v hintereinander schreibt. Wir können ein Dyck-Wort $w \in \{a, b\}^*$ durch die folgenden beiden Bedingungen definieren.

(a) $|w|_a = |w|_b$,

(b) $|u|_a \geq |u|_b$ für alle Präfixe u von w.

Mit D_n bezeichnen wir in diesem Abschnitt die Menge der Dyck-Wörter der Länge $2n$. Also gilt:

$$D_n = \left\{ w \in \{a, b\}^{2n} \ \middle| \ |w|_a = n, \ \forall u \leq w : |u|_a \geq |u|_b \right\}$$

Satz 4.39. *Die Anzahl der Dyck-Wörter der Länge n ist C_n, also:*

$$|D_n| = \frac{1}{n+1} \binom{2n}{n}$$

Beweis. Sei $E_n = D_n b$ die Menge der Wörter wb, wobei w ein Dyck-Wort der Länge n ist. Klar ist $|E_n| = |D_n|$. Schließlich sei W_n die Menge der Wörter der Länge $2n + 1$, die an genau $n + 1$ Positionen ein b enthalten. Dann gilt

$$|W_n| = \binom{2n+1}{n+1} = \frac{2n+1}{n+1} \binom{2n}{n}$$

Zu zeigen ist also noch $(2n + 1)|E_n| = |W_n|$. Ein Wort aus W_n kann durch ein *Klammergebirge* visualisiert werden: Bei einem a gehen wir nach oben und bei einem b nach unten. Beginnen wir auf dem Niveau Null, so enden wir auf dem Niveau -1. Die Wörter aus E_n erkennen wir daran, dass wir bis auf den letzten Schritt oberhalb der Null-Linie bleiben. Allgemein gilt dies nicht, so entspricht dem Wort $w = abbabaaabbb \in W_5$ das folgende Klammergebirge:

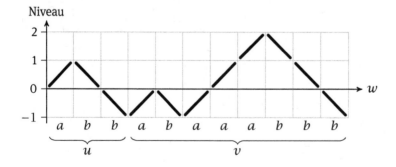

Betrachten wir jetzt für ein Wort w aus W_n den kürzesten Präfix u, der eine tiefste Position erreicht und zerlegen das Wort w in uv, so ist die zyklische Vertauschung vu ein Wort aus E_n. In dem Beispiel ist $u = abb$ und $vu = abaaabbbabb$

$\in E_5$. Man überzeugt sich leicht, dass *abb* der einzige Präfix ist, der auf diese Weise zu einem Wort aus E_5 führt. Diese Beobachtung lässt sich auf jedes Wort in W_n anwenden.

Zunächst ist klar, dass u nicht leer ist und mit einem b enden muss. Angenommen, es existieren zwei Präfixe u_1 und u_1u_2 von $w = u_1u_2v \in W_n$ mit $0 < |u_1| < |u_1u_2| \le |w|$, so dass u_2vu_1 und vu_1u_2 beides Wörter aus E_n sind. Betrachten wir das Niveau ℓ am Ende von u_2, wenn wir bei Niveau 0 beginnen. Aus $u_2vu_1 \in E_n$ folgt $\ell \ge 0$, da das Niveau von Wörtern aus E_n bis vor dem letzten Zeichen nie negativ ist. Aus $vu_1u_2 \in E_n$ folgt $\ell < 0$, da das Niveau von vu_1u_2 mit dem letzten Zeichen aus u_2 negativ wird und vorher nie negativ war. Dies ist ein Widerspruch. Damit entsprechen jedem Wort aus E_n genau $2n + 1$ zyklische Vertauschungen in W_n, und jedes Wort aus W_n lässt sich durch zyklische Vertauschung eindeutig einem Wort aus E_n zuordnen. Wir erhalten wie gewünscht $(2n + 1)|E_n| = |W_n|$. □

4.9.2 Binärbäume und Catalan-Zahlen

Sucht man einen Namen in einem Telefonbuch, so sieht ein typisches Vorgehen etwa wie folgt aus. Das Buch wird irgendwo aufgeschlagen und anhand des oben stehenden Buchstabens entscheidet sich, ob man weiter vorne oder weiter hinten sucht. Dies ist der erste Schritt eines allgemeinen Prinzips – der *binären Suche*. Die Idee ist, einer linear geordneten Menge eine Datenstruktur zuzuordnen, die schnelles Aufsuchen, Einfügen und Löschen erlaubt. Die Basisstruktur hierfür sind Binärbäume.

Wir definieren zunächst *saturierte Binärbäume* induktiv. Ein einzelner Knoten v definiert einen saturierten Binärbaum mit der Knotenmenge $\{v\}$. Der Knoten von v ist zugleich *Wurzel* und *Blatt*. Die *Höhe* des Baumes ist 0, und die Menge der inneren Knoten ist leer. Seien jetzt B_1 und B_2 saturierte Binärbäume mit Knotenmengen V_1 und V_2, wobei wir $V_1 \cap V_2 = \emptyset$ annehmen. Sei v ein neuer Knoten, $v \notin V_1 \cup V_2$. Dann definieren wir einen saturierten Binärbaum B mit Knotenmenge $\{v\} \cup V_1 \cup V_2$ wie folgt. Die Wurzel ist v, diese hat als *linkes Kind* die Wurzel von B_1 und als *rechtes Kind* die Wurzel von B_2. Die Menge der Blätter ist die Vereinigung der Blätter von B_1 und B_2. Damit ist v also kein Blatt. Die Menge der inneren Knoten besteht jetzt aus v und den inneren Knoten von B_1 und B_2. Hat B_i die Höhe h_i für $i = 1,2$, so erhält B die Höhe $\max\{h_1, h_2\} + 1$.

In einem saturierten Binärbaum gibt es also genau eine Wurzel, die inneren Knoten haben genau zwei Kinder, während die Blätter die Knoten ohne Kinder sind. Gibt es n Blätter, so gibt es genau $n - 1$ innere Knoten. Dies folgt mit Induktion und der Beobachtung $1 + (n - 1) + (m - 1) = (n + m) - 1$. Die Knotenzahl bei n inneren Knoten ist also $2n + 1$ und damit insbesondere ungerade.

Einen allgemeinen Binärbaum erhalten wir, indem wir bei einem saturierten Binärbaum alle Blätter entfernen. Dies verringert die Höhe um 1 und die Knoten haben

nun 0, 1 oder 2 Kinder. Neu ist also, dass der Baum leer sein kann und dass es Knoten mit nur einem Kind geben kann. Wir übernehmen die alte Wurzel (sofern sie über-lebt hat) und teilen die Knoten wieder in innere Knoten (Knoten mit einem oder zwei Kindern) sowie Blätter (Knoten ohne Kinder) ein.

saturierter Binärbaum   allgemeiner Binärbaum

Ausgehend von einem Binärbaum können wir den saturierten Binärbaum rekonstru-ieren, indem wir Blätter anfügen. Zwischen der Menge der Binärbäume mit n Knoten und der Menge der saturierten Binärbäume mit $2n + 1$ Knoten gibt es also eine natür-liche Bijektion. Deren Anzahl wird genau wie die Anzahl der Dyck-Wörter durch die Catalan-Zahlen beschrieben.

Satz 4.40. *Die Anzahl der Binärbäume mit n Knoten (bzw. die Anzahl der saturierten Binärbäume mit n inneren Knoten) ist $C_n = \frac{1}{n+1}\binom{2n}{n}$.*

Beweis. Nach den Vorbemerkungen reicht es, die Anzahl der saturierten Binärbäume mit $2n + 1$ Knoten zu bestimmen. Diese codieren wir durch Wörter der Länge $2n + 1$ über einem Alphabet mit zwei Buchstaben. Die informelle Beschreibung ist eine Tie-fensuche von links nach rechts: Besuchen wir in dieser Tiefensuche das erste Mal einen inneren Knoten, so schreiben wir ein a, bei einem Blatt schreiben wir ein b. Insgesamt schreiben wir also n mal ein a und $n + 1$ mal ein b. Bis wir das letzte Blatt besucht haben, haben wir zu jedem Zeitpunkt mindestens so viele innere Knoten wie Blätter besucht. In den Bezeichnungen aus dem letzten Abschnitt ist das geschriebe-ne Wort eine Zeichenkette aus $E_n = D_n b$. Dem saturierten Binärbaum aus der obigen Skizze entspricht das Wort *abaaabbbabb*.

Aufgrund von Satz 4.39 müssen wir daher nur die Menge $\mathcal{B}_n$ der saturierten Bi-närbäume mit n inneren Knoten in Bijektion mit E_n setzen. Hierfür formalisieren wir die Tiefensuche durch eine Abbildung code : $\mathcal{B}_n \to \{a,b\}^{2n+1}$. Diese wird induktiv definiert:

Für $n = 0$ setzen wir code$(B) = b$ für den einzigen Baum $B \in \mathcal{B}_0$. Sei jetzt $B \in \mathcal{B}_n$ mit $n > 0$ und v die Wurzel von B. Sei L der linke Teilbaum unterhalb von v und R der rechte. Dann setzen wir

$$\text{code}(B) = a \cdot \text{code}(L) \cdot \text{code}(R)$$

Mit Induktion sind code(L) und code(R) Dyck-Wörter gefolgt von einem b, also ist auch code(B) ein solches Wort; und damit gilt code$(B) \in E_n$. Da code(B) mit ei-

nem a beginnt, ist $a \cdot \mathrm{code}(L)$ der kürzeste nichtleere Präfix, der ein Dyck-Wort ist. Damit können wir B rekonstruieren, wenn wir $\mathrm{code}(B)$ kennen. Die Abbildung code ist also injektiv.

Betrachten wir ein beliebiges Wort $w \in E_n$, so gibt es genau eine Zerlegung $w = aubvb$ für die u und v Dyck-Wörter sind. Mit Induktion gilt $ub = \mathrm{code}(L)$ und $vb = \mathrm{code}(R)$ für saturierte Binärbäume, also ist $w = \mathrm{code}(B)$ für ein $B \in \mathcal{B}_n$. Die Abbildung ist damit auch surjektiv und insgesamt bijektiv. □

Aus der Bildungsvorschrift für Binärbäume ergibt sich Korollar 4.41.

Korollar 4.41. *Die Catalan-Zahlen erfüllen das folgende Bildungsgesetz:*

$$C_0 = 1, \quad C_{n+1} = \sum_k C_k C_{n-k} \quad \text{für } n \in \mathbb{N}$$

4.10 Die mittlere Höhe binärer Suchbäume

Wir setzen in diesem Abschnitt die Untersuchung der binären Suche fort. Wir untersuchen folgende Fragestellung: Wie lange müssen wir in einem zufällig erzeugten Binärbaum suchen, um einen Eintrag zu finden? „Zufällig erzeugt" bedeutet für uns, dass die Elemente in einer beliebigen Reihenfolge in den Suchbaum eingefügt werden und dass dabei jede mögliche Reihenfolge gleich wahrscheinlich ist. Die Dauer einer Suche entspricht der Länge des Suchpfades. Es wird sich herausstellen, dass wir durchschnittlich nur logarithmisch lange (in der Anzahl der Elemente) suchen müssen.

Es existieren verschiedene Ansätze, bei denen die Elemente nicht ganz naiv in den Suchbaum eingefügt werden, und die diese Komplexität auch im schlechtesten Fall erreichen. Tatsächlich zeigt die folgende Abschätzung, dass man sich den zusätzlichen Aufwand beim Einfügen ersparen kann, wenn die Reihenfolge der Daten zufällig ist.

Sei π eine Permutation der Elemente $\{1, \ldots, n\}$, dann schreiben wir auch $\pi = (\pi(1), \pi(2), \ldots, \pi(n))$. Diese Schreibweise ist nicht zu verwechseln mit der Zykelschreibweise aus Abschnitt 4.6.2. Für $I \subseteq \{1, \ldots, n\}$ sei $B_I(\pi)$ der binäre Suchbaum, der durch Einfügen der Elemente $i \in I$ in den zunächst leeren Baum entsteht, wenn die Reihenfolge durch π (von links nach rechts) gegeben wird. Für $I = \{1, \ldots, n\}$ schreiben wir $B(\pi)$ statt $B_{\{1,\ldots,n\}}(\pi)$.

Betrachten wir ein Beispiel: Für $n = 7$, $\pi = (3, 2, 6, 1, 5, 7, 4)$ und $I = \{2, 3, 4, 5, 7\}$ ist $B_{\{2,3,4,5,7\}}(\pi)$ der Baum, der entsteht, wenn in den anfänglich leeren Baum der Reihe nach die Elemente 3, 2, 5, 7 und 4 eingefügt werden. Beachte $\pi^{-1}(3) < \pi^{-1}(2) < \pi^{-1}(5) < \pi^{-1}(7) < \pi^{-1}(4)$.

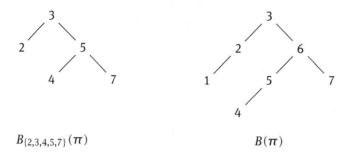

$$B_{\{2,3,4,5,7\}}(\pi) \qquad\qquad\qquad B(\pi)$$

Wir definieren folgende Zufallsvariablen:

$$R_i(\pi) = \text{„Wurzel von } B(\pi) \text{ ist } i\text{“}$$
$$X_I(\pi) = \text{„Höhe von } B_I(\pi)\text{“}$$
$$Y_I(\pi) = 2^{X_I(\pi)}$$

Wir wollen den Erwartungswert $E[X_n]$ abschätzen, wobei $X_n = X_{\{1,\dots,n\}}$ meint. Dabei gehen wir von einer Gleichverteilung aller Permutationen aus. Es stellt sich heraus, dass es geschickter und einfacher ist, zuerst $E[Y_n]$ zu betrachten. Zunächst zeigen wir, dass $E[Y_n] = E[2^{X_n}]$ durch ein Polynom 3. Grades abgeschätzt werden kann. Es gilt $E[Y_1] = 1$. Sei jetzt $n \geq 2$. Wegen

$$Y_n(\pi) = 2 \sum_{i=1}^{n} R_i(\pi) \cdot \max\left\{ Y_{\{1,\dots,i-1\}}(\pi), Y_{\{i+1,\dots,n\}}(\pi) \right\}$$

gilt:

$$E[Y_n] = 2 \sum_{i=1}^{n} E\left[R_i \cdot \max\{Y_{\{1,\dots,i-1\}}, Y_{\{i+1,\dots,n\}}\} \right]$$

Man beachte, dass $R_i(\pi) = 1$, falls $\pi(1) = i$ ist, und $R_i(\pi) = 0$ andernfalls. Für $i \notin I$ sind die Zufallsvariablen R_i und Y_I unabhängig. Also gilt:

$$E[Y_n] = 2 \sum_{i=1}^{n} E[R_i] \cdot E\left[\max\{Y_{\{1,\dots,i-1\}}, Y_{\{i+1,\dots,n\}}\} \right]$$

Es gilt $E[R_i] = \frac{1}{n}$ und $\max\{Y_I, Y_J\} \leq Y_I + Y_J$ und damit

$$E[Y_n] \leq \frac{2}{n} \sum_{i=1}^{n} \left(E[Y_{\{1,\dots,i-1\}}] + E[Y_{\{i+1,\dots,n\}}] \right)$$

Aufgrund der Linearität der Erwartungswerte sowie der Eigenschaft

$$E[Y_I] = E[Y_{\{1,\dots,|I|\}}] = E[Y_{|I|}]$$

erhalten wir

$$E[Y_n] \leq \frac{4}{n} \sum_{i=1}^{n} E[Y_{i-1}] = \frac{4}{n} \sum_{i=0}^{n-1} E[Y_i],$$

da jeder Term zweimal gezählt wird. Wir zeigen nun, dass $E[Y_n] \le \frac{1}{4}\binom{n+3}{3}$ gilt. Für $n = 1$ gilt $E[Y_1] = 1 = \frac{1}{4}\binom{4}{3}$. Sei ab jetzt $n \ge 2$. Dann gilt mit $E[Y_0] = 0$:

$$E[Y_n] \le \frac{4}{n}\sum_{i=0}^{n-1} E[Y_i] \overset{\text{IV}}{\le} \frac{4}{n}\sum_{i=0}^{n-1}\frac{1}{4}\binom{i+3}{3}$$

$$= \frac{1}{n}\sum_{i=0}^{n-1}\binom{i+3}{3} = \frac{1}{n}\cdot\binom{n+3}{4} = \frac{1}{4}\cdot\binom{n+3}{3}$$

Die vorletzte Gleichung folgt mit Satz 4.7 zur oberen Summation. Jetzt haben wir also eine Abschätzung für $E[Y_n]$. Um daraus $E[X_n]$ ableiten zu können, wenden wir die Jensen'sche Ungleichung nach Korollar 3.6 mit der konvexen Funktion $f : x \mapsto 2^x$ an. Wir erhalten

$$2^{E[X_n]} \le E[2^{X_n}] = E[Y_n] \le \frac{1}{4}\binom{n+3}{3} \le cn^3 + c$$

für eine geeignete Konstante $c \in \mathbb{R}$. Daraus erhalten wir den Satz 4.42 über die durchschnittliche Höhe X_n binärer Suchbäume mit n Knoten.

Satz 4.42.
$$E[X_n] \le 3\log_2 n + \mathcal{O}(1) \in \mathcal{O}(\log n)$$

Aufgaben

4.1. Seien A, B und C beliebige Mengen und sei A^B die Menge der Abbildungen von B nach A.

(a) Zeigen Sie, dass die Mengen $C^{(A\times B)}$ und $(C^B)^A$ bijektiv aufeinander abgebildet werden können.

(b) Zeigen Sie, dass die Mengen $C^{A\cup B}$ und $C^A \times C^B$ bijektiv aufeinander abgebildet werden können, falls $A \cap B = \varnothing$ gilt.

(c) Zeigen Sie, dass die Mengen A und 2^A nicht bijektiv aufeinander abgebildet werden können, da es keine Surjektion von A auf 2^A gibt.

4.2. Wie viele 9-stellige Zahlen gibt es, in welchen jede Ziffer zwischen 0 und 9 höchstens einmal vorkommt, die 0 aber mindestens einmal vorkommt?

4.3. Aus einer Menge von 15 Frauen und 12 Männern soll eine Kommission mit 8 Mitgliedern gewählt werden.

(a) Wie viele Wahlmöglichkeiten gibt es, wenn die Kommission gleich viele Männer wie Frauen enthalten soll?

(b) Wie viele Wahlmöglichkeiten gibt es, wenn die Kommission mindestens 2 Männer enthalten soll?

(c) Wie viele Wahlmöglichkeiten gibt es, wenn die Kommission mehr Männer als Frauen enthalten soll?

4.4. In dem Spiel *Carcassone* sind die Karten quadratisch. Die vier Seiten entsprechen einer Straße (s), einer befestigten Stadt (b) oder einer Wiese (w). Wie viele Muster gibt es? Mit einem Muster meinen wir etwa, dass sich zwei Straßen- und zwei Wiesenseiten gegenüber liegen.

4.5. Zeigen Sie, dass jede nichtleere endliche Menge gleich viele Teilmengen mit einer geraden Anzahl von Elementen wie Teilmengen mit einer ungeraden Anzahl von Elementen enthält.

4.6. Zeigen Sie:

(a) $\displaystyle\sum_{k=m}^{n} \binom{k}{m}\binom{n}{k} = \binom{n}{m} \cdot 2^{n-m}$

(b) $\displaystyle\sum_{k,\ell} \binom{n}{k}\binom{k}{\ell}\ell = n \cdot 3^{n-1}$

(c) $\displaystyle\sum_{i}\sum_{j} \binom{n}{i}\binom{n+i}{j} = 6^n$

(d) $\displaystyle\sum_{k} \binom{m-k}{n}\binom{m+k}{n} = \binom{2m+1}{2n+1}$

(e) $\displaystyle\sum_{k=1}^{m} \left(\binom{m+1}{k}\sum_{i=1}^{n} i^k\right) = (n+1)^{m+1} - (n+1)$

4.7. Die Fibonacci-Zahlen sind $F_0 = 0$, $F_1 = 1$ und $F_n = F_{n-1} + F_{n-2}$ für $n \geq 2$. Zeigen Sie für $n \in \mathbb{N}$:

(a) $\displaystyle F_{n+1} = \sum_{k \leq n} \binom{n-k}{k}$

(b) $\displaystyle F_{2n} = \sum_{i} \binom{n}{i}F_i$

(c) $\displaystyle F_{3n} = \sum_{i} \binom{n}{i}2^i F_i$

(d) $\displaystyle 0 = \sum_{i} \binom{n}{i}(-1)^i F_{n+i}$

4.8. Sei $n \geq 3$. Mit $G^{(3)}(n)$ bezeichnen wir die Anzahl aller Teilmengen $A \subseteq \{1, \ldots, n\}$ mit $|A| = 3$ und $\mathrm{sum}(A)$ gerade. Dabei ist $\mathrm{sum}(A) = \sum_{a \in A} a$. Geben Sie eine Formel für $G^{(3)}(n)$ an.

4.9. Zeigen Sie:

(a) $\displaystyle\sum_k \binom{n}{k}\left\{{k \atop m}\right\} = \left\{{n+1 \atop m+1}\right\}$

(b) $\displaystyle\sum_k \left[{n \atop k}\right]k = \left[{n+1 \atop 2}\right]$

(c) $\displaystyle\sum_k \left[{n \atop k}\right]\binom{k}{m} = \left[{n+1 \atop m+1}\right]$

4.10. Die hundert Schlümpfe wurden vom König gefangen genommen, da er neidisch auf ihre Klugheit war. Der König hat einen Schrank mit 100 Schubladen, die die Nummern 1 bis 100 tragen und in welche er die Ausweise der Schlümpfe so legt, dass jede Schublade genau einen Ausweis enthält. Die Schlümpfe bekommen noch eine letzte Chance, aus der Gefangenschaft entlassen zu werden. Der König erklärt ihnen sein Spiel: In einer zufälligen Reihenfolge werden die Schlümpfe nacheinander in das Zimmer mit dem Schubladenschrank gelassen. Sie werden nicht wissen, wie viele vor ihnen an der Reihe waren. In dem Zimmer dürfen Sie 50 Schubladen eine nach der anderen öffnen und hinein schauen, aber nicht die Ausweise berühren. Danach werden die Schubladen wieder verschlossen. Am Ende sieht alles vollkommen unverändert aus. Wenn ein Schlumpf dabei seinen Ausweis sah, darf er zurück in die Zelle und dort warten. Ansonsten bricht das Spiel ab und der König erklärt, dass er schon bei einem Versagen eines Einzelnen alle Schlümpfe auf ewige Zeit in ihren Zellen festhalten wird. Der König verpflichtet sich jedoch die Schlümpfe freizulassen, wenn jeder der hundert Schlümpfe seinen Ausweis während des Besuches im Zimmer gesehen hat. Er rechnet sich aus, dass ein Schlumpf nur mit Wahrscheinlichkeit $1/2$ seinen Ausweis finden kann, egal welche Strategie er im Kopf hat. Dies ist richtig! Sein Irrtum ist zu glauben, dass er sie nur mit der Wahrscheinlichkeit 2^{-100} entlassen muss. Daher dürfen sich die Schlümpfe kurz zum letzten Mal besprechen. Danach kommt jeder Schlumpf in Einzelhaft und jegliche Kommunikation unter ihnen wird verhindert. Das Spiel beginnt.

(a) Zeigen Sie, dass die schlauen Schlümpfe eine Strategie haben, mit einer Wahrscheinlichkeit von mehr als 31% frei zu kommen.

(b) Zeigen Sie, dass es keine Strategie gibt, bei der die Schlümpfe mit einer Wahrscheinlichkeit von 32% oder mehr frei kommen.

Hinweis: Betrachten Sie das folgende Spiel, welches für die Schlümpfe nicht schwieriger ist. Ohne Einschränkung seien die Schlümpfe durchnummeriert von 1 bis 100. Am Anfang befinden sich alle Schlümpfe im Raum und Schlumpf Nummer 1 beginnt, Schubladen zu öffnen. Dies tut er so lange, bis er seinen Ausweis gefunden hat. Die Schubladen, welche er öffnet, werden nicht wieder verschlossen. Als Nächstes kommt der Schlumpf mit der kleinsten Nummer an

die Reihe, welcher seinen Ausweis noch in keiner geöffneten Schublade vorfindet. Auch er öffnet wieder Schubladen (ohne diese wieder zu verschließen), bis er seinen Ausweis findet. Dies wird solange wiederholt, bis alle Schubladen geöffnet sind. Jeder kann jederzeit in die offenen Schubladen hineinsehen. Am Ende kommen die Schlümpfe frei, wenn kein Schlumpf mehr als 50 Schubladen geöffnet hat.

4.11. Wir betrachten folgendes Ratespiel mit Spielern Alice und Bob: Zunächst einigen sich Alice und Bob auf Zahlen $n, r \in \mathbb{N}$. Dann wählt Alice eine beliebige Menge $R \subseteq \{1, \ldots, n\}$. Das Ziel von Bob ist es, R zu bestimmen. Dazu darf er bis zu r Fragen der Form

$$\text{„Ist } R \cap M = \varnothing ?\text{“}$$

für beliebige Mengen $M \subseteq \{1, \ldots, n\}$ stellen. Alice antwortet auf diese Fragen wahrheitsgemäß. Nach maximal r Fragen benennt Bob eine Menge R und hat gewonnen, falls seine Behauptung korrekt ist.

Beispiel: Es ist $(n, r) = (5, 4)$ und das Spiel verläuft wie folgt:

Bob:	Ist $R \cap \{1, 2, 3\} = \varnothing$?	Alice:	Nein.
Bob:	Ist $R \cap \{4, 5\} = \varnothing$?	Alice:	Ja.
Bob:	Ist $R \cap \{1, 2\} = \varnothing$?	Alice:	Ja.
Bob:	R ist gleich $\{3\}$!		

Damit hat Bob bereits nach drei Fragen gewonnen.

(a) Geben Sie eine Gewinnstrategie für Bob an, die mit n Fragen auskommt.

(b) Zeigen Sie, dass dies optimal ist, d.h., zu jeder Strategie gibt es eine Ratemenge, für die die Strategie mehr als $n - 1$ Fragen benötigt.

(c) Ist ein Spiel mit $r = n - 1$ fair?

4.12. Zeigen Sie:

(a) $C_n = \dbinom{2n}{n} - \dbinom{2n}{n+1}$

(b) $C_n = \dfrac{1}{n+1} \sum_k \dbinom{n}{k}^2$

(c) $C_{n+1} = \dfrac{2(2n+1)}{n+2} C_n$

4.13. Zeigen Sie, dass es genau C_{n-2} viele Möglichkeiten gibt, ein regelmäßiges n-Eck mit $n \geq 3$ Knoten in Dreiecke zu unterteilen (triangulieren). Die folgende Abbil-

dung zeigt die $C_4 = 14$ Möglichkeiten, um ein Hexagon zu triangulieren:

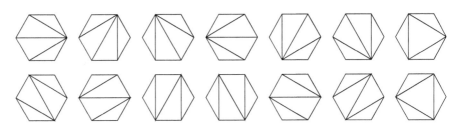

4.14. Eine Familie $\mathcal{A} \subseteq 2^{\{1,\dots,n\}}$ heißt *Antikette*, falls ihre Mitglieder paarweise unvergleichbar sind, d. h., aus $M \subseteq N$ folgt $M = N$. Zeigen Sie:

(a) (Satz von Sperner) Jede Antikette enthält höchstens $\binom{n}{\lfloor n/2 \rfloor}$ Teilmengen.

   Hinweis: Jedes Mitglied eine Antikette kommt in einer maximalen Kette (vergleiche Abschnitt 7.1) vor und maximale Ketten enthalten keine zwei verschiedenen Mitglieder der selben Antikette.

(b) Die obige Schranke ist scharf.

Zusammenfassung

Begriffe

- gleichmächtig
- Abbildungen B^A
- Fakultät $n!$
- fallende Faktorielle $n^{\underline{k}}$
- Potenzmenge 2^A
- charakteristische Abbildung
- Binomialkoeffizienten $\binom{n}{k}$
- k-elementige Teilmengen $\binom{A}{k}$
- Multinomialkoeffizienten $\binom{n}{k_1,\dots,k_d}$
- Bubble-Sort
- Fehlstellung
- Rencontres-Zahlen $R_n, R_{n,m}$
- Fixpunkt
- Stirling-Zahlen erster Art $\left[{n \atop k} \right]$
- Stirling-Zahlen zweiter Art $\left\{ {n \atop k} \right\}$

- Partition, Klassen
- Zykel, Zykelschreibweise
- steigende Faktorielle $n^{\overline{k}}$
- harmonische Zahlen H_n
- Bell-Zahlen B_n
- arithmetische Partitionszahlen $P(n, k)$
- summatorische Partitionszahlen $P(n)$
- Ferrers-Diagramm
- untere Partitionszahlen $p(n, k)$
- Catalan-Zahlen C_n
- Dyck-Wort
- Klammergebirge
- (saturierter) Binärbaum

Methoden und Resultate

- kombinatorische Interpretation, bijektiver Beweis, Polynommethode
- $|B^A| = |B|^{|A|}$, $|2^A| = 2^{|A|}$, $\left|\binom{A}{k}\right| = \binom{|A|}{k}$
- Es gibt $n!$ Permutationen auf $\{1, \ldots, n\}$.
- Es gibt $n^{\underline{k}}$ injektive Abbildungen von $\{1, \ldots, k\}$ nach $\{1, \ldots, n\}$.
- $\frac{1}{e^{n-1}} \le \frac{n!}{n^n} \le \frac{n}{e^{n-1}}$
- $\binom{n}{k} = \binom{n}{n-k}$ für $n \in \mathbb{N}$ und $k \in \mathbb{Z}$
- Additionstheorem: $\binom{x}{k} = \binom{x-1}{k} + \binom{x-1}{k-1}$
- Binomialsatz: $(x+y)^r = \sum_k \binom{r}{k} x^k y^{r-k}$ für $|x| < |y|$ oder für $r \in \mathbb{N}$
- Trinomiale Revision: $\binom{x}{m}\binom{m}{k} = \binom{x}{k}\binom{x-k}{m-k}$
- Binomialinversion: $f_i = \sum_k \binom{i}{k} g_k$ für $0 \le i \le n \Rightarrow g_n = \sum_k (-1)^{n-k}\binom{n}{k} f_k$
- Urnenmodell: Ziehen mit/ohne Zurücklegen und mit/ohne Reihenfolge
- Gauß-Formel: $\sum_{k=0}^n k = \binom{n+1}{2}$
- Obere Summation: $\binom{n+1}{m+1} = \sum_{0 \le k \le n} \binom{k}{m}$
- Parallele Summation: $\binom{x+n+1}{n} = \sum_{k \le n} \binom{x+k}{k}$
- Vandermonde'sche Identität: $\binom{x+y}{n} = \sum_k \binom{x}{k}\binom{y}{n-k}$
- $|\{(e_1, \ldots, e_\ell) \in \mathbb{N}^\ell \mid \sum_{1 \le k \le \ell} e_k \le t\}| = \binom{t+\ell}{\ell}$
- Multinomialsatz: $(x_1 + \cdots + x_d)^n = \displaystyle\sum_{k_i \ge 0,\, k_1 + \cdots + k_d = n} \binom{n}{k_1, \ldots, k_d} x_1^{k_1} \cdots x_d^{k_d}$
- Bubble-Sort benötigt im Durchschnitt $\Theta(n^2)$ Vergleiche.
- Siebformel von Sylvester: $|A_1 \cup \cdots \cup A_n| = \displaystyle\sum_{k \ge 1} (-1)^{k+1} \sum_{1 \le r_1 < \cdots < r_k \le n} |A_{r_1} \cap \cdots \cap A_{r_k}|$
- $R_n = n! \sum_{k=0}^n \frac{(-1)^k}{k!}$, $R_{n,m} = \frac{n!}{m!} \sum_{k=0}^{n-m} (-1)^k \frac{1}{k!}$
- Additionstheorem für Stirling-Zahlen zweiter Art: $\left\{\begin{matrix} n \\ k \end{matrix}\right\} = \left\{\begin{matrix} n-1 \\ k-1 \end{matrix}\right\} + k\left\{\begin{matrix} n-1 \\ k \end{matrix}\right\}$
- Es gibt $m!\left\{\begin{matrix} n \\ m \end{matrix}\right\}$ Surjektionen von $\{1, \ldots, n\}$ auf $\{1, \ldots, m\}$.
- $\left\{\begin{matrix} 2n \\ n \end{matrix}\right\} \ge n^n$
- $x^n = \sum_k \left\{\begin{matrix} n \\ k \end{matrix}\right\} \cdot x^{\underline{k}}$
- $m!\left\{\begin{matrix} n \\ m \end{matrix}\right\} = \sum_k (-1)^{m-k}\binom{m}{k} k^n$
- Stirling'scher Schmetterling
- $n! = \sum_k \left[\begin{matrix} n \\ k \end{matrix}\right]$, $\left\{\begin{matrix} n \\ k \end{matrix}\right\} \le \left[\begin{matrix} n \\ k \end{matrix}\right]$, $\sum_k \left\{\begin{matrix} n \\ k \end{matrix}\right\} \le n!$
- Additionstheorem für Stirling-Zahlen erster Art: $\left[\begin{matrix} n \\ k \end{matrix}\right] = \left[\begin{matrix} n-1 \\ k-1 \end{matrix}\right] + (n-1)\left[\begin{matrix} n-1 \\ k \end{matrix}\right]$

- $x^{\overline{n}} = \sum_k \left[{n \atop k} \right] x^k, \; x^{\underline{n}} = \sum_k (-1)^{n-k} \left[{n \atop k} \right] x^k$

- Für $0 \le m \le n$ gilt $\sum_k (-1)^{n-k} \binom{n}{k} k^m = \sum_k (-1)^{n-k} \left[{n \atop k} \right] m^k = \begin{cases} n!, & n = m \\ 0, & m < n \end{cases}$

- $\left[{n \atop k} \right] = \left\{ {-k \atop -n} \right\}$

- $\left[{n \atop 2} \right] = (n-1)! \, H_{n-1}$ für $n \ge 1$

- $(\frac{n}{2})^{\frac{n}{2}} \le B_n = \sum_k \left\{ {n \atop k} \right\} \le n!$

- $B_0 = 1$ und $B_{n+1} = \sum_k \binom{n}{k} B_k$

- Dobiński-Formel: $B_n = \frac{1}{e} \sum_{k \ge 0} \frac{k^n}{k!}$

- $P(n) = \sum_k P(n, k)$

- Ferrers-Spiegelung

- $P(n, k) = P(n-1, k-1) + P(n-k, k) = \sum_{j \le k} P(n-k, j) = \sum_{j \ge 0} P(n - jk, k-1)$

- $p(n, k) = p(n, n) + \sum_{j \ge k} p(n-j, j)$ für $k \ge 1$

- $C_n = \frac{1}{n+1} \binom{2n}{n} =$ Anzahl Dyck-Wörter der Länge $2n =$ Anzahl Binärbäume mit n Knoten = Anzahl saturierter Binärbäume mit n inneren Knoten

- $C_0 = 1$ und $C_{n+1} = \sum_k C_k C_{n-k}$

- Die mittlere Höhe von binären Suchbäumen mit n Knoten ist in $\mathcal{O}(\log n)$.

5 Erzeugende Funktionen

Häufig untersucht man Folgen a_n von reellen oder komplexen Zahlen mit $|a_n| \in 2^{O(n)}$. Dies bedeutet, die Absolutbeträge wachsen höchstens einfach exponentiell; und es gibt eine positive reelle Zahl r mit $|a_n| \leq r^n$ für alle $n \in \mathbb{N}$. In diesem Fall können wir die unendliche Reihe $a(z) = \sum_{n \geq 0} a_n z^n$ bilden, die für $|z| < 1/r$ absolut konvergiert. Dies liegt daran, dass die geometrische Reihe $\sum_{n \geq 0} z^n$ genau dann konvergiert, wenn $|z| < 1$ ist. Insbesondere sehen wir, dass es zwischen dem Wachstum der Folge $(a_n)_{n \geq 0}$ und dem Konvergenzradius von $\sum_{n \geq 0} a_n z^n$ einen engen Zusammenhang gibt. Ein typisches Beispiel ist $f(z) = \sum_{n \geq 0} F_n z^n$. Da die Fibonacci-Zahlen F_n das Wachstum $F_n \sim \Phi^n / \sqrt{5}$ mit $\Phi = (1 + \sqrt{5})/2$ aufweisen, ist der Konvergenzradius von $f(z)$ gerade $1/\Phi$.

Wir können $a(z)$ in dem Konvergenzbereich unendlich oft gliedweise differenzieren und erhalten für die k-te Ableitung:

$$a^{(k)}(z) = k! \sum_{n \geq k} \binom{n}{k} a_n z^{n-k}$$

Insbesondere gilt $a_n = \frac{a^{(n)}(0)}{n!}$ und damit „kennt" die analytische Funktion $a(z)$ die Folge $(a_n)_{n \in \mathbb{N}}$. Indem man erzeugende Funktionen manipuliert, kann man dann nach anschließendem Koeffizientenvergleich zu nichttrivialen Aussagen über die Ausgangsfolgen kommen. Wir wollen uns die erzeugenden Funktionen zu verschiedenen Zahlenfolgen ansehen, die wir im vorigen Kapitel behandelt haben. Dies sind die Fibonacci-Zahlen, Stirling-Zahlen, Catalan-Zahlen, Partitionszahlen und die Bell-Zahlen. Die Vielseitigkeit von erzeugenden Funktionen ist vor allem dadurch begründet, dass sich hier Techniken aus verschiedenen Zweigen der Mathematik kombinieren lassen. Wir illustrieren dies an Hand des Pentagonalzahlensatzes. Einige Folgen wachsen zu schnell und dann konvergiert die Reihe $\sum_{n \geq 0} a_n z^n$ nicht für $z > 0$. Dies führt auf exponentielle erzeugende Funktionen, die wir für die Stirling-Zahlen der ersten Art und die Bell-Zahlen heranziehen werden. Zunächst beschäftigen wir uns mit den gewöhnlichen erzeugenden Funktionen, welche die obige Definition zugrunde legen.

5.1 Gewöhnliche erzeugende Funktionen

Die unendliche Reihe

$$a(z) = \sum_{n \geq 0} a_n z^n$$

heißt die (gewöhnliche) *erzeugende Funktion* der Folge $(a_n)_{n \in \mathbb{N}}$. Wir interpretieren die Reihe als analytische Funktion, sofern sie einen positiven Konvergenzradius r hat. Dies bedeutet, dass der Wert $a(z)$ für alle $z \in \mathbb{C}$ mit $|z| < r$ definiert ist. Ansonsten lesen wir $a(z)$ als *formale Potenzreihe*. In beiden Interpretationen können

wir die Reihen addieren, multiplizieren und beliebig oft differenzieren. Die Formel für die k-te Ableitung steht schon oben. Addition und Multiplikation ergeben sich durch:

$$\sum_{n\geq 0} a_n z^n + \sum_{n\geq 0} b_n z^n = \sum_{n\geq 0} (a_n + b_n) z^n$$

$$\left(\sum_{n\geq 0} a_n z^n\right) \cdot \left(\sum_{n\geq 0} b_n z^n\right) = \sum_{n\geq 0} \left(\sum_{k+\ell=n} a_k b_\ell\right) z^n$$

Ferner gilt stets $a(0) = a_0$. Wichtig ist, dass wir für $a_0 \neq 0$ die multiplikativ inverse Reihe $a^{-1}(z)$ bilden können. Dies geschieht rein formal. Wir setzen zunächst $b_0 = \frac{1}{a_0}$ und für $n \geq 1$ bestimmen wir b_n induktiv durch Lösen der Gleichung

$$\sum_{k=0}^{n} a_k b_{n-k} = 0$$

Es ist dann $(\sum_{n\geq 0} a_n z^n) \cdot (\sum_{n\geq 0} b_n z^n) = 1$ und für $|a_n| \in 2^{O(n)}$ gilt auch $|b_n| \in 2^{O(n)}$. Betrachten wir die folgenden Beispiele. Wenn $a_n = 1$ für alle $n \in \mathbb{N}$ gilt, dann ist $a(z) = \sum_{n\geq 0} z^n = \frac{1}{1-z}$ die geometrische Reihe mit Konvergenzradius 1. Falls $a_n = n$ für alle $n \in \mathbb{N}$ ist, dann gilt

$$a(z) = \sum_{n\geq 0} n z^n = z \cdot \sum_{n\geq 1} n z^{n-1} = z \cdot \left(\sum_{n\geq 0} z^n\right)' = z \cdot \left(\frac{1}{1-z}\right)' = \frac{z}{(1-z)^2}$$

und der Konvergenzradius ist wieder 1. Ist $a_n = n!$ für alle $n \in \mathbb{N}$, so hat $a(z) = \sum_{n\geq 0} n! z^n$ keinen positiven Konvergenzradius. Die formale Reihe $a^{-1}(z)$ kann dennoch gebildet werden. Wir berechnen nun die erzeugenden Funktionen für einige konkrete Beispiele.

5.1.1 Fibonacci-Zahlen

Wir haben bereits gesehen, dass erzeugende Funktionen häufig durch ihre Inversen auf einfache Weise dargestellt werden können. Wir wenden dieses Vorgehen im Folgenden auf die Fibonacci-Zahlen F_n und andere Beispiele an. Es sei $f(z) = \sum_{n\geq 0} F_n \cdot z^n$ die erzeugende Funktion der Fibonacci-Zahlen F_n. Wegen $F_n \in \Theta(\Phi^n)$ konvergiert die Reihe $f(z) = \sum_{n\geq 0} F_n z^n$ für $|z| < \Phi^{-1}$ und insbesondere für $|z| < \frac{1}{2}$.

Satz 5.1.

$$f(z) = \frac{z}{1-z-z^2}$$

Beweis. Mittels der Rekursionsformel erhalten wir

$$f(z) = z + \sum_{n\geq 2} F_{n-1} z^n + \sum_{n\geq 2} F_{n-2} z^n = z + z f(z) + z^2 f(z)$$

Hieraus folgt $f(z) = \frac{z}{1-z-z^2}$. □

Die erzeugende Funktion liefert nun die schon bekannte explizite Darstellung der Fibonacci-Zahlen und den Zusammenhang zum goldenen Schnitt $\Phi = \frac{1+\sqrt{5}}{2}$ und $\hat{\Phi} = -\Phi^{-1} = \frac{1-\sqrt{5}}{2}$. Für die Fibonacci-Zahlen gilt:

$$F_n = \frac{1}{\sqrt{5}} \left(\left(\frac{1+\sqrt{5}}{2} \right)^n - \left(\frac{1-\sqrt{5}}{2} \right)^n \right)$$

Die Nullstellen des Nenners der erzeugenden Funktion $f(z)$ sind $\frac{-1}{2}(1 \pm \sqrt{5})$, also $-\Phi$ und $-\hat{\Phi}$. Damit erhalten wir durch eine Partialbruchzerlegung

$$f(z) = \frac{1}{\sqrt{5}} \left(\frac{1}{1 - \Phi z} - \frac{1}{1 - \hat{\Phi} z} \right)$$

Die Summanden in der Klammer sind nun Grenzfunktionen der geometrischen Reihen $\sum_{n \geq 0} \Phi^n z^n$ und $\sum_{n \geq 0} \hat{\Phi}^n z^n$. Wir erhalten $f(z) = \frac{1}{\sqrt{5}} \sum_{n \geq 0} (\Phi^n - \hat{\Phi}^n) z^n$. Ein Koeffizientenvergleich liefert schließlich die Behauptung.

Dieses Vorgehen liefert ein allgemeines Schema für das Lösen von einfachen Rekursionsgleichungen: (1) Stelle die erzeugende Funktion als rationale Funktion dar. (2) Zerlege die rationale Funktion mittels Partialbruchzerlegung, so dass alle Nenner linear sind. (3) Verwende die Formel für die geometrische Reihe, um den Koeffizienten von z^n zu ermitteln.

5.1.2 Catalan-Zahlen

Die erzeugende Funktion für die Catalan-Zahlen $C_n = \frac{1}{n+1} \binom{2n}{n}$ wird sich als Lösung einer quadratischen Gleichung ergeben. Wir setzen $c(z) = \sum_{n \geq 0} C_n z^n$. Wegen $C_n \leq 4^n$ hat die Reihe mindestens den Konvergenzradius $1/4$.

Satz 5.2.
$$c(z) = \frac{1 - \sqrt{1 - 4z}}{2z}$$

Beweis. Wir bilden $c^2(z) = \sum_{n \geq 0} (\sum_k C_k \cdot C_{n-k}) z^n$. Es ist $C_0 = 1$ und nach Korollar 4.41 gilt $C_{n+1} = \sum_k^n C_k \cdot C_{n-k}$. Also erhalten wir

$$z \cdot c^2(z) = \sum_{n \geq 0} C_{n+1} z^{n+1} = -1 + \sum_{n \geq 0} C_n z^n = c(z) - 1$$

Daher muss entweder $c(z) = (1 + \sqrt{1 - 4z})/2z$ oder $c(z) = (1 - \sqrt{1 - 4z})/2z$ gelten. Im Gegensatz zu $(1 + \sqrt{1 - 4z})/2z$ hat $c(z)$ jedoch keinen Pol für $z = 0$, also bleibt nur $c(z) = (1 - \sqrt{1 - 4z})/2z$. $\qquad\square$

5.1.3 Stirling-Zahlen zweiter Art

Als nächstes Beispiel betrachten wir die erzeugende Funktion für die Stirling-Zahlen $\left\{{n \atop k}\right\}$ der zweiten Art. Für ein festes $k \in \mathbb{N}$ sei $S_k(t) = \sum_{n \geq 0} \left\{{n \atop k}\right\} t^n$ die entsprechende erzeugende Funktion. Dann gilt insbesondere $S_0(z) = \left\{{0 \atop 0}\right\} = 1$ und $S_1(z) = -1 + \sum_{n \geq 0} z^n = z/(1-z)$. Für $k \geq 0$ und $n < 0$ ist $\left\{{n \atop k}\right\} = 0$, daher gilt $S_k(z) = \sum_{n \in \mathbb{Z}} \left\{{n \atop k}\right\} z^n$. Nach dem Additionstheorem 4.20 gilt $\left\{{n \atop k}\right\} = \left\{{n-1 \atop k-1}\right\} + k \left\{{n-1 \atop k}\right\}$. Für die erzeugenden Funktionen heißt dies für $k \geq 1$:

$$S_k(z) = z\, S_{k-1}(z) + kz\, S_k(z)$$

Wir können also $S_k(z)$ rekursiv bestimmen: $S_0(z) = 1$ und $S_k(z) = \frac{zS_{k-1}(z)}{1-kz}$ für $k \geq 1$. Außerdem erhalten wir induktiv Konvergenz für $|z| < \frac{1}{k}$.

Satz 5.3.

$$S_k(z) = \prod_{1 \leq i \leq k} \frac{z}{(1-iz)} = \frac{z^k}{(1-z)\cdots(1-kz)}$$

Beweis. Dies folgt sofort aus der Rekursion $S_0(z) = 1$ und $S_k(z) = \frac{zS_{k-1}(z)}{1-kz}$ für $k \geq 1$. $\qquad\square$

5.1.4 Partitionszahlen

Seien $u_1, \ldots, u_k$ paarweise verschiedene Unbestimmte, dann gibt es $\binom{k}{n}$ Möglichkeiten, n dieser k Unbestimmten zu wählen. Nehmen wir eine Unbestimmte z hinzu, so können wir das Polynom $(1 + u_1 z) \cdots (1 + u_k z)$ bilden und nach Ausmultiplizieren erscheint dann vor z^n als Koeffizient die Summe

$$\sum_{1 \leq i_1 < \cdots < i_n \leq k} u_{i_1} \cdots u_{i_n}$$

Dies ist die Summe aller n-Kombinationen von Zahlen aus $\{u_1, \ldots, u_k\}$. Setzen wir nun $u_i = 1$ für alle i, so ergibt sich $\binom{k}{n}$ als der Koeffizient vor z^n. Dies zeigt, wie man manchmal die erzeugende Funktion direkt angeben und dann die Koeffizienten bestimmen kann.

Der nächste Satz verallgemeinert dies auf *Multimengen*. Eine Multimenge über A ist eine Abbildung $M : A \to \mathbb{N}$. Multimengen sind also Elemente in $\mathbb{N}^A$. Die Idee ist, dass jedes Element $a \in A$ genau $M(a)$ mal in der Multimenge vorkommt. Entsprechend definieren wir die Größe von M als $|M| = \sum_{a \in A} M(a)$. Es ist üblich, Multimengen als formale Summen zu beschreiben. Hierfür lesen wir die Elemente $a \in A$ als Unbestimmte und schreiben $M = \sum_{a \in A} M(a) \cdot a$. Terme mit $M(a) = 0$ schreibt man meistens gar nicht hin. Ist etwa $A = \{a, b, c, d\}$ mit $M(a) = 27$, $M(b) = 0$, $M(c) = 1$ und $M(d) = 14$, so ist $M = 27a + c + 14d$ und hat die Größe 42. Eine

Teilmenge von A ist der Spezialfall mit $M(A) \subseteq \{0, 1\}$. Daher gibt man manchmal die erlaubten Vielfachheiten $N_a \subseteq \mathbb{N}$ für jedes $a \in A$ vor. Wir fragen also nach der Anzahl der Multimengen über A bei vorgeschriebenen Vielfachheiten, die eine gewisse Größe n haben. Die Antwort kann durch ihre erzeugende Funktion beschrieben werden.

Satz 5.4. *Die erzeugende Funktion für die Anzahl der Multimengen der Größe n über* $\{1, \ldots, k\}$*, bei denen die Vielfachheit von j aus einer gegebenen Zahlenmenge N_j* $\subseteq \mathbb{N}$ *stammt, ist*

$$\prod_{j=1}^{k} \left(\sum_{i \in N_j} z^i \right)$$

Beweis. Seien zunächst $u_1, \ldots, u_k$ paarweise verschiedene Unbestimmte. Beim Ausmultiplizieren von

$$\left(\sum_{i \in N_1} u_1^i z^i \right) \cdots \left(\sum_{i \in N_k} u_k^i z^i \right)$$

erhalten wir eine Reihe, wobei der Koeffizient von z^n die Summe über alle Terme der Form $u_1^{i_1} \cdots u_k^{i_k}$ ist mit $i_1 + \cdots + i_k = n$ und $i_\ell \in N_j$. Setzen wir wieder alle $u_i = 1$, so gibt der so entstehende Koeffizient die Anzahl der Multimengen, wie in dem Satz behauptet, an. $\square$

Wir bestimmen mit diesem Ansatz die Anzahl aller Multimengen der Größe 4 mit Elementen aus $\{a, b, c, d\}$ mit $N_a = \{0, 1, 2\}$, $N_b = \{2\}$, $N_c = \mathbb{N}$ und $N_d = \{0\}$. Wir erhalten als erzeugende Funktion

$$(1 + z + z^2) \cdot z^2 \cdot \sum_{i \geq 0} z^i = z^2 + 2z^3 + 3z^4 + 3z^5 + \cdots$$

Die Anzahl der zulässigen Multimengen ist also 3. Als formale Summe geschrieben sind dies $2b + 2c$, $a + 2b + c$ und $2a + 2b$.

Korollar 5.5. *Die erzeugende Funktion für die Anzahl aller Multimengen der Größe n über einer festen Menge mit k Elementen ist*

$$\prod_{j=1}^{k} \left(\sum_{i \geq 0} z^i \right) = \frac{1}{(1-z)^k} = \sum_{n \geq 0} \binom{n+k-1}{k-1} z^n$$

Beweis. Nach Satz 5.4 ist die erzeugende Funktion gerade $\prod_{j=1}^{k} (\sum_{i \geq 0} z^i) = \frac{1}{(1-z)^k}$. Beim Ausmultiplizieren von $\prod_{j=1}^{k} (\sum_{i \geq 0} z^i)$ ergibt sich der Koeffizient vor z^n als die Anzahl der Folgen $(i_1, \ldots, i_k)$ mit $i_1 + \cdots + i_k = n$. Wir haben in Satz 4.10 gesehen, dass diese Zahl der Binomialkoeffizient $\binom{n+k-1}{k-1}$ ist. $\square$

Für eine Menge M von positiven natürlichen Zahlen sei $Z_M(n)$ die Anzahl aller Zerlegungen von n in Summanden aus M (ohne Berücksichtigung der Reihenfolge):

$$Z_M(n) = \left| \left\{ (r_m)_{m \in M} \mid n = \sum_{m \in M} r_m m \right\} \right|$$

Satz 5.6. *Die erzeugende Funktion der Zahlen $Z_M(n)$ ist:*

$$\prod_{m \in M} \frac{1}{1 - z^m}$$

Beweis. Die Menge M bestehe etwa aus den Zahlen $1 \le m_1 < m_2 < m_3 < \cdots$. Es ist $\frac{1}{1-z^m} = \sum_{n \ge 0}(z^m)^n$; daher gilt

$$\prod_{m \in M} \frac{1}{1 - z^m} = (1 + z^{m_1} + z^{2m_1} + \cdots)(1 + z^{m_2} + z^{2m_2} + \cdots) \cdots$$

Beim Ausmultiplizieren tritt z^n so oft auf, wie es Sequenzen $(r_1, r_2, \ldots)$ von Zahlen aus $\mathbb{N}$ gibt mit

$$n = r_1 m_1 + r_2 m_2 + \cdots$$

und dies entspricht gerade der Definition von $Z_M(n)$. □

Beispiel 5.7. Zur Zeit der D-Mark gab es Münzen in Werten von 1, 2, 5, 10 und 50 Pfennigen. Auf wie viele Weisen konnte man eine D-Mark in kleinere Münzen wechseln? Sei c_n die Anzahl bei einem Betrag von n Pfennigen. Wir suchen also c_{100} in der folgenden Reihe:

$$\sum_{n \ge 0} c_n z^n = \frac{1}{(1 - z)(1 - z^2)(1 - z^5)(1 - z^{10})(1 - z^{50})}$$

$$= (1 + z + z^2 + \cdots)(1 + z^2 + z^4 + \cdots) \cdots (1 + z^{50} + z^{100} + \cdots)$$

Um c_{100} auszurechnen, kann man sukzessiv vorgehen. Ausgehend von $\frac{1}{1-z} = 1 + z + z^2 + \cdots$ dividiert man die anderen Polynome nacheinander gemäß der Methode:

$$(a_0 + a_1 z + a_2 z^2 + \cdots) : (1 - z^k) = b_0 + b_1 z + b_2 z^2 + \cdots$$

$$\text{mit } b_n = \begin{cases} a_n & \text{für } n = 0, \ldots, k - 1 \\ a_n + b_{n-k} & \text{für } n \ge k \end{cases}$$

Es folgt $c_{100} = 2498$. ◇

Die summatorischen Partitionszahlen $P(n)$ (bzw. $P(n, k)$) sind erklärt durch die Anzahl der Zerlegungen von n in positive Summanden (bzw. in k positive Summanden), siehe Abschnitt 4.8. Mit Hilfe der Ferrers-Spiegelung wie in Abbildung 4.2 hatten wir gesehen, dass $P(n, k)$ auch die Anzahl der Zerlegungen von n in positive Summanden ist, wenn der größte Summand k ist. Wir können jetzt die erzeugenden Funktionen für die Folgen $(P(n))_{n \in \mathbb{N}}$ und $(P(n, k))_{n \in \mathbb{N}}$ angeben. Die erzeugende Funktion der summatorischen Partitionszahlen $P(n)$ wurde von Euler entdeckt.

Korollar 5.8. *(a) Die erzeugende Funktion für $P(n)$ ist:*

$$\prod_{m \ge 1} \frac{1}{1 - z^m}$$

(b) *Die erzeugende Funktion für die Anzahl der Zerlegungen von n in natürliche Summanden, die nicht größer als k sind, ist*

$$\prod_{m=1}^{k} \frac{1}{1-z^m}$$

(c) *Die erzeugende Funktion für $P(n,k)$ ist:*

$$\prod_{m=1}^{k} \frac{z}{1-z^m}$$

Beweis. Die ersten beiden Aussagen sind ein Spezialfall von Satz 5.6. Für die erste können wir $M = \mathbb{N}$ und für die zweite $M = \{1,\ldots,k\}$ wählen. Die erzeugende Funktion für $P(n,k)$ erhalten wir unter Beachtung, dass $P(n,k)$ gleich der Anzahl aller Zerlegungen von $n-k$ in natürliche Summanden ist, die nicht größer als k sind. Also erhalten wir die dritte Aussage aus der zweiten unter Beachtung von $z^k \cdot \prod_{m=1}^{k} \frac{1}{1-z^m} = \prod_{m=1}^{k} \frac{z}{1-z^m}$. $\qquad\square$

Wir definieren nun die Menge der Partitionen von n, in denen die Summanden n_i ungerade (engl. *odd*) bzw. paarweise verschieden (engl. *different*) sind:

$$\mathcal{P}_o(n) = \Big\{ (n_1,\ldots,n_m) \in \mathcal{P}(n) \;\Big|\; n_i \text{ ungerade} \Big\}$$

$$\mathcal{P}_d(n) = \Big\{ (n_1,\ldots,n_m) \in \mathcal{P}(n) \;\Big|\; n_i \neq n_j \text{ für } i \neq j \Big\}$$

Wir setzen $P_o(n) = |\mathcal{P}_o(n)|$ und $P_d(n) = |\mathcal{P}_d(n)|$ für $n \in \mathbb{Z}$. Insbesondere ist $P_o(n) = P_d(n) = 0$ für $n < 0$ und $P_o(0) = P_d(0) = 1$. Erstaunlicher Weise stimmen $P_o(n)$ und $P_d(n)$ überein.

Satz 5.9. *Die erzeugenden Funktionen von $P_d(n)$ und $P_o(n)$ sind beide:*

$$\prod_{m \geq 1} (1 + z^m)$$

Insbesondere gilt $P_d(n) = P_o(n)$ für alle $n \in \mathbb{N}$.

Beweis. Die erzeugende Funktion für die Partitionszahlen $P_d(n)$ mit paarweise verschiedenen Summanden ist durch das Produkt $\prod_{m \geq 1}(1+z^m)$ gegeben. Nach Satz 5.6 ist die erzeugende Funktionen für die Partitionszahlen $P_o(n)$ mit ungeraden Summanden das Produkt $\prod_{m \text{ ungerade}} \frac{1}{1-z^m}$. Nun gilt $1 - z^{2m} = (1 + z^m)(1 - z^m)$, also ist

$$\prod_{m \geq 1} (1 + z^m) = \prod_{m \geq 1} \frac{1 - z^{2m}}{1 - z^m} = \prod_{m \text{ ungerade}} \frac{1}{1 - z^m}$$

Vielleicht ist einem nicht ganz wohl bei der unbefangenen Art, hier in unendlichen Produkten unendlich oft zu kürzen. Aber wir können die Rechnung rein mit Polynomen durchführen. Um $P_d(n)$ und $P_o(n)$ zu berechnen reicht es, $1 \leq m \leq n$ zu betrachten. Auch in den geometrischen Reihen $\frac{1}{1-z^m} = 1 + z^m + z^{2m} + \cdots$ können wir alle Summanden ignorieren deren Exponent größer als n ist. $\qquad\square$

5.1.5 Das Wachstum der Partitionszahlen

Das asymptotische Wachstum der Partitionszahlen $P(n)$ ist bekannt:

$$P(n) \sim \frac{1}{4n\sqrt{3}} e^{\pi\sqrt{\frac{2n}{3}}}$$

Diese erstaunliche Formel wurde 1918 von Hardy und Ramanujan gefunden (Godfrey Harold Hardy, 1877–1947 und Srinivasa Ramanujan, 1887–1920). Wir begnügen uns mit der schwächeren Aussage 5.10, für die ein elementarer Beweis existiert.

Satz 5.10.
$$\log_2 P_d(n) \geq \sqrt{n} \quad \text{für } n \geq 32$$
$$\log P(n) \in \Theta(\sqrt{n})$$

Beweis. Die Herleitung der unteren Schranke $\sqrt{n} \leq \log_2 P_d(n)$ ist einfach. Betrachte hierfür die Teilmengen von $\{1,\ldots,\lceil\sqrt{n}\rceil\}$. Deren Anzahl ist mindestens $2^{\sqrt{n}}$. Summieren wir die Elemente einer solchen Teilmenge I auf, so ist die Summe der Elemente kleiner als $(\sqrt{n}+1)(\sqrt{n}+2)/2 = n/2 + 3\sqrt{n}/2 + 1$. Ist n groß genug ($n \geq 32$), so können wir ein weiteres Element zwischen $\sqrt{n}+1$ und n hinzunehmen und erhalten eine durch die Teilmenge I eindeutig bestimmte Zerlegung von n in paarweise verschiedene Summanden. Dies ergibt $\sqrt{n} \leq \log_2 P_d(n)$ für n genügend groß.

Wir zeigen jetzt die Schranke $\log P(n) \in \mathcal{O}(\sqrt{n})$ und folgen für diesen Teil der Beweisführung in [28]. Wir wissen $\prod_{m\geq 1}\frac{1}{1-z^m} = \sum_{n\in\mathbb{N}} P(n)z^n$. Da Faktoren mit $m > n$ auf der rechten Seite nicht zum Koeffizienten von z^n beitragen, gilt für alle $0 < x < 1$ die Abschätzung:

$$P(n)x^n \leq \prod_{m=1}^{n} \frac{1}{1-x^m}$$

Wir erhalten

$$\ln P(n) \leq -n\ln x - \sum_{m=1}^{n} \ln(1-x^m)$$

Wegen $-\ln(1-y) = \sum_{i\geq 1}\frac{y^i}{i}$ gilt

$$-\sum_{m=1}^{n}\ln(1-x^m) = \sum_{m=1}^{n}\sum_{i\geq 1}\frac{x^{mi}}{i} = \sum_{i\geq 1}\sum_{m=1}^{n}\frac{x^{mi}}{i} \leq \sum_{i\geq 1}\frac{x^i}{(1-x^i)i}$$

Die Summanden dürfen vertauscht werden, da die Reihe $\sum_{i\geq 1}\frac{x^{mi}}{i}$ für ein festes m absolut konvergiert. Wir benutzen die geometrische Reihe und erhalten wegen $0 < x < 1$ für alle $i \geq 1$ die Abschätzung:

$$(1-x^i) = (1-x)(1+x+\cdots+x^{i-1}) \geq (1-x)ix^{i-1}$$

Damit ergibt sich

$$-\sum_{m=1}^{n} \ln(1-x^m) \le \frac{x}{1-x}\sum_{i\ge1}\frac{1}{i^2}$$

Die Reihe $\sum_{i\ge1}\frac{1}{i^2}$ konvergiert, also gilt $\sum_{i\ge1}\frac{1}{i^2}\in\mathcal{O}(1)$. Wir setzen jetzt $y=\frac{x}{1-x}$, also $\frac{1}{x}=1+\frac{1}{y}$ mit $0<\frac{1}{y}$, und sehen damit

$$\ln P(n)\le n\ln\left(1+\frac{1}{y}\right)+y\sum_{i\ge1}\frac{1}{i^2}$$

Es gilt $\ln(1+\frac{1}{y})\le\frac{1}{y}$; und wir erhalten

$$\ln P(n)\le\frac{n}{y}+y\sum_{i\ge1}\frac{1}{i^2}$$

Wählen wir $y=\sqrt{n}$, so ergibt sich $\log P(n)\in\mathcal{O}(\sqrt{n})$, also auch $\log P(n)\in\Theta(\sqrt{n})$, da ja $P_d(n)\le P(n)$. Dies zeigt den Satz. $\qquad\square$

Betrachten wir nochmals die letzten Beweisschritte. Setzen wir $y=c\sqrt{n}$ für eine Konstante c, so erhalten wir $\ln P(n)\le\sqrt{n}(1/c+c\sum_{i\ge1}\frac{1}{i^2})$. Um die Konstante c geschickt wählen zu können, müssen wir $\sum_{i\ge1}\frac{1}{i^2}$ berechnen. Dies gelang Euler um 1735 und er fand heraus

$$\sum_{i\ge1}\frac{1}{i^2}=\frac{\pi^2}{6}$$

Damit können wir $c=\pi/\sqrt{6}$ optimal wählen und erhalten

$$\ln P(n)\le\pi\sqrt{\frac{2n}{3}}$$

Die untere Schranke für $\log P_d(n)$ können wir mit den vorgestellten Techniken ebenfalls etwas verfeinern. Unser Ansatz zeigt sogar für jedes $c<2$ die Abschätzung $c\sqrt{n}\le\log_2 P_d(n)$ für fast alle n. Denn für eine Teilmenge $I\subseteq\{1,\dots,\lceil c\sqrt{n}\rceil\}$ oder für ihr Komplement liegt die Summe über ihre Elemente in $c^2 n/4+\mathcal{O}(\sqrt{n})$, denn beide Summen zusammen ergeben genau $(\lceil c\sqrt{n}\rceil)(\lceil c\sqrt{n}\rceil+1)/2$. Wir finden also mindestens $2^{c\sqrt{n}-1}$ Teilmengen, deren Summen über ihre Elemente jeweils in $c^2 n/4+\mathcal{O}(\sqrt{n})$ liegen. Ist n genügend groß, dann finden wir in dem Bereich von $c\sqrt{n}$ bis n eine ganze Zahl, so dass wir jede dieser Teilmengen zu einer Partition von n mit paarweise verschiedenen Summanden ergänzen können.

5.1.6 Der Pentagonalzahlensatz

Pentagonalzahlen erweitern die Konstruktion der Dreiecks- und Quadratzahlen auf regelmäßige Fünfecke. Die j-te Pentagonalzahl (oder *Fünfeckszahl*) entspricht der Anzahl der Kugeln, die man zum Legen eines Musters ineinandergeschachtelter regelmäßiger Fünfecke benötigt, die eine gemeinsame Ecke haben.

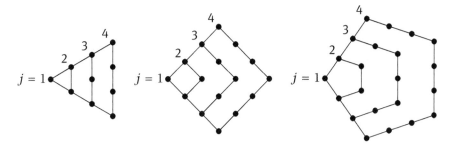

Die ersten Pentagonalzahlen sind

$$0, 1, 5, 12, 22, 35, 51, 70, 92, 117, 145, \ldots$$

Sie lassen sich nach der Formel $\frac{3j^2 - j}{2}$ berechnen. Wir erweitern die Definition für alle $j \in \mathbb{Z}$ und setzen

$$f(j) = \frac{3j^2 + j}{2}$$

Die j-te Pentagonalzahl ist in dieser Bezeichnung $f(-j)$, was zu etwas übersichtlicheren Formeln weiter unten führen wird. Wir bemerken, dass $f(j) \in \mathbb{N}$ für alle $j \in \mathbb{Z}$ gilt. Als Nächstes erkennen wir $f(i) \neq f(j)$ für alle $i \neq j$. Denn sei $i \neq j$, dann erhalten wir die folgenden Äquivalenzen:

$$f(i) = f(j) \iff 3i^2 - 3j^2 = j - i \iff 3(i + j) = -1$$

Multiplizieren wir das Produkt $\prod_{m \in \mathbb{N}} (1 - z^m)$ aus, so ergibt sich etwas Erstaunliches. Die Anfangsterme sehen wie folgt aus:

$$\prod_{m \geq 1} (1 - z^m) = 1 - z - z^2 + z^5 + z^7 - z^{12} - z^{15} + z^{22} + z^{26} - \cdots$$

Die ersten Koeffizienten beschränken sich auf ± 1 und treten nur bei den Exponenten der Form $f(j)$ auf. Dies ist kein Zufall und wurde von Euler durch seinen Pentagonalzahlensatz bestätigt.

Satz 5.11 (Pentagonalzahlensatz).

$$\prod_{m \geq 1} (1 - z^m) = \sum_{j \in \mathbb{Z}} (-1)^j z^{f(j)}$$

Beweis. In der Beweisführung folgen wir [2]. Zunächst erinnern wir daran, dass $f(i) \in \mathbb{N}$ und $f(i) \neq f(j)$ für alle $i \neq j$ gilt; also steht rechts wirklich eine Potenzreihe mit Koeffizienten ± 1 oder 0. Nach Korollar 5.8 hat die erzeugende Funktion der summatorischen Partitionszahlen $P(n)$ die Produktdarstellung

$$p(z) = \prod_{m \geq 1} \frac{1}{1 - z^m}$$

Wir können das Produkt $1/p(z)$ als eine formale Potenzreihe entwickeln

$$\prod_{m \geq 1} (1 - z^m) = \sum_{n \in \mathbb{N}} c(n) z^n$$

Die Koeffizienten $c(n)$ in dieser Reihe ergeben sich eindeutig durch $c(0) = 1$ und die Forderung, dass für alle $n \geq 1$ gilt:

$$\sum_{k=0}^{n} c(k) P(n - k) = 0$$

Wir müssen also zeigen, dass die Folge $c(k) = 0$ für $k \neq f(j)$, $c(k) = 1$ für $k = f(j)$ und j gerade, sowie $c(k) = -1$ für $k = f(j)$ und j ungerade alle diese Gleichungen löst. Wegen $P(m) = 0$ für $m < 0$ ist dies äquivalent mit der Forderung, dass für alle $n \geq 1$ gilt:

$$\sum_{j \in \mathbb{Z}} (-1)^j P(n - f(j)) = 0$$

Zu zeigen ist also

$$\sum_{j \text{ gerade}} P(n - f(j)) = \sum_{j \text{ ungerade}} P(n - f(j))$$

Dies ruft geradezu nach einem bijektiven Beweis! Wir suchen also eine Bijektion ψ zwischen den Mengen $\bigcup_{j \text{ gerade}} \mathcal{P}(n - f(j))$ und $\bigcup_{j \text{ ungerade}} \mathcal{P}(n - f(j))$. Die folgende Involution φ von $\bigcup_{j \in \mathbb{Z}} \mathcal{P}(n - f(j))$ wurde von David Bressoud und Doron Zeilberger (geb. jeweils 1950) gefunden; die gesuchte Bijektion ψ ist die Restriktion von φ auf die Teilmenge $\bigcup_{j \text{ gerade}} \mathcal{P}(n - f(j))$.

Wir starten mit einer Partition $d = (d_1, \ldots, d_m) \in \mathcal{P}(n - f(j))$. Man beachte, es kann $n = f(j)$ sein. Wegen $n \geq 1$ ist dann allerdings $j \neq 0$. Für $n = f(j)$ ist d die leere Folge, also $m = 0$. In diesem Fall setzen wir $d_1 = 0$. Es gilt also stets $d_1 + \cdots + d_m + f(j) = n$; und für $m \geq 1$ gilt auch $d_1 \geq \cdots \geq d_m \geq 1$. Aufgrund von $f(j) - f(j-1) = 3j - 1$ erscheint dieser Wert $3j - 1$ in den folgenden Definitionen.

– Falls $m \geq d_1 - 3j$, dann setze

$$\varphi(d) = (m + 3j - 1, d_1 - 1, \ldots, d_m - 1)$$

Summanden, die Null geworden sind, lassen wir fort. Es gilt $\varphi(d) \in \mathcal{P}(n - f(j - 1))$. Die maximale Zahl der Terme in $\varphi(d)$ ist $m + 1$ und dies sind zu wenige, um in diesem Fall zu bleiben. Wir geraten in die andere Situation.

– Falls $m < d_1 - 3j$, dann setze

$$\varphi(d) = (d_2 + 1, \ldots, d_m + 1, \underbrace{1, \ldots, 1}_{d_1 - m - 3j - 1})$$

Jetzt gilt $\varphi(d) \in \mathcal{P}(n - f(j + 1))$. Die neue Zahl der Terme ist $d_1 - 3j - 2 \geq d_2 - 3j - 2 \geq (d_2 + 1) - 3(j + 1)$. Damit sind wir zurück in der ersten Situation.

Eine einfache Rechnung zeigt jetzt $\varphi(\varphi(d)) = d$ und damit ist der Pentagonalzahlensatz bewiesen.   □

Die ganz unterschiedliche Entwicklung der Koeffizienten in den beiden Reihen

$$\prod_{m \geq 1} (1 + z^m) = \sum_{n \in \mathbb{N}} P_d(n)\, z^n = 1 + z + z^2 + 2z^3 + 2z^4 + 3z^5 + 4z^6 + \cdots$$

und

$$\prod_{m \geq 1} (1 - z^m) = \sum_{j \in \mathbb{Z}} (-1)^j z^{f(j)} = 1 - z - z^2 + z^5 + z^7 - z^{12} - z^{15} + \cdots$$

lässt sich kombinatorisch interpretieren. Hierfür betrachten wir zunächst diejenigen
Partitionen in $\mathcal{P}_d(n)$, die aus genau k Summanden bestehen, und nennen diese Men-
ge $\mathcal{P}_d(n, k)$. Sie besteht also aus den Folgen $(d_1, \ldots, d_k) \in \mathbb{N}^k$ mit $d_1 + \cdots + d_k = n$,
wobei $d_1 > \cdots > d_k \geq 1$. Als Nächstes wählen wir neben z eine weitere Unbestimm-
te u und bilden die formale Reihe

$$\prod_{m \geq 1} (1 + z^m u) = \sum_{n \in \mathbb{N}} \left(\sum_{k \in \mathbb{N}} |\mathcal{P}_d(n, k)| u^k \right) z^n$$

Schließlich fassen wir die Partitionen mit einer geraden und die mit einer ungeraden
Anzahl von Termen zusammen und definieren

$$E_d(n) = \sum_{k \text{ gerade}} |\mathcal{P}_d(n, k)| \qquad O_d(n) = \sum_{k \text{ ungerade}} |\mathcal{P}_d(n, k)|$$

Die Bezeichnungen E_d und O_d setzen sich aus $E = $ *Even*, $O = $ *Odd* und $d = $ *different* zusammen. Man beachte, $E_d(0) = 1$ und $O_d(0) = 0$. Spezialisieren wir u zu $+1$ beziehungsweise zu -1, so erhalten wir:

$$\prod_{m \geq 1} (1 + z^m) = \sum_{n \in \mathbb{N}} (E_d(n) + O_d(n))\, z^n$$

$$\prod_{m \geq 1} (1 - z^m) = \sum_{n \in \mathbb{N}} (E_d(n) - O_d(n))\, z^n$$

Damit haben auch die Koeffizienten der beiden Reihen eine kombinatorische Interpretation und der Pentagonalzahlensatz liefert jetzt eine überraschende Identität für
die Differenz $E_d(n) - O_d(n)$:

Korollar 5.12.

$$E_d(n) - O_d(n) = \begin{cases} 1 & \text{für } n = f(j) \text{ und } j \in \mathbb{Z} \text{ gerade} \\ -1 & \text{für } n = f(j) \text{ und } j \in \mathbb{Z} \text{ ungerade} \\ 0 & \text{sonst, also falls } n \neq f(j) \text{ für alle } j \in \mathbb{Z} \end{cases}$$

5.2 Exponentielle erzeugende Funktionen

Wollen wir die Anzahl von kombinatorischen Objekten der Größe n in einer Folge $a_n \in \mathbb{N}$ festhalten, so wachsen diese Werte häufig sehr schnell. Dies passiert insbesondere dann, wenn die Objekte mit Permutationen in Verbindung stehen, die Terme der Form $n!$ ins Spiel bringen. Vielfach trifft man dann auf die Situation $a_n \in 2^{\omega(n)}$. In diesem Fall divergiert die Reihe $\sum_{n\geq 0} a_n z^n$ für jedes positive $z > 0$ und eine gewöhnliche erzeugende Funktion hilft nicht weiter.

Ein offensichtlicher Fall für $a_n \in 2^{\omega(n)}$ ist $a_n = n!$, was gerade die Anzahl der Permutationen von $\{1, \ldots, n\}$ ist. Dies führt auf den Begriff der *exponentiellen erzeugenden Funktion*. Diese ist für eine Folge von reellen oder komplexen Zahlen $(a_n)_{n\in\mathbb{N}}$ zunächst als die formale Potenzreihe definiert:

$$\tilde{a}(z) = \sum_{n\geq 0} \frac{a_n}{n!} z^n$$

Die Regeln zur Addition, Multiplikation und Ableitung ergeben sich wie folgt.

$$\sum_{n\geq 0} \frac{a_n}{n!} z^n + \sum_{n\geq 0} \frac{b_n}{n!} z^n = \sum_{n\geq 0} \frac{a_n + b_n}{n!} z^n$$

$$\sum_{n\geq 0} \frac{a_n}{n!} z^n \cdot \sum_{n\geq 0} \frac{b_n}{n!} z^n = \sum_{n\geq 0} \sum_{k=0}^{n} \frac{a_k \cdot b_{n-k}}{k! \, (n-k)!} z^n$$

$$= \sum_{n\geq 0} \frac{1}{n!} \left(\sum_{k=0}^{n} \binom{n}{k} a_k \cdot b_{n-k} \right) z^n$$

$$\left(\sum_{n\geq 0} \frac{a_n}{n!} z^n \right)' = \sum_{n\geq 0} \frac{a_{n+1}}{n!} z^n$$

Können wir in einer Folge $(a_n)_{n\in\mathbb{N}}$ die Absolutbeträge der Folgenglieder durch $|a_n| \in 2^{n \log_2 n + O(n)}$ abschätzen, so gibt es ein $r > 0$ mit $|a_n| \leq (rn)^n$ für fast alle n. Damit konvergiert die Reihe $\sum_{n\geq 0} \frac{a_n}{n!} z^n$ absolut für alle $z < 1/re$ (denn $\frac{n^n}{n!} \leq ne^n$). Insbesondere hat die Reihe einen positiven Konvergenzradius und über die Ableitungen sehen wir erneut, dass die entsprechende analytische Funktion alle Koeffizienten a_n eindeutig bestimmt.

Wir betrachten einige einfache Beispiele. Für $a_n = 1$ ist $\tilde{a}(z) = \exp(z) = e^z$ die Exponentialfunktion in der Reihendarstellung $\exp(z) = \sum_{n\geq 0} \frac{z^n}{n!}$. Für $a_n = n!$ ist die exponentielle erzeugende Funktion $\tilde{a}(z) = \frac{1}{1-z}$ die geometrische Reihe $\sum_{n\geq 0} z^n$. Für $n, m \in \mathbb{N}$ sei $I(n, m)$ die Anzahl der Injektionen einer n-elementigen Menge in eine m-elementige Menge, also $I(n, m) = m^{\underline{n}} = n! \binom{m}{n}$. Wählen wir m fest und betrachten $a_n = I(n, m)$ als Folge in n, so ergibt sich die exponentielle erzeugende Funktion als das Polynom

$$\tilde{a}(z) = \sum_{n\geq 0} \frac{I(m, n)}{n!} z^n = \sum_{n} \binom{m}{n} z^n = (1+z)^m$$

5.2.1 Stirling-Zahlen erster Art

Ein erstes nichttriviales Beispiel für eine exponentielle erzeugende Funktion liefert die Folge der Zykelzahlen $\begin{bmatrix} n \\ 1 \end{bmatrix} = (n-1)!$. Die gewöhnliche erzeugende Funktion konvergiert nicht, aber die exponentielle erzeugende Funktion hat die Gestalt $\sum_{n \geq 0} \frac{1}{n!} \cdot \begin{bmatrix} n \\ 1 \end{bmatrix} z^n = \sum_{n \geq 0} \frac{z^n}{n} = -\ln(1-z)$. Dies wiederum ist ein Spezialfall für die exponentielle erzeugende Funktion der Stirling-Zahlen erster Art. Wegen $\sum_k \begin{bmatrix} n \\ k \end{bmatrix} = n!$ ist die zugehörige summatorische exponentielle erzeugende Funktion gerade die schon eben betrachtete geometrische Reihe. Interessant ist jedoch die exponentielle erzeugende Funktion für die Folge $\begin{bmatrix} n \\ k \end{bmatrix}$ der Stirling-Zahlen erster Art, wenn k fest ist. Zunächst beobachten wir, dass diese Folge langsamer wächst als $n!$, aber nach der Abschätzung (4.6) mindestens das Wachstum $(n-k)!$ hat. Die zugehörige exponentielle erzeugende Funktion hat also einen positiven Konvergenzradius, während die gewöhnliche erzeugende Funktion für $z > 0$ nicht konvergiert.

Wir untersuchen jetzt für $k \in \mathbb{N}$ die Reihe $\sum_n \frac{1}{n!} \begin{bmatrix} n \\ k \end{bmatrix} z^n$. Hierzu benutzen wir den bereits bekannten allgemeinen Binomialsatz 4.11. Seien $r, z \in \mathbb{C}$ mit $|z| < 1$. Dann gilt

$$(1+z)^r = \sum_n \binom{r}{n} z^n = \sum_n \frac{r^{\underline{n}}}{n!} z^n$$

Nach Korollar 4.29 wissen wir, dass $r^{\underline{n}} = r(r-1) \cdots (r-n+1)$ ein Polynom in r vom Grad n mit den Koeffizienten $s(n,k) = (-1)^{n-k} \begin{bmatrix} n \\ k \end{bmatrix}$ ist. Es gilt also $r^{\underline{n}} = \sum_k s(n,k) r^k$ und hieraus folgt:

$$e^{r \ln(1+z)} = (1+z)^r = \sum_k \left(\sum_n \frac{s(n,k)}{n!} z^n \right) r^k$$

Ein Koeffizientenvergleich für jedes feste z mit $|z| < 1$ zeigt nun:

$$\sum_n \frac{s(n,k)}{n!} z^n = \frac{(\ln(1+z))^k}{k!}$$

Die letzte Formel liefert also für jedes feste k die exponentielle erzeugende Funktion für die *Stirling-Zahlen der ersten Art mit Vorzeichen* $s(n,k)$. Gehen wir zurück zu $\begin{bmatrix} n \\ k \end{bmatrix} = (-1)^{n-k} s(n,k)$ und ersetzen z durch $-z$, so erhalten wir die exponentielle erzeugende Funktion für die Stirling-Zahlen der ersten Art $\begin{bmatrix} n \\ k \end{bmatrix}$:

$$\sum_{n \geq 0} \frac{1}{n!} \begin{bmatrix} n \\ k \end{bmatrix} z^n = \frac{(-\ln(1-z))^k}{k!}$$

Der Konvergenzradius dieser Reihen ist jeweils 1.

5.2.2 Bell-Zahlen

Wir behandeln jetzt ein weiteres schönes Beispiel für eine exponentielle erzeugende Funktion, nämlich die für die Bell-Zahlen B_n. Wir wissen schon aus den Ungleichungen in (4.7), dass zwar $B_n \in 2^{\omega(n)}$ aber auch $B_n \leq n!$ gilt; im Gegensatz zur gewöhnlichen erzeugenden Funktion hat die exponentielle erzeugende Funktion also mit Sicherheit einen positiven Konvergenzradius.

Sei $\tilde{b}(z) = \sum_{n \geq 0} \frac{B_n}{n!} z^n$ die exponentielle erzeugende Funktion für die Bell-Zahlen B_n. Wir hatten gesehen, dass $B_{n+1} = \sum_k \binom{n}{k} B_k$ für alle $n \in \mathbb{N}$ gilt. Mit e^z bezeichnen wir die übliche Exponentialfunktion zur Basis e, die etwa durch die Reihe $e^z = \sum_{n \geq 0} \frac{z^n}{n!}$ definiert ist. Unter Benutzung der Rechenregeln von oben gilt damit die Differentialgleichung:

$$\tilde{b}'(z) = \sum_{n \geq 0} \frac{B_{n+1}}{n!} z^n = \sum_{n \geq 0} \left(\frac{1}{n!} \sum_k \binom{n}{k} B_k \right) z^n$$

$$= \left(\sum_n \frac{z^n}{n!} \right) \cdot \left(\sum_n \frac{B_n}{n!} z^n \right) = \tilde{b}(z) \cdot e^z$$

Diesen Typ von Differentialgleichung wollen wir lösen. Die Funktion e^{e^z} erfüllt die Differentialgleichung. Seien nun $b_1(z)$ und $b_2(z)$ zwei Lösungen dieser Differentialgleichung mit $b_1(z) > 0$ und $b_2(z) > 0$ für alle $0 < z \in \mathbb{R}$. Dann gilt für die logarithmischen Ableitungen $(\ln \circ b_1)'(z) = \frac{b_1'(z)}{b_1(z)} = \frac{b_2'(z)}{b_2(z)} = e^z$. Also unterscheiden sich $b_1(z)$ und $b_2(z)$ nur um einen konstanten Faktor, denn die Ableitung von $b_1(z)/b_2(z)$ ist Null. Damit gilt $\tilde{b}(z) = c e^{e^z}$ für ein $c > 0$. Wegen $\tilde{b}(0) = 1$ folgt $c = \frac{1}{e}$. Damit ergibt sich:

$$\tilde{b}(z) = e^{e^z - 1}$$

Insbesondere sehen wir, dass die Reihe $\tilde{b}(z) = \sum_{n \geq 0} \frac{B_n}{n!} z^n$ überall konvergiert. Wir erhalten einen neuen Beweis für die schon aus Satz 4.34 bekannte Dobiński-Formel: Durch Reihenentwicklung gilt $\tilde{b}(z) = e^{e^z - 1} = \frac{1}{e} \sum_k (\frac{1}{k!} \sum_n \frac{z^n k^n}{n!}) = \sum_n (\frac{1}{e} \sum_k \frac{k^n}{k!}) \frac{z^n}{n!}$. Koeffizientenvergleich liefert die Dobiński-Formel

$$B_n = \frac{1}{e} \sum_{k \geq 0} \frac{k^n}{k!}$$

Aufgaben

5.1. Sei F_n die n-te Fibonacci-Zahl. Zeigen Sie, dass die unendliche Summe $\sum_{n \geq 0} 10^{-n} F_n$ gegen eine rationale Zahl konvergiert.

5.2. Gegeben seien $c_1, c_2 \in \mathbb{R}$ mit $c_1 \neq 0 \neq c_2$ und $c_1^2 + 4c_2 > 0$. Wir setzen

$$\lambda_1 = \frac{c_1}{2} + \sqrt{\left(\frac{c_1}{2}\right)^2 + c_2} \quad \text{und} \quad \lambda_2 = \frac{c_1}{2} - \sqrt{\left(\frac{c_1}{2}\right)^2 + c_2}$$

Sei $(a_n)_{n\geq 0}$ eine Folge, die rekursiv definiert ist durch $a_0 = 0$, $a_1 = 1$ und $a_n = c_1 a_{n-1} + c_2 a_{n-2}$ für $n \geq 2$. Zeigen Sie:

(a) Die erzeugende Funktion $a(z)$ der Folge $(a_n)_{n\geq 0}$ ist $a(z) = \frac{z}{1-c_1 z - c_2 z^2}$.

(b) $a_n = \frac{1}{2 \cdot \sqrt{(\frac{c_1}{2})^2 + c_2}}(\lambda_1^n - \lambda_2^n)$.

5.3. Die Zahlen a_n sind induktiv definiert durch $a_0 = 2$, $a_1 = 5$ und $a_{n+2} = 5a_{n+1} - 6a_n$. Bestimmen Sie die erzeugende Funktion der Zahlen a_n, und zeigen Sie $a_n = 2^n + 3^n$.

5.4. Wir definieren die Zahlenfolge $(a_n)_{n\geq 0}$ durch $a_0 = 0$, $a_1 = 1$ und $a_n = 3a_{n-1} - 2a_{n-2} + 2^{n-1}$ für $n \geq 2$. Bestimmen Sie die erzeugende Funktion von $(a_n)_{n\geq 0}$, und zeigen Sie $a_n = 1 + (n-1)2^n$.

5.5. Sei $H_n = \sum_{k=1}^{n} 1/k$ die Folge der harmonischen Zahlen. Bestimmen Sie deren erzeugende Funktion $h(z)$.

Hinweis: Sie können $-\ln(1-z) = \sum_{n\geq 1} z^n/n$ als bekannt voraussetzen.

5.6. Bestimmen Sie die erzeugende Funktion von $(F_{2n})_{n\geq 0}$. Hierbei ist F_n die n-te Fibonacci-Zahl.

5.7. Sei $a_0 = 1$ und $a_n = \sum_{i=0}^{n-1}(n-i)a_i$. Bestimmen Sie die erzeugende Funktion von $(a_n)_{n\geq 0}$.

5.8. Bestimmen Sie die exponentielle erzeugende Funktion der Rencontres-Zahlen R_n.

5.9. Ein *Automat* über dem endlichen Alphabet Σ ist ein 4-Tupel $\mathcal{A} = (Q, \delta, q_0, F)$ mit einer endlichen Zustandsmenge Q, einem Startzustand $q_0 \in Q$, einer Menge von Endzuständen $F \subseteq Q$ und einer Übergangsfunktion $\delta : Q \times \Sigma \to Q$. Man kann die Übergangsfunktion δ auf Sequenzen von Elementen aus Σ ausdehnen, indem man $\delta(q, \varepsilon) = q$ und $\delta(q, wa) = \delta(\delta(q, w), a)$ setzt; hierbei ist ε die leere Sequenz (das leere Wort), $a \in \Sigma$ und w eine beliebige Sequenz. Die Menge aller endlichen Sequenzen über Σ bezeichnen wir mit Σ^*. Die von $\mathcal{A}$ akzeptierte Sprache ist $L(\mathcal{A}) = \{w \in \Sigma^* \mid \delta(q_0, w) \in F\}$. Wir sind an der Frage interessiert, wie viele Wörter der Länge n von $\mathcal{A}$ akzeptiert werden.

(a) Für einen Zustand q sei $L_q = \{w \in \Sigma^* \mid \delta(q_0, w) = q\}$ die Menge der Wörter, die nach q führen. Zeigen Sie: Es gilt $L_{q_0} = \{\varepsilon\} \cup \bigcup_{\delta(p,a)=q_0} L_p \cdot a$ sowie $L_q = \bigcup_{\delta(p,a)=q} L_p \cdot a$ für $q \neq q_0$.

(b) Sei a_n^q die Anzahl der Wörter der Länge n in L_q und sei $a^q(z)$ die erzeugende Funktion von $(a_n^q)_{n\geq 0}$. Dann gilt $a^{q_0}(z) = 1 + \sum_{\delta(p,a)=q_0} z a^p(z)$ sowie $a^q(z) = \sum_{\delta(p,a)=q} z a^p(z)$ für $q \neq q_0$.

(c) Sei b_n die Anzahl der Wört der Länge n in $L(\mathcal{A})$. Dann ist die erzeugende Funktion von $(b_n)_{n\geq 0}$ gegeben durch $\sum_{q\in F} a^q(z)$.

(d) Sei $\Sigma = \{a, b\}$, $Q = \{q_0, q_1, q_2\}$, $F = \{q_0, q_1\}$ und δ gegeben durch

q	c	$\delta(q, c)$
q_0	a	q_1
q_0	b	q_0
q_1	a	q_2
q_1	b	q_0
q_2	a	q_2
q_2	b	q_2

Bestimmen Sie die erzeugenden Funktion für die Anzahl der Wörter der Länge n in der von diesem Automaten akzeptierten Sprache. Wieviele Wörter der Länge n akzeptiert der Automat?

Zusammenfassung

Begriffe

- gewöhnliche erzeugende Funktion
- analytische Funktion
- Konvergenzradius
- formale Potenzreihe
- Multimenge
- n in Summanden aus M, $Z_M(n)$

- Partitionen $P_o(n)$
- Partitionen $P_d(n)$
- Pentagonalzahl, Fünfeckzahl
- Partitionszahlen $E_d(n)$
- Partitionszahlen $O_d(n)$
- exponentielle erzeugende Funktion

Methoden und Resultate

- Gewöhnliche erzeugende Funktionen: Zusammenhang zwischen asymptotischem Wachstum und Konvergenzradius
- Rechnen mit formalen Potenzreihen
- Invertieren von formalen Potenzreihen
- Erzeugende Funktion der Fibonacci-Zahlen: $f(z) = \frac{z}{1-z-z^2}$
- Lösen von einfachen Rekursionsgleichungen mittels erzeugender Funktionen
- Erzeugende Funktion der Catalan-Zahlen: $c(z) = \frac{1-\sqrt{1-4z}}{2z}$
- Erzeugende Funktion der Stirling-Zahlen zweiter Art: $S_k(z) = \prod_{1 \le i \le k} \frac{z}{1-iz}$
- Erzeugende Funktion für Anzahl Multimengen über $\{1, \ldots, k\}$ mit Randbedingungen $N_j \subseteq \mathbb{N}$: $\prod_{j=1}^{k} (\sum_{i \in N_j} z^i)$
- Erzeugende Funktion für die Anzahl der Multimengen über $\{1, \ldots, k\}$: $\frac{1}{(1-z)^k}$
- Erzeugende Funktion für $Z_M(n)$: $\prod_{m \in M} \frac{1}{1-z^m}$

- Erzeugende Funktion für $P(n)$: $\prod_{m \geq 1} \frac{1}{1-z^m}$
- Erzeugende Funktion für $P(n,k)$: $\prod_{m=1}^{k} \frac{z}{1-z^m}$
- Erzeugende Funktion für $P_d(n)$ und für $P_o(n)$: $\prod_{m \geq 1}(1+z^m)$
- Für $n \geq 32$ gilt $\log_2 P_d(n) \geq \sqrt{n}$
- $\log P(n) \in \Theta(\sqrt{n})$
- Pentagonalzahlensatz: $\prod_{m \geq 1}(1-z^m) = \sum_{j \in \mathbb{Z}}(-1)^j z^{f(j)}$
- $E_d(n) - O_d(n) = \begin{cases} 1 & \text{für } n = f(j) \text{ und } j \in \mathbb{Z} \text{ gerade} \\ -1 & \text{für } n = f(j) \text{ und } j \in \mathbb{Z} \text{ ungerade} \\ 0 & \text{sonst, also falls } n \neq f(j) \text{ für alle } j \in \mathbb{Z} \end{cases}$
- Rechenregeln für exponentielle erzeugende Funktionen
- $e^z = \sum_{n \geq 0} \frac{z^n}{n!}$
- Exponentielle erzeugende Funktion für $\begin{bmatrix} n \\ k \end{bmatrix}$: $\sum_{n \geq 0} \frac{1}{n!}\begin{bmatrix} n \\ k \end{bmatrix}z^n = \frac{(-\ln(1-z))^k}{k!}$
- Exponentielle erzeugende Funktion für Bell-Zahlen: $\sum_{n \geq 0} \frac{B_n}{n!}z^n = e^{e^z - 1}$

6 Graphentheorie

Graphen dienen der Beschreibung und Veranschaulichung von Relationen zwischen Objekten. Die Idee ist es, Objekte durch Punkte darzustellen und falls zwischen zwei Objekten eine Verbindung besteht, eine Linie zwischen ihnen zu ziehen. Die Objekte bezeichnet man als Knoten und die Verbindungen zwischen ihnen als Kanten. Manchmal benutzt man Beschriftungen, um die Art der Verbindung zu kennzeichnen oder um die Objekte in Klassen einzuteilen. Durch Abstraktion kann man sehr viele Zusammenhänge durch Graphen darstellen. Beispielsweise könnte man als Knoten alle Städte und als Kanten das Straßennetz nehmen, d. h., wir zeichnen eine Kante zwischen zwei Städten, falls diese durch eine Straße verbunden sind. Ein anderes Beispiel ergibt sich mit Filmen und Schauspielern als Knotenmenge. Wir ziehen eine Kante zwischen einem Schauspieler x und einem Film y, falls x in y mitgewirkt hat. Wir können auch die Spielstellungen eines Spiels als Knotenmenge auffassen; eine Kante ziehen wir von einer Spielstellung x zu einer Spielstellung y, falls sich x durch einen Zug nach y überführen lässt. Mit Graphen lassen sich die unterschiedlichsten Sachverhalte modellieren. Allgemein lassen sich Relationen $R \subseteq A \times B$ als Graph auffassen, indem man $A \cup B$ als Knotenmenge wählt und eine Kante zwischen $x, y \in A \cup B$ zieht, falls $(x, y) \in R$ gilt. Diese Art der Darstellung hat mehrere Vorteile. Der erste ist, dass Graphen sich sehr gut „*graphisch*" veranschaulichen lassen. Der andere Vorteil ist, dass man zur Lösung von Problemen bereits existierende Resultate und Verfahren aus der Graphentheorie heranziehen kann. Des Weiteren erlaubt die Graphentheorie einheitliche Begriffsbildungen.

6.1 Grundbegriffe

Es existieren mehrere verschiedene Modelle für Graphen. Im allgemeinsten Fall besteht ein Graph $G = (V, E, \sigma, \tau)$ aus einer Menge von *Knoten* V (engl. *vertex*), einer Menge von *Kanten* E (engl. *edge*) und zwei Abbildungen $\sigma, \tau : E \to V$. Die Abbildung σ ordnet jeder Kante aus E einen Startknoten (engl. *source*) zu, und die Abbildung τ gibt den Zielknoten (engl. *target*) an. Zwei Knoten, die durch eine Kante verbunden sind, heißen *adjazent* (oder *benachbart*). Ein Knoten x und eine Kante e sind *inzident*, wenn x Start- oder Zielknoten von e ist. Eine Kante $e \in E$ lässt sich graphisch durch einen Pfeil $\sigma(e) \longrightarrow \tau(e)$ veranschaulichen. Dieses Modell erlaubt mehrere Kanten zwischen zwei Knoten – sogenannte *Mehrfachkanten*. Zudem sind auch Kanten von einem Knoten zu sich selbst möglich; solche Kanten nennt man *Schlingen*.

Mehrfachkanten   Schlingen

Graphen ohne Schlingen und Mehrfachkanten bezeichnet man als *einfach*. In vielen Anwendungen kann man Schlingen und Mehrfachkanten durch Einführen von weiteren Knoten vermeiden. Eine Kante $x \longrightarrow y$ kann man beispielsweise durch $x \longrightarrow z \longrightarrow y$ ersetzen, wobei z ein neuer Knoten ist. Für einfache Graphen existieren auch einfachere Beschreibungsmodelle. In den meisten Fällen sind die hier behandelten Graphen einfach. Ein *gerichteter* Graph ist ein Paar (V, E), wobei V eine beliebige Menge ist und $E \subseteq V \times V$. Jedes Paar $(x, y) \in E$ stellt eine Kante vom Knoten x zum Knoten y dar. Hierbei ist x der Startknoten und y der Zielknoten. Falls wir nicht zwischen Startknoten und Zielknoten unterscheiden, erhalten wir *ungerichtete* Graphen. Bei ungerichteten Graphen verbindet jede Kante zwei Knoten, ohne dabei zwischen diesen beiden Knoten einen Unterschied zu machen. Als Modelle für ungerichtete Graphen ergeben sich Paare (V, E) mit $E \subseteq \binom{V}{2}$, d. h., die Kanten bestehen aus zweielementigen Teilmengen $\{x, y\}$ mit $x, y \in V$. Alternativ könnte man bei gerichteten Graphen zusätzlich fordern, dass $(x, y) \in E$ genau dann gilt, wenn $(y, x) \in E$ ist. Damit entsprechen Kanten bei ungerichteten Graphen einer symmetrischen Relation. Sowohl bei gerichteten als auch bei ungerichteten Graphen benutzen wir die Schreibweise $xy \in E$ für Kanten. Hierbei ist zu beachten, dass bei ungerichteten Graphen xy und yx dieselbe Kante beschreiben. Man kann jedem gerichteten Graphen einen ungerichteten Graphen zuordnen, indem man die Orientierung der Kanten vergisst. Umgekehrt kann man jeden ungerichteten Graphen orientieren, indem man für jede ungerichtete Kante $\{x, y\}$ eine Richtung festlegt: (x, y) oder (y, x). Die *Orientierung* eines ungerichteten Graphen ist nicht eindeutig. Bei orientierten Graphen ist höchstens eine der beiden Kanten (x, y) und (y, x) in E enthalten.

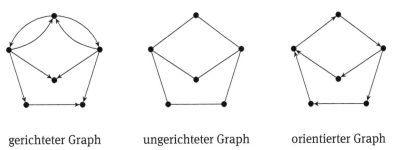

gerichteter Graph   ungerichteter Graph   orientierter Graph

Ein Graph ist *endlich*, falls er nur endlich viele Knoten besitzt. Wenn nicht anders angegeben, meinen wir im Folgenden mit „Graph" stets einen endlichen, ungerichteten Graph ohne Schlingen und Mehrfachkanten. Bei „gerichteten Graphen" meinen

wir endliche gerichtete Graphen ohne Mehrfachkanten. Ein *Teilgraph* $G' = (V', E')$ des Graphen $G = (V, E)$ ist ein Graph mit $V' \subseteq V$ und $E' \subseteq E$. Ein *induzierter* Teilgraph $G' = (V', E')$ des Graphen $G = (V, E)$ ist ein Teilgraph, für den gilt $E' = \binom{V'}{2} \cap E$; das heißt, in G' sind alle Kanten aus G enthalten, welche Knoten aus V' verbinden. Ein induzierter Teilgraph ist bereits durch eine Teilmenge der Knoten eindeutig spezifiziert. Manchmal bezeichnet man induzierte Teilgraphen auch als *Untergraphen*.

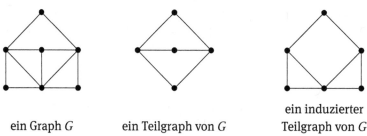

ein Graph G ein Teilgraph von G ein induzierter Teilgraph von G

Wir behandeln im Rest dieses Abschnitts nur ungerichtete Graphen. Die meisten Begriffsbildungen lassen sich aber leicht auf beliebige Graphen (wie zum Beispiel gerichtete Graphen mit Mehrfachkanten) übertragen. Sei $G = (V, E)$ ein Graph. Eine Folge von Knoten $x_0 x_1 \cdots x_n$ ist ein *Pfad* (oder ein *Weg*), falls je zwei aufeinander folgende Knoten durch eine Kante verbunden sind; das heißt, für alle $0 \leq i < n$ gilt $x_i x_{i+1} \in E$. Wir nennen n die *Länge* des Pfads; es ist die Anzahl der Kanten. Ein Pfad $x_0 \cdots x_n$ ist *einfach*, falls die Knoten $x_0, \ldots, x_n$ alle verschieden sind. Der Knoten x_0 ist der *Startpunkt* des Pfads und x_n sein *Endpunkt*. Falls bei einem Pfad der Länge $n \geq 3$ der Startpunkt und der Endpunkt identisch sind, so sprechen wir von einem *Kreis*. Ein Kreis $x_0 \cdots x_{n-1} x_0$ ist *einfach*, falls die Knoten $x_0, \ldots, x_{n-1}$ alle verschieden sind. Der folgende Graph wird *Petersen-Graph* (Julius Peter Christian Petersen, 1839–1910) genannt.

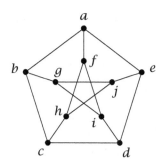

Petersen-Graph

Ein Pfad der Länge 8 im Petersen-Graph ist z. B. *abgjeafhc*; ein einfacher Pfad der Länge 9 ist *abcdejhfig*. Ein Kreis ist *abgjhcba*; und ein einfacher Kreis ist beispielsweise *abcdejgifa*.

Ein Graph G heißt *zusammenhängend*, falls je zwei Knoten durch einen Pfad verbunden sind. Eine *Zusammenhangskomponente* eines Graphen ist eine maximale Teilmenge von Knoten, welche paarweise durch einen Pfad verbunden sind. Die *Distanz* (oder der *Abstand*) zwischen zwei Knoten x und y ist die Länge eines kürzesten Pfads mit Startpunkt x und Endpunkt y. Falls x und y in verschiedenen Zusammenhangskomponenten liegen, ist ihr Abstand unendlich.

Häufig ist man nicht an der exakten Struktur eines Graphen interessiert, sondern nur an seinem „Aussehen". Die Namen der Knoten spielen in vielen Fällen eine untergeordnete Rolle. Dies führt auf den Begriff der Isomorphie. Zwei Graphen $G = (V, E)$ und $G' = (V', E')$ sind *isomorph*, falls eine bijektive Abbildung $\varphi : V \to V'$ existiert mit

$$xy \in E \iff \varphi(x)\varphi(y) \in E'$$

Isomorphie von Graphen bedeutet also, dass die Graphen durch Umbenennung der Knoten in einander überführt werden können. Die folgenden beiden Graphen über der Knotenmenge $\{1, 2, 3, 4, 5, 6\}$ sind isomorph:

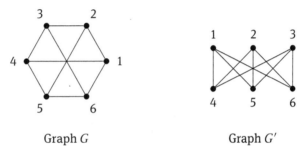

Graph G Graph G'

Ein möglicher Isomorphismus von G nach G' ist gegeben durch $1 \mapsto 1, 2 \mapsto 4, 3 \mapsto 2$, $4 \mapsto 5, 5 \mapsto 3$ und $6 \mapsto 6$. Ein nichttrivialer Isomorphismus von G auf sich selbst ist die Drehung gegen den Uhrzeigersinn $1 \mapsto 2, 2 \mapsto 3, 3 \mapsto 4, 4 \mapsto 5, 5 \mapsto 6, 6 \mapsto 1$. Die Komposition dieser beiden Isomorphismen $G \to G$ und $G \to G'$ liefert einen weiteren Isomorphismus $G \to G'$.

Beispiel 6.1. Sei $V = \{M_1, \ldots, M_n\}$ eine endliche Familie von Mengen M_i. Wir können aus V einen Graph $G = (V, E)$ konstruieren, indem wir für $i \neq j$ genau dann eine Kante $\{M_i, M_j\}$ zeichnen, wenn $M_i \cap M_j \neq \varnothing$ gilt.

Umgekehrt lässt sich jeder Graph auf diese Weise darstellen: Sei $G = (V, E)$ ein beliebiger Graph mit $V = \{x_1, \ldots, x_n\}$. Für $1 \leq i \leq n$ setzen wir $M_i = \{x_i\} \cup \{e \in E \mid x_i \in e\}$. Nun definiert die Familie $\{M_1, \ldots, M_n\}$ einen zu G isomorphen Graphen.   ◇

Der *komplementäre* Graph von $G = (V, E)$ ist $\overline{G} = (V, \overline{E})$ mit $\overline{E} = \binom{V}{2} \setminus E$. Der komplementäre Graph $\overline{G}$ von G enthält also genau die Kanten, welche G nicht hat.

| ein Graph | komplementärer Graph |

Ein Graph G heißt *selbstkomplementär*, falls G und sein komplementärer Graph $\overline{G}$ isomorph sind. Im Folgenden geben wir einige Beispiele selbstkomplementärer Graphen an:

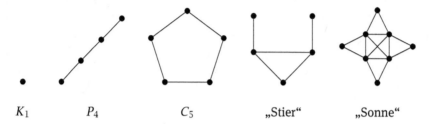

| K_1 | P_4 | C_5 | „Stier" | „Sonne" |

Wir wollen nun die Namen einiger spezieller Graphen vereinbaren. Sei hierzu $V_n = \{v_1, \dots, v_n\}$ eine Menge mit n Elementen. Der *vollständige* Graph mit n Knoten ist $K_n = (V_n, \binom{V_n}{2})$; das heißt, der Graph K_n enthält alle möglichen Kanten. Der Graph K_n hat genau $\binom{n}{2} = \frac{n(n-1)}{2}$ viele Kanten. Dies ist die Maximalzahl an Kanten, die ein ungerichteter Graph ohne Schlingen und Mehrfachkanten haben kann. Insbesondere ist die Anzahl der Kanten jedes Graphen mit n Knoten in $\mathcal{O}(n^2)$.

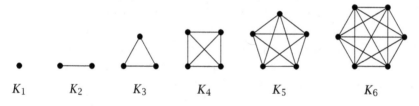

| K_1 | K_2 | K_3 | K_4 | K_5 | K_6 |

Der *leere* Graph mit n Knoten ist $(V_n, \varnothing)$. Er enthält keine Kanten und ist der komplementäre Graph des K_n. Mit P_n bezeichnen wir den Graphen, der genau aus einem einfachen Pfad der Länge $n - 1$ besteht, das heißt $P_n = (V_n, \{v_i v_{i+1} \mid 1 \leq i < n\})$. Der Graph P_n enthält $n - 1$ Kanten.

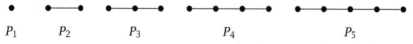

| P_1 | P_2 | P_3 | P_4 | P_5 |

Der Graph C_n ist der einfache Kreis mit n Knoten. Man erhält den Graph C_n, indem man bei einem P_n den ersten mit dem letzten Knoten verbindet; das heißt $C_n = (V_n, \{v_i v_{i+1} \mid 1 \leq i < n\} \cup \{v_n v_1\})$. Für $n \geq 3$ besitzt der Graph C_n genau n Kanten.

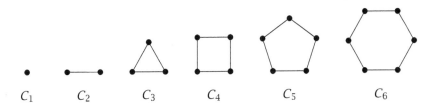

$C_1 \qquad C_2 \qquad C_3 \qquad C_4 \qquad C_5 \qquad C_6$

Ein Graph $G = (V, E)$ heißt *bipartit*, wenn $V = A \cup B$ mit $A \cap B = \varnothing$ und $E \subseteq \{\{a, b\} \mid a \in A, b \in B\}$ gilt. Dies bedeutet, wir können die Knoten in zwei Klassen A und B einteilen, so dass jede Kante einen Endpunkt in A und den anderen Endpunkt in B hat; das heißt, es gibt keine Kanten zwischen Knoten aus A und keine Kanten zwischen Knoten aus B. Beispielsweise sind die Graphen C_n für $n \geq 2$ genau dann bipartit, wenn n gerade ist (dann kommen die geraden Knoten nach A und die ungeraden Knoten nach B). Seien $A_m = \{a_1, \dots, a_m\}$ und $B_n = \{b_1, \dots, b_n\}$ disjunkte Mengen. Der *vollständig bipartite* Graph über A_m und B_n ist $K_{m,n} = (A_m \cup B_n, \{\{a, b\} \mid a \in A_m, b \in B_n\})$. Der Graph $K_{m,n}$ besitzt genau mn viele Kanten.

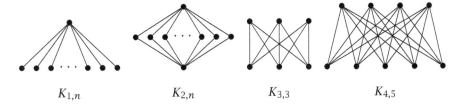

$$K_{1,n} \qquad\qquad K_{2,n} \qquad\qquad K_{3,3} \qquad\qquad K_{4,5}$$

Die naheliegendste Maßzahl eines Knotens x ist die Anzahl der inzidenten Kanten. Wir nennen dies den *Knotengrad* (oder kurz: den *Grad*) d_x von x, d. h. $d_x = |\{e \in E \mid x \in e\}|$ für einen Graphen (V, E) und seinen Knoten $x \in V$. Im vollständigen Graph K_n haben alle Knoten den Grad $n - 1$; und beim Kreis C_n mit $n \geq 3$ gilt $d_x = 2$ für alle Knoten. Bei Pfaden P_n mit $n \geq 2$ haben genau zwei Knoten (die Endpunkte) den Grad 1, und alle übrigen Knoten haben den Grad 2. Eine erste Beobachtung zu Graden liefert Satz 6.2 für Graphen $G = (V, E)$.

Satz 6.2 (Handschlaglemma).

$$\sum_{x \in V} d_x = 2|E|$$

Beweis. Wir zählen die Anzahl der *Kantenenden* auf zwei verschiedene Weisen. Jede Kante verbindet zwei Knoten und besitzt deshalb zwei Enden. Dies entspricht der rechten Seite. Andererseits lässt sich jedes Kantenende eindeutig einem Knoten zuordnen; auf diese Weise werden die Kantenenden auf der linken Seite gezählt. Der Name des Satzes ist dadurch motiviert, dass man sich eine Kante xy als das Händeschütteln von x und y vorstellt; um die Gleichung zu erhalten, zählt man auf zwei verschiedene Arten, wie viele Hände geschüttelt werden (d. h. die Anzahl der Kantenenden). □

Die deutsche Sprache hilft, sich Satz 6.2 leicht zu merken: „Die Summe der Grade ist gerade." Dies kann auch über die Knoten mit ungeradem Grad formuliert werden.

Korollar 6.3. *Die Anzahl der Knoten mit ungeradem Grad ist gerade.*

6.2 Eulerkreise und Hamiltonkreise

Als Euler in Königsberg war, wurde ihm die folgende Frage gestellt: Ist es möglich, durch die Stadt zu gehen, jede der sieben Brücken über den Fluss Pregel genau einmal zu überqueren, und am Ende wieder dort anzukommen, wo man gestartet ist; siehe Abbildung 6.1. In der Sprache der Graphentheorie lässt sich dies wie folgt formulieren: Besitzen die folgenden Graphen jeweils einen Kreis, der jede Kante genau einmal verwendet?

Im rechten Graph vermeiden wir die Mehrfachkanten des linken durch zusätzliche Knoten und Kanten. Euler zeigte 1736, dass solche Kreise nicht existieren. Dieses Ereignis wird häufig als die Geburtsstunde der Graphentheorie betrachtet. Einen Kreis in einem Graphen nennt man *Eulerkreis*, wenn jede Kante genau einmal besucht wird. Der Satz von Euler (Satz 6.4) liefert eine leicht zu überprüfende Charakterisierung derjenigen Graphen, die einen Eulerkreis besitzen.

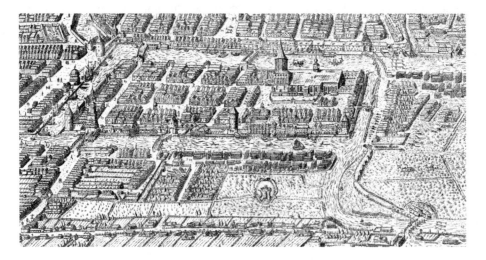

Abb. 6.1. Königsberg (Ausschnitt eines Stichs von *Joachim Bering*, 1613).

Satz 6.4 (Euler 1736). *Ein zusammenhängender Graph hat genau dann einen Eulerkreis, wenn alle Knoten einen geraden Grad haben.*

Beweis. Wenn ein Graph einen Eulerkreis besitzt, dann hat ein Knoten x, der auf dem Kreis k-mal vorkommt, den Grad $d_x = 2k$.

Sei umgekehrt $G = (V, E)$ ein Graph, bei dem jeder Knoten einen geraden Grad hat. Sei $P = v_0 \cdots v_n$ ein Pfad maximaler Länge, der jede Kante höchstens einmal benutzt. Da der Grad von v_n gerade ist, muss $v_0 = v_n$ gelten (andernfalls könnte man P verlängern). Falls eine Kante existiert, welche bei P nicht verwendet wird, dann gibt es auch eine Kante $v_i x$, die P nicht verwendet (da G zusammenhängend ist). Nun ist aber $v_i \cdots v_n v_1 \cdots v_i x$ ein längerer Pfad als P, der jede Kante nur einmal verwendet. Dies ist ein Widerspruch. Also ist P ein Eulerkreis. □

Bemerkung 6.5. Ein Eulerkreis in einem gerichteten Graphen $G = (V, E)$ mit $E \subseteq V \times V$ durchläuft alle gerichteten Kanten $xy \in E$ von x nach y. Auch Schleifen xx sind hier kein Problem. Der obige Beweis lässt sich leicht an gerichtete Graphen anpassen. Die zu zeigende Aussage ist hier, dass ein zusammenhängender gerichteter Graph genau dann einen Eulerkreis besitzt, wenn bei jedem Knoten x der Eingangsgrad (d. h., die Anzahl der Kanten der Form yx) gleich dem Ausgangsgrad (d. h., die Anzahl der Kanten der Form xy) ist. Die einzige Modifikation im Beweis von Satz 6.4 ist, dass man unterscheiden muss, ob eine Kante xv_i oder eine Kante $v_i x$ existiert; erstere hängt man bei $v_i \cdots v_n v_1 \cdots v_i$ vorne an und letztere hinten. ◊

Ein *Eulerweg* ist ein Pfad, der jede Kante eines Graphen genau einmal verwendet. Jeder Eulerkreis ist auch ein Eulerweg, aber ein Eulerweg kann bei einem anderen Knoten enden, als er beginnt. Aus Satz 6.4 lässt sich leicht herleiten, dass ein Graph genau dann einen Eulerweg besitzt, wenn höchstens zwei Knoten einen ungeraden Grad haben: Man zeichnet zwischen den beiden Knoten mit ungeradem Grad einen Pfad der Länge 2 mit einem neuen Knoten in der Mitte ein (nach Korollar 6.3 kann es nicht nur einen einzigen Knoten mit ungeradem Grad geben); in dem entstandenen Graphen hat jeder Knoten einen geraden Grad und es lässt sich Satz 6.4 anwenden. Wir sehen auch, dass jeder Eulerweg in diesem Fall bei einem Knoten mit ungeradem Grad beginnt und bei dem anderen Knoten mit ungeradem Grad endet. Dies führt sofort zu einer Lösung des Kinderrätsels *Haus vom Nikolaus*: Lässt sich das folgende Bild zeichnen, ohne den Stift abzusetzen und ohne eine Linie mehrfach zu malen, d. h., ein Strich für jede der 8 Silben des Satzes „*Das ist das Haus vom Nikolaus*".

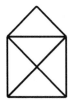

Haus vom Nikolaus

Ein ganz ähnliches Konzept zum Eulerkreis ist der Hamiltonkreis (nach Sir William Rowan Hamilton, 1805–1865). Ein *Hamiltonkreis* ist ein Kreis $v_0 \cdots v_n$, der jeden Knoten genau einmal besucht (wenn man Start und Ende $v_0 = v_n$ nur einmal zählt). Beispielsweise besitzt der vollständige Graph K_n für $n \geq 3$ stets einen Hamiltonkreis. Der Petersengraph hingegen besitzt keinen Hamiltonkreis. Ein Hamiltonkreis des Dodekaeders ist in der folgenden Zeichnung durch dicke Kanten angedeutet:

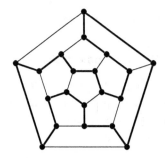

Hamiltonkreis im Dodekaeder

Im Gegensatz zu Eulerkreisen ist für Hamiltonkreise kein einfaches Kriterium bekannt, mit dem man überprüfen kann, ob ein Graph einen Hamiltonkreis besitzt (ein solches Kriterium würde das sogenannte P-NP-Problem lösen). Eine hinreichende Bedingung liefert der Satz von Ore (Satz 6.6, Øystein Ore, 1899–1968).

Satz 6.6 (Ore 1960). *Wenn in einem Graph $G = (V, E)$ mit $|V| \geq 3$ je zwei nicht benachbarte Knoten x, y die Bedingung $d_x + d_y \geq |V|$ erfüllen, dann besitzt G einen Hamiltonkreis.*

Beweis. Angenommen es existiert ein Graph mit $n \geq 3$ Knoten, welcher die Voraussetzung des Satzes erfüllt, aber keinen Hamiltonkreis hat. Sei $G = (V, E)$ ein solcher Graph mit n Knoten und mit einer maximalen Anzahl von Kanten. Sei $xy \notin E$. Nach Maximalität von E besitzt der Graph $(V, E \cup \{xy\})$ einen Hamiltonkreis, und dieser Kreis verwendet die Kante xy. Wenn wir die Kante xy weglassen, dann liefert dies einen Pfad $v_1 \cdots v_n$ in G von $x = v_1$ nach $y = v_n$, der jeden Knoten genau einmal besucht. Sei $X = \{v_i \mid v_{i+1}x \in E\}$ und $Y = \{v_i \mid v_i y \in E\}$. Es gilt $y \notin X$ und $y \notin Y$, sowie $|X| = d_x$ und $|Y| = d_y$. Mit $d_x + d_y \geq n$ folgt daraus $X \cap Y \neq \emptyset$. Sei

$v_i \in X \cap Y$. Die Knotenfolge

$$v_1 \cdots v_i v_n \cdots v_{i+1} v_1$$

definiert einen Hamiltonkreis, denn es gilt $v_i v_n = v_i y \in E$ und $v_{i+1} v_1 = v_{i+1} x$ $\in E$. Dies ist ein Widerspruch. Also existiert kein Graph G, der die Voraussetzung des Satzes erfüllt, aber keinen Hamiltonkreis besitzt. $\square$

6.3 Bäume

Ein *Baum* ist ein nichtleerer zusammenhängender Graph ohne einfache Kreise. Ein Knoten eines Baums ist ein *Blatt*, falls er höchstens den Grad 1 hat; Knoten, die keine Blätter sind, nennt man *innere Knoten*. Das folgende Bild zeigt alle Bäume mit 5 Knoten.

Die drei Bäume mit 5 Knoten

In manchen Fällen zeichnet man einen Knoten eines Baums aus und nennt ihn *Wurzel*. Man spricht auch von *gewurzelten* Bäumen. Die Idee ist, dass die Wurzel der Stelle entspricht, an der der Baum beginnt. Satz 6.7 fasst einige Eigenschaften von (nicht gewurzelten) Bäumen zusammen:

Satz 6.7. *Sei $G = (V, E)$ ein nichtleerer Graph. Die folgenden Eigenschaften sind äquivalent:*

(a) G ist ein Baum.

(b) Zwischen je zwei Knoten aus V gibt es genau einen einfachen Pfad in G.

(c) G ist zusammenhängend und $|E| = |V| - 1$.

(d) G ist zusammenhängend, aber durch Entfernen von jeder beliebigen Kante aus E wird der Graph unzusammenhängend.

Beweis. (a) $\Rightarrow$ (b): Angenommen, zwei Knoten $x, y \in V$ sind durch zwei verschiedene Pfade $x v_1 \cdots v_{m-1} y$ und $x w_1 \cdots w_{n-1} y$ verbunden. Wir wählen die Knoten x und y so, dass $m + n$ minimal ist. Dann ist $\{v_1, \ldots, v_{m-1}\} \cap \{w_1, \ldots, w_{n-1}\}$ $= \emptyset$. Daher ist $x v_1 \cdots v_{m-1} y w_{n-1} \cdots w_1 x$ ein einfacher Kreis. Dies ist ein Widerspruch, denn G ist ein Baum. Also folgt, dass zwischen je zwei Knoten höchstens ein Pfad existiert. Da G zusammenhängend ist, existiert auch mindestens ein Pfad.

(b) $\Rightarrow$ (c): Für jeden Knoten gibt es genau eine ausgehende Kante, die zur Wurzel führt; und jede Kante erfüllt diese Aufgabe für irgendeinen Knoten. Es gibt $|V| - 1$ Knoten, welche nicht die Wurzel sind. Dies ist nach der Vorüberlegung auch die Anzahl der Kanten. Im Folgenden ist diese Idee genauer beschrieben.

Wir wählen einen beliebigen Knoten $r \in V$ zur Wurzel (engl. *root*) und ordnen jedem Knoten $x \in V \setminus \{r\}$ die erste Kante $e_x = xv_1 \in E$ zu, die auf dem eindeutigen Pfad $xv_1 \cdots v_{m-1}r$ von x zur Wurzel liegt. Weiter definieren wir die Höhe $h(x)$ von x als die Länge des Pfads von x zu r, d. h. $h(x) = m$. Falls $y \neq x$ auf dem Pfad von x zur Wurzel liegt, dann gilt $h(y) < h(x)$. Angenommen, es gilt $e_x = e_y$ für Knoten $x \neq y$. Seien $p = xv_1 \cdots v_{m-1}r$ und $q = yw_1 \cdots w_{n-1}r$ zwei einfache Pfade in G. Dann gilt $v_1 = y$ und $w_1 = x$. Es folgt $h(x) < h(y)$ und $h(y) < h(x)$. Dies ist ein Widerspruch; also gilt $e_x \neq e_y$ für $x \neq y$. Nun gilt für $E_V = \{e_x \in E \mid x \in V \setminus \{r\}\}$, dass $|E_V| = |V \setminus \{r\}| = |V| - 1$. Angenommen, es existiert eine Kante $xy \in E \setminus E_V$. Ohne Einschränkung sei $h(y) \leq h(x)$. Falls y auf dem Pfad von x zur Wurzel liegt, dann folgt aus $e_x \neq xy$, dass $h(y) \leq h(x) - 2$ gilt. In jedem Fall liefert deshalb der Pfad mit der Kante xy über den Knoten y einen neuen einfachen Pfad von x zur Wurzel, was ein Widerspruch zu (b) ist. Also gilt $E = E_V$ und damit $|E| = |V| - 1$.

(c) $\Rightarrow$ (d): Sei $e \in E$. Der Graph $G' = (V, E \setminus \{e\})$ hat nur $|V| - 2$ Kanten, aber $|V| - 2$ Kanten können höchstens $|V| - 1$ Knoten miteinander verbinden. Also ist G' unzusammenhängend.

(d) $\Rightarrow$ (a): Wenn G einen einfachen Kreis enthalten würde, dann könnten wir jede beliebige Kante auf diesem Kreis entfernen, und der entstandene Graph wäre immer noch zusammenhängend. Dies ist ein Widerspruch zu (d). Also enthält G keine einfachen Kreise. $\qquad\qquad\square$

Wenn wir mit einem beliebigen zusammenhängenden Graphen starten, können wir nach Satz 6.7 (d) so lange Kanten entfernen, bis wir einen Baum erhalten. Deshalb enthält jeder zusammenhängende Graph (mindestens) einen sogenannten *Spannbaum*. Wir formulieren dies im Korollar 6.8.

Korollar 6.8. *Jeder nichtleere zusammenhängende Graph $G = (V, E)$ besitzt einen Baum mit Knoten V als Teilgraph.*

Insbesondere folgt aus Korollar 6.8, dass jeder zusammenhängende Graph mit n Knoten mindestens $n - 1$ Kanten besitzt. Spannbäume sind ein einfaches, aber sehr vielseitiges Hilfsmittel in der Graphentheorie. Wir betrachten hierzu das folgende Beispiel: Sei G ein zusammenhängender Graph mit n Knoten. Ein Durchlauf eines Spannbaums von G mittels Tiefensuche zeigt, dass G einen (nicht einfachen) Kreis der Länge $2(n - 1)$ besitzt, welcher jeden Knoten mindestens einmal besucht.

Natürlich lässt sich das Konzept der Spannbäume auch auf nicht zusammenhängende Graphen verallgemeinern, indem man die Zusammenhangskomponenten einzeln betrachtet. Ein weiteres einfaches Korollar aus Satz 6.7 belegt die gewählten Begriffsbildungen.

Korollar 6.9. *Jeder Baum besitzt Blätter.*

Beweis. Angenommen, jeder Knoten des Baums $G = (V, E)$ hätte mindestens den Grad 2. Aus Satz 6.2 folgt $|E| \geq |V|$. Dies ist ein Widerspruch zu Satz 6.7 (c). Also besitzt G mindestens ein Blatt. □

Tatsächlich folgt aus Korollar 6.9, dass jeder Baum mit mindestens zwei Knoten auch mindestens zwei Blätter besitzt. Die Beweistechnik ist typisch: Man *pflückt die Blätter*. Für zwei Knoten gilt die Behauptung. Bei einem Baum mit mehr als zwei Knoten können wir ein Blatt x entfernen ("pflücken"). Dieses Blatt existiert nach Korollar 6.9. Der entstandene Baum besitzt nach Induktionsvoraussetzung zwei Blätter y und z. Da x nur mit einem Knoten verbunden war, ist y oder z auch ein Blatt im ursprünglichen Baum. Zusammen mit x sind dies zwei Blätter. Für alle $n \geq 2$ ist der Pfad P_n ein Beispiel eines Baums mit n Knoten und genau zwei Blättern.

6.4 Die Cayley-Formel

Die Cayley-Formel (nach Arthur Cayley, 1821–1895) bestimmt die Anzahl der Spannbäume in einem vollständigen Graphen mit n Knoten. Dies ist etwas anderes, als die Anzahl der Bäume mit n Knoten bis auf Isomorphie zu bestimmen. So gibt es nur einen Baum mit drei Knoten, nämlich den Graph P_3. Aber es gibt drei Spannbäume mit der Knotenmenge $\{1, 2, 3\}$.

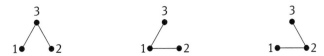

Wir betrachten den vollständigen Graphen K_n mit der Knotenmenge $V = \{1, \ldots, n\}$. Die Kantenmenge E besteht also aus der Menge $\binom{V}{2}$ und hat $\binom{n}{2}$ Kanten. Ein Spannbaum besteht aus $n-1$ Kanten, damit gibt es $\binom{m}{n-1}$ potentielle Kandidaten für Spannbäume, wobei $m = \binom{n}{2}$ ist. Es ist klar, dass die Anzahl der Spannbäume viel geringer ist. Die genaue Anzahl findet sich in dem folgenden Satz. Es gibt diverse Beweise für diesen klassischen Satz der abzählenden Kombinatorik. Wir verwenden die Methode, Bäume mittels *Prüfer-Codes* darzustellen. Diese Codierungen sind nach Ernst Paul Heinz Prüfer (1896–1934) benannt, mit deren Hilfe er 1918 einen sehr eleganten und einfachen Beweis für Satz 6.10 gefunden hatte. Diesen Beweis stellen wir unten vor.

Satz 6.10 (Cayley-Formel). *Sei $n \geq 2$. Die Anzahl der Spannbäume in einem vollständigen Graphen mit n Knoten ist n^{n-2}.*

Beweis. Sei $K_n = (V, E)$ der vollständige Graph mit der n-elementigen Knotenmenge V. Wir nehmen an, dass die Knoten in V linear angeordnet sind. Des Weiteren sei (V, T) mit $T \subseteq E$ ein Spannbaum von K_n. Wir codieren T durch eine Folge in V^{n-2}. Für $n = 2$ ist dies die leere Folge und dies entspricht dem einzigen Spannbaum mit $T = E$ in diesem Spezialfall. Sei jetzt $n \geq 3$. Sei $b_1 \in V$ das kleinste Blatt von

(V, T). Der wesentliche Trick ist, den Nachbarn p_1 von b_1 zu notieren und nicht das Blatt b_1 selbst. Es gilt also $b_1 p_1 \in T$. Wir setzen $V' = V \setminus \{b_1\}$ und $T' = T \setminus \{b_1 p_1\}$. Dann ist (V', T') ein Spannbaum für einen vollständigen Graphen mit $n - 1$ Knoten. Nach Induktion existiert eine Folge $(p_2, \ldots, p_{n-2})$, die (V', T') codiert. Wir definieren die Kodierung von (V, T) durch die Folge $(p_1, p_2, \ldots, p_{n-2})$ und nennen diese Folge den *Prüfer-Code* von (V, T).

Mit Induktion erkennt man, dass $\{p_1, \ldots, p_{n-2}\}$ genau die Menge der inneren Knoten von T ist. Insbesondere können einige der p_i's gleich sein. Daher können wir nun aus der Folge $(p_1, \ldots, p_{n-2})$ das Blatt b_1 herauslesen. Es ist das kleinste Element in $V \setminus \{p_1, \ldots, p_{n-2}\}$. Wir wissen damit $b_1 p_1 \in T$. Induktiv können wir nun aus $\{p_2, \ldots, p_{n-2}\}$ den Spannbaum T' mit Knotenmenge $V' = V \setminus \{b_1\}$ rekonstruieren. Wir erhalten $T = T' \cup \{b_1 p_1\}$. Die Zuordnung, die jedem T den Prüfer-Code zuordnet ist also eine injektive Abbildung von der Menge aller Spannbäume von (V, E) in die Menge V^{n-2}.

Es verbleibt noch zu zeigen, dass jede Folge $(p_1, \ldots, p_{n-2})$ Prüfer-Code eines Spannbaums ist. Sei b_1 das kleinste Element in $V \setminus \{p_1, \ldots, p_{n-2}\}$. Mit Induktion ist die Restfolge $(p_2, \ldots, p_{n-2})$ der Prüfer-Code eines Spannbaums T' von $V' = V \setminus \{b_1\}$. Also ist (V', T') zusammenhängend und T' hat $n - 2$ Kanten. Setzen wir $T = T' \cup \{b_1 p_1\}$, so ist (V, T) zusammenhängend und T hat $n - 1$ Kanten. Also ist (V, T) ein Spannbaum mit Prüfer-Code $(p_1, \ldots, p_{n-2})$. $\qquad\square$

Das Verfahren, den Prüfer-Code eines Spannbaums (V, T) zu erzeugen, kann wie folgt beschrieben werden. Üblicherweise startet man mit der Knotenmenge $V = \{1, \ldots, n\}$. Dann wird das kleinste Blatt gepflückt und sein Nachbar notiert. Nun fährt man mit dem kleineren Baum fort, bis nur noch zwei Knoten übrig bleiben. Der Abbruch passiert also genau dann, wenn der Baum keine inneren Knoten mehr hat. Es werden nur innere Knoten aus T notiert. Umgekehrt kann man den Baum zu einem so gebildeten Prüfer-Code $(p_1, \ldots, p_{n-2})$ dadurch rekonstruieren, dass man zunächst $V = \{1, \ldots, n\}$ setzt (n ist die Länge der Folge plus zwei). Nun findet man heraus, welches Blatt zuerst gepflückt wurde. Es ist der kleinste Knoten in $V \setminus \{p_1, \ldots, p_{n-2}\}$. Man weiß, dass der verbleibende Baum über der Knotenmenge $V' = V \setminus \{b_1\}$ gebildet wurde, und sein Prüfer-Code ist $(p_2, \ldots, p_{n-2})$. Induktiv kann man daraus den Baum (V', T') ermitteln, und zusammen mit dem Knoten b_1 und der Kante $b_1 p_1$ ist der so konstruierte Graph genau der gesuchte Baum zum Prüfer-Code $(p_1, \ldots, p_{n-2})$. Die Rekursion bricht ab, sobald $n = 2$ gilt und der Prüfer-Code leer ist. Dann ist der zu bildende Baum eindeutig durch die beiden Knoten aus der Knotenmenge bestimmt. Das folgende Diagramm gibt ein Beispiel an.

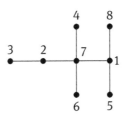

Baum mit Prüfer-Code $(2, 7, 7, 1, 7, 1)$

Bäume sind sicherlich diejenige Graphenklasse, welche am häufigsten anzutreffen ist, sei es als Binärbäume (Maximalgrad 3, Wurzel hat Maximalgrad 2), als Wahrscheinlichkeitsbäume, als Suchbäume für geordnete Mengen, oder, wie in Abschnitt 4.9.2, zur Darstellung von korrekt geklammerten Ausdrücken, wie man sie etwa in der Struktur von XML-Dokumenten (Extensible Markup Language) vorfindet.

6.5 Der Heiratssatz

Seien A und B disjunkt, und sei $G = (A \cup B, E)$ ein bipartiter Graph; das heißt, jede Kante in E verbindet einen Knoten aus A mit einem Knoten aus B. Eine Teilmenge $M \subseteq E$ ist ein *Matching*, falls keine zwei Kanten in M einen gemeinsamen Knoten haben. Wir sagen, M ist ein *perfektes Matching* für A, wenn jeder Knoten aus A auf einer Kante aus M liegt.

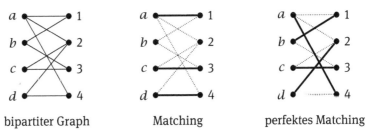

bipartiter Graph   Matching   perfektes Matching

Der Heiratssatz gibt eine Bedingung dafür an, wann genau ein perfektes Matching für A existiert. Sei hierzu

$$N_G(a) = \{b \in B \mid ab \in E\}$$

die Menge der Nachbarn von $a \in A$ im Graphen G. Diese Notation lässt sich durch $N_G(X) = \bigcup_{a \in X} N_G(a)$ auf Teilmengen $X \subseteq A$ erweitern. Die *Heiratsbedingung* sagt, dass $|N_G(X)| \geq |X|$ für alle Teilmengen $X \subseteq A$ gilt. Wenn $|B| < |A|$ ist, dann kann kein perfektes Matching existieren. Ebenso muss die Heiratsbedingung gelten, wenn ein perfektes Matching existiert. Der Heiratssatz sagt nun, dass auch die Umkehrung gilt: Wenn die Heiratsbedingung gilt, dann existiert ein perfektes Matching für A.

In dieser Form wurde der Heiratssatz 1935 von Philip Hall (1904–1982) bewiesen. In einer leicht anderen Formulierung wurde er bereits 1931 in zwei unabhängigen Ar-

beiten von Kőnig (Dénes Kőnig, 1884–1944) und Egerváry (Jenő Egerváry, 1891–1958) gezeigt. Noch etwas früher, bereits 1929, wurde der Satz von Menger bewiesen; und der Heiratssatz lässt sich leicht aus dem Satz von Menger herleiten. Dennoch ist es inzwischen üblich, Hall den Heiratssatz zuzuschreiben.

Namensgeber für den Heiratssatz ist die folgende Situation: Man kann sich A (wie *Alice*) als eine Menge von Frauen und B (wie *Bob*) als eine Menge von Männern vorstellen. Eine Kante zwischen Frau und Mann existiert, wenn eine Heirat möglich ist. Ein perfektes Matching bedeutet, dass es möglich ist, alle Personen aus der Gruppe A (in diesem Fall die Frauen) zu verheiraten, ohne dass dabei ein Mann mit zwei Frauen verheiratet wird.

Satz 6.11 (Heiratssatz). *Sei $G = (A \cup B, E)$ ein bipartiter Graph. Es gibt genau dann ein perfektes Matching für A, wenn $|N_G(X)| \geq |X|$ für alle $X \subseteq A$ gilt.*

Beweis. Sei M ein perfektes Matching, dann erfüllt der Graph $(A \cup B, M)$ die Heiratsbedingung. Also erfüllt auch G die Heiratsbedingung.

Sei jetzt umgekehrt G ein Graph, der die Heiratsbedingung erfüllt. Falls für jede echte Teilmenge $\emptyset \neq X \subsetneq A$ die Abschätzung $|N_G(X)| > |X|$ gilt, dann können wir eine beliebige Kante e aus G entfernen. Der verbleibende Graph erfüllt immer noch die Heiratsbedingung und besitzt deshalb mit Induktion nach der Kantenzahl ein perfektes Matching. Dieses ist auch ein perfektes Matching von G.

Wenn der obige Fall nicht eintritt, dann existiert eine nichtleere Menge $X \subsetneq A$ mit $|N_G(X)| = |X|$. Sei G_1 der von $X \cup N_G(X)$ induzierte Untergraph von G, und sei G_2 der von den verbliebenen Knoten (außerhalb von $X \cup N_G(X)$) induzierte Untergraph. Der Graph G_1 erfüllt die Heiratsbedingung, da $N_{G_1}(X') = N_G(X')$ für alle $X' \subseteq X$ gilt. Wir müssen noch zeigen, dass G_2 die Heiratsbedingung erfüllt. Mit Induktion besitzen dann sowohl G_1 als auch G_2 perfekte Matchings, und deren Vereinigung ist ein perfektes Matching von A. Sei $G_2 = (A' \cup B', E')$ mit $A' \subseteq A$ und $B' \subseteq B$. Für $X' \subseteq A'$ gilt

$$|N_{G_2}(X')| + |N_G(X)| \geq |N_G(X' \cup X)| \geq |X' \cup X| = |X'| + |X|$$

und damit $|N_{G_2}(X')| \geq |X'|$. Also erfüllt G_2 die Heiratsbedingung. □

Eine häufige Anwendung des Heiratssatzes 6.11 ist mit $|A| = |B|$. Dann ist jedes perfekte Matching von A im bipartiten Graph $G = (A \cup B, E)$ auch ein perfektes Matching von B.

6.6 Stabile Heirat

Der Nobelpreis für Wirtschaftswissenschaften, genauer der *Preis der Reichsbank Schwedens für die ökonomische Wissenschaft zum Andenken an Alfred Nobel*, wurde 2012 an Alvin Elliot Roth (geb. 1951) und Lloyd Stowell Shapley (geb. 1923) verliehen.

Damit wurde ihr Beitrag zur Theorie stabiler Zuordnungen und zur Gestaltung bestimmter Märkte gewürdigt. Zentral ist dabei der Gale-Shapley-Algorithmus, der bereits 50 Jahre vorher von David Gale (1921–2008) und Shapley entwickelt wurde. Die
später durchgeführten empirischen Arbeiten von Roth belegten dann die Bedeutung
der Stabilität bei realen Bedingungen.

Wir wollen den zugrunde liegenden Algorithmus von Gale und Shapley in diesem Abschnitt vorstellen. Es seien A und B Mengen von je n Personen. Ohne Einschränkung seien A die Frauen und B die Männer. Jede Person hat eine Präferenzliste
der Personen vom anderen Geschlecht. Für $a \in A$ können wir uns die Präferenz-
liste P_a als eine lineare Ordnung $b_{a(1)} > \cdots > b_{a(n)}$ mit $B = \{b_{a(1)}, \ldots, b_{a(n)}\}$
vorstellen. Wenn $b_{a(i)}$ in der Ordnung vor $b_{a(j)}$ steht, so bevorzugt a den Mann $b_{a(i)}$
vor $b_{a(j)}$. Analog verfügt jeder Mann $b \in B$ über eine Präferenzliste P_b. Eine Heirat
(oder das Matching $M \subseteq A \times B$) ist *stabil*, wenn alle Frauen verheiratet sind und es
zu keinen Scheidungen kommt. Es kommt zu einer Scheidung, wenn es zwei Paare
$(a, b'), (a', b)$ gibt mit $P_a(b) > P_a(b')$ und $P_b(a) > P_b(a')$. Dann lassen sich nämlich a und b von ihren Partnern scheiden und bilden ein neues Paar (a, b). Wenn
sich anschließend a' und b' zusammentun, sind wieder alle verheiratet. Die Zufrie-
denheit von a und b ist gestiegen, während die von a' und b' gesunken sein kann.

Die Situation von zwei Paaren ist leicht zu analysieren. Es gibt also zwei Frauen
a, a' und zwei Männer b, b'. Betrachten wir zunächst den Fall bei dem es ein Paar
gibt, welches sich wechselseitig die höchste Präferenz gibt. Ohne Einschränkung ist
dies das Paar (a, b). Es gilt also $P_a(b) > P_a(b')$ und $P_b(a) > P_b(a')$. Dann ist
$(a, b), (a', b')$ stabil, und es ist die einzige stabile Heirat. Wenn kein solches Paar
existiert, so überkreuzen sich die Präferenzen jeweils. Sagen wir a hat als Favorit b,
aber b favorisiert a', die nun b' bevorzugt, für den schließlich die Frau a die höhere
Präferenz hat. Wir erhalten eine kreisförmige Anordnung der höchsten Präferenzen.

In diesem Fall sind beide möglichen Paarbildungen stabil; sie unterscheiden sich dadurch, dass entweder die beiden Männer oder die beiden Frauen ihre Favoriten heiraten. Immerhin ist auch in komplizierteren Situationen eine stabile Heirat möglich,
allerdings nur unter der Bevorzugung einer Partei.

Im Allgemeinen stellt sich keine stabile Heirat ein, wenn Paare in einer zufälligen
Reihenfolge aufeinander treffen und sich bei entsprechenden Präferenzen von ihrem
aktuellen Partner trennen und neu heiraten. In der vorherigen Situation könnten zuerst (a, b) verheiratet sein, dann (a', b), dann (a', b'), dann (a, b') und schließlich
wieder (a, b). Die übrigen beiden Personen sind hier jeweils unverheiratet. In unserer
Betrachtung einer stabilen Heirat gibt es keine unverheirateten Personen, also fügen
wir eine zusätzliche Frau a_0 und einen zusätzlichen Mann b_0 hinzu, welche jeweils
die niedrigste Präferenz bei den bisherigen Männern und Frauen haben. Personen

welche im obigen Beispiel unverheiratet sind, werden nun mit a_0 oder b_0 verheiratet. Dadurch liefert das obige Beispiel eine unendliche Folge von nichtstabilen Heiraten, bei denen jeweils alle Personen verheiratet sind. Ein Durchlauf ist im folgenden Bild veranschaulicht; die gestrichelte Kante widerspricht jeweils der Stabilität. Nach nur vier Neuorientierungen der Paare sind wir wieder in der Ausgangssituation.

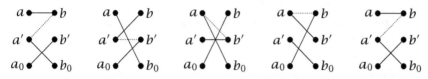

Beim Berechnen einer stabilen Heirat ist es üblich, verloben und heiraten von einander abzugrenzen, um keine Paare scheiden zu müssen. *Verloben* meint das vorläufige Auswählen eines Partners, während *heiraten* endgültig ist. Der *Gale-Shapley-Algorithmus* berechnet wie folgt eine Heirat.

(1) Zu Anfang ist niemand verlobt oder verheiratet.

(2) Solange noch ein unverlobter Mann $b \in B$ existiert, macht b derjenigen Dame $a \in A$ einen Antrag, die er noch nicht vorher gefragt hatte und die für ihn unter diesen Frauen die höchste Präferenz hat.

Die Frau a nimmt den Antrag an und verlobt sich mit b, wenn sie noch keinen Partner hat, oder sie den Antragsteller b ihrem derzeitigen Verlobten vorzieht. Gegebenenfalls wird dabei eine Verlobung gelöst, um eine andere einzugehen.

(3) Sind alle Männer verlobt, so heiratet jeder seine Verlobte.

Jeden Durchlauf von Schritt (2) bezeichnen wir im Folgenden als *Runde*.

Satz 6.12 (Gale, Shapley 1962). *Der Gale-Shapley-Algorithmus berechnet in maximal n^2 Runden eine stabile Heirat.*

Beweis. Verlobte Frauen bleiben in jeder Runde verlobt. Eine Frau verlobt sich nur dann neu, wenn sie sich verbessert. Insbesondere verlobt sich jede Frau maximal n-mal. Spätestens nach n^2 Runden hat jede Frau einen Antrag erhalten, und alle Frauen (und damit auch alle Männer) sind verlobt.

Angenommen, die vom Gale-Shapley-Algorithmus berechnete Heirat wäre instabil, das heißt, es gibt zwei verheiratete Paare (a, b') und (a', b) mit $P_a(b) > P_a(b')$ und $P_b(a) > P_b(a')$. Dann hatte jedoch a vor der Verlobung von b mit a' entweder einen Antrag von b abgelehnt oder ihn verlassen. Der Grund hierfür war eine Verlobung mit einem Mann b'' mit $P_a(b'') > P_a(b)$. Da sich Frauen im Verlauf des Verfahrens nur verbessern, gilt $P_a(b') \geq P_a(b'')$. Dies ist ein Widerspruch zu $P_a(b) > P_a(b')$. □

Insbesondere existiert stets eine stabile Heirat. Die vom Gale-Shapley-Algorithmus berechnete Heirat lässt sich noch etwas genauer beschreiben.

Bemerkung 6.13. Der Gale-Shapley-Algorithmus ist für die Männer optimal. Sie erhalten jeweils diejenige Partnerin, welche die höchste Präferenz unter allen Frauen hat, mit denen überhaupt eine stabile Paarbildung möglich ist.

Hierzu genügt es zu zeigen, dass ein Paar $(a, b') \in A \times B$ in keiner stabilen Heirat realisierbar ist, wenn die Frau a im Gale-Shapley-Verfahren einen Antrag von b' abgelehnt oder b' verlässt. Mit Widerspruch betrachten wir den ersten Zeitpunkt t, zu dem eine Frau a einen Mann b' ablehnt oder verlässt, obwohl eine stabile Heirat M mit $(a, b') \in M$ existiert. Der Grund ist in beiden Fällen ein Mann b mit $P_a(b) > P_a(b')$. Es gilt $(a', b) \in M$ für eine Frau $a' \neq a$. Wäre $P_b(a) < P_b(a')$, dann hätte b von a' bereits vor t eine Absage erhalten oder wäre von a' verlassen worden. Dies ist nach Wahl von (a, b') nicht möglich. Also gilt $P_b(a) > P_b(a')$. Treffen nun (a, b') und (a', b) aufeinander, so verlassen a und b ihre Partner und bilden ein neues Paar (a, b). Damit ist M nicht stabil, ein Widerspruch. ◇

Aufgrund der Bemerkung 6.13 kommt es im Gale-Shapley-Verfahren nicht auf die Reihenfolge der Anträge an. Die Männer können sich sogar Zeit lassen. Dies ändert sich, wenn Frauen ihre Präferenzen nur partiell geordnet haben. Dann ist es wichtig, wer zuerst den Antrag stellt. Frauen werden insoweit benachteiligt, als dass sie die Partner niedrigster Präferenz bekommen, mit denen eine stabile Heirat möglich ist; siehe Aufgabe 6.16.

6.7 Der Satz von Menger

Der Satz von Menger (Karl Menger, 1902–1985) stellt einen Zusammenhang her zwischen einem Parameter, der maximiert wird, und einem Parameter, der minimiert wird. Derartige Aussagen sind typisch für die Graphentheorie. Genauer geht es bei dem einen Parameter um die minimale Anzahl von Knoten, die man braucht, um zwei (nicht notwendigerweise disjunkte) Knotenmengen A und B in einem gegebenen Graphen G zu trennen. Der andere Parameter ist die maximale Anzahl von disjunkten Pfaden in G von A nach B. Der Satz von Menger besagt, dass diese beiden Parameter übereinstimmen.

Sei $G = (V, E)$ ein gerichteter oder ein ungerichteter Graph und $A, B \subseteq V$. Ein *AB-Pfad* ist ein Pfad $x_0 \cdots x_n$ mit $x_0 \in A$ und $x_n \in B$; des Weiteren gilt für die inneren Knoten $x_1, \ldots, x_{n-1}$, dass $\{x_1, \ldots, x_{n-1}\} \cap (A \cup B) = \emptyset$. Bei einem AB-Pfad ist $x_0 = x_n \in A \cap B$ möglich. Ein *AB-Separator* ist eine Teilmenge $C \subseteq V$, so dass jeder AB-Pfad einen Knoten auf C hat. Zwei Pfade sind *disjunkt*, wenn sie keine gemeinsamen Knoten besitzen.

Satz 6.14 (Menger 1929; gerichtete Graphen, disjunkte Pfade). *Sei $G = (V, E)$ ein gerichteter Graph und $A, B \subseteq V$. Dann ist die Größe k eines kleinsten AB-Separators gleich der maximalen Anzahl von paarweise disjunkten AB-Pfaden.*

Beweis. Wenn C ein AB-Separator ist, dann gibt es höchstens $|C|$ disjunkte AB-Pfade. Es verbleibt zu zeigen, dass k disjunkte AB-Pfade existieren. Wenn $E = \varnothing$ gilt, dann ist $|A \cap B| = k$ und die Knoten in $A \cap B$ bilden k disjunkte AB-Pfade der Länge 0. Sei jetzt $e = xy$ eine gerichtete Kante in E. Wir entfernen die Kante e und erhalten den Graph $G' = (V, E \setminus \{e\})$. Wenn der kleinste AB-Separator in G' die Größe k hat, dann erhalten wir mit Induktion k disjunkte AB-Pfade in G'. Diese sind auch disjunkt in G.

Sei also C ein AB-Separator in G' mit $|C| \leq k - 1$. Sowohl $S = C \cup \{x\}$ als auch $T = C \cup \{y\}$ sind AB-Separatoren in G. Es folgt $|S| = k = |T|$. Sowohl jeder AS-Separator in G' als auch jeder TB-Separator in G' ist ein AB-Separator in G; dies verwendet die Orientierung der Kante e von x nach y. Mit Induktion existieren sowohl k disjunkte AS-Pfade $\mathcal{P}$ in G' als auch k disjunkte TB-Pfade $\mathcal{Q}$ in G'. Die Pfade aus $\mathcal{P}$ und die Pfade aus $\mathcal{Q}$ schneiden sich nur in C, denn sonst gäbe es einen AB-Pfad, der C gar nicht schneidet. In jedem Knoten aus S endet ein AS-Pfad aus $\mathcal{P}$ und in jedem Knoten aus T beginnt ein TB-Pfad aus $\mathcal{Q}$. Wir können damit die Pfade aus $\mathcal{P}$ und aus $\mathcal{Q}$ aneinander hängen und erhalten k disjunkte AB-Pfade in G; den Pfad nach x setzen wir hierbei mit dem Pfad ab y fort; dies ist möglich, da $xy \in E$ ist. $\qquad\square$

Der obige Beweis von Satz 6.14 erschien im Jahre 2000 in einer Arbeit von Frank Göring [21].

Satz 6.15 (Menger 1929; ungerichtete Graphen, disjunkte Pfade). *Sei $G = (V, E)$ ein Graph und $A, B \subseteq V$. Dann ist die Größe eines kleinsten AB-Separators gleich der maximalen Anzahl von paarweise disjunkten AB-Pfaden.*

Beweis. Dies folgt aus der gerichteten Version des Satzes von Menger 6.14, wenn wir anstelle einer ungerichteten Kante $\{x, y\}$ die beiden Orientierungen (x, y) und (y, x) aufnehmen. $\qquad\square$

Den Heiratssatz 6.11 kann man als ein Korollar des Satzes von Menger erhalten: Die Heiratsbedingung besagt, dass A ein minimaler AB-Separator ist, und der Satz von Menger liefert nun das perfekte Matching als disjunkte AB-Pfade der Länge 1.

6.8 Maximale Flüsse

Bevor wir *Flüsse* einführen, betrachten wir eine Variante von Satz 6.14, welche die maximale Anzahl von kantendisjunkten Pfaden zwischen zwei Knoten charakterisiert. Hierzu benötigen wir noch ein paar Begriffsbildungen. In der folgenden Variante des Satzes von Menger erlauben wir Mehrfachkanten. Daher betrachten wir Kantenzüge anstelle von Pfaden. Sei $G = (V, E, \sigma, \tau)$ ein Graph, so dass $\sigma(e)$ der Startknoten der Kante $e \in E$ und $\tau(e)$ ihr Endknoten ist. Ein *Kantenzug* ist eine Folge von Kanten $\pi = e_1 \cdots e_n$ mit $\tau(e_i) = \sigma(e_{i+1})$ für alle $1 \leq i < n$. Wir sagen, π ist ein *st-Kantenzug*, wenn $\sigma(e_1) = s$ und $\tau(e_n) = t$ gilt. Der Kantenzug π definiert den Pfad $\sigma(e_1) \cdots \sigma(e_n)\tau(e_n)$. Zwei Kantenzüge $\pi_1 = e_1 \cdots e_m$ und $\pi_2 = f_1 \cdots f_n$ sind

disjunkt, falls $e_i \neq f_j$ für alle i, j gilt. Seien $s, t \in V$ zwei verschiedene Knoten von $G = (V, E, \sigma, \tau)$. Ein *st-Schnitt* ist ein Paar (A, B) mit $s \in A \subseteq V$ und $t \in B = V \setminus A$. Ein st-Schnitt (A, B) zerschneidet also V in zwei nichtleere disjunkte Teile, und jeder Kantenzug von s nach t muss A verlassen und B betreten. Das *Gewicht* von (A, B) ist die Anzahl der Kanten $e \in E$ mit $\sigma(e) \in A$ und $\tau(e) \in B$. Dies sind genau jene Kanten, die einen Übergang von A nach B ermöglichen.

Satz 6.16 (Menger 1929; gerichtete Graphen, disjunkte Kantenzüge). *Sei $G = (V, E, \sigma, \tau)$ ein gerichteter Graph, bei dem Mehrfachkanten erlaubt sind, und seien $s, t \in V$ mit $s \neq t$. Dann ist das minimale Gewicht eines st-Schnitts gleich der maximalen Anzahl von disjunkten st-Kantenzügen.*

Beweis. Wir verwenden eine Konstruktion, die *Kantengraph* (oder auch *Liniengraph*) genannt wird. Der Kantengraph $L(G)$ von G hat die Knotenmenge E und die Kanten-menge

$$F = \{(e, f) \in E \times E \mid \tau(e) = \sigma(f)\}$$

d. h., die Kanten des ursprünglichen Graphen G sind die Knoten des Kantengraphen, und es wird eine Kante von e nach f gezeichnet, wenn der Zielknoten von e gleich dem Startknoten von f ist. Man beachte, dass der Kantengraph $L(G)$ niemals Mehr-fachkanten besitzt. Jeder Pfad $e_1 \cdots e_m$ in $L(G)$ übersetzt sich deshalb in einen Kan-tenzug $e_1 \cdots e_m$ in G und umgekehrt.

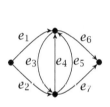 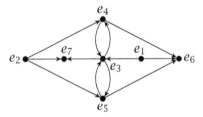

gerichteter Graph G dazugehöriger Kantengraph $L(G)$

Die Aussage des Satzes folgt nun, wenn wir die gerichtete Version des Satzes von Men-ger 6.14 auf den Kantengraphen $L(G)$ mit $A = \{e \mid \sigma(e) = s\}$ und $B = \{f \mid \tau(f) = t\}$ anwenden. □

6.8.1 Der Satz von Ford und Fulkerson

Wir wollen uns nun mit einer quantitativen Version des vorigen Satzes befassen, dem sogenannten *Max-Flow-Min-Cut-Theorem*. Ein Beweis davon wurde zuerst in einem technischen Bericht von 1954 von Ford und Fulkerson (Lester Randolph Ford junior, geb. 1927, und Delbert Ray Fulkerson, 1924–1976) veröffentlicht [19]. Parallel zu des-sen Zeitschriftenpublikation [20] im Jahr 1956 haben Peter Elias (1923–2001), Amiel

Feinstein (geb. 1930) und Claude Elwood Shannon (1916–2001) einen weiteren Beweis gefunden [17]. Die Formulierung des Max-Flow-Min-Cut-Theorems führt uns in die Welt der Netzwerke und Flüsse und bedarf einiger Vorbereitung.

Ein *Flussnetzwerk* besteht aus einer endlichen Knotenmenge V mit zwei ausgezeichneten Knoten, einem Startknoten s (*source*) und einem Zielknoten t (*target*) sowie einer Kapazitätsfunktion $c : V \times V \to \mathbb{R}_{\geq 0}$ in die nicht negativen reellen Zahlen. Die Kapazitätsfunktion definiert einen gewichteten gerichteten Graphen, wobei wir nur die Kanten mit positiver Kapazität zeichnen. Wir können uns jede Kante (x, y) als ein Rohr- oder Leitungsstück mit der Kapazität $c(x, y)$ vorstellen. Entsprechend ist ein Flussnetzwerk (V, c, s, t) ein System von Rohren oder Leitungen mit einer Quelle s und einem Ziel t. Ein typisches Problem der kombinatorischen Optimierung ist es, einen maximalen Fluss von der Quelle zum Ziel zu berechnen, der die Kapazitäten einhält. Formal ist ein *Fluss* eine Abbildung $f : V \times V \to \mathbb{R}$, die den drei folgenden Bedingungen genügt:

- (Schiefsymmetrie) Für alle $x, y \in V$ gilt $f(x, y) = -f(y, x)$.
- (Flusserhaltung) Für alle $u \in V$ mit $s \neq u \neq t$ gilt $\sum_{v \in V} f(u, v) = 0$.
- (Kapazitätsbedingung) Für alle $x, y \in V$ gilt $f(x, y) \leq c(x, y)$.

Die Schiefsymmetrie können wir uns so vorstellen, dass ein positiver Fluss von x nach y das Gleiche ist wie ein negativer Fluss von y nach x. Es folgt $f(x, x) = 0$ und damit ist es keine Einschränkung $c(x, x) = 0$ für alle Knoten x zu fordern. Flussnetzwerke haben also keine Schlingen. Die Flusserhaltung besagt, dass bei inneren Knoten genau so viel hineinfließt, wie auch wieder herauskommt. Bei der Quelle und dem Ziel braucht dies nicht zu gelten. Wir messen den *Wert* $\|f\|$ eines Flusses $f : V \times V \to \mathbb{R}$ an der Quelle durch

$$\|f\| = \sum_{y \in V} f(s, y)$$

Wie bei Flussnetzwerken zeichnen wir auch bei Flüssen nur Kanten xy mit einem positiven Wert $f(x, y)$ ein.

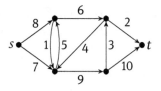

Flussnetzwerk

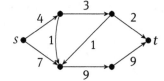

Fluss mit Wert 11

Die Definition des Wertes $\|f\|$ berücksichtigt die Quelle und nicht das Ziel. Das nächste Lemma zeigt uns, dass man einen identischen Wert auch beim Ziel vorfindet. Etwas allgemeiner betrachten wir dabei beliebige st-Schnitte. Wie bei Graphen ist ein st-Schnitt (A, B) eines Flussnetzwerks (V, c, s, t) durch die Bedingungen $s \in A \subseteq V$ und $t \in B = V \setminus A$ charakterisiert ist.

Lemma 6.17. *Sei $f : V \times V \to \mathbb{R}$ ein Fluss und (A, B) ein st-Schnitt. Dann gilt*

$$\|f\| = \sum_{x \in A,\, y \in B} f(x, y)$$

Insbesondere ist $\|f\| = \sum_{y \in V} f(s, y) = \sum_{x \in V} f(x, t)$.

Beweis. Wir zeigen die Aussage mit Induktion nach $|A|$. Für $A = \{s\}$ ist die Aussage trivial. Sei jetzt $A = A' \cup \{u\}$ mit $s \in A'$ und $u \notin A'$. Mit Induktion ist $\|f\| = \sum_{x \in A',\, y \in B'} f(x, y)$ für $B' = B \cup \{u\}$. Es gilt

$$\sum_{x \in A,\, y \in B} f(x, y) = \sum_{x \in A',\, y \in B'} f(x, y) + \sum_{y \in B} f(u, y) - \sum_{x \in A} f(x, u)$$

Nun ist $s \neq u \neq t$ und daher liefert uns die Schiefsymmetrie zusammen mit der Flusserhaltung die Behauptung:

$$\sum_{y \in B} f(u, y) - \sum_{x \in A} f(x, u) = \sum_{y \in B} f(u, y) + \sum_{x \in A} f(u, x) = \sum_{v \in V} f(u, v) = 0$$

Wegen $f(t, t) = 0$ ist die Aussage $\|f\| = \sum_{x \in V} f(x, t)$ gerade der Spezialfall mit $B = \{t\}$. □

Als *Kapazität* $c(A, B)$ eines st-Schnitts (A, B) bezeichnen wir die nicht negative reelle Zahl

$$c(A, B) = \sum_{x \in A,\, y \in B} c(x, y)$$

Aufgrund der Kapazitätsbedingung besagt Lemma 6.17, dass die Kapazität eines st-Schnitts eine obere Schranke für den Wert eines Flusses ist. Für alle st-Schnitte (A, B) und alle Flüsse f gilt:

$$\|f\| \leq c(A, B) \tag{6.1}$$

Der Satz von Ford und Fulkerson (Satz 6.18) stellt fest, dass diese Grenze scharf ist. Wir werden im nächsten Abschnitt 6.8.3 eine allgemeinere und zugleich algorithmische Version des Max-Flow-Min-Cut-Theorems behandeln. Vorab wollen wir zeigen, dass man dieses Theorem als Korollar zum Satz von Menger 6.16 auffassen kann.

Satz 6.18 (Max-Flow-Min-Cut-Theorem). *Sei $N = (V, c, s, t)$ ein Flussnetzwerk. Der maximale Wert $\|f\|$ eines st-Flusses f ist gleich der minimalen Kapazität $c(A, B)$ eines st-Schnitts (A, B). Sind ferner alle Kapazitäten natürliche Zahlen, so kann der maximale Fluss f ganzahlig gewählt werden.*

Beweis. Nach Gleichung (6.1) ist nur die Existenz eines Flusses zu zeigen, dessen Wert gleich der Kapazität eines st-Schnitts ist. Seien zunächst alle Kapazitäten ganzzahlig. Dann ersetzen wir jede Kante (x, y) durch $c(x, y)$ Kopien; dadurch erhalten wir einen ungewichteten gerichteten Graphen G mit Mehrfachkanten. Von x nach y sind jetzt $c(x, y)$ Kanten gezogen. Sei (A, B) ein minimaler st-Schnitt in G. Das Gewicht k

von (A, B) ist gleich $c(A, B)$. Mit Satz 6.16 finden wir k disjunkte st-Kantenzüge π_1, ..., π_k in G. Für zwei Knoten x, y sei $w(x, y)$ die Anzahl der Kanten von x nach y, die in einem der Wege π_i vorkommen. Wir definieren einen Fluss f durch $f(x, y) = w(x, y) - w(y, x)$. Damit ist $\|f\| = k = c(A, B)$. Dies beweist insbesondere den zweiten Teil des Satzes.

Die Erweiterung auf rationale Kapazitäten ist einfach: Wir multiplizieren mit dem Hauptnenner der Kapazitäten. Dies reduziert den Fall von rationalen Kapazitäten auf ganzzahlige Kapazitäten. Treten irrationale Kapazitäten auf, so lässt sich dieser Fall durch Stetigkeitsargumente auf rationale Kapazitäten zurückführen. Wir belassen es für reelle Kapazitäten bei dieser Beweisskizze, da wir im nächsten Abschnitt einen anderen Beweis kennen lernen werden. ☐

6.8.2 Residualgraphen und Verbesserungspfade

Wir betrachten noch einen anderen Zugang zu Satz 6.18, da wir uns der algorithmischen Lösung nähern möchten. Für einen beliebigen Fluss f (etwa den Nullfluss) ist der *Residualgraph* (V, R_f) definiert durch die Kantenmenge

$$R_f = \{ (x, y) \in V \times V \mid c(x, y) > f(x, y) \}$$

Sei $A \subseteq V$ die Menge der im Residualgraphen von s aus erreichbaren Knoten. Setzen wir $B = V \setminus A$, so gilt

$$\sum_{x \in A,\, y \in B} (c(x, y) - f(x, y)) = 0$$

Ist nun $t \notin A$, so definiert (A, B) einen st-Schnitt mit $\|f\| = c(A, B)$. Für $t \in A$ können wir den Wert des Flusses entlang eines Pfades von s nach t im Residualgraphen erhöhen.

Als Nächstes untersuchen wir die Erhöhung des Wertes von f entlang eines Pfades im Residualgraphen genauer. Es kommt entscheidend darauf an, geschickt vorzugehen, wenn man einen effizienten Algorithmus zur Lösung des Flussproblems implementieren will. Wir betrachten hierzu den in der Flusstheorie zentralen Begriff eines *Verbesserungspfades*. Dies ist ein gerichteter Pfad π im Residualgraphen von s nach t; insbesondere erfüllen alle Kanten xy auf dem Pfad π die Ungleichung $c(x, y) > f(x, y)$. Ein Verbesserungspfad wird in der Literatur auch *augmentierender Pfad* genannt. Wenn ein Verbesserungspfad existiert, dann ist die Kapazität nicht ausgenutzt und wir können den Wert des Flusses f entlang des Verbesserungspfades erhöhen und erhalten einen neuen Fluss mit dem Wert

$$\|f\| + \min \{ c(x, y) - f(x, y) \mid xy \text{ ist Kante auf dem Pfad } \pi \}$$

Diese Argumentation ist im Prinzip schon der Originalbeweis von Ford und Fulkerson, der auch sofort einen Algorithmus liefert: Starte mit dem Nullfluss $f(x, y) = 0$ für alle (x, y). Berechne den Residualgraphen und einen Verbesserungspfad π, und

erhöhe den Fluss entlang des Pfades um den positiven Wert, der zur vollen Ausnutzung der Kapazität einer Kante von π führt. Damit wird der Wert des Flusses echt vergrößert und auch immer um eine natürliche Zahl erhöht, wenn die Kapazitäten in $\mathbb{N}$ liegen. Der Algorithmus terminiert auch, wenn alle Kapazitäten in $\mathbb{Q}$ liegen, aber bei binärer Eingabe kann bei ungünstiger Wahl der Verbesserungspfade eine exponentielle Laufzeit auftreten. Schlimmer, sind einige Kapazitäten irrational, so kann der Ford-Fulkerson-Algorithmus immer kleinere Wertzuwächse wählen. Dann ist nicht einmal die Konvergenz gegen einen maximalen Fluss gesichert.

Beispiel 6.19. Sei $\varphi = \frac{-1+\sqrt{5}}{2} \approx 0{,}6180339887 \cdots$ eine Lösung der quadratischen Gleichung $x^2 + x - 1 = 0$. Insbesondere gilt $\varphi + \varphi^2 = 1$. Damit ist $1 + \varphi + \varphi^2 = 2$ der maximale Fluss im folgenden Flussnetzwerk.

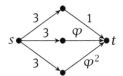

Es gibt drei Pfade von s nach t. Wir beginnen mit dem obersten davon als Verbesserungspfad. Danach haben die anderen beiden Pfade die freien Kapazitäten φ und φ^2. Als Nächstes konstruieren wir eine unendliche Folge von Verbesserungspfaden. Hierzu nehmen wir nach $n - 1$ weiteren Verbesserungspfaden die folgende Situation an: Ein Pfad hat keine freie Kapazität, ein Pfad hat die freie Kapazität φ^n, und der dritte Pfad hat noch φ^{n+1} an Kapazität frei. Als Nächstes betrachten wir den folgenden Verbesserungspfad: wir gehen von s nach t über den Pfad mit freier Kapazität φ^{n+1}, dann gehen wir nach s zurück über den Pfad ohne freie Kapazität (man beachte, dass auf diesem Pfad die Kanten in Richtung von t nach s tatsächlich im Residualgraphen vorhanden sind), schließlich gehen wir zu t zurück über den Pfad mit freier Kapazität φ^n. Dieser Verbesserungspfad bringt eine Verbesserung von φ^{n+1}. Die freien Kapazitäten ändern sich wie folgt: Der Pfad von s nach t ohne freie Kapazität hat jetzt φ^{n+1} frei; der Pfad mit ursprünglich freier Kapazität φ^n hat jetzt noch $\varphi^n - \varphi^{n+1} = \varphi^{n+2}$ frei; und der Pfad, welcher ursprünglich φ^{n+1} Kapazität frei hatte, hat nun nichts mehr frei. Die Situation ist also wie vor dem Verbesserungspfad, nur dass n durch $n + 1$ ersetzt wurde. Wenn wir so weiter verfahren, dann erreichen wir niemals den Fluss mit Wert 2, da $\varphi^n > 0$ für alle $n \in \mathbb{N}$ gilt.

Wenn wir noch eine direkte Kante von s nach t mit Kapazität $c > 0$ hinzufügen, dann ist $2 + c$ der maximale Fluss, aber die Sequenz obiger Verbesserungspfade konvergiert gegen 2. $\diamond$

Es hat fast zwanzig Jahre gedauert, bevor Jack Edmonds (geb. 1934) und Richard Karp (geb. 1935) ihre Heuristik vorstellten, den Wert immer entlang eines kürzesten Pfades im Residualgraphen zu erhöhen. Wir bezeichnen mit $n = |V|$ die Anzahl der Knoten und mit m die Anzahl der Kanten (x, y) mit $c(x, y) + c(y, x) > 0$. Damit

ist $m \le n^2$ und höchstens doppelt so groß, wie die Anzahl der Kanten mit einer positiven Kapazität. Die Heuristik von Edmonds und Karp führt zu der polynomiellen Laufzeit $\mathcal{O}(m^2 n)$, welche insbesondere unabhängig von den Kapazitäten ist [16]. Etwa zeitgleich, aber unabhängig von Edmonds und Karp, die in den USA forschten, fand Yefim Dinitz (geb. 1949) in der damaligen UdSSR ein ähnliches Verfahren zur Berechnung eines maximalen Flusses mit einer besseren Laufzeit von $\mathcal{O}(mn^2)$, siehe [14]. Den Algorithmus von Dinitz stellen wir jetzt vor. Die Kenntnis von Satz 6.18 wird nicht benötigt.

6.8.3 Der Algorithmus von Dinitz

Der Algorithmus von Dinitz (engl. *Dinic's algorithm*) startet mit dem Nullfluss und arbeitet in Phasen. Vor jeder Phase ist bereits ein Fluss f berechnet. Ist f noch kein maximaler Fluss, so werden innerhalb einer Phase verschiedene Verbesserungspfade gefunden, bis der Abstand von s zu t im Residualgraphen größer geworden ist. Innerhalb jeder Phase wird spätestens in jedem n-ten Schritt eine von maximal m Kanten gelöscht; damit terminiert eine Phase nach $\mathcal{O}(mn)$ Schritten. Am Ende der Phase hat der Fluss einen echt größeren Wert. Der entscheidende Punkt ist allerdings, dass höchstens n Phasen ausgeführt werden, denn nach jeder Phase wird sich der Abstand von s zu t im Residualgraphen um mindestens eins erhöht haben. Dieses Argument liefert schließlich die Laufzeitschranke $\mathcal{O}(mn^2)$.

Wir haben bereits einen Fluss f vorliegen und beschreiben nun den Ablauf einer Phase. Sei (V, R_f) der Residualgraph von f. Aus $(x, y) \in R_f$ folgt über die Schiefsymmetrie, dass $c(x, y) + c(y, x) > 0$ gilt. Also hat R_f für jeden möglichen Fluss höchstens m Kanten. Die Länge eines kürzesten Pfades von x nach y im aktuellen Residualgraphen bezeichnen wir mit $d_f(x, y)$. Wir setzen $d_f(x, y) = \infty$, falls es keinen Pfad gibt. Der Wert $d_f(x, y)$ ist die aktuelle *Distanz* von x nach y und hängt von f ab. Am Anfang einer Phase konstruieren wir den *Levelgraphen*. In den Levelgraphen werden gewisse Knoten und Kanten von (V, R_f) aufgenommen. Für $d \ge 0$ setzen wir $L_d = \{x \in V \mid d_f(s, x) = d\}$ und für $d \ge 1$ setzen wir:

$$E_d = \Big\{ (x, y) \in L_{d-1} \times L_d \ \Big| \ (x, y) \in R_f \Big\}$$

Der Levelgraph (L, E) ist gegeben durch $L = \bigcup_{d \ge 0} L_d$ und $E = \bigcup_{d \ge 1} E_d$. Mit einer Breitensuche können wir (L, E) in $\mathcal{O}(m)$ Schritten berechnen. Innerhalb einer Phase ändern sich die Knotenmengen L_d nicht. Insbesondere entfernen wir keine Knoten aus L, lediglich Kanten aus E werden gestrichen. Der Residualgraph wird im weiteren Verlauf der Phase zu keinem Zeitpunkt mehr explizit berechnet.

Befindet sich t nicht in L, so ist $(L, V \setminus L)$ ein st-Schnitt mit Wert $\|f\| = c(L, V \setminus L)$. Nach Gleichung (6.1) ist $\|f\|$ maximal, und wir sind fertig. Im Folgenden sei daher $t \in L$ und $k = d_f(s, t)$ zu Beginn der Phase. Damit liegt t in L_k. Wir entfernen nun Kanten aus dem Levelgraphen, solange es noch von s ausgehende Kanten gibt. Dies bedeutet, wir werden (L, E) dynamisch verändern und Flüsse augmentieren. Hierfür

definieren wir drei Invarianten, die für alle in dieser Phase aktuellen Graphen (L, E) und Residualgraphen (V, R_f) erhalten bleiben:

(1) Es gilt $E \subseteq R_f$.

(2) Aus $p \in L_a$ und $q \in L_b$ folgt $d_f(p, q) \geq b - a$.

(3) Jeden Pfad in (V, R_f) von s nach t der Länge k gibt es auch in (L, E).

Zu Beginn der Phase sind die drei Invarianten erfüllt. Wir beginnen damit, die Kanten E_{k+1} zu entfernen, denn diese kommen auf keinem Pfad von s nach t der Länge k vor. Insbesondere hat t danach den Ausgangsgrad Null. Die Invarianten wurden nicht verletzt. Wir starten jetzt eine Tiefensuche von s aus und stoppen, wenn wir einen Knoten mit Ausgangsgrad Null finden. Die Tiefensuche liefert damit einen Pfad $\pi = (p_0, \ldots, p_\ell)$ mit $p_0 = s$, $(p_{d-1}, p_d) \in E_d$ für $1 \leq d \leq \ell$ und p_ℓ hat keine ausgehende Kante in (L, E). Die Tiefensuche kostet $\mathcal{O}(n)$ Zeit. Wir unterscheiden jetzt die Fälle $p_\ell \neq t$ und $p_\ell = t$.

Im ersten Fall sei $p_\ell \neq t$. Dann gibt es in (L, E) keinen Pfad von s nach t, der p_ℓ benutzt, denn p_ℓ hat Ausgangsgrad Null. Also gibt es nach der dritten Invariante auch in (V, R_f) keinen Pfad der Länge k von s nach t, welcher p_ℓ benutzt. Wir entfernen die Kante $(p_{\ell-1}, p_\ell)$ aus E. Dies verändert R_f nicht und die Invarianten bleiben erhalten. Danach starten wir eine neue Tiefensuche bei s (oder setzen die alte bei $p_{\ell-1}$ fort).

Der zweite Fall ist etwas subtiler. Es sei jetzt $p_\ell = t$. Nach der ersten Invariante ist $\pi = (p_0, \ldots, p_\ell)$ ein Verbesserungspfad. Wir bezeichnen mit E_π die Menge der Kanten von π:

$$E_\pi = \big\{ (p_{d-1}, p_d) \mid 1 \leq d \leq \ell \big\}$$

Für $\rho = \min\{c(e) - f(e) \mid e \in E_\pi\}$ gilt $\rho > 0$. Wir definieren einen neuen Fluss f_π vermöge

$$f_\pi(x, y) = \begin{cases} f(x, y) + \rho & \text{für } (x, y) \in E_\pi \\ f(x, y) - \rho & \text{für } (y, x) \in E_\pi \\ f(x, y) & \text{sonst} \end{cases}$$

Dadurch erhöhen wir den Fluss entlang des Pfades π um den Wert ρ und haben mindestens eine der Kanten $e \in E_\pi$ *gesättigt*; dies bedeutet, für den veränderten Fluss f_π gilt $f_\pi(e) = c(e)$ für eine Kante $e \in E_\pi$. Wir durchlaufen den Pfad π nochmals und entfernen alle gesättigten Kanten aus E. Danach starten wir eine neue Tiefensuche bei s.

Da sich der Fluss verändert hat, müssen wir die Invarianten bezüglich R_{f_π} überprüfen. Hierfür müssen wir untersuchen, wie sich der Residualgraph verändert hat. Die gesättigten Kanten e vom Pfad π sind in R_{f_π} nicht mehr vorhanden. Das Löschen war also problemlos. Alle ungesättigten Kanten (p_{d-1}, p_d) auf π sind weiterhin in R_{f_π} und auch in (L, E) vorhanden. Wir müssen dennoch vorsichtig sein, denn wenn $f(x, y)$ zu $f(x, y) + \rho$ vergrößert wird, so verringert sich gleichzeitig $f(y, x)$ zu

$f(y,x) - \rho$. Solche Kanten (y,x) können in R_{f_π} neu hinzukommen. Sie sind aber nicht in (L, E) aufgenommen worden. Die erste Invariante bleibt trivialerweise erhalten. Aber die zweite und dritte müssen nachgewiesen werden. Die neuen Kanten (y,x) haben alle die Eigenschaft, dass $y \in L_{d+1}$ und $x \in L_d$ für ein $d \geq 0$ gilt. Wir definieren jetzt eine Menge R, die wir schrittweise in R_{π_f} transformieren und die ebenfalls die drei Invarianten erfüllt, wenn wir bei den Bedingungen R anstelle von R_f einsetzen. Die Distanz in R bezeichnen wir mit d_R. Zu Anfang besteht R aus den Kanten R_f ohne die gesättigten Kanten von E_π. Die Invarianten sind erfüllt. Sei jetzt $R' = R \cup \{(y,x)\}$ für ein beliebiges Paar $(y,x) \in L_{d+1} \times L_d$. Wir wollen zeigen, dass die Invarianten für R' mit entsprechender Distanz $d_{R'}$ gelten. Die erste Invariante bleibt erhalten, weil wir eine Kante in R' hinzunehmen. Betrachte jetzt $p \in L_a$ und $q \in L_b$. Benutzt ein kürzester Pfad von p nach q in R' nicht die Kante (y,x), so gilt $b - a \leq d_R(p,q) = d_{R'}(p,q)$. Benutzt dieser Pfad die Kante, so können wir $d_{R'}(p,q)$ wie folgt errechnen:

$$d_{R'}(p,q) = d_R(p,y) + 1 + d_R(x,q) \geq (d + 1 - a) + 1 + (b - d) = b - a + 2$$

Dies zeigt die zweite Invariante. Um schließlich die dritte Invariante zu zeigen, betrachten wir einen Pfad der Länge k von s nach t in R'. Benutzt er nicht die Kante (y,x), so ist es ein Pfad in R und damit auch in (L, E), da die Invarianten für R gelten. Benutzt er die Kante, so erhalten wir den folgenden Widerspruch:

$$k = d_{R'}(s,t) = d_R(s,y) + 1 + d_R(x,t) \geq (d + 1) + 1 + (k - d) = k + 2$$

Dies zeigt die dritte Invariante. Über mehrere dieser Schritte kann man von R aus die Menge R_{f_π} erhalten. Daher gelten die Invarianten für R_{f_π}.

Nach jeweils $\mathcal{O}(n)$ Schritten verliert E eine Kante, also gibt es nach $\mathcal{O}(mn)$ Schritten keine Ausgangskante bei s. Dies beendet die Phase. Die Distanz von s nach t im aktuellen Residualgraphen ist nach der zweiten Invariante mindestens k. Andererseits gibt es in (L, E) keinen Pfad von s nach t. Nach der dritten Invariante gibt es dann auch in R_f keinen Pfad der Länge k. Daher muss im aktuellen Residualgraphen $d_f(s,t) \geq k + 1$ gelten. Nach höchstens n Phasen gilt $d_f(s,t) = \infty$. Dann haben wir einen maximalen Fluss berechnet, und die Laufzeit ist insgesamt durch $\mathcal{O}(mn^2)$ beschränkt.

Beispiel 6.20. Der Algorithmus von Dinitz berechnet auf dem Flussnetzwerk von Seite 137 den dort dargestellten Fluss nach drei Phasen. Wir betrachten noch ein anderes Flussnetzwerk:

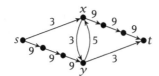

Die erste Phase des Algorithmus von Dinitz beginnen wir mit dem Nullfluss. Der Fluss, sein zugehöriger Residualgraph und der resultierende Levelgraph am Anfang dieser Phase sind wie folgt:

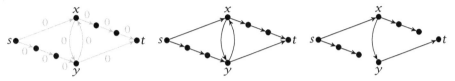

Nach Hinzufügen des Verbesserungspfades (s, x, y, t) mit Wert 3 ergibt sich zu Beginn der zweiten Phase das folgende Bild:

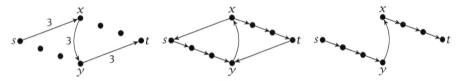

Der Fluss der Kante (y, x) ist -3; daher hat sie noch freie Kapazität 8. Also hat der – in diesem Fall eindeutige – Verbesserungspfad den Wert 8. Nach Hinzufügen dieses Pfades zum Fluss erhalten wir:

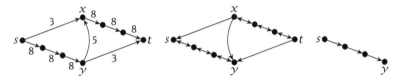

Zu Beginn der dritten Phase ist nun t nicht im Levelgraphen enthalten. Der berechnete Fluss mit Wert 11 ist somit maximal. Der zugehörige minimale Schnitt wird durch die drei Kanten (s, x), (y, x) und (y, t) definiert. ◊

Die Laufzeit zur Berechnung maximaler Flüsse wurde in den letzten Jahren weiter verbessert und ist Gegenstand aktueller Forschung. Der Algorithmus von Dinitz ist robust und einfach zu implementieren; er spielt daher auch in der Praxis weiterhin eine wichtige Rolle. Von Dinitz stammt auch eine lesenswerte Betrachtung über die historische Entwicklung der Fluss-Algorithmen [15].

6.9 Planare Graphen

Ein bekanntes Rätsel ist das folgende. Gegeben seien drei Versorgungsstationen Strom, Gas und Wasser, sowie drei Häuser A, B und C. Kann man jede Versorgungsstation mit jedem Haus so verbinden, dass sich keine zwei Versorgungsleitungen schneiden?

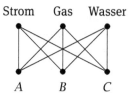

Ein Graph heißt *planar*, wenn man ihn so in die Ebene zeichnen kann, dass sich die Kanten nicht schneiden. Es ist gleichwertig, ob sich ein Graph kreuzungsfrei in die Ebene oder auf die Kugeloberfläche zeichnen lässt. Das obige Rätsel lässt sich nun wie folgt graphentheoretisch formulieren: Ist der Graph $K_{3,3}$ planar?

Wie das folgende Beispiel zeigt, ist der Graph K_4 planar.

Eine *Facette* eines planaren Graphen ist eine maximale zusammenhängende Fläche in der Ebene, welche keine Kanten und keine Knoten enthält. Häufig werden Facetten auch als *Flächen* oder *Gebiete* bezeichnet. Bei zusammenhängenden Graphen mit mehr als zwei Knoten wird jede Facette von einem (nicht notwendigerweise einfachen) Kreis umrandet. Insbesondere umrandet die Außenseite eines planaren Graphen eine unbeschränkte Facette. Die vollständigen Graphen K_1 und K_2 besitzen jeweils nur eine Facette. Der folgende Graph besitzt 4 Facetten.

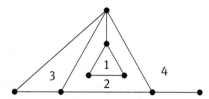

Die Facetten Nummer 1 und 3 werden jeweils von einem einfachen Kreis umrandet, während die Facetten 2 und 4 nicht von einfachen Kreisen umrandet werden.

Manchmal unterscheidet man zwischen den Begriffen *planar* und *plättbar*. Bei *plättbaren* Graphen meint man dann solche Graphen, die sich kreuzungsfrei in der Ebene zeichnen lassen. Im Gegensatz dazu ist im strengeren Sinne bei einem *planaren* Graphen eine kreuzungsfreie Einbettung in die Ebene gegeben. Diese Unterscheidung ist für uns hier nicht wichtig. Wir verwenden deshalb nur den Term *planar* (auch dann, wenn wir nicht von einer gegebenen kreuzungsfreien Einbettung ausgehen). Damit folgen wir dem englischen Sprachgebrauch, der einen kreuzungsfrei eingebetteten Graphen als *plane graph* bezeichnet. Der Unterschied ist, dass Graphen isomorph sein können, aber verschiedene kreuzungsfreie Einbettungen besitzen. Die

folgenden beiden Graphen sind isomorph, aber die erste Zeichnung hat zwei Facetten, welche von einem Kreis der Länge 5 umrandet werden, während bei der zweiten Zeichnung gar keine solche Facette existiert.

Zwei Facetten der Länge 5   Keine Facette der Länge 5

6.9.1 Die Eulerformel

Um die Darstellung im Folgenden knapp zu halten, legen wir einige Bezeichner für diesen Abschnitt fest: Mit n meinen wir stets die Anzahl der Knoten eines Graphen G, mit m bezeichnen wir die Anzahl der Kanten, und wenn G planar ist, dann sei f die Anzahl der Facetten (inklusive der äußeren Facette). Die *Eulerformel* (auch *Euler'sche Polyederformel* genannt) gibt einen Zusammenhang zwischen den Größen n, m und f an. Insbesondere folgt aus der Eulerformel, dass alle kreuzungsfreien Zeichnungen eines gegebenen planaren Graphen (in der Ebene) dieselbe Anzahl von Facetten haben.

Satz 6.21 (Eulerformel; Euler 1758). *In nichtleeren, zusammenhängenden, planaren Graphen gilt*

$$n - m + f = 2$$

Beweis. Sei G ein zusammenhängender planarer Graph mit mindestens einem Knoten. Wenn G keine einfachen Kreise enthält, dann ist G ein Baum; und es gilt die Eulerformel, da $m = n - 1$ sowie $f = 1$ ist. Sei nun G kein Baum und sei e eine Kante auf einem einfachen Kreis in G. Entfernen wir diese Kante, so werden zwei Facetten zu einer verschmolzen (wenn wir die kreuzungsfreie Zeichnung von G zugrunde legen), denn auf den beiden Seiten von e liegen unterschiedliche Facetten. Die Differenz zwischen Kanten und Facetten hat sich also nicht verändert. Die Behauptung folgt mit Induktion nach m. □

Aus der Eulerformel lassen sich einige weitere interessante Eigenschaften von planaren Graphen herleiten.

Korollar 6.22. *Sei G ein nichtleerer planarer Graph.*

(a) Wenn $n \geq 3$ ist, dann gilt $m \leq 3n - 6$.

(b) Die Graphen K_5 und $K_{3,3}$ sind nicht planar.

(c) Es gibt einen Knoten x mit Grad $d_x \leq 5$.

Beweis. (a) Durch Hinzufügen von Kanten können wir annehmen, dass G zusammenhängend ist. Wegen $n \geq 3$ wird jede Facette von einem (nicht notwendigerweise einfachen) Kreis der Länge mindestens 3 umrandet. An jeder Seite einer Kante liegen maximal 2 Facetten. Dies zeigt

$$3f \leq 2m$$

Mit der Eulerformel folgt

$$6 = 3n - 3m + 3f \leq 3n - 3m + 2m$$

und damit die Behauptung.

(b) Der K_5 hat 10 Kanten. Dies widerspricht der Abschätzung aus (a). Also ist der K_5 nicht planar. Um zu zeigen, dass der Graph $K_{3,3}$ nicht planar ist, geben wir eine stärkere Schranke für die Kantenzahl in bipartiten planaren Graphen an.

Sei $n \geq 4$ und sei G ein zusammenhängender, bipartiter, planarer Graph mit n Knoten. Jede Facette wird von einem Kreis der Länge mindestens 4 umrandet (Länge 3 ist nicht möglich, da G bipartit ist). Dies liefert $4f \leq 2m$ und

$$4 = 2n - 2m + 2f \leq 2n - 2m + 1m$$

Also gilt $m \leq 2n - 4$. Der $K_{3,3}$ hat 9 Kanten und widerspricht damit dieser Abschätzung. Also ist der Graph $K_{3,3}$ nicht planar.

(c) Indem wir uns auf eine Zusammenhangskomponente beschränken, können wir annehmen, dass G zusammenhängend ist. Außerdem habe G mindestens 7 Knoten, sonst ist nichts zu zeigen. Sei $\bar{d} = (\sum_{x \in V} d_x)/n$ der durchschnittliche Knotengrad. Wegen $\sum_x d_x = 2m$ folgt mit der Eulerformel, dass

$$\bar{d} = \frac{2m}{n} \leq \frac{6n - 12}{n} < 6$$

Da der Durchschnittsgrad kleiner als 6 ist, muss ein Knoten mit Grad höchstens 5 existieren. □

Die Graphen K_5 und $K_{3,3}$ sind nicht planar, und der Satz von Kuratowski (Kazimierz Kuratowski, 1896–1980) sagt, dass jeder nicht planare Graph in einem gewissen Sinn einen K_5 oder einen $K_{3,3}$ enthält, siehe z. B. [13]. Daher sind K_5 und $K_{3,3}$ die einzigen Archetypen von nicht planaren Graphen.

Im Ikosaeder hat jeder Knoten Grad 5. Dies zeigt, dass sich die Abschätzung in Korollar 6.22 (c) im Allgemeinen nicht verbessern lässt.

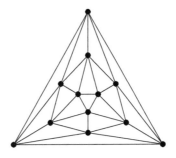

Ikosaeder als planarer Graph

6.9.2 Färbungen von planaren Graphen

Eine *C-Färbung* eines Graphen $G = (V, E)$ ist eine Abbildung $c : V \to C$ mit $c(x) \neq c(y)$ für alle $xy \in E$. Hierbei ist C die Menge der Farben (engl. *colors*). Wir sagen, G ist *k-färbbar*, wenn eine C-Färbung mit $|C| = k$ existiert. Der berühmte Vierfarbensatz von Kenneth Appel (geb. 1932) und Wolfgang Haken (geb. 1928) besagt, dass jeder planare Graph 4-färbbar ist [4, 5]. Wir werden in diesem Abschnitt eine schwächere Aussage beweisen, nämlich dass jeder planare Graph 5-färbbar ist.

Wir überlegen uns zunächst, dass jeder planare Graph 6-färbbar ist: Sei v ein Knoten mit Grad höchstens 5; dieser existiert nach Korollar 6.22 (c). Der Graph ohne den Knoten v ist nach Induktion 6-färbbar. Da v maximal 5 Nachbarn hat, verbrauchen diese Nachbarn auch maximal 5 Farben, so dass wir v wieder hinzunehmen können und eine der 6 Farben für v übrig ist. Dieselbe Idee verfolgen wir im Beweis des nächsten Satzes, nur dass wir eine Farbe einsparen.

Satz 6.23 (Fünffarbensatz). *Jeder planare Graph ist 5-färbbar.*

Beweis. Mit Induktion nach der Anzahl der Knoten zeigen wir, dass sich jeder planare Graph mit den Farben C mit $|C| = 5$ färben lässt. Sei $G = (V, E)$ ein planarer Graph. Nach Korollar 6.22 (c) existiert ein Knoten v mit Grad $d_v \leq 5$.

Falls $d_v \leq 4$ gilt, dann betrachten wir den von $V \setminus \{v\}$ induzierten Untergraphen von G. Dieser besitzt nach Induktion eine C-Färbung c. Wir fügen den Knoten v wieder hinzu und setzen $c(v)$ auf eine der Farben, die von den höchstens 4 Nachbarn von v nicht verwendet wurde. Die Farbe der übrigen Knoten wird nicht verändert. Dies liefert eine Färbung c von G.

Sei nun $d_v = 5$ und seien a, b, c, d, e die Nachbarn von v so gewählt, dass die Knoten in der Zeichnung von v aus gesehen im Uhrzeigersinn angeordnet sind.

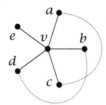

Es können nicht beide Kanten ac und bd vorhanden sein, da sie sich sonst schneiden müssten. Ohne Einschränkung sei $ac \notin E$. Wir entfernen aus G den Knoten v und alle seine Kanten. Dann schieben wir die Knoten a und c zusammen und verschmelzen sie zu einem einzigen Knoten $z_{ac} \notin V$. Dies liefert einen planaren Graphen mit zwei Knoten weniger. Mit Induktion existiert eine 5-Färbung c' für den so entstandenen Graphen. Wir konstruieren daraus eine Färbung $c : V \to C$ von G, indem die Knoten a und c beide die Farbe $c'(z_{ac})$ bekommen. Die übrigen Knoten aus G behalten ihre Farbe. Wir müssen noch den Knoten v färben: Da die fünf Nachbarn von v höchstens 4 Farben verbrauchen, bleibt eine Farbe übrig, die wir für v verwenden können. $\square$

6.9.3 Planare Separatoren

Ein *Separator* eines Graphen $G = (V, E)$ ist eine Menge von Knoten C, so dass sich V in Teilmengen A, B, C einteilen lässt mit der Eigenschaft, dass zwischen A und B keine Kanten verlaufen. Die Idee ist, dass das effiziente Finden von kleinen Separatoren einen sogenannten *Teile-und-Herrsche* Ansatz für algorithmische Probleme auf Graphen erlaubt: Berechne einen Separator C, so dass die Teilmengen A und B in etwa gleich groß sind (*Teilen*); finde dann rekursiv Lösungen für die von A und B induzierten Untergraphen und setze diese mit C als „Adapter" zu einer Lösung auf G zusammen (*Herrschen*). Nun existiert leider nicht in jedem Graphen ein geeigneter Separator C. Bei planaren Graphen hingegen ist die Situation sehr viel angenehmer; hier kann man stets einen Separator C mit $|C| \in O(\sqrt{n})$ finden, so dass grob mindestens ein Drittel aller Knoten in A und ein Drittel in B ist. Das *planare Separator-Theorem* fasst dies etwas genauer:

Satz 6.24 (Lipton, Tarjan 1979). *Sei $G = (V, E)$ ein planarer Graph. Dann gibt es disjunkte Knotenmengen A, B, C mit $A \cup B \cup C = V$ und*
- *$|A| < 2n/3$ und $|B| < 2n/3$,*
- *$|C| \leq \sqrt{8n}$, und*
- *es gibt keine Kanten zwischen Knoten aus A und Knoten aus B.*

Beweis. Ohne Einschränkung sei $n \geq 3$. Durch Hinzufügen von weiteren Kanten können wir annehmen, dass in der Zeichnung von G alle Facetten von Dreiecken begrenzt

werden (auch die äußere Facette). Sei $k = \lfloor \sqrt{2n} \rfloor$. Jeder einfache Kreis, welcher ein Teilgraph von G ist, definiert auch einen Kreis in der planaren Einbettung von G. Für einen einfachen Kreis C von G sei $V(C)$ die Menge der Knoten auf dem Kreis C, $A(C)$ die Menge der Knoten außerhalb von C und $B(C)$ die Menge der Knoten innerhalb von C.

Sei C ein Kreis, welcher die folgenden drei Bedingungen erfüllt:

(1) C hat höchstens $2k$ Knoten,

(2) $|A(C)| < 2n/3$, und

(3)   unter den Bedingungen (1) und (2) ist $|B(C)| - |A(C)|$ minimal.

Ein solcher Kreis C existiert, da das Dreieck um die äußere Facette die Bedingungen (1) und (2) erfüllt.

Wir nehmen an, dass $|B(C)| \geq 2n/3$ gilt und führen dies im Rest des Beweises zu einem Widerspruch. Sei D der von $B(C) \cup V(C)$ induzierte Untergraph von G. Für $x, y \in V(C)$ sei $c(x, y)$ die minimale Anzahl von Kanten auf einem Pfad von x nach y im Graphen C, und $d(x, y)$ sei die minimale Anzahl von Kanten auf einem Pfad von x nach y in D.

Behauptung 6.25. *Für alle $x, y \in V(C)$ gilt $c(x, y) = d(x, y)$.*

Beweis von Behauptung 6.25. Da C ein Teilgraph von D ist, gilt $d(x, y) \leq c(x, y)$. Angenommen, es existieren Knoten $x, y \in V(C)$ mit $d(x, y) < c(x, y)$; dann betrachten wir ein Paar x, y solcher Knoten, bei denen $d(x, y)$ minimal ist. Sei P ein Pfad in D von x nach y mit $d(x, y)$ vielen Kanten. Aufgrund der Minimalität von $d(x, y)$ sind x und y die einzigen Knoten von P auf dem Kreis C. Der Graph $C \cup P$ bestehend aus dem Kreis C zusammen mit dem Pfad P enthält drei einfache Kreise: den Kreis C selbst, sowie die Kreise C_1 und C_2 mit P als Teilstück. Ohne Einschränkung sei $|B(C_1)| \geq |B(C_2)|$. Es gilt $A(C_1) < 2n/3$, denn

$$
\begin{aligned}
n - |A(C_1)| &= |B(C_1)| + |V(C_1)| \\
&> \frac{1}{2}\left(|B(C_1)| + |B(C_2)| + |V(P)| - 2\right) \\
&= \frac{1}{2}|B(C)| \geq n/3.
\end{aligned}
$$

Deshalb erfüllt C_1 die Bedingung (2), und wegen $d(x, y) < c(x, y)$ erfüllt C_1 die Bedingung (1). Aus Bedingung (3) und $B(C_1) \subseteq B(C)$ folgt $B(C) = B(C_1)$. Also hat P keine inneren Knoten, und es gilt $c(x, y) \leq 1$, was ein Widerspruch zu $d(x, y) < c(x, y)$ ist. Dies zeigt die Behauptung 6.25. □

Behauptung 6.26. *C hat genau $2k$ Knoten.*

Beweis von Behauptung 6.26. Angenommen $|V(C)| < 2k$. Sei $e = xy$ eine beliebige Kante auf C. Da G trianguliert ist, liegt e an einem Dreieck in D an. Sei z der dritte Knoten dieses Dreiecks. Wegen $|B(C)| \geq 2n/3$ gilt insbesondere $B(C) \neq \emptyset$.

Mit Behauptung 6.25 folgt $z \in B(C)$, denn andernfalls hätte der Kreis C eine Sehne in D. Wie können nun den Kreis C' betrachten, der aus C entsteht, indem wir die Kante e löschen und dann den Knoten z sowie die Kanten xz und zy aufnehmen. Dieser Kreis C' erfüllt die Bedingungen (1) und (2), allerdings ist $B(C') \subsetneq B(C)$ und $A(C') = A(C)$, was einen Widerspruch zu Bedingung (3) für C bedeutet. Dies zeigt die Behauptung 6.26. □

Seien $x_0, \ldots, x_{2k-1}$ die Knoten auf dem Kreis C in genau dieser Reihenfolge. Um die Notation im Folgenden zu vereinfachen, setzen wir $x_{2k} = x_0$. Sei $S = \{x_0, \ldots, x_k\}$ und $T = \{x_k, \ldots, x_{2k}\}$.

Behauptung 6.27. *In D gibt es $k + 1$ disjunkte Pfade von S nach T.*

Beweis von Behauptung 6.27. Nach dem Satz von Menger (Satz 6.15) gibt es entweder $k + 1$ disjunkte ST-Pfade oder es gibt einen ST-Separator der Größe kleiner gleich k. Sei P ein ST-Separator mit $|P| \le k$. Wegen $S \cap T = \{x_0, x_k\}$ gilt $x_0, x_k \in P$. Sei Q die Zusammenhangskomponente von x_0 in dem von P induzierten Untergraph von D. Dann kann Q nicht x_k enthalten, denn sonst wäre die Distanz $d(x_0, x_k)$ zwischen x_0 und x_k kleiner als $|P|$ und damit kleiner als k im Widerspruch zu $d(x_0, x_k) = c(x_0, x_k) = k$. Sei R die Menge von Knoten außerhalb von Q, die einen Nachbarn in der Komponente Q haben und die einen Pfad zu x_k besitzen, welcher keine Knoten aus Q verwendet (in R befinden sich die Nachbarn des äußeren Rands von Q von x_0 aus betrachtet). Nach Definition von Q sind die Knotenmengen R und P disjunkt. Da G trianguliert ist, induzieren die Knoten aus R in D einen Pfad von S nach T, denn dieser Pfad beginnt auf C auf der einen Seite zwischen x_0 und x_k und endet auf C auf der anderen Seite zwischen diesen beiden Knoten. Also ist P kein ST-Separator, ein Widerspruch. Dies zeigt die Behauptung 6.27. □

Da G planar ist, können sich die $k + 1$ disjunkten Pfade $\pi_0, \ldots, \pi_k$ aus Behauptung 6.27 nicht kreuzen. Deshalb können wir annehmen, dass der Pfad π_i bei Knoten x_i beginnt und bei Knoten x_{2k-i} endet. Mit Behauptung 6.25 sehen wir, dass π_i mindestens $c(x_i, x_{2k-i}) + 1$ viele Knoten enthält. Es gilt $\sum_{i=0}^{k} \min\{i, k - i\} = \lfloor k/2 \rfloor \cdot \lceil k/2 \rceil \ge (k^2 - 1)/4$ und hieraus folgt

$$n \ge \sum_{i=0}^{k} |V(\pi_i)| \ge \sum_{i=0}^{k} \min\{2i + 1, 2(k - i) + 1\} \ge \frac{(k + 1)^2}{2}$$

Dies ist ein Widerspruch zur Definition von $k = \lfloor \sqrt{2n} \rfloor$. Also gilt $|B(C)| < 2n/3$. Damit erfüllen die Knotenmengen $A(C), B(C), V(C)$ die Aussage des Satzes. □

Lipton und Tarjan (Richard Jay Lipton und Robert Endre Tarjan, geb. 1948) haben den Satz 6.24 im Jahr 1979 veröffentlicht [26]. Weiter haben sie gezeigt, dass sich der Separator C in Linearzeit berechnen lässt. Damit sind Separatoren in planaren Graphen für viele Berechnungsprobleme ein Hilfsmittel, welches zu effizienten Algo-

rithmen führt. Der hier vorgestellte Beweis folgt einer Arbeit von Alon, Seymour und Thomas [3] (Noga Alon, geb. 1956, Paul D. Seymour, geb. 1950, und Robin Thomas).

6.10 Der Satz von Ramsey

Erreicht eine Population von Lemmingen eine gewisse Größe und Dichte, so begibt sich eine große Zahl von ihnen auf eine Wanderung mit manchmal ungewissem Ausgang. Ein Problem sind breite Flüsse, so dass bei ihrer Überquerung unter Umständen nur ein kleiner Teil der Lemminge das andere Ufer erreicht. Angenommen, nach der Überquerung von k Flüssen sollen noch mindestens n Lemminge vorhanden sein. Was ist zu tun? Eine vernünftige Strategie ist, die Wanderung mit einer genügend großen Zahl zu beginnen. Dies tun die Lemminge, und die Konstruktion der Ramsey-Zahlen (Frank Plumpton Ramsey, 1903–1930) verfolgt eine ähnliche Vorgehensweise.

Abb. 6.2. Berglemminge (aus *Brehms Tierleben*, 1927).

Eine *Färbung* der Kanten eines Graphen $G = (V, E)$ mit Farben C ist eine Abbildung $f : E \to C$. Häufig färbt man die Kanten eines vollständigen Graphen. Färbungen beliebiger Graphen lassen sich dadurch realisieren, dass man denjenigen Kanten, die nicht in E sind, eine spezielle Farbe gibt. Beispielsweise ist die charakteristische Funktion $\chi_E : \binom{V}{2} \to \{0, 1\}$ von E mit

$$\chi_E(e) = 1 \iff e \in E$$

eine oft verwendete Färbung. Es ist eine natürliche Verallgemeinerung in diesem Abschnitt, nicht nur 2-elementige Teilmengen zu färben, sondern die Kanten von vollständigen k-Hypergraphen. Ein k-*Hypergraph* ist ein Paar (V, E) von Knoten V und k-*Hyperkanten* $E \subseteq \binom{V}{k}$. Wir nennen k die *Dimension* des Hypergraphen. Deshalb ist in diesem Abschnitt eine *Färbung* stets eine Abbildung von der Form

$$f : \binom{V}{k} \to C$$

für $k \geq 1$. Eine Teilmenge $X \subseteq V$ heißt *monochromatisch*, wenn eine Farbe $b \in C$ existiert mit $f(K) = b$ für alle $K \in \binom{X}{k}$; das heißt, alle Hyperkanten innerhalb von X haben dieselbe Farbe. Im Falle der charakteristischen Funktion eines Graphen G ist eine Teilmenge X von Knoten genau dann monochromatisch, wenn X eine Clique bildet (d. h., alle Kanten zwischen Knoten aus X sind in G vorhanden) oder wenn X unabhängig ist (d. h., in G gibt es keine Kanten zwischen Knoten aus X).

Die wesentliche Aussage der Ramsey-Theorie ist, dass große monochromatische Teilmengen X für sehr, sehr große V garantiert werden können. Im Spezialfall der gewöhnlichen Graphen bedeutet dies damit das Folgende: Für jedes $n \in \mathbb{N}$ gibt es eine kleinste Zahl $R(n)$ mit der folgenden Eigenschaft: Ist (V, E) ein Graph mit mindestens $R(n)$ Knoten, so gibt es eine Teilmenge $X \subseteq V$ mit $|X| \geq n$ und X ist entweder eine Clique oder eine unabhängige Menge.

Beispiel 6.28. In einem Restaurant sitzen 6 Leute. Dann gibt es 3 Personen, die einander alle kennen, oder es gibt 3 Personen, die sich gegenseitig nicht kennen.

Der zugrunde liegende Graph hat die 6 Leute als Knoten, und eine Kante zwischen zwei Personen x und y wird gezeichnet, wenn sich x und y gegenseitig kennen. Nehmen wir zuerst an, dass Person A drei Leute kennt. Wenn sich diese drei Leute nicht kennen sind wir fertig; andernfalls kennen sich 2 Leute und zusammen mit A ergibt dies 3 Leute, die sich kennen. Wenn A keine 3 Leute kennt, dann gibt es 3 Leute, die A nicht kennt. Wenn man die Eigenschaften *sich kennen* und *sich nicht kennen* vertauscht, dann ist die Situation symmetrisch zum vorigen Fall. Dies zeigt die obige Aussage.

Bei 5 Gästen können wir diese Situation nicht immer erzwingen. Es reicht einen runden Tisch zu betrachten, an dem 5 Personen sitzen, die genau die beiden Tischnachbarn kennen. Dann gibt es weder eine Clique der Größe 3 noch eine unabhängige Menge der Größe 3. Insgesamt erhalten wir $R(3) = 6$. ◊

Etwas allgemeiner garantieren die *Ramsey-Zahlen* $R_{k,c}(n)$ bei jeder Knotenmenge V mit $|V| \geq R_{k,c}(n)$ und jeder Färbung $f : \binom{V}{k} \to C$ mit $|C| = c$ eine monochromatische Teilmenge von V der Größe n. Die Existenz und die formale Definition dieser Zahlen werden durch den Satz 6.29 bereitgestellt. Die Ramsey-Zahlen $R(n)$ für gewöhnliche Graphen (mit der charakteristischen Funktion der Kanten als Färbung) ergeben sich dann durch $R(n) = R_{2,2}(n)$.

Satz 6.29 (Ramsey 1930; endliche Version). *Für alle $k, c, n \in \mathbb{N}$ gibt es eine kleinste Zahl $R_{k,c}(n) \in \mathbb{N}$ mit folgender Eigenschaft: Ist V eine Menge mit $|V| \geq R_{k,c}(n)$ und $f : \binom{V}{k} \to C$ eine Färbung mit $|C| = c$, so gibt es eine monochromatische Teilmenge $X \subseteq V$ mit $|X| = n$.*

Beweis. Wir machen eine Induktion nach der Dimension k. Die Fälle $k = 0$ oder $c \leq 1$ oder $n = 0$ sind trivial und uninteressant. Sei daher $k \geq 1$, $c \geq 2$ und $n \geq 1$. Für

$k = 1$ gilt nach einem *Schubfachschluss* $R_{1,c}(n) = c(n-1) + 1$, denn ab dieser Zahl muss eine der c Farben für mindestens n Knoten vergeben werden.

Sei jetzt $k \geq 1$ und $r_k = R_{k,c}(n)$ schon definiert. Wir geben eine obere Schranke für $r_{k+1} = R_{k+1,c}(n)$ an. Hierfür betrachten wir eine endliche Menge V zusammen mit einer Färbung $f : \binom{V}{k+1} \to C$. Die Menge V sei sehr groß; sie enthalte weit mehr als r_k Elemente. Eine ausreichende Größe wird sich später ergeben. Die Idee ist, V extrem auszudünnen; dabei gehen wir auf der Restmenge zu einer Färbung g der k-elementigen Teilmengen über. Enthält die Restmenge noch r_k Elemente, so gibt es bezüglich g eine monochromatische Teilmenge. Diese wird sich auch bezüglich f als monochromatisch herausstellen.

Um den Ausdünnungsprozess zu starten, legen wir auf V zunächst eine lineare Ordnung $<$ fest. Jede nichtleere Teilmenge hat damit eindeutig bestimmte kleinste und größte Elemente. Die Elemente aus $\binom{V}{k+1}$ schreiben wir jetzt als Paare (K, b) mit $K \in \binom{V}{k}$, $b \in V$ und $\max(K) < b$. Wir wollen V so ausdünnen, dass $g(K) = f(K, b)$ unabhängig von b gilt. Für $m \in \mathbb{N}$ mit $m \leq r_k$ definieren wir induktiv Teilmengen $A_m, B_m \subseteq V$ und eine Färbung $g : \binom{A_m}{k} \to C$, die den folgenden Eigenschaften genügen sollen:

(1) A_m enthält m Elemente und B_m enthält „genügend viele" Elemente.

(2) Für alle $a \in A_m$ und alle $b \in B_m$ gilt $a < b$.

(3) Für alle $K \in \binom{A_m}{k}$ und alle $b \in B_m$ gilt $f(K, b) = g(K)$.

Für $m = 0$ setzen wir $A_0 = \varnothing$ und $B_0 = V$. Die Färbung g ist dann die leere Abbildung; Farben bezüglich g sind damit noch nicht vergeben. Die Färbung g wird auf immer größere Mengen erweitert und erhält deshalb keinen Index.

Sei jetzt $m \geq 0$ und $B_m \neq \varnothing$. Wir definieren $a_{m+1} = \min(B_m)$ und setzen $A_{m+1} = A_m \cup \{a_{m+1}\}$. Damit gilt $|A_{m+1}| = m + 1$. Als Nächstes erweitern wir die Färbung g zu einer Färbung $g : \binom{A_{m+1}}{k} \to C$. Die noch nicht gefärbten Elemente $K \in \binom{A_{m+1}}{k}$ enthalten alle das Element a_{m+1}. Hiervon gibt es also $\binom{m}{k-1}$ Stück. Wir suchen eine Teilmenge $B_{m+1} \subseteq B_m \setminus \{a_{m+1}\}$ maximaler Größe mit der Eigenschaft, dass für alle $K \in \binom{A_{m+1}}{k}$ und alle $b \in B_{m+1}$ der Wert $f(K, b)$ identisch ist. Die Menge B_{m+1} muss existieren, kann aber leer sein. Nachdem wir ein solches B_{m+1} gefunden haben, erweitern wir die Färbung g, indem wir für jedes $K \in \binom{A_{m+1}}{k}$ mit $a_{m+1} \in K$ diejenige Farbe $y \in C$ wählen, für die für alle $b \in B_{m+1}$ die Bedingung $f(K, b) = y$ erfüllt ist (wenn B_{m+1} leer ist, dann wählen wir $y \in C$ beliebig). Es gelten erneut alle drei Bedingungen, denn genau so wurde B_{m+1} konstruiert.

Nehmen wir zunächst an, die Mengen B_m wären stets groß genug und wir hätten die Menge A_{r_k} konstruiert. Nach Definition von r_k finden wir für A_{r_k} mit Färbung g eine monochromatische Teilmenge X mit n Elementen. Es gibt also eine Farbe $y \in C$ mit $g(K) = y$ für alle $K \in \binom{X}{k}$. Betrachten wir jetzt ein Element aus $\binom{X}{k+1}$, so können wir dieses als Paar (K, b) mit $K \in \binom{X}{k}$ und $\max(K) < b$ schreiben. Aus den

Invarianten (2) und (3) erhalten wir

$$f(K, b) = g(K) = \gamma$$

Dies zeigt, dass X auch für die Färbung f monochromatisch ist.

Eine Frage bleibt: Wie groß muss B_m sein, um eine genügend große Menge B_{m+1} zu garantieren? Angenommen, wir wollen nur ein einziges $K \in \binom{A_{m+1}}{k}$ färben und danach noch r Elemente in B_{m+1} garantieren. Nach dem Schubfachschluss reichen $c(r-1)+1$ Elemente in $B_m \setminus \{a_{m+1}\}$. Da $c \geq 2$ ist, ist jede Menge B_m mit $|B_m| \geq cr$ ausreichend groß. Nun haben wir nicht nur ein Element, sondern $\binom{m}{k-1}$ Elemente zu färben. Soll also B_{m+1} am Ende noch r Elemente enthalten, so genügt es, wenn B_m mindestens $c^{\binom{m}{k-1}}r$ Elemente enthält. Um die Invariante (1) für $m = r_k$ zu gewährleisten, muss V so groß sein, dass $B_{r_{k}-1} \neq \emptyset$ garantiert werden kann. Nach obiger Argumentation genügt hierfür

$$|V| \geq \prod_{m < r_k} c^{\binom{m}{k-1}} = c^{\binom{r_k}{k}}$$

Die zweite Abschätzung ist hierbei gerade die Formel zur oberen Summation, Satz 4.7. Deshalb gilt $R_{k+1,c}(n) \leq c^{\binom{r_k}{k}}$. $\square$

Über das genaue Wachstum der Ramsey-Zahlen ist wenig bekannt. Die hier angegebenen oberen Schranken wachsen exorbitant schnell. Es scheint allerdings kaum möglich, sie substantiell zu verbessern. Der wichtigste Fall ist sicher $k = 2$, denn dies betrifft die Färbung von Graphen. Hier liefert unsere Herleitung die Schranke:

$$R_{2,c}(n) \leq 2^{nc \log c}$$

Satz 6.30 zeigt, dass diese Abschätzung schon ganz moderat ist.

Satz 6.30 (Erdős 1947). *Für $n \geq 2$ gilt $R_{2,2}(n) \geq 2^{\frac{n}{2}}$.*

Beweis. Wegen Beispiel 6.28 können wir $n \geq 4$ annehmen. Wir betrachten einen vollständigen Graph K_m mit einer festen Knotenmenge. Es existieren $2^{\binom{m}{2}}$ Färbungen der Kanten von K_m mit 2 Farben. Sei K_n ein fester Untergraph von K_m. Dann existieren mindestens $2^{\binom{m}{2}-\binom{n}{2}+1}$ Färbungen, bei denen K_n monochromatisch ist. Es gibt $\binom{m}{n}$ viele vollständige Untergraphen von K_m, deren Größe n ist. Damit bei jeder Färbung eine einfarbige Teilmenge der Größe n existiert, muss deshalb $\binom{m}{n} \cdot 2^{\binom{m}{2}-\binom{n}{2}+1} \geq 2^{\binom{m}{2}}$ gelten. Für $m < 2^{\frac{n}{2}}$ ergibt sich aber:

$$\frac{\binom{m}{n} \cdot 2^{\binom{m}{2}-\binom{n}{2}+1}}{2^{\binom{m}{2}}} \leq \frac{m^n}{2^{n-1}} \cdot 2^{-\binom{n}{2}+1} \qquad \text{da } n! \geq 2^{n-1}$$

$$< \frac{2^{\frac{n^2}{2}}}{2^{n-1}} \cdot 2^{-\frac{n^2}{2}+\frac{n}{2}+1} \qquad \text{da } m < 2^{\frac{n}{2}}$$

$$= 2^{-n+\frac{n}{2}+2}$$

$$\leq 1 \qquad \text{da } n \geq 4$$

Also muss $m \geq 2^{\frac{n}{2}}$ gelten, damit jede Färbung des K_m einen monochromatischen Untergraph K_n enthält. $\qquad\qquad\qquad\qquad\qquad\qquad\qquad\qquad\qquad\qquad\qquad$ □

Da die Ramsey-Zahlen so riesig sind, ergeben sich Anwendungen häufig erst für unendliche Objekte. Wir beschließen diesen Abschnitt daher mit Aussage 6.31.

Satz 6.31 (Ramsey 1930; unendliche Version). *Sei V eine unendliche Menge und f : $\binom{V}{k} \to C$ eine Färbung mit endlich vielen Farben. Dann gibt es eine unendliche monochromatische Teilmenge von V.*

Beweis. Aus dem Satz von Ramsey 6.29 folgt sofort, dass es beliebig große monochromatische Teilmengen gibt, aber diese könnten alle nebeneinander liegen. Es ist nicht unmittelbar klar, dass es eine unendliche monochromatische Teilmenge gibt. Betrachten wir daher nochmals den Beweis von Satz 6.29. Da V unendlich ist, können wir eine unendliche Folge $A_0 \subsetneq A_1 \subsetneq \cdots$ definieren und finden damit eine unendliche Teilmenge $A = \bigcup_m A_m$. Bei Invariante (1) wird hier gewährleistet, dass B_m unendlich ist. Mit Induktion nach k gibt es bezüglich g eine unendliche monochromatische Teilmenge $X \subseteq A$ und diese ist mit demselben Argument wie oben auch monochromatisch für f. $\qquad\qquad\qquad\qquad\qquad\qquad\qquad\qquad\qquad\qquad\qquad$ □

Aufgaben

6.1. Wieviele Graphen mit der Knotenmenge $\{1, \ldots, n\}$ gibt es?

6.2. Für $n \geq 1$ sei V_n die Menge aller Teilmengen von $\{1, \ldots, n\}$. Sei G_n der Graph mit der Knotenmenge V_n, für den zwei Knoten A und B genau dann durch eine Kante verbunden sind, wenn $A \cap B = \varnothing$ gilt.

(a) Zeichnen Sie den Graphen G_3.

(b) Wie viele Knoten und wie viele Kanten hat der Graph G_n?

6.3. Sei $G = (V, E)$ ein Graph, bei dem jeder Knoten mindestens den Grad 4 hat.

(a) Zeigen Sie, dass $|E| \geq 2|V|$ gilt.

(b) Zeigen Sie, dass für alle $n \geq 5$ ein Graph mit n Knoten existiert, so dass jeder Knoten den Grad 4 hat.

6.4. Sei $G = (V, E)$ ein zusammenhängender Graph mit $|V| \geq 3$, der nicht vollständig ist. Zeigen Sie, dass es dann drei Knoten $u, v, w \in V$ gibt mit $uv \in E$, $vw \in E$ und $uw \notin E$.

6.5. Sei $G = (V, E)$ ein beliebiger ungerichteter Graph. Zeigen Sie, dass G oder der komplementäre Graph $\overline{G}$ zusammenhängend ist.

6.6. Sei $G = (V, E)$ ein endlicher Graph. Mit $\ell(G)$ bezeichnen wir die Länge eines längsten einfachen Weges, d. h., für alle einfachen Wege $v_0 \cdots v_\ell$ in G gilt $\ell \leq \ell(G)$.

Zeigen Sie, dass für einen zusammenhängenden Graphen G zwei einfache Wege der Länge $\ell(G)$ stets einen gemeinsamen Knoten haben.

6.7. Zeigen Sie, dass in einem ungerichteten Graph $G = (V, E)$ entweder weniger als zwei Knoten vorhanden sind oder mindestens zwei Knoten den gleichen Grad haben.

6.8. Eine *Brücke* in einem zusammenhängenden Graphen $G = (V, E)$ ist eine Kante $e \in E$, für die der Graph $G' = (V, E \setminus \{e\})$ nicht mehr zusammenhängend ist. Zeigen Sie, dass ein Graph, in dem alle Knoten geraden Grad haben, keine Brücke enthält.

6.9. (De Bruijn-Folgen) Sei Σ ein endliches Alphabet. Zeigen Sie: Es existiert ein Wort $w \in \Sigma^*$ der Länge $|w| = |\Sigma|^k + k - 1$, welches jedes Wort der Länge k genau einmal als Faktor enthält.

6.10. Sei G ein zusammenhängender Graph, bei dem jeder Knoten geraden Grad hat. Geben Sie einen Algorithmus mit linearer Laufzeit an, der auf Eingabe von G einen Eulerkreis berechnet.

6.11. Sei φ eine bijektive Funktion, die den 12 Kanten eines Würfels ein Gewicht aus der Menge $\{1, \ldots, 12\}$ zuordnet. Zeigen Sie, dass es stets zwei Ecken gibt, für die die Gewichtssumme der inzidenten Kanten verschieden ist.

6.12. Seien $n \geq 0$ und $d_1, \ldots, d_n \in \mathbb{N} \setminus \{0\}$. Zeigen Sie, dass genau dann ein Baum mit n Knoten und den Knotengraden $d_1, \ldots, d_n$ existiert, wenn $\sum_{i=1}^{n} d_i = 2n - 2$ gilt.

6.13. Ein Automorphismus eines Graphen (V, E) ist eine bijektive Abbildung $\varphi : V \to V$ mit

$$\forall x, y \in V \colon \{x, y\} \in E \;\Leftrightarrow\; \{\varphi(x), \varphi(y)\} \in E$$

Sei $G = (V, E)$ ein Baum und sei φ ein Automorphismus von G. Zeigen Sie, dass dann φ einen Knoten von G oder eine Kante von G auf sich selbst abbildet.

6.14. Seien $M = \{1, \ldots, n\}$, $1 \leq \ell \leq n$ und $A_1, \ldots, A_\ell$ paarweise verschiedene Teilmengen von M. Zeigen Sie, dass ein Element $x \in M$ existiert, so dass die Mengen $A_1 \setminus \{x\}, \ldots, A_\ell \setminus \{x\}$ paarweise verschieden sind.

6.15. Seien $\{P_1, \ldots, P_m\}$ und $\{Q_1, \ldots, Q_m\}$ zwei Partitionen einer endlichen Menge M, so dass alle Klassen P_i und Q_i jeweils genau k Elemente enthalten. Zeigen Sie, dass ein gemeinsames Vertretersystem für die beiden Partitionen existiert.

Hinweis: Elemente $v_1, \ldots, v_m \in M$ bilden ein *Vertretersystem* einer Partition $\{P_1, \ldots, P_m\}$ von M, falls für jede Klasse P_i genau ein Vertreter v_j mit $v_j \in P_i$ existiert.

6.16. Zeigen Sie, dass das Gale-Shapley-Verfahren aus Satz 6.12 für die Menge A der Frauen denkbar ungünstig ist. Ist $a \in A$ am Ende mit b verheiratet, so hat b aus

ihrer Sicht die niedrigste Präferenz unter allen Partnern, die in einer stabilen Heirat möglich sind.

6.17. Sei G ein zusammenhängender planarer Graph mit n Knoten und f Facetten (Gebieten).

(a) Zeigen Sie: $f \leq 2n - 4$.

(b) Sei n gerade. Zeigen Sie: Wenn die eine Hälfte der Knoten den Grad d und die andere Hälfte der Knoten den Grad $2d$ hat, dann gilt $d \leq 3$.

(c) Geben Sie einen zusammenhängenden planaren Graphen ohne Schlingen und Mehrfachkanten mit 12 Knoten an, so dass 6 Knoten den Grad 3 und 6 Knoten den Grad 6 haben.

6.18. Sei G ein einfacher planare Graph mit n Knoten, m Kanten, f Facetten und z Zusammenhangskomponenten. Zeigen Sie: $n - m + f - z = 1$.

6.19. In der Ebene (bzw. auf der Kugeloberfläche) können bereits die Graphen K_5 und $K_{3,3}$ nicht ohne Kantenüberschneidung gezeichnet werden. Zeigen Sie, dass man auf der Torusoberfläche sogar die Graphen K_7 und $K_{4,4}$ kreuzungsfrei einbetten kann.

6.20. Zeigen Sie, dass es bis auf Isomorphie nur endlich viele Graphen G gibt, so dass sowohl G als auch $\overline{G}$ planar sind.

6.21. (Satz von Wernicke, 1904) Sei $G = (V, E)$ ein planarer Graph mit Minimalgrad $d_x \geq 5$ für alle $x \in V$. Zeigen Sie, dass eine Kante $xy \in E$ existiert mit $d_x = 5$ und $d_y \leq 6$.

6.22. Ein *Turnier* ist ein gerichteter Graph $G = (V, E)$ ohne Schlingen und Mehrfachkanten, der für alle $u, v \in V$ mit $u \neq v$ genau eines der beiden Paare (u, v) und (v, u) als Kante enthält. Es gilt also jeweils $(u, v) \subset E \Leftrightarrow (v, u) \notin E$. Ist V eine Menge von Spielern und hat am Ende eines Turniers jeder gegen jeden gespielt, so soll $(u, v) \in E$ bedeuten, dass u gegen v gewonnen hat. Zeigen Sie für ein Turnier $G = (V, E)$:

(a) (Rédei 1934) Jedes Turnier besitzt einen gerichteten Hamiltonpfad. Es existiert also ein Pfad in G, der jeden Knoten genau einmal besucht.

(b) Es existiert ein Knoten $v \in V$, genannt der *Lion King*, von dem aus jeder andere Knoten durch einen Pfad der Länge ≤ 2 erreicht werden kann.

6.23. Sei $M \subseteq \{1, \ldots, 2n\}$ eine Teilmenge mit $n + 1$ Elementen. Zeigen Sie:

(a) M enthält ein Paar aufeinander folgender Zahlen.

(b) M enthält zwei Zahlen, deren Summe $2n + 1$ ist.

(c) M enthält zwei Zahlen k und ℓ, so dass k ein Teiler von ℓ ist.

6.24. Zeigen Sie:

(a) Es gibt einen Graph mit 8 Knoten ohne eine Clique der Größe 3 und ohne eine unabhängige Menge der Größe 4.

(b) Jeder Graph mit 9 Knoten hat eine Clique der Größe 3 oder eine unabhängige Menge der Größe 4.

6.25. Sei $c : \mathbb{N} \to \{1, \ldots, k\}$ eine beliebige Färbung mit k Farben. Zeigen Sie, dass $x, y, z \in \mathbb{N}$ existieren mit $x + y = z$ und $c(x) = c(y) = c(z)$.

6.26. Sei $0 < p < 1$ beliebig, und sei V eine Knotenmenge mit n Knoten. Im Folgenden betrachten wir das Zufallsexperiment, bei dem zwischen je zwei verschiedenen Knoten $x, y \in V$ eine Kante xy (unabhängig von anderen Kanten) mit Wahrscheinlichkeit p gesetzt wird. Sei $G = (V, E)$ der resultierende Graph. Zeigen Sie: Für $n \to \infty$ gilt

$$\Pr\left[G \text{ ist zusammenhängend}\right] \to 1$$

Zusammenfassung

Begriffe

–	Graph G	–	Spannbaum	–	Fluss
–	Knoten V	–	bipartit	–	Flussnetzwerk
–	Kanten E	–	(perfektes) Matching	–	planar
–	Pfad	–	Heiratsbedingung	–	Facette
–	zusammenhängend	–	stabile Heirat	–	Färbung
–	Grad d_x	–	Separator	–	Ramsey-Zahl $R_{k,c}(n)$
–	Eulerkreis	–	Kantenzug	–	monochromatisch
–	Hamiltonkreis	–	Kantengraph	–	Clique
–	Baum	–	Schnitt	–	unabhängige Menge

Methoden und Resultate

- Handschlaglemma: $\sum_{x \in V} d_x = 2|E|$

- Eulerkreis $\Leftrightarrow$ zusammenhängend und alle Knoten haben geraden Grad

- Ore: Für alle $x \neq y$ mit $xy \notin E$ gilt $d_x + d_y \geq |V| \Rightarrow$ Hamiltonkreis

- Baum $\Leftrightarrow$ zusammenhängend und $|E| = |V| - 1$

- Heiratssatz: Heiratsbedingung $\Leftrightarrow$ perfektes Matching

- Bei n Frauen und n Männern gibt es stets eine stabile Heirat.

- Der Gale-Shapley-Algorithmus berechnet eine stabile Heirat in $\mathcal{O}(n^2)$ Schritten.

- Menger: Minimale Größe eines AB-Separators = maximale Anzahl disjunkter AB-Pfade

- Max-Flow-Min-Cut-Theorem: Minimales Gewicht eines st-Schnitts = maximaler Wert eines st-Flusses
- Der Algorithmus von Dinitz berechnet einen maximalen Fluss in $\mathcal{O}(mn^2)$ Schritten.
- Eulerformel: $G \neq \varnothing$ zusammenhängend und planar $\Rightarrow n - m + f = 2$
- G planar und $n \geq 3 \Rightarrow m \leq 3n - 6$
- K_5 und $K_{3,3}$ sind nicht planar
- $G \neq \varnothing$ planar $\Rightarrow \exists x \in V : d_x \leq 5$
- Fünffarbensatz: Planare Graphen sind 5-färbbar.
- Planares Separator-Theorem: G planar $\Rightarrow \exists$ Zerlegung $V = A \cup B \cup C$ mit $|A| < 2n/3$, $|B| < 2n/3$, $|C| \leq \sqrt{8n}$ und $A \times B \cap E = \varnothing$
- Ramsey: $\forall k, c, n \in \mathbb{N} \; \exists R_{k,c}(n) \in \mathbb{N}$: Wenn $|V| \geq R_{k,c}(n)$, dann besitzt jede Färbung von $\binom{V}{k}$ eine monochromatische Teilmenge $X \subseteq V$ mit $|X| = n$
- $2^{\frac{n}{2}} \leq R_{2,2}(n) \leq 2^{2n}$
- Ramsey (unendliche Version): V unendlich $\Rightarrow$ Jede Färbung von $\binom{V}{k}$ mit endlich vielen Farben besitzt eine unendliche monochromatische Teilmenge $X \subseteq V$.

7 Ordnungsstrukturen und Verbände

In diesem Kapitel untersuchen wir Halbordnungen. Dies sind Strukturen, welche besonders gut dafür geeignet sind, Reihenfolgen und Kausalitäten sowie Enthaltenseinsbeziehungen und Größenvergleiche abstrakt zu analysieren. Ein besonderes Augenmerk legen wir auf vollständige Halbordnungen und Verbände. Wir beweisen den Fixpunktsatz von Kleene und zeigen eine Anwendung für die Semantik von Programmiersprachen; und wir beweisen den Fixpunktsatz von Knaster und Tarski für vollständige Verbände. Danach untersuchen wir allgemeine boolesche Verbände und zeigen unter anderem den Satz von Stone, dass jeder endliche boolesche Verband als ein Potenzmengenverband realisiert werden kann. Insbesondere hat jeder endliche boolesche Verband genau 2^n Elemente für ein $n \in \mathbb{N}$. Zum Abschluss leiten wir noch den allgemeinen Darstellungssatz von Stone her, der jeden booleschen Verband als Mengenverband realisiert.

7.1 Halbordnungen

Eine *Halbordnung* (oder *partielle Ordnung*) ist eine Menge M zusammen mit einer Relation $R \subseteq M \times M$, die reflexiv, transitiv und antisymmetrisch ist. Es gelten also für alle $x, y, z \in M$ die Beziehungen:

- $(x, x) \in R$,
- wenn $(x, y) \in R$ und $(y, z) \in R$, dann ist auch $(x, z) \in R$,
- wenn $(x, y) \in R$ und $(y, x) \in R$, dann gilt $x = y$.

Ist (M, R) eine Halbordnung, so ist es auch (M, R^{-1}), wobei R^{-1} wie üblich die inverse Relation $R^{-1} = \{(y, x) \in M \times M \mid (x, y) \in R\}$ bezeichnet. Häufig schreiben wir $\leq$ für die Relation R und dann steht $x < y$ für $x \leq y \land x \neq y$. Zwei Elemente heißen *unvergleichbar*, wenn weder $x \leq y$ noch $y \leq x$ gilt. Eine Halbordnung $(M, \leq)$ heißt *lineare* oder *totale* Ordnung, wenn je zwei Elemente von M vergleichbar sind, also stets $x \leq y$ oder $y \leq x$ gilt.

Eine Halbordnung $(M, \leq)$ heißt *wohlfundiert*, wenn jede nichtleere Teilmenge $X \subseteq M$ minimale Elemente enthält. Ein Element $x \in X$ ist *minimal* in X, wenn es kein $y \in X$ mit $y < x$ gibt. Analog werden *maximale* Elemente definiert. Eine *Wohlordnung* ist eine wohlfundierte lineare Ordnung. Insbesondere enthält jede nichtleere Teilmenge einer Wohlordnung genau ein minimales Element. Wir benutzen den *Wohlordnungssatz* der Mengenlehre. Dieser ist äquivalent zum *Auswahlaxiom* und besagt, dass sich jede Menge M mit einer Wohlordnung versehen lässt. Für abzählbare Mengen ist der Wohlordnungssatz trivial, da die natürlichen Zahlen wohlgeordnet sind.

Beispiel 7.1. Die Zahlenbereiche $\mathbb{N}$, $\mathbb{Z}$, $\mathbb{Q}$ und $\mathbb{R}$ sind lineare Ordnungen mit der üblichen Relation „$\leq$". Hierunter ist nur das Paar $(\mathbb{N}, \leq)$ eine Wohlordnung. Auf $\mathbb{N}$ erhalten wir durch die Teilerrelation $m \mid n$ eine Halbordnung mit 1 als kleinstem und 0

als größtem Element. Für eine beliebige Menge M definiert die Inklusion „$\subseteq$" auf der Potenzmenge 2^M eine Halbordnung.

Seien (M_i, R_i) Halbordnungen für $i \in I$, wobei I eine beliebige Indexmenge ist. Dann definiert man auf dem kartesischen Produkt $\prod_{i \in I} M_i$ eine Halbordnung R wie folgt:

$$R = \{ ((x_i)_{i \in I}, (y_i)_{i \in I}) \mid \forall i \in I : (x_i, y_i) \in R_i \}$$

Wie eben seien (M_i, R_i) Halbordnungen für $i \in I$. Wir nehmen an, dass fast alle (also alle bis auf endlich viele Ausnahmen) der M_i ein kleinstes Element $\perp$ haben. Dann definieren wir

$$\coprod_{i \in I} M_i = \left\{ (x_i) \in \prod_{i \in I} M_i \mid x_i = \perp \text{ für fast alle } i \in I \right\}$$

Damit ist $\coprod_{i \in I} M_i$ eine Halbordnung. Dies ist eine Unterhalbordnung von $\prod_{i \in I} M_i$. Beide Halbordnungen haben ein kleinstes Element, wenn alle M_i ein kleinstes Element besitzen.

Beschreibt man eine positive natürliche Zahl n durch ihre Primfaktorzerlegung $\prod p^{n_p}$, wobei p die Primzahlen $\mathbb{P}$ durchläuft, so wird die Teilbarkeitsrelation auf $\mathbb{N} \setminus \{0\}$ zur partiellen Ordnung $\coprod_{p \in \mathbb{P}} (\mathbb{N}, \leq)$. Hierfür beachte man, dass 1 das kleinste Element von $(\mathbb{N}, \mid)$ ist und dass bei einer Primfaktorzerlegung fast alle Exponenten n_p gleich Null sind. $\diamond$

Zur graphischen Darstellung einer endlichen nichtleeren Halbordnung $(M, \leq)$ verwendet man häufig ihr *Hasse-Diagramm*. Diese Darstellung geht auf den Zahlentheoretiker Helmut Hasse (1898–1979) zurück und spiegelt die Nachbarschaftsrelation einer Halbordnung wider. Hierbei heißt x *unterer Nachbar* von y (in Zeichen $x \lessdot y$), wenn $x < y$ gilt und es kein $z \in M$ mit $x < z < y$ gibt. Entsprechend ist y ein *oberer Nachbar* von x wenn $x \lessdot y$ gilt.

Das Hasse-Diagramm erhalten wir wie folgt: Für jedes Element zeichnen wir einen Punkt in der Ebene und markieren ihn gegebenenfalls mit dem zugeordneten Element. Punkte x, y werden genau dann auch durch eine Strecke verbunden, wenn $x \lessdot y$ gilt. Um die Ordnung eindeutig aus ihrem Hasse-Diagramm rekonstruieren zu können, vereinbaren wir, dass y oberhalb von x zu zeichnen ist, wenn $x < y$ gilt. Damit erhalten wir $x < y$ genau dann, wenn man im Diagramm von x aus von unten nach oben über eine Folge von Strecken zu y gelangen kann. Für unendliche Halbordnungen ist nicht gesagt, dass überhaupt Paare benachbarter Elemente existieren, betrachte etwa $(\mathbb{Q}, \leq)$.

Wir betrachten nun zwei Beispiele für Hasse-Diagramme. Für drei Elemente haben wir fünf mögliche Hasse-Diagramme:

Sei $M = \{1, 2, \ldots, 10\}$ versehen mit der Teilbarkeitsordnung. Wir erhalten folgendes Hasse-Diagramm:

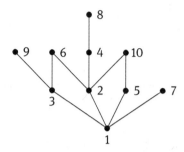

Sei $(M, \leq)$ eine Halbordnung. Eine *Kette K* ist eine linear geordnete Teilmenge von M. Sie heißt maximal, wenn benachbarte Elemente in K auch benachbart in M sind. In unendlichen Halbordnungen können x und y vergleichbar sein, ohne dass eine maximale Kette zwischen ihnen existiert. Als Länge einer Kette K bezeichnen wir die Anzahl der Kanten in ihrem Hasse-Diagramm; die Länge ist also $|K| - 1$. Besitzt die Halbordnung ein kleinstes Element $\perp$ und ist x ein Element der Halbordnung, so heißt die Länge einer längsten Kette von $\perp$ nach x die *Dimension* von x und wird mit $\dim(x)$ bezeichnet. Man setzt $\dim(x) = \infty$, falls beliebig lange Ketten existieren. Formal gilt:

$$\dim(x) = \sup \{ \, |K| - 1 \mid K \text{ ist eine Kette von } \perp \text{ nach } x \, \} \in \mathbb{N} \cup \{\infty\}$$

Die Dimension von $\perp$ ist 0. Im folgenden Beispiel gilt $\dim(\perp) = 0$, $\dim(a) = 1$, $\dim(b) = \dim(c) = 5$ und $\dim(d) = 6$, und es gibt drei maximale Ketten von $\perp$ nach d der Längen 4, 5 und 6.

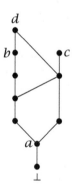

Sind R und R' Halbordnungen auf einer Menge M, so sagen wir, dass R' eine *Verfeinerung* von R ist, falls $R \subseteq R'$ gilt. Eine *topologische Sortierung* oder *topologische Reihenfolge* einer endlichen Halbordnung (M, R) ist eine Verfeinerung zu einer linearen Ordnung $(M, \leq)$. Häufig schreibt man einfach $M = \{x_0, \ldots, x_n\}$ und verlangt $R \subseteq \{(x_i, x_j) \mid 0 \leq i \leq j \leq n\}$.

Aus dem Auswahlaxiom folgt, dass für alle Halbordnungen lineare Verfeinerungen existieren. Wir führen dies hier nicht aus. Dafür beweisen wir ein anderes Resultat, welches zeigt, dass sich jede abzählbare Halbordnung nach $(\mathbb{Q}, \leq)$ einbetten lässt. Naiv betrachtet ist dies ein sehr erstaunliches Resultat, vor allem wenn man an die schier unüberschaubaren Möglichkeiten denkt, abzählbare Mengen partiell zu ordnen.

Satz 7.2. *Sei $(M, \leq)$ eine abzählbare Halbordnung. Dann gibt es eine injektive Abbildung $f : M \to [0, 1] \subseteq \mathbb{Q}$ mit $f(x) < f(y)$ für alle $x < y$.*

Beweis. Wir können Elemente $\bot$ und $\top$ neu hinzunehmen. Deshalb können wir annehmen, dass M ein kleinstes Element $\bot$ und ein größtes Element $\top$ enthält und $|M| \geq 2$ gilt. Wir setzen $f(\bot) = 0$ und $f(\top) = 1$ und schreiben $M = \{x_1, x_2, \ldots\}$ mit $x_1 = \bot$ und $x_2 = \top$ und nehmen an, dass die x_i paarweise verschieden sind. Sei $n \geq 3$. Induktiv seien $n - 1$ paarweise verschiedene Zahlen $f(x_j)$ im Intervall $[0, 1]$ für $1 \leq j < n$ schon definiert, und in diesem Bereich impliziere $x_i < x_j$ die Anordnung $f(x_i) < f(x_j)$. Wir definieren jetzt $f(x_n) \in [0, 1]$ wie folgt. Zunächst setzen wir:

$$a_n = \max\left\{ f(x_i) \mid x_i < x_n, 1 \leq i < n \right\}$$
$$b_n = \min\left\{ f(x_j) \mid x_n < x_j, 1 \leq j < n \right\}$$

Für $1 \leq i, j < n$ folgt aus $x_i < x_n$ und $x_n < x_j$, dass $x_i < x_j$ und $f(x_i) < f(x_j)$ gilt. Dies bedeutet $0 \leq a_n < b_n \leq 1$ und wir können für $f(x_n)$ eine Zahl c_n wählen, die $a_n < c_n < b_n$ erfüllt und die von allen $f(x_0), \ldots, f(x_{n-1})$ verschieden ist. Damit ist $f : M \to [0, 1]$ überall definiert und injektiv. Die Bedingung $f(x) < f(y)$ für alle $x < y$ gilt nach Konstruktion. $\square$

Wenden wir das Verfahren aus dem obigen Beweis auf eine endliche Halbordnung $(M, \leq)$ an und multiplizieren die Zahlen in $f(M)$ mit ihrem Hauptnenner, so erhalten wir eine Verfeinerung von $(M, \leq)$ in die natürlichen Zahlen und damit eine topologische Sortierung. Eine andere Methode besteht darin, zunächst ein Hasse-Diagramm zu zeichnen und dieses um einen möglicherweise unsichtbaren ε-Winkel nach links zu kippen. Wir nummerieren danach von unten nach oben. Dies ist am folgenden Beispiel illustriert.

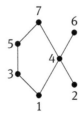

7.2 Vollständige Halbordnungen

In diesem Abschnitt bezeichnen $(M, \leq)$ und $(M', \leq)$ stets Halbordnungen. Für Teilmengen $D \subseteq M$ bezeichnen wir mit sup D die kleinste obere Schranke (Supremum), sofern sie existiert. Analog bezeichnen wir mit inf D die größte untere Schranke (Infimum). Die leere Menge hat in M genau dann ein Supremum bzw. Infimum, wenn M ein eindeutiges minimales Element $\perp$, bzw. ein eindeutiges maximales Element $\top$, enthält. Gibt es diese beiden Elemente, so gilt also sup $\varnothing = \perp$ und inf $\varnothing = \top$. Eine Teilmenge $D \subseteq M$ heißt *gerichtet*, falls für alle $x, y \in D$ ein $z \in D$ existiert mit $x \leq z$ und $y \leq z$. Die leere Menge und Ketten sind gerichtete Teilmengen, ebenso alle Teilmengen, die genau ein maximales Element enthalten. In linearen Ordnungen sind alle Teilmengen gerichtet. Eine Halbordnung $(M, \leq)$ heißt *vollständig*, wenn jede gerichtete Teilmenge $D \subseteq M$ eine kleinste obere Schranke hat. Es wird dabei nicht verlangt, dass das Supremum sup D in D enthalten ist. Insbesondere enthalten vollständige Halbordnungen ein eindeutig bestimmtes kleinstes Element sup $\varnothing = \perp$. Eine vollständige Halbordnung wird in der Literatur auch als *CPO* bezeichnet (engl. *complete partial order*).

Die lineare Ordnung $(\mathbb{N}, \leq)$ ist keine vollständige Halbordnung, aber $(\mathbb{N} \cup \{\infty\}, \leq)$ ist eine mit $\perp = 0$ und $\top = \infty$. Endliche nichtleere gerichtete Teilmengen von M besitzen genau ein maximales Element. Insbesondere sind endliche Halbordnungen genau dann vollständig, wenn sie ein eindeutig bestimmtes kleinstes Element besitzen. Das kartesische Produkt $M \times M'$ vollständiger Halbordnungen mit komponentenweiser Ordnung ist eine vollständige Halbordnung. Sei Σ ein Alphabet und $(\Sigma^*, \leq)$ die Menge der Wörter mit der Präfixordnung. Dann ist das leere Wort das kleinste Element, aber $(\Sigma^*, \leq)$ ist nicht vollständig. Indem wir die unendlichen Sequenzen hinzunehmen erhalten wir eine vollständige Halbordnung $(\Sigma^\infty, \leq)$. Hierbei ist $\Sigma^\infty = \Sigma^* \cup \Sigma^\omega$ die Mengen aller endlichen und aller unendlichen Wörter über dem Alphabet Σ.

Seien $(M, \leq)$ und $(M', \leq)$ vollständige Halbordnungen. Eine Abbildung $f : M \to M'$ zwischen Halbordnungen heißt *monoton*, falls $f(x) \leq f(y)$ für alle $x \leq y$ gilt. Ist f monoton und $D \subseteq M$ gerichtet, so ist auch $f(D)$ gerichtet. Sind M und M' jeweils vollständige Halbordnungen, so gilt dann sup $f(D) \leq f(\text{sup } D)$, da $x \leq$ sup D für alle $x \in D$. Die Abbildung $f : M \to M'$ heißt *stetig*, falls f monoton ist und sup $f(D) = f(\text{sup } D)$ für jede nichtleere gerichtete Teilmenge $D \subseteq M$ gilt. Wir verlangen also nicht, dass stetige Abbildungen kleinste Elemente aufeinander abbilden. Bei einigen Autoren sind gerichtete Mengen niemals leer, dann wird dieser Hinweis überflüssig; aber im Gegenzug muss man dann betonen, dass eine vollständige Halbordnung ein kleinstes Element hat.

Ist M eine vollständige Halbordnung, so reicht die Stetigkeit einer Abbildung $f : M \to M$, um durch eine iterierte Anwendung von f auf das kleinste Element $\perp$ eine monoton wachsende Folge zu definieren, deren Supremum der dann eindeutig bestimmte kleinste Fixpunkt ist. Dies ist der Inhalt des Satzes 7.3. Dabei heißt wie üb-

lich $x \in M$ ein *Fixpunkt*, falls $f(x) = x$ gilt. Er ist benannt nach Stephen Cole Kleene (1909–1994).

Satz 7.3 (Fixpunktsatz von Kleene). *Es sei $(M, \leq)$ eine vollständige Halbordnung und $f : M \to M$ stetig. Dann ist $x_f = \sup\{f^i(\bot) \mid i \geq 0\}$ der eindeutig bestimmte kleinste Fixpunkt von f.*

Beweis. Da $\bot$ das kleinste Element ist, gilt zunächst $\bot \leq f(\bot)$. Aufgrund der Monotonie von f folgt mit Induktion die Aussage $f^i(\bot) \leq f^{i+1}(\bot)$ für alle $i \in \mathbb{N}$. Also ist $\bot = f^0(\bot) \leq f^1(\bot) \leq f^2(\bot) \leq \cdots$ eine Kette, die in der vollständige Halbordnung ein Supremum $x_f = \sup\{f^i(\bot) \mid i \geq 0\}$ besitzt. Jetzt nutzen wir die Stetigkeit von f aus und erhalten:

$$f(x_f) = f\left(\sup\{f^i(\bot) \mid i \geq 0\}\right) = \sup\{f^{i+1}(\bot) \mid i \geq 0\} = x_f$$

Also ist x_f ein Fixpunkt von f. Sei jetzt y ein weiterer Fixpunkt von f. Dann gilt $\bot \leq y$ und damit $f^i(\bot) \leq f^i(y) = y$ für alle $i \in \mathbb{N}$. Also gilt $x_f = \sup\{f^i(\bot) \mid i \geq 0\} \leq y$. $\square$

7.3 Denotationale Semantik

In der Semantik von Programmiersprachen interessiert man sich nicht für die syntaktische Beschreibung eines Programms, sondern versucht „zu verstehen", was Computer-Programme „tun". Die Bedeutung eines Programms wird dann zur Vorschrift, die gegebene Eingaben in die durch das Programm berechneten Ausgaben überführt. Da Programme nicht immer terminieren, ist deren *Semantik* eine nur partiell definierte Funktion. Tatsächlich ist das sogenannte *Halteproblem* unentscheidbar. Dies bedeutet, es gibt kein algorithmisches Verfahren, das aus der syntaktischen Beschreibung eines Programms in einer höheren Programmiersprache (wie etwa Java) bestimmt, ob das Programm terminiert.

Man kann also nicht für alle Programme die Bedeutung herleiten, aber man kann die Bedeutung beliebig genau approximieren. Die Idee, mit Hilfe einer *denotationalen Semantik* die Bedeutung von Programmen zu definieren, geht wesentlich auf Dana Scott (geb. 1932) zurück. Er entwickelte Ende der 1960er Jahre die Bereichstheorie und zeigte, wie man mittels kleinster Fixpunkte Berechnungen beliebig genau approximieren kann. Im Folgenden erklären wir nur, wie man die Bedeutung der *while-Schleife* durch einen kleinsten Fixpunkt erklären kann. Dies ist der entscheidende Schritt in der Theorie von Scott.

Wir starten mit einer abstrakten Menge von Daten Σ und stellen uns vor, dass gewisse syntaktisch definierte Ausdrücke partiell definierte Funktionen von Σ nach Σ beschreiben. Die Menge dieser sich ergebenden Funktionen sei $\mathcal{F}$. Bezeichnet $(\Sigma \to_p \Sigma)$ die Menge der partiell definierten Abbildungen von Σ nach Σ, so gilt $\mathcal{F} \subseteq (\Sigma \to_p \Sigma)$. Es geht uns hier nicht um eine genaue Festlegung der Funktionen in $\mathcal{F}$,

aber wir nehmen an, dass $\mathcal{F}$ eine genügend große Anzahl von Grundfunktionen enthält und dass $\mathcal{F}$ unter Komposition abgeschlossen ist. Wir interessieren uns für den Abschluss von $\mathcal{F}$ unter der Hinzunahme von *while-Schleifen*. Eine while-Schleife besteht aus einer booleschen Bedingung und einem Rumpf. Beim Betreten der Schleife wird zunächst die boolesche Bedingung ausgewertet. Liefert die Auswertung das Ergebnis *falsch*, so wird die Schleife übersprungen, andernfalls wird der Rumpf einmal ausgeführt und danach die Schleife erneut betreten. Die Schleife terminiert also nur, wenn die Auswirkung der booleschen Bedingung schließlich *falsch* wird. Zur Definition einer while-Schleife benötigen wir neben $\mathcal{F}$ also auch eine Menge berechenbarer boolescher Funktionen $\mathcal{B} \subseteq (\Sigma \to \mathbb{B})$, die uns im Weiteren zur Verfügung stehe. Hierbei bezeichnet $(\Sigma \to \mathbb{B})$ die Menge der (totalen) Abbildungen von Σ nach $\mathbb{B} = \{0, 1\}$. Unter $\mathcal{B}$ sollte man sich einfache Auswertungen vorstellen, etwa ob sich ein arithmetischer Ausdruck wie $X \cdot Y - Z$ zu einer positiven Zahl auswertet. Für $\sigma \in \Sigma$ und $b \in \mathcal{B}$ ist also ein Wahrheitswert $b(\sigma) = 0$ (falsch) oder $b(\sigma) = 1$ (wahr) definiert.

Sei $b \in \mathcal{B}$ und $c \in \mathcal{F}$, dann erweitern wir $\mathcal{F}$ um das folgende Konstrukt:

$$\textbf{while } b \textbf{ do } c \textbf{ od}$$

Für $w = \textbf{while } b \textbf{ do } c \textbf{ od}$ müssen wir den Definitionsbereich $\mathrm{dom}(w)$ von w und die Bedeutung $w(\sigma)$ für $\sigma \in \mathrm{dom}(w) \subseteq \Sigma$ erklären. Dies basiert auf einer Fallunterscheidung und einer Rekursion. Für $b(\sigma) = 0$ wertet sich die Bedingung zu „falsch" aus und die Schleife w wird nicht betreten. Also definieren wir $w(\sigma) = \sigma$ für $b(\sigma) = 0$. Im anderen Fall ist $b(\sigma) = 1$ und die Schleife wird betreten und zunächst wird der Funktionswert $c(\sigma)$ berechnet. Ist $c(\sigma)$ nicht definiert, so wird auch $w(\sigma)$ nicht definiert. Ist $c(\sigma) = \sigma'$ definiert, so erklären wir $w(\sigma)$ durch $w(\sigma')$, falls $\sigma' \in \mathrm{dom}(w)$ gilt.

Wir können dies auch wie folgt ausdrücken. Wir erhalten $\sigma \in \mathrm{dom}(w)$ genau dann, wenn ein $t \in \mathbb{N}$ existiert mit $b(c^t(\sigma)) = 0$ und $b(c^k(\sigma)) = 1$ sowie $c^k(\sigma) \in \mathrm{dom}(c)$ für alle $0 \le k < t$. Hierbei ist, wie üblich, $c^0(\sigma) = \sigma$ und

$$c^k(\sigma) = \underbrace{c \circ \cdots \circ c}_{k \text{ mal}}(\sigma)$$

Der Parameter t misst die „Zeit" oder genauer die Anzahl der Schleifendurchläufe von w, wenn wir bei $\sigma \in \Sigma$ beginnen. Gilt $\sigma \in \mathrm{dom}(w)$, so ist t eindeutig definiert und wir können $w(\sigma) = c^t(\sigma)$ setzen.

Die denotationale Semantik nach Scott geht wie folgt vor. Zunächst wird $(\Sigma \to_p \Sigma)$ mit einer partiellen Ordnung versehen. Wir setzen $f \sqsubseteq g$, falls die folgenden beiden Bedingungen gelten:

- Der Definitionsbereich $\mathrm{dom}(f)$ von f ist eine Teilmenge von $\mathrm{dom}(g)$.
- Für alle $x \in \mathrm{dom}(f)$ gilt $f(x) = g(x)$.

Dies bedeutet, es gilt $f \sqsubseteq g$ genau dann, wenn der Graph von f eine Teilmenge des Graphen von g ist. Man beachte, ist f total definiert und gilt $f \sqsubseteq g$, so muss schon

$f = g$ gelten. Allgemeiner stellen wir fest: Gilt $f \sqsubseteq g$ und $\mathrm{dom}(g) \subseteq \mathrm{dom}(f)$, so ist $f = g$.

Die Halbordnung der partiell definierten Abbildungen $((\Sigma \rightarrow_p \Sigma), \sqsubseteq)$ ist vollständig: Sei $D = \{f_i \mid i \in I\} \subseteq (\Sigma \rightarrow_p \Sigma)$ gerichtet. Dann ergibt sich $f = \sup D$ mit $f : \Sigma \rightarrow_p \Sigma$ wie folgt. Wir setzen $\mathrm{dom}(f) = \bigcup_{i \in I} \mathrm{dom}(f_i)$. Für $x \in \mathrm{dom}(f)$ wähle ein $i \in I$ mit $x \in \mathrm{dom}(f_i)$. Setze $f(x) = f_i(x)$. Diese Wahl hängt nicht von i ab. Denn sei $x \in \mathrm{dom}(f_i)$ und $x \in \mathrm{dom}(f_j)$, dann wähle $f_k \in D$ mit $f_i \sqsubseteq f_k$ und $f_j \sqsubseteq f_k$. Es gilt $x \in \mathrm{dom}(f_k)$ und $f_i(x) = f_k(x) = f_j(x)$. Wegen $f_i(x) = f_j(x)$ ist $f(x)$ wohldefiniert.

Für $b \in \mathcal{B}$ und $c \in \mathcal{F}$ definieren wir einen Operator $\Gamma_{b,c}$ von $(\Sigma \rightarrow_p \Sigma)$ nach $(\Sigma \rightarrow_p \Sigma)$ wie folgt:

$$\Gamma_{b,c}(g)(\sigma) = \begin{cases} \sigma & \text{falls } b(\sigma) = 0 \\ g(c(\sigma)) & \text{sonst} \end{cases}$$

Eine leichte Rechnung zeigt, dass der Operator $\Gamma_{b,c}$ stetig ist. Wir wissen also nach dem Kleene'schen Fixpunktsatz 7.3, dass $\Gamma_{b,c}$ einen kleinsten Fixpunkt in der vollständigen Halbordnung $(\Sigma \rightarrow_p \Sigma)$ hat. Dieser Fixpunkt ist genau die Semantik der while-Schleife. Dies wird in Satz 7.4 gezeigt, der aufgrund seiner fundamentalen Bedeutung für die Entwicklung dieser Theorie als *Hauptsatz der denotationalen Semantik* gelten kann.

Satz 7.4. *Sei* $c \in \mathcal{F}$ *und* $w = $ **while** b **do** c **od**, *dann ist die partiell definierte Funktion* $w \in (\Sigma \rightarrow_p \Sigma)$ *der kleinste Fixpunkt des Operators* $\Gamma_{b,c}$.

Beweis. Die Definition $w \in \mathcal{F}$ wurde oben angegeben und zeigt:

$$w(\sigma) = \begin{cases} \sigma & \text{falls } b(\sigma) = 0 \\ w(c(\sigma)) & \text{sonst} \end{cases}$$

Also gilt $\Gamma_{b,c}(w) = w$ und w ist ein Fixpunkt von $\Gamma_{b,c}$. Für den Satz müssen wir nur zeigen, dass die Inklusion der Definitionsbereiche $\mathrm{dom}(w) \subseteq \mathrm{dom}(g)$ gilt, wenn g der kleinste Fixpunkt von $\Gamma_{b,c}$ ist. Nach dem Kleene'schen Fixpunktsatz ist $\mathrm{dom}(g) = \bigcup_{i \in \mathbb{N}} \mathrm{dom}(\Gamma_{b,c}^i(\bot))$. Sei $\sigma \in \mathrm{dom}(w)$. Wir zeigen $\sigma \in \mathrm{dom}(\Gamma_{b,c}^i(\bot))$ für ein $i \geq 0$.

Wegen $\sigma \in \mathrm{dom}(w)$ gibt es nach der Definition von $w(\sigma)$ ein eindeutig bestimmtes $t \in \mathbb{N}$ (dies ist die Zahl der „Schleifendurchläufe") mit $b(c^t(\sigma)) = 0$ und $b(c^k(\sigma)) = 1$ sowie $c^k(\sigma) \in \mathrm{dom}(c)$ für alle $0 \leq k < t$. Wir beweisen $\sigma \in \mathrm{dom}(\Gamma_{b,c}^{t+1}(\bot))$ mit Induktion nach t. Für $t = 0$ ist $b(\sigma) = 0$ und $\sigma \in \mathrm{dom}(\Gamma_{b,c}(\bot)) = \{\sigma \in \Sigma \mid b(\sigma) = 0\}$.

Sei jetzt $t \geq 1$. Insbesondere gilt $b(\sigma) = 1$ und $\sigma' = c(\sigma)$ ist definiert. Ferner gilt $w(\sigma') = w(\sigma)$ und $\sigma' \in \mathrm{dom}(w)$. Mit Induktion nach t erhalten wir $\sigma' \in \mathrm{dom}(\Gamma_{b,c}^t(\bot))$. Wegen $b(\sigma) = 1$ ist $\Gamma_{b,c}^{t+1}(\bot)(\sigma) = \Gamma_{b,c}(\Gamma_{b,c}^t(\bot)(\sigma)) = \Gamma_{b,c}^t(\bot)(c(\sigma)) = \Gamma_{b,c}^t(\bot)(\sigma')$. Also ist $\sigma \in \mathrm{dom}(\Gamma_{b,c}^{t+1}(\bot))$ und der Satz ist bewiesen. $\square$

Betrachten wir die Fakultätsfunktion als Beispiel mit $\Sigma = \mathbb{N}^{\{X,Y\}}$.

$w = $ **while** $X > 0$ **do**

      $Y := Y \cdot X;$

      $X := X - 1$

  od

Die Semantikfunktion von w berechnet also die Wirkung auf zwei Programm-parameter X und Y. Wir fassen Σ daher als die Menge $\mathbb{N}^2$ auf. Hierbei bedeutet ein Paar (m, n), dass X den Wert m und Y den Wert n hat. Für das obige Programm behaupten wir $w(m, n) = (0, n \cdot m!)$. Die Fakultät $m!$ wird demnach durch $w(m, 1)$ berechnet, wenn am Ende der Wert von Y ausgegeben wird. Zunächst sehen wir, dass die Semantik der Schleife w total definiert ist, denn die Schleife terminiert auf allen Eingaben. Betrachte jetzt die Funktion $f : \Sigma \to \Sigma$ mit $f(m, n) = (0, n \cdot m!)$. Da w überall definiert ist, folgt die Behauptung, wenn wir $\Gamma_{b,c}(f) = f$ zeigen können. Die boolesche Bedingung b ist hierbei die Auswertung von $X > 0$ und es gilt $c(m, n) = (m - 1, nm)$. Für $b(m, n) = 0$ ist $m = 0$. Also ist $(\Gamma_{b,c}(f))(0, n) = (0, n) = (0, n \cdot 0!) = f(0, n)$. Für $m > 0$ ist $(\Gamma_{b,c})(f)(m, n) = f(c(m, n)) = f(m - 1, nm) = (0, nm \cdot (m - 1)!) = (0, n \cdot m!)$. Dies zeigt, dass das Programm w in der Tat Fakultät berechnet.

Programmstücke der Form **while** b **do** c **od** werden übersprungen, wenn sich die Bedingung b stets zu falsch ergibt. Ein Problem ist, dass wir diese Eigenschaft arithmetischer Ausdrücke algorithmisch gar nicht überprüfen können. Dies ergibt sich aus dem berühmten *Gödel'schen Unvollständigkeitssatz*. Kurt Friedrich Gödel (1906–1978) veröffentlichte dieses Resultat in seiner vielleicht wichtigsten Arbeit, die er bereits im Alter von 25 Jahren mit dem Titel *Über formal unentscheidbare Sätze der Principia mathematica und verwandter Systeme* verfasste.

7.4 Kleinste Fixpunkte für monotone Abbildungen

Der Kleene'sche Fixpunktsatz liefert die Existenz kleinster Fixpunkte für stetige Abbildungen in vollständigen Halbordnungen. Er sagt uns aber mehr, denn der Fixpunktsatz von Kleene liefert auch die Approximation, dass der kleinste Fixpunkt das Supremum der Menge $\{f^i(\bot) \mid i \in \mathbb{N}\}$ ist. Ist $(M, \leq)$ vollständig und die Abbildung $f : M \to M$ monoton, aber nicht stetig, so kann die letzte Aussage falsch sein. Erweitere etwa die natürliche Ordnung $(\mathbb{N}, \leq)$ zu $(M, \leq) = (\mathbb{N} \cup \{\omega_1, \omega_2\}, \leq)$, wobei $n < \omega_1 < \omega_2$ für alle $n \in \mathbb{N}$ gelte. Dann ist $(M, \leq)$ eine vollständige Halbordnung, und $f(n) = n + 1$, $f(\omega_1) = f(\omega_2) = \omega_2$ definiert eine monotone Abbildung. Es gilt $\bot = 0$ und $\omega_1 = \sup\{f^i(0) \mid i \in \mathbb{N}\}$, aber ω_1 ist kein Fixpunkt, sondern der einzige Fixpunkt ist ω_2.

Sei jetzt $(M, \leq)$ eine beliebige vollständige Halbordnung mit kleinstem Element $\bot$ und $f : M \to M$ eine monotone Abbildung. Setze $f^0 = \bot$ und $f^{i+1} = f(f^i)$ für $i \in \mathbb{N}$. Dann gilt $f^i \leq f^{i+1}$ und daher existiert $f^\omega = \sup\{f^i \mid i \geq 0\}$. Gilt

$f(f^\omega) = f^\omega$, so ist f^ω ein kleinster Fixpunkt. Im anderen Fall gilt $f^\omega < f(f^\omega)$, und wir können eine neue Fixpunktiteration bei f^ω starten, denn es gilt $f^\omega \leq x$ für alle Fixpunkte x von f. Auch nach der zweiten Iteration müssen wir keinen Fixpunkt gefunden haben. Dann starten wir die dritte und so fort. *Transfinite Induktion* erlaubt nun, das obige Verfahren beliebig oft zu iterieren, um auf diesem Wege zu einem kleinsten Fixpunkt zu gelangen. Wir halten dies in Satz 7.5 fest. Der hier geführte Beweis von Satz 7.5 benutzt den Wohlordnungssatz. Damit ist die Beweisführung eine Standardroutine für Wohlordnungen analog zum Beweis des Fixpunktsatzes von Kleene.

Satz 7.5. *Sei $(M, \leq)$ eine vollständige Halbordnung und $f : M \to M$ eine monotone Abbildung. Dann existiert ein eindeutig bestimmter kleinster Fixpunkt.*

Beweis. Sei Ω zunächst eine beliebige wohlgeordnete Menge. Das heißt, Ω ist linear geordnet und jede nichtleere Teilmenge von Ω hat ein eindeutig bestimmtes minimales Element. Als Nächstes definieren wir eine Abbildung $\Omega \to M$, $\alpha \mapsto f^\alpha$ und später werden wir Ω so groß wählen, dass diese Abbildung nicht injektiv sein kann. Im Augenblick spielt dies noch keine Rolle. Für $\alpha \in \Omega$ setzen wir

$$f^\alpha = f \left(\sup \{ f^\beta \mid \beta < \alpha \} \right)$$

Als Erstes müssen wir zeigen, dass f^α wohldefiniert ist, denn a priori ist nicht klar, dass die Schranke $\sup\{f^\beta \mid \beta < \alpha\}$ existiert. Wir zeigen mehr und behaupten die folgenden vier Aussagen für alle $\alpha, \beta \in \Omega$:

(a) f^α ist definiert.

(b) $\beta < \alpha$ impliziert $f^\beta \leq f^\alpha$.

(c) $f^\alpha \leq f(f^\alpha)$.

(d) Für alle $x \in M$ mit $f(x) = x$ gilt $f^\alpha \leq x$.

Der Beweis dieser Behauptung erfolgt durch Widerspruch. Angenommen, eine der obigen Aussagen wäre für ein $\alpha \in \Omega$ falsch. Dann gibt es ein minimales α mit dieser Eigenschaft und alle vier Aussagen gelten für alle γ mit $\gamma < \alpha$. Der Widerspruch ergibt sich, da wir zeigen, dass alle vier Aussagen auch für α gelten.

Zu (a): Alle f^β für $\beta < \alpha$ sind definiert und für $\beta \leq \gamma < \alpha$ gilt $f^\beta \leq f^\gamma$. Also ist $\{f^\beta \mid \beta < \alpha\}$ linear geordnet und damit eine gerichtete Teilmenge. Damit existiert $f^\alpha = f(\sup\{f^\beta \mid \beta < \alpha\})$, da M eine vollständige Halbordnung ist. Insbesondere ist $f^\perp = f(\perp)$. Zu (b): Es gilt $f^\alpha = f(\sup\{f^\beta \mid \beta < \alpha\})$ und $f^\beta \leq f(f^\beta)$ für alle $\beta < \alpha$. Also gilt für alle $\beta < \alpha$ aufgrund der Monotonie von f auch

$$f^\beta \leq f \left(f^\beta \right) \leq f \left(\sup \{ f^\beta \mid \beta < \alpha \} \right) = f^\alpha$$

Zu (c): Wir betrachten die folgende Rechnung.

$$
\begin{aligned}
f^\alpha &= f\left(\sup\left(\{\,f^\beta \mid \beta < \alpha\,\}\right)\right) && \text{nach Definition} \\
&\le f\left(\sup\{\,f\,(f^\beta)\mid \beta < \alpha\,\}\right) && \text{da } f^\beta \le f\,(f^\beta) \\
&\le f\left(f\left(\sup\{\,f^\beta \mid \beta < \alpha\,\}\right)\right) && \text{da } f \text{ monoton ist} \\
&= f\,(f^\alpha) && \text{nach Definition}
\end{aligned}
$$

Zu (d): Sei $x \in M$ mit $f(x) = x$. Wir wissen $f^\beta \le x$ für alle $\beta < \alpha$. Also gilt $\sup\{f^\beta \mid \beta < \alpha\} \le x$. Hieraus folgt

$$
f^\alpha = f\left(\sup\{\,f^\beta \mid \beta < \alpha\,\}\right) \le f(x) = x
$$

Wir haben damit die Behauptung gezeigt, dass alle vier Aussagen für alle $\alpha \in \Omega$ gelten.

Wir betrachten jetzt die Potenzmenge $\Omega = 2^M$ von M und versehen diese mit einer Wohlordnung. Hier benutzen wir den Wohlordnungssatz, der, wie oben erwähnt, zum Auswahlaxiom äquivalent ist. Potenzmengen haben stets eine gößere Kardinalität als die Ausgangsmenge. Es gibt also keine Injektion von Ω nach M. Das obige Verfahren angewendet auf die Menge Ω erzwingt nun die Existenz von $\alpha, \beta \in \Omega$ mit $f^\beta = f^\alpha$ und $\beta < \alpha$. Nun gilt $f^\beta \le \sup\{f^\beta \mid \beta < \alpha\}$ also $f(f^\beta) \le f^\alpha$ und insgesamt

$$
f^\beta \le f\,(f^\beta) \le f^\alpha = f^\beta
$$

Damit ist f^β ein Fixpunkt von f und nach der vierten Aussage ist f^β der kleinste Fixpunkt. $\qquad\qquad\square$

7.5 Verbände

Eine nichtleere Halbordnung $(V, \le)$ heißt *Verband*, wenn für alle x und y aus V sowohl eine kleinste obere Schranke $x \vee y$ als auch eine größte untere Schranke $x \wedge y$ existiert. Alle linearen nichtleeren Ordnungen sind Verbände. In Verbänden sind minimale und maximale Elemente eindeutig bestimmt, sofern sie existieren. Insbesondere besitzen alle endlichen Verbände ein kleinstes Element $\bot$ und ein maximales Element $\top$ und in ihnen haben alle Teilmengen ein Supremum und ein Infimum. Wir betrachten nun einige Beispiele.

Sei $(\mathbb{N}, |)$ die Menge der natürlichen Zahlen mit der Teilerrelation als Halbordnung definiert. Für $m, n \in \mathbb{N}$ ist die kleinste obere Schranke das kleinste gemeinsame Vielfache $\mathrm{kgV}(m, n)$ und die größte untere Schranke der größte gemeinsame Teiler $\mathrm{ggT}(m, n)$. Die Potenzmenge mit der Teilmengenbeziehung $(2^M, \subseteq)$ ist ein Verband. Die Vereinigung definiert das Supremum und der Durchschnitt liefert das Infimum. Ist $\{(V_i, \le) \mid i \in I\}$ eine Familie von Verbänden, so ist das kartesische Produkt $(\prod_{i \in I} V_i, \le)$ mit der komponentenweisen Ordnung ebenfalls ein Verband. Bis

auf Umbenennung der Elemente gibt es zehn Verbände mit höchstens 5 Elementen. Die Hasse-Diagramme dieser zehn Verbände sind wie folgt.

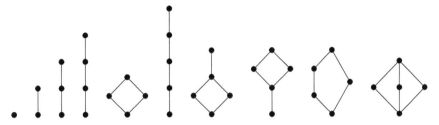

In jedem Verband $(V, \le)$ gelten die folgenden vier Rechenregeln:

$$(V1) \quad x \wedge x = x \qquad\qquad\qquad\qquad \text{(Idempotenz)}$$
$$x \vee x = x$$

$$(V2) \quad x \wedge y = y \wedge x \qquad\qquad\qquad \text{(Kommutativität)}$$
$$x \vee y = y \vee x$$

$$(V3) \quad x \wedge (y \wedge z) = (x \wedge y) \wedge z \qquad \text{(Assoziativität)}$$
$$x \vee (y \vee z) = (x \vee y) \vee z$$

$$(V4) \quad x \wedge (x \vee y) = x \vee (x \wedge y) = x \quad \text{(Absorption)}$$

Die Forderung der Idempotenz ist redundant, denn (V1) ergibt sich aus den beiden Absorptionsgesetzen (V4). Dies sieht man wie folgt. Mit $y = x \vee x$ erhalten wir:

$$x = x \vee (x \wedge y) = x \vee (x \wedge (x \vee x)) = x \vee x$$

Die Rechnung für $x = x \wedge x$ ist vollkommen analog. Verbände sind spezielle Halbordnungen; aber es sind zugleich auch algebraische Strukturen mit zwei inneren Verknüpfungen $\wedge$ und $\vee$. Innere Verknüpfungen, die (V1) bis (V4) (beziehungsweise äquivalent (V2) bis (V4)) erfüllen, charakterisieren Verbände, wie Satz 7.6 zeigt.

Satz 7.6. *Sei V eine nichtleere Menge mit zwei inneren Verknüpfungen $\vee$ und $\wedge$, die den drei Eigenschaften (V2) bis (V4) genügen. Definieren wir für $x, y \in V$ die Relation $x \le y$ durch $x = x \wedge y$, so ist $(V, \le)$ ein Verband; hier haben $\wedge$ und $\vee$ die Bedeutung von Infimum und Supremum.*

Beweis. Wir haben gesehen, dass wir aufgrund von (V4) die Idempotenz (V1) hinzunehmen dürfen. Wir zeigen zunächst, dass $\le$ eine partielle Ordnung definiert. Die Reflexivität $x \le x$ folgt aus der Idempotenz (V1). Die Antisymmetrie folgt aus der Kommutativität (V2), denn für $x \le y$ und $y \le x$ gilt $x = x \wedge y = y \wedge x = y$. Die Transitivität folgt aus der Assoziativität (V3), denn für $x \le y$ und $y \le z$ gilt $x = x \wedge y = x \wedge (y \wedge z) = (x \wedge y) \wedge z = x \wedge z$.

Wir zeigen, dass $x \wedge y$ die größte untere Schranke ist. Zunächst ist $(x \wedge y) \wedge x = x \wedge y$, also ist $x \wedge y$ eine untere Schranke für x und aus Symmetrie auch für y. Sei

jetzt $z \le x$ und $z \le y$, also $z = x \wedge z$ und $z = y \wedge z$. Zu zeigen ist $z = (x \wedge y) \wedge z$. Dies folgt aus $(x \wedge y) \wedge z = x \wedge (y \wedge z) = x \wedge z = z$.

Der Beweis bis hierhin verwendet nur die Axiome (V1) bis (V3), denn die Absorptionsgesetze (V4) wurden nur zur Herleitung von (V1) benutzt. Als Nächstes zeigen wir mit Hilfe von (V4) die Dualität

$$x = x \wedge y \Leftrightarrow y = x \vee y \qquad (7.1)$$

In der Tat, (V4) besagt $y = y \vee (y \wedge x)$. Also folgt aus $x = x \wedge y = y \wedge x$ die Beziehung $y = y \vee x = x \vee y$. Ist umgekehrt $y = x \vee y$, so folgt erneut mit (V4) die Beziehung $x = x \wedge (x \vee y) = x \wedge y$.

Die Dualität in Gleichung 7.1 liefert, dass $x \vee y$ genau dann die kleinste obere Schranke von x und y ist, wenn $x \wedge y$ die größte untere Schranke ist. Aber wir wissen bereits, dass $x \wedge y$ die größte untere Schranke von x und y ist. $\qquad \square$

Sei V ein Verband und $V' \subseteq V$. Dann heißt V' ein *Unterverband* von V, wenn V' bezüglich der inneren Verknüpfungen $\wedge$ und $\vee$ von V abgeschlossen ist. Nicht leere Ketten in Verbänden sind Unterverbände. Eine Teilmenge $V' \subseteq V$ kann sehr wohl bezüglich der von V induzierten Halbordnung einen Verband bilden, ohne Unterverband von V zu sein; solche Teilmengen heißen *Teilverbände*. Der Unterschied zwischen Unter- und Teilverbänden findet sich etwa in Beispiel 7.7.

Beispiel 7.7. Sei der Verband $(V, \le)$ mit $V = \{\bot, a, b, c, \top\}$ gegeben durch:

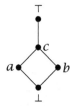

Hier ist $\{\bot, a, b, c\}$ ein Unterverband, aber $\{\bot, a, b, \top\}$ ist nur ein Teilverband, denn es gilt $a \vee b = c \notin \{\bot, a, b, \top\}$. $\qquad \Diamond$

7.6 Vollständige Verbände

Ein *vollständiger Verband* ist eine Halbordnung $(V, \le)$, in der jede Teilmenge ein Supremum hat. Insbesondere existiert sup $\emptyset = \bot$ und ein vollständiger Verband ist niemals leer und besitzt ein kleinstes Element $\bot$. Damit ist jeder vollständige Verband eine vollständige Halbordnung. Viele vollständige Halbordnungen sind keine vollständigen Verbände, da in vollständigen Halbordnungen nur gerichtete Teilmengen ein Supremum haben müssen. So ist etwa die Menge der endlichen und unendlichen Wörter Σ^∞ mit der Präfixordnung eine vollständige Halbordnung, aber kein vollständiger Verband, falls Σ mindestens zwei Buchstaben enthält. In einem vollständigen

Verband V hat jede Teilmenge auch ein Infimum, denn es gilt

$$\inf D = \sup\{x \in V \mid \forall y \in D : x \le y\}$$

Insbesondere ist ein vollständiger Verband ein Verband mit kleinstem und größtem Element. Jeder endliche Verband ist vollständig. Für eine beliebige Menge M ist der Potenzmengenverband $(2^M, \subseteq)$ vollständig.

Wir zeigen jetzt den wichtigen Fixpunktsatz von Knaster und Tarski (nach Bronisław Knaster, 1893–1980, und Alfred Tarski, 1901–1983) und benutzen hierfür Satz 7.5, der kleinste Fixpunkte in vollständigen Halbordnungen garantiert.

Satz 7.8 (Fixpunktsatz von Knaster und Tarski). *Es sei V ein vollständiger Verband und $f : V \to V$ eine monotone Abbildung. Dann bildet die Menge $P(f) = \{y \in V \mid f(y) = y\}$ der Fixpunkte einen vollständigen Teilverband. Insbesondere existieren eindeutig bestimmte kleinste und größte Fixpunkte.*

Beweis. Sei $Y \subseteq P(f)$. Zu zeigen ist, dass Y ein Supremum in $P(f)$ hat. Zu zeigen ist also, dass es einen eindeutig bestimmten kleinsten Fixpunkt in $P(f)$ gibt, der mindestens genauso groß wie alle $y \in Y$ ist. Betrachte hierfür $V_Y = \{x \in V \mid \sup Y \le x\}$. Dann ist V_Y ein vollständiger Verband mit kleinstem Element $\bot_Y = \sup Y$. Für $y \in Y$ und $x \in V_Y$ ist $y \le \sup Y \le x$, also auch $f(y) = y \le f(\sup Y) \le f(x)$, da f monoton ist. Damit gilt $\sup Y \le f(\sup Y) \le f(x)$, und f induziert eine monotone Abbildung von V_Y nach V_Y.

Als vollständiger Verband ist V_Y eine vollständige Halbordnung. Also existiert nach dem Satz 7.5 ein kleinster Fixpunkt $x_{f,Y} \in P(f)$ von f innerhalb von V_Y. Dies ist in Bezug auf $P(f)$ die kleinste obere Schranke von Y. $\square$

Sei V ein vollständiger Verband und $f : V \to V$ monoton. Dann ist der Teilverband $P(f) = \{y \in V \mid f(y) = y\}$ im Allgemeinen kein Unterverband von V. Betrachte hierfür etwa den vollständigen Verband $\{\bot, a, b, c, \top\}$ aus Beispiel 7.7. Gilt $f(c) = \top$ und lässt f alle anderen Elemente invariant, so ist f monoton und die Menge der Fixpunkte $P(f)$ ist der Teilverband $\{\bot, a, b, \top\}$.

7.7 Modulare und distributive Verbände

In jedem Verband V gelten die Distributivitätsungleichungen:

$$x \vee (y \wedge z) \le (x \vee y) \wedge (x \vee z)$$
$$x \wedge (y \vee z) \ge (x \wedge y) \vee (x \wedge z)$$

Unter der Nebenbedingung $x \le z$ folgt dann die Modularungleichung:

$$x \vee (y \wedge z) \le (x \vee y) \wedge z$$

Modulare und distributive Verbände sind dadurch definiert, dass die jeweiligen Ungleichungen durch Gleichheiten ersetzt werden können.

Ein Verband V heißt *distributiv*, falls er eine der beiden äquivalenten Bedingungen für alle $x, y, z \in V$ erfüllt:

(D) $x \vee (y \wedge z) = (x \vee y) \wedge (x \vee z)$

(D') $x \wedge (y \vee z) = (x \wedge y) \vee (x \wedge z)$

Die Äquivalenz der Bedingungen (D) und (D') ergibt sich aus der folgenden Überlegung. Es genügt, etwa (D) $\Rightarrow$ (D') zu zeigen, die Umkehrung liefert dann das Dualitätsprinzip. Betrachte hierfür $a, b, c \in V$ und setze $x = a \wedge b$, $y = a$ und $z = c$. Zu zeigen ist $a \wedge (b \vee c) = (a \wedge b) \vee (a \wedge c)$. Dies erkennen wir wie folgt:

$$
\begin{aligned}
(a \wedge b) \vee (a \wedge c) &= ((a \wedge b) \vee a) \wedge ((a \wedge b) \vee c) && \text{nach (D)} \\
&= a \wedge ((a \wedge b) \vee c) && \text{Absorption} \\
&= a \wedge (a \vee c) \wedge (b \vee c) && \text{nach (D)} \\
&= a \wedge (b \vee c) && \text{Absorption}
\end{aligned}
$$

Ein Verband V heißt *modular*, falls für alle $x, y, z \in V$ die folgende Implikation erfüllt ist:

(M) $x \leq z \Rightarrow x \vee (y \wedge z) = (x \vee y) \wedge z$

Jeder distributive Verband ist modular. Unter den zehn Verbänden mit höchstens fünf Elementen, die in Abschnitt 7.5 dargestellt sind, sind genau die ersten acht distributiv. Der neunte Verband ist nicht modular, während der zehnte Verband modular aber nicht distributiv ist. Da diese beiden Verbände eine zentrale Rolle im Rest dieses Abschnitts spielen, bezeichnen wir sie kurz mit N_5 und M_5. Wir halten noch fest, dass alle Verbände mit höchstens vier Elementen distributiv sind.

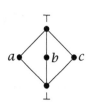 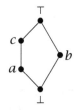

Verband M_5 Verband N_5

(modular, nicht distributiv)   (nicht modular)

Der Satz von Dedekind (Satz 7.9, Julius Wilhelm Richard Dedekind, 1831–1916) sagt aus, dass der Verband N_5 der Archetyp eines nichtmodularen Verbands ist.

Satz 7.9 (Dedekind). *Für jeden Verband V sind die folgenden Aussagen äquivalent:*

(a) V ist modular.

(b) Für alle $x, y, z \in V$ gilt die modulare Kürzungsregel:
Wenn $x \leq y$ und $z \wedge x = z \wedge y$ und $z \vee x = z \vee y$ gilt, dann ist $x = y$.

(c) V enthält keinen zu N_5 isomorphen Unterverband.

Beweis. Ist V modular, so gilt die Kürzungsregel aufgrund von:

$$x = x \vee (x \wedge z) = x \vee (z \wedge y) \overset{(M)}{=} (x \vee z) \wedge y = (z \vee y) \wedge y = y$$

Klar ist auch, dass in N_5 die Kürzungsregel nicht gilt. Betrachte hierzu das vorige Bild von N_5 mit $x = a$, $y = c$ und $z = b$. Sei jetzt V nicht modular. Dann gibt es $a, b, c \in V$ mit $a \leq c$ und

$$a \vee (b \wedge c) < (a \vee b) \wedge c \tag{7.2}$$

Wir zeigen, dass die durch $x = a \vee (b \wedge c)$, $y = b$, $z = (a \vee b) \wedge c$, $n = b \wedge c$ und $e = a \vee b$ definierten Elemente von V einen zu N_5 isomorphen Unterverband bilden:

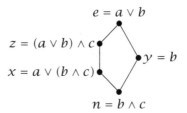

Es ist $n \leq x < z \leq e$ sowie $n \leq y \leq e$. Es kann nicht $x \leq y$ gelten, denn dann wäre $a \leq b$ und zusammen mit $a \leq c$ würde in Gleichung (7.2) links und rechts jeweils $b \wedge c$ stehen. Es kann nicht $y \leq z$ gelten, denn dann wäre $b \leq c$ und zusammen mit $a \leq c$ würde in 7.2 links und rechts jeweils $a \vee b$ stehen. Hieraus folgt $y \notin \{n, x, z, e\}$ und, dass y weder mit x noch mit z vergleichbar ist. Insbesondere sind die fünf Elemente n, x, y, z, e paarweise verschieden. Weiter gilt:

$$x \vee y = a \vee (b \wedge c) \vee b = a \vee b = e$$
$$z \wedge y = (a \vee b) \wedge c \wedge b = b \wedge c = n$$

Dies bedeutet einerseits, dass y weder mit x noch mit z vergleichbar ist und andererseits, dass $\{n, x, y, z, e\}$ bezüglich $\wedge$ und $\vee$ abgeschlossen und isomorph zu N_5 ist. Damit ist der Satz bewiesen. $\square$

Wir hatten gesehen, dass der Verband M_5 modular, aber nicht distributiv ist. Erscheinen weder M_5 noch N_5 als Unterverbände, so ist der Verband distributiv. Dies charakterisiert also distributive Verbände gemäß dem Satz 7.10 von Garrett Birkhoff (1911–1996). Birkhoff gilt als Begründer der *universellen Algebra* und war Sohn von George David Birkhoff (1884–1944), der unter anderem für seine 1933 entworfene *mathematische Theorie der Ästhetik* [6] auch außerhalb der Mathematik Bekanntheit erlangte.

Satz 7.10 (Birkhoff). *Für jeden Verband V sind die folgenden Aussagen äquivalent:*

(a) V ist distributiv.

(b) Für alle $x, y, z \in V$ gilt die Kürzungsregel:
   Wenn $x \wedge z = y \wedge z$ und $x \vee z = y \vee z$ gilt, dann ist $x = y$.

(c) V enthält weder einen zu M_5 noch zu N_5 isomorphen Unterverband.

Beweis. Ist V distributiv und $x \wedge z = y \wedge z$ sowie $x \vee z = y \vee z$, so gilt die Kürzungsregel aufgrund von:

$$
\begin{aligned}
x &= x \vee (x \wedge z) && \text{Absorption} \\
&= x \vee (y \wedge z) && \text{da } x \wedge z = y \wedge z \\
&= (x \vee y) \wedge (x \vee z) && \text{Distributivität} \\
&= (x \vee y) \wedge (y \vee z) && \text{da } x \vee z = y \vee z \\
&= y \vee (x \wedge z) && \text{Distributivität} \\
&= y \vee (y \wedge z) && \text{da } x \wedge z = y \wedge z \\
&= y && \text{Absorption}
\end{aligned}
$$

Weder in M_5 noch in N_5 gilt die (distributive) Kürzungsregel: Betrachte erneut das obige Bild von M_5 und N_5 mit $\{x, y, z\} = \{a, b, c\}$ für M_5 und, wie eben, mit $x = a$, $y = c$ und $z = b$ für N_5.

Sei V nicht distributiv. Wir nehmen an, dass V keinen zu N_5 isomorphen Unterverband enthält. Dann ist V modular nach Satz 7.9. Da V nicht distributiv ist, gibt es $a, b, c \in V$ mit $(a \wedge b) \vee (a \wedge c) < a \wedge (b \vee c)$. Wir definieren

$$
\begin{aligned}
n &= (a \wedge b) \vee (a \wedge c) \vee (b \wedge c) \\
e &= (a \vee b) \wedge (a \vee c) \wedge (b \vee c) \\
x &= (a \wedge e) \vee n \\
y &= (b \wedge e) \vee n \\
z &= (c \wedge e) \vee n
\end{aligned}
$$

Es gilt $n \leq x, y, z$. Mit (M) sehen wir

$$
\begin{aligned}
a \wedge n &= a \wedge ((a \wedge b) \vee (a \wedge c) \vee (b \wedge c)) \\
&= ((a \wedge b) \vee (a \wedge c)) \vee (a \wedge (b \wedge c)) \\
&= (a \wedge b) \vee (a \wedge c)
\end{aligned}
$$

Zusammen mit $a \wedge e = a \wedge (b \vee c)$ folgt nun $n < e$. Daraus erhalten wir $x, y, z \leq e$: Um beispielsweise $x \leq e$ einzusehen, beachte man dass $a \wedge e \leq e$ und $n < e$ gilt. Um zu zeigen, dass $\{n, e, x, y, z\}$ einen zu M_5 isomorphen Unterverband von V bilden, reicht es zu zeigen, dass $x \wedge y = x \wedge z = y \wedge z = n$ und $x \vee y = x \vee z = y \vee z = e$ ist. Wir zeigen dies exemplarisch für die Identität $x \wedge y = n$. Die anderen Fälle sind

analog. Es ist

$$
\begin{aligned}
x \wedge y &= ((a \wedge e) \vee n) \wedge ((b \wedge e) \vee n) && \text{Definition} \\
&= ((a \wedge e) \wedge ((b \wedge e) \vee n)) \vee n && \text{Modularität} \\
&= ((a \wedge e) \wedge ((b \vee n) \wedge e)) \vee n && \text{Modularität} \\
&= ((a \wedge e) \wedge e \wedge (b \vee n)) \vee n && \text{Kommutativität} \\
&= ((a \wedge e) \wedge (b \vee n)) \vee n && \text{Idempotenz} \\
&= (a \wedge (b \vee c) \wedge (b \vee (a \wedge c))) \vee n && \text{Absorption} \\
&= (a \wedge (b \vee ((b \vee c) \wedge (a \wedge c)))) \vee n && \text{Modularität} \\
&= (a \wedge (b \vee (a \wedge c))) \vee n && a \wedge c \leq c \leq b \vee c \\
&= (a \wedge b) \vee (a \wedge c) \vee n && \text{Modularität} \\
&= n && \text{Idempotenz}
\end{aligned}
$$

Damit ist der Satz bewiesen. ☐

Es sei V ein Verband. Ein Element $a \in V$ heißt *irreduzibel* (genauer $\vee$-*irreduzibel*), falls es kleinere Elemente gibt, aber a nicht das Supremum von zwei echt kleineren Elementen ist. Dies bedeutet, a ist kein minimales Element, und für alle $b, c \in V$ folgt aus $a = b \vee c$ schon $a = b$ oder $a = c$. Mit $\mathcal{J}(V)$ bezeichnen wir die Menge der irreduziblen Elemente von V.

Ist M eine Menge, so ist $(2^M, \subseteq)$ ein vollständiger und distributiver Verband, und die Menge der irreduziblen Elemente sind die einelementigen Teilmengen. Wir können also $\mathcal{J}(2^M)$ mit der Grundmenge M identifizieren. Wir sagen, dass V ein *Mengenverband* ist, wenn V isomorph zu einem Unterverband eines Potenzmengenverbandes $(2^M, \subseteq)$ ist. Jeder Mengenverband ist distributiv, denn Distributivität vererbt sich auf Unterverbände.

Jeder endliche Verband mit mehr als einem Element enthält irreduzible Elemente. Es gibt unendliche Verbände ohne irreduzible Elemente. Ein Beispiel hierfür ist $\mathbb{Z} \times \mathbb{Z}$ mit komponentenweisem Vergleich. In Ketten ist jedes von $\perp$ verschiedene Element irreduzibel. Im folgenden Hasse-Diagramm sind alle Elemente bis auf $\perp$ irreduzibel.

Im nächsten Beispiel sind die irreduziblen Elemente eingekreist.

Satz 7.11. *Sei V ein endlicher, distributiver Verband. Dann ist V ein Mengenverband vermöge der Zuordnung $\rho : V \to 2^{\mathcal{J}(V)}$, die für $a \in V$ wie folgt definiert ist:*

$$\rho(a) = \{x \in \mathcal{J}(V) \mid x \le a\}$$

Insbesondere ist ρ injektiv und erfüllt die Gleichungen $\rho(x \vee y) = \rho(x) \cup \rho(y)$ und $\rho(x \wedge y) = \rho(x) \cap \rho(y)$. Ferner gilt $\rho(\bot) = \varnothing$ und $\rho(\top) = \mathcal{J}(V)$.

Beweis. Die Abbildung ρ erfüllt $\rho(\bot) = \varnothing$ und $\rho(\top) = \mathcal{J}(V)$, und sie überführt die Ordnung $\le$ in Teilmengenbeziehungen $\subseteq$. Wir können jedes $a \in V$ als ein Supremum $a = \sup J$ mit $J = \{x \in V \mid \bot < x \le a\}$ darstellen. Falls $b \in J \setminus \mathcal{J}(V)$ existiert, so schreibe $b = c \vee d$ für echt kleinere Elemente $c < b$ und $d < b$. Insbesondere gilt $\bot \notin \{c, d\} \subseteq J$. Streichen wir in J das Element b, bilden also $J' = J \setminus \{b\}$, so gilt weiterhin $a = \sup J'$. Da V endlich ist, terminiert dieser Streichungsprozess und wir erhalten

$$a = \sup\{x \in \mathcal{J}(V) \mid x \le a\} = \sup \rho(a)$$

Insbesondere ist ρ injektiv. Zu zeigen ist für $a, b \in V$ nur noch $\rho(a \vee b) = \rho(a) \cup \rho(b)$ und $\rho(a \wedge b) = \rho(a) \cap \rho(b)$. Es gilt

$$\rho(a) \cap \rho(b) = \{x \in \mathcal{J}(V) \mid x \le a \text{ und } x \le b\} = \rho(a \wedge b)$$

Für die Vereinigung gilt zunächst

$$\rho(a) \cup \rho(b) = \{x \in \mathcal{J}(V) \mid x \le a \text{ oder } x \le b\} \subseteq \rho(a \vee b) \in \rho(V)$$

Sei umgekehrt $x \in \rho(a \vee b)$. Dann folgt $x = x \wedge (a \vee b) = (x \wedge a) \vee (x \wedge b)$, da V distributiv ist. Nun ist $x \in \mathcal{J}(V)$ und damit irreduzibel, also gilt $x = x \wedge a$ oder $x = x \wedge b$. Dies bedeutet $x \le a$ oder $x \le b$; damit erhalten wir $x \in \rho(a) \cup \rho(b)$. Dies liefert $\rho(a \vee b) = \rho(a) \cup \rho(b)$ und der Satz ist bewiesen. □

7.8 Boolesche Verbände

Es sei V ein Verband mit einem kleinsten Element $\bot$ und einem größten Element $\top$. Ist V endlich, so existieren $\bot$ und $\top$ automatisch. Zwei Elemente $x, y \in V$ heißen *komplementär* zueinander, falls $x \wedge y = \bot$ und $x \vee y = \top$ gilt. Die Elemente $\bot$ und $\top$ sind stets komplementär zueinander. Ist V eine Kette, so sind $\bot$ und $\top$ die einzigen Elemente, die komplementäre Elemente besitzen. Ein Verband V heißt *komplementär*, falls alle Elemente komplementäre Elemente besitzen. Ein Potenzmengenverband $(2^M, \subseteq)$ ist vollständig, distributiv und komplementär. Hier sind A und $M \setminus A$ komplementär zueinander. Ein Verband V mit einem kleinsten Element $\bot$ und einem größten Element $\top$ heißt *boolescher Verband*, wenn er distributiv und komplementär ist. Die booleschen Verbände sind nach George Boole (1815–1864) benannt, da sie auf dessen Logikkalküle von 1847 zurückgehen. Den Namen prägte Henry Maurice Sheffer (1882–1964) im Jahre 1913.

In einem distributiven Verband sind komplementäre Elemente eindeutig bestimmt, sofern sie existieren. Dies folgt aus der Kürzungsregel in Satz 7.10, kann aber natürlich auch direkt verifiziert werden. Denn seien $x, y_1, y_2 \in V$ mit $x \wedge y_i = \bot$ und $x \vee y_i = \top$ für $i = 1,2$. Dann gilt aufgrund der Distributivität:

$$y_1 = y_1 \wedge (x \vee y_2) = (y_1 \wedge x) \vee (y_1 \wedge y_2) = y_1 \wedge y_2$$

Dies zeigt $y_1 = y_1 \wedge y_2$ und vermöge Symmetrie erhalten wir $y_1 = y_2 = y_1 \wedge y_2$. Insbesondere gibt es in einem booleschen Verband zu jedem Element a genau ein Komplement. Wir verwenden hierfür im Folgenden die Bezeichnung $\overline{a}$. Aufgrund der Eindeutigkeit der Komplemente gilt $\overline{\overline{a}} = a$ und zusammen mit Dualitätsprinzip erhalten wir die Regeln von de Morgan.

$$\overline{a \wedge b} = \overline{a} \vee \overline{b} \qquad \text{und} \qquad \overline{a \vee b} = \overline{a} \wedge \overline{b}$$

Die Regeln sind nach Augustus de Morgan (1806–1871) benannt, waren aber spätestens seit dem Mittelalter durch das Handbuch der Logik *Summa Logicae* von Wilhelm von Occam (ca. 1288–1347) bekannt. Sein Name wird häufig mit dem *Sparsamkeitsprinzip* in der Scholastik, *Occams Rasiermesser*, in Verbindung gebracht.

Im Folgenden wollen wir die endlichen booleschen Verbände bestimmen und zeigen, dass sie zu Potenzmengenverbänden $(2^M, \subseteq)$ isomorph sind. Es ist üblich, die Elemente der Dimension 1 als *Atome* zu bezeichnen. Ein Atom ist also ein Element $a \neq \bot$ einer Halbordnung V mit kleinstem Element $\bot$, für welches kein $b \in V$ mit $\bot < b < a$ existiert.

Lemma 7.12. *Sei $(V, \leq)$ ein boolescher Verband und $a \in V$. Dann ist a genau dann irreduzibel, wenn a ein Atom ist.*

Beweis. Atome sind irreduzibel. Sei nun $\bot \neq b \in V$ kein Atom. Dann existiert ein $a \in V$ mit $\bot < a < b$, und es folgt:

$$a \vee (b \wedge \overline{a}) = (a \vee b) \wedge (a \vee \overline{a}) = b \wedge \top = b$$

Es reicht jetzt zu zeigen, dass $b \wedge \overline{a} < b$ gilt, denn dann ist b nicht irreduzibel. Angenommen, es wäre $b \wedge \overline{a} = b$. Nach der Regel von de Morgan gilt dann $\overline{b} \vee a = \overline{b}$, also auch $a \leq \overline{b}$. Damit folgt $a \leq \overline{b} \wedge b = \bot$ im Widerspruch zu $a \neq \bot$. Also ist $b \wedge \overline{a} < b$, und das Lemma ist bewiesen. ☐

Satz 7.13 (Stone). *Jeder endliche boolesche Verband V kann als ein Potenzmengenverband realisiert werden. Genauer gilt, dass $(V, \leq)$ und $(2^A, \subseteq)$ kanonisch isomorph sind, wenn A die Menge der Atome von V ist.*

Beweis. Nach dem Lemma 7.12 ist die Menge der irreduziblen Elemente $\mathcal{J}(V)$ genau die Menge A der Atome von V. Als boolescher Verband ist V distributiv und somit nach Satz 7.11 isomorph zu dem Mengenverband $(\rho(V), \subseteq)$ mit $\rho(V) = \{\rho(a) \mid a \in V\}$ und $\rho(a) = \{x \in \mathcal{J}(V) \mid x \leq a\}$. Da für ein einzelnes Atom a gerade $\rho(a) = \{a\}$

gilt, gehören alle einelementigen Mengen zu $\rho(V)$. Außerdem gilt $\rho(a_1 \vee \cdots \vee a_n) = \{a_1, \ldots, a_n\}$. Damit induziert die Inklusion $A \subseteq V$ einen kanonischen Isomorphismus zwischen den Verbänden $(2^A, \subseteq)$ und $(V, \leq)$. $\qquad\qquad\square$

Wir wissen also insbesondere: Es gibt genau dann einen endlichen booleschen Verband mit k Elementen, wenn $k = 2^n$ eine Zweierpotenz ist. Dieser Verband ist dann bis auf Isomorphie eindeutig bestimmt. Der Satz von Stone, auch bekannt als Darstellungssatz für boolesche Algebren, ist 1936 von Marshall Harvey Stone (1903–1989) entdeckt worden. Dieser Satz gilt unter Ausnutzung des Auswahlaxioms viel allgemeiner, wie wir im übernächsten Abschnitt sehen werden. Der dort geführte Beweis benutzt das technische Konzept von Ultrafiltern.

7.9 Boolesche Ringe

Die Potenzmenge 2^M einer Menge M ist nicht nur ein boolescher Verband, sondern ja auch die Menge der Abbildungen von M nach $\{0, 1\}$. Wir identifizieren hierfür eine Teilmenge $A \subseteq M$ mit ihrer *charakteristischen Abbildung* $\chi_A : M \to \{0, 1\}$, die die Elemente von A auf 1 schickt und die anderen auf 0. Lesen wir $\{0, 1\}$ als booleschen Verband $\mathbb{B}$ mit $0 < 1$, so ist der Verband 2^M gerade das kartesische Produkt $\mathbb{B}^M$. Wir können $\{0, 1\}$ aber auch als den Körper $\mathbb{F}_2 = \mathbb{Z}/2\mathbb{Z}$ interpretieren. Dann wird 2^M zu der $\mathbb{F}_2$-Algebra $\mathbb{F}_2^M$, wobei die Addition und die Multiplikation komponentenweise erklärt sind. Insbesondere ist $2^M = \mathbb{F}_2^M$ ein kommutativer Ring mit $2x = 0$ und idempotenter Multiplikation $x^2 = x$. Wenn wir die Elemente in diesem Ring wieder als Teilmengen von M lesen, ergibt sich die Multiplikation als Durchschnitt

$$A \cdot B = A \cap B$$

und die Addition wird zur *symmetrischen Differenz*

$$A + B = A \bigtriangleup B = (A \cup B) \setminus (A \cap B)$$

Ein Unterring von $\mathbb{F}_2^M$, genauer von $(2^M, \bigtriangleup, \cap, \varnothing, M)$, wird auch *Mengenring* genannt. Ein Ring $R = (R, +, \cdot, 0, 1)$ heißt *boolescher Ring*, falls $x^2 = x$ für alle $x \in R$ gilt. Wir werden in diesem Abschnitt zeigen, dass boolesche Verbände und boolesche Ringe äquivalente Begriffe sind. Im darauffolgenden Abschnitt beweisen wir (mit dem Konzept von Ultrafiltern) den allgemeinen Darstellungssatz von Stone, der besagt, dass die booleschen Ringe genau die Mengenringe sind. Natürlich sind alle Mengenringe boolesch. Diese Richtung ist also trivial.

Proposition 7.14. *Sei R ein boolescher Ring. Dann ist R kommutativ und es gilt $2x = 0$ für alle $x \in R$. Anders ausgedrückt, boolesche Ringe sind genau die kommutativen $\mathbb{F}_2$-Algebren mit idempotenter Multiplikation.*

Beweis. Wir zeigen zuerst, dass $2x = 0$ für jedes $x \in R$ gilt. Aus der booleschen Eigenschaft folgt

$$2x = (2x)^2 = 4x^2 = 4x$$

Damit gilt wie behauptet $0 = 4x - 2x = 2x$. Nun zeigen wir $xy = yx$. Wir betrachten $(x + y)^2$ und rechnen wie folgt:

$$x + y = (x + y)^2 = x^2 + xy + yx + y^2 = x + xy + yx + y$$

Also ist $xy + yx = 0$ und nach Addition von yx folgt jetzt $xy = yx$. □

Ist R ein boolescher Ring, so definieren wir $x \leq y$ durch die Bedingung $x = xy$. Die Relation $\leq$ ist eine partielle Ordnung. Sie ist reflexiv, da $x^2 = x$ gilt. Sie ist antisymmetrisch, da R kommutativ ist. Schließlich erhalten wir die Transitivität durch $xz = xyz = xy = x$ für $x = xy$ und $y = yz$.

Proposition 7.15. *Sei $R(+, \cdot, 0, 1)$ ein boolescher Ring und sei $x \leq y$ durch $x = xy$ definiert. Dann ist $(R, \leq)$ ein boolescher Verband. In dem Verband gilt:*

(a) $\perp = 0$ und $\top = 1$

(b) $x \wedge y = xy$

(c) $x \vee y = x + y + xy$

(d) $\overline{x} = 1 + x$

Beweis. Wir hatten schon gesehen, dass $(R, \leq)$ eine Halbordnung ist. Klar ist auch $\perp = 0$ und $\top = 1$. Zu (b): Wegen $xyx = xy = xyy$ ist xy eine untere Schranke für $\{x, y\}$. Sei jetzt z wegen $z = zx$ und $z = zy$ eine andere untere Schranke für $\{x, y\}$. Dann gilt $z = zy = zxy$, also gilt $z \leq xy$ und damit ist $x \wedge y = xy$ die größte untere Schranke. Zu (c): Es gilt $x(x + y + xy) = x^2 + xy + x^2y = x + 2xy = x$ und analog $y(x + y + xy) = y$, also ist $x + y + xy$ eine obere Schranke von $\{x, y\}$. Ist z eine andere obere Schranke, so gilt $x = xz$ und $y = yz$. Dann folgt $(x + y + xy)z = xz + yz + xyz = x + y + xy$, also ist $x + y + xy \leq z$ und $x \vee y = x + y + xy$. Zu (d): Wegen $x + (1 + x) = 1$ und $x(1 + x) = 0$ sind x und $1 + x$ komplementär zueinander. Dies zeigt, dass $(R, \leq)$ ein komplementärer Verband ist. Das Distributivgesetz ist erfüllt:

$$(x \vee y) \wedge z = (x + y + xy)z = xz + yz + xyz = (x \wedge z) \vee (y \wedge z)$$

Also ist $(R, \leq)$ ein boolescher Verband. □

Proposition 7.16. *Sei $(V, \leq)$ ein boolescher Verband. Setze $0 = \perp$ und $1 = \top$ und definiere Addition und Multiplikation durch:*

(a) $x + y = (x \wedge \overline{y}) \vee (\overline{x} \wedge y)$

(b) $x \cdot y = x \wedge y$

Dann ist $(V, +, \cdot, 0, 1)$ ein boolescher Ring und $x \leq y$ äquivalent mit $x = xy$.

Beweis. Offenbar gilt $x + x = 0$, $x + 0 = x$, $x \cdot x = x$ und $x \cdot 1 = x$. Ferner ist $x + y = y + x$ und $xy = yx$ sowie $(xy)z = x(yz)$. Wir zeigen als nächstes die Assoziativität der Addition. Dies gilt wegen:

$$(x + y) + z = (x \wedge \overline{y} \wedge \overline{z}) \vee (\overline{x} \wedge y \wedge \overline{z}) \vee (\overline{x} \wedge \overline{y} \wedge z) \vee (x \wedge y \wedge z)$$
$$= x + (y + z)$$

Die Rechnung kann durch ein Venn-Diagramm veranschaulicht werden, indem man sich x, y und z als Mengen vorstellt, bei denen $\vee$ die Vereinigung und $+$ die symmetrische Differenz ist. Es verbleibt noch, das Distributivgesetz $(x + y)z = xz + yz$ zu zeigen. Es ist $(x + y)z = ((x \wedge \overline{y}) \vee (\overline{x} \wedge y)) \wedge z = (x \wedge \overline{y} \wedge z) \vee (\overline{x} \wedge y \wedge z)$. Außerdem ergibt sich

$$xz + yz = (x \wedge z \wedge \overline{y \wedge z}) \vee (y \wedge z \wedge \overline{x \wedge z})$$
$$= (x \wedge z \wedge (\overline{y} \vee \overline{z})) \vee (y \wedge z \wedge (\overline{x} \vee \overline{z}))$$
$$= (x \wedge \overline{y} \wedge z) \vee (\overline{x} \wedge y \wedge z)$$

Dies beweist $(x + y)z = xz + yz$ und damit die Proposition. □

Die Konzepte „boolescher Verband", „boolesche Algebra" und „boolescher Ring" sind also vollkommen äquivalent. Wir sprechen von einem Verband, wenn wir die Halbordnung hervorheben wollen, und leiten hieraus die inneren Verknüpfungen $\wedge$ und $\vee$ ab. Wir sprechen dann von einer booleschen Algebra. Starten wir mit einer booleschen Algebra mit den inneren Verknüpfungen $\wedge$ und $\vee$, dann erhalten wir einen booleschen Verband, indem wir $x \le y$ durch $x = x \wedge y$ definieren. Wenn wir in der booleschen Algebra von der „Oder-Funktion" $\vee$ zu dem „exklusiven Oder" $+$ übergehen, so erhalten wir mittels $x + y = (x \wedge \overline{y}) \vee (\overline{x} \wedge y)$ und $xy = x \wedge y$ gerade einen booleschen Ring. Schließlich führt uns der Ring mit der Festlegung $x \le y \Leftrightarrow x = xy$ zurück zum Ausgangsverband, bzw. zur Ausgangsalgebra. Dieser Zusammenhang liefert eine zweite Version des Satzes 7.11 von Stone.

Satz 7.17. *Jeder endliche boolesche Ring R ist isomorph zu einem Potenzmengenring 2^M.*

Lässt man die Endlichkeit weg, so erhält man den allgemeinen Darstellungssatz von Stone, der im nächsten Abschnitt behandelt wird.

7.10 Der allgemeine Darstellungssatz von Stone

Eine *Mengenalgebra* ist ein Unterverband V in einem Potenzmengenverband $(2^M, \subseteq)$ mit $\emptyset, M \in V$ und V enthält für alle $A, B \in V$, neben $A \cup B$ und $A \cap B$ auch das Komplement $M \setminus A$.

Beispiel 7.18. Sei X ein topologischer Raum, dann bildet die Familie der Mengen, die gleichzeitig offen und abgeschlossen sind, eine Mengenalgebra. Mit einem topologischen Raum meint man eine Menge X zusammen mit einer *Topologie*. Dies ist

eine Familie von Teilmengen von X, die abgeschlossen gegenüber beliebiger Vereinigung und endlichem Durchschnitt ist. Die Teilmengen von X, die zur Topologie gehören, nennt man *offen*, deren Komplemente nennt man *abgeschlossen*. Die Mengen $\emptyset$ und X sind gleichzeitig offen und abgeschlossen.

Sei Σ ein endliches Alphabet, dann ist die in der Informatik häufig anzutreffende Familie der regulären Sprachen (auch bekannt als Typ-3 Sprachen) über Σ eine Mengenalgebra. $\diamond$

Wir beweisen in diesem Abschnitt den allgemeinen Darstellungssatz von Stone, der besagt, dass jeder boolesche Ring isomorph zu einem Mengenring ist. Dies liefert dann automatisch die Aussage, dass jeder boolesche Verband isomorph zu einer Mengenalgebra ist. Der Beweis benutzt das Auswahlaxiom und basiert auf der Existenz von Ultrafiltern. Mit dem Konzept der Ultrafilter ist der Beweis des Darstellungssatzes sehr einfach. Die Schwierigkeit besteht nun darin, sich mit Ultrafiltern vertraut zu machen und deren Existenz „anzunehmen". Der Begriff *Filter* wurde von Henri Paul Cartan (1904–2008) geprägt, der auch Gründungsmitglied der Mathematikergruppe Nicolas Bourbaki war. Diese Gruppe begann ab 1934 mit der Veröffentlichung von grundlegenden Lehrbüchern der Mathematik, den *Éléments de mathématique*, und nahm damit entscheidenden Einfluss auf die moderne Entwicklung dieser Wissenschaft.

Filter haben in Halbordnungen $(R, \le)$ die folgende anschauliche Interpretation. Es ist eine nichtleere Teilmenge $F \subseteq R$, in der genügend große, aber nicht alle, Elemente „ausgefiltert" werden, sie bleiben in F „hängen". Insbesondere gilt $F \ne R$, da der Filter einiges durchlassen soll. Ist ferner x in F und y größer als x, so gilt auch $y \in F$. Sind x und y in F, so gibt es einen gemeinsamen Grund hierfür, dies ist ein $z \in F$ mit $z \le x$ und $z \le y$. Hat die Halbordnung ein kleinstes Element $\bot$, so gilt wegen $F \ne R$ also $\bot \notin F$. Einige Autoren verzichten auf die Forderung $F \ne R$ und sprechen von *eigentlichen Filtern*, wenn $F \ne R$ gilt.

Im Folgenden benutzen wir Filter nur in booleschen Verbänden und übersetzen die obige anschauliche Definition für diesen Spezialfall gleich in die äquivalente Sprache der booleschen Ringe. Hierdurch werden die Formeln etwas kompakter. Statt von einem Verband R gehen wir von einem booleschen Ring R aus. Sei also $R = (R, +, \cdot, 0, 1)$ ein boolescher Ring und wie üblich $x \le y$ genau dann, wenn $x = xy$. Eine Teilmenge $F \subseteq R$ heißt jetzt *Filter*, falls die folgenden vier Bedingungen erfüllt sind:

(a) $F \ne \emptyset$

(b) $0 \notin F$

(c) $\forall x \in F \; \forall y \in R : x = xy \Rightarrow y \in F$

(d) $\forall x, y \in F : xy \in F$

Aus der vorletzten Forderung folgt insbesondere $1 \in F$. Ist $0 \ne x \in R$ ein Element, so bildet die Menge der Ringelemente $F_x = \{ y \in R \mid x \le y \}$ einen Filter, einen so-

genannten *Hauptfilter*. Ist $\{F_i \mid i \in I\}$ eine Familie von Filtern für eine linear geordnete Indexmenge I und gilt $F_i \subseteq F_j$ für alle $i \leq j$, so ist offenbar deren Vereinigung $\bigcup\{F_i \mid i \in I\}$ erneut ein Filter. In der Menge der Filter von R haben also Ketten $\{F_i \mid i \in I\}$ ein Supremum $\bigcup\{F_i \mid i \in I\}$ und nach dem Lemma von Zorn (welches, wie der Wohlordnungssatz, äquivalent zum Auswahlaxiom ist) liegt damit jeder Filter in einem maximalen Filter. Dies ist ein Filter $U \subseteq R$ mit der Eigenschaft, dass jeder Filter F mit $U \subseteq F \subseteq R$ schon gleich U ist. Die maximalen Filter nennen wir *Ultrafilter*.

Kein Filter $F \subseteq R$ kann sowohl x als auch $\overline{x} = 1 + x$ enthalten, denn dann müsste auch $0 = x(1 + x)$ in F liegen, was ausgeschlossen ist. Diese Beobachtung liefert schon die Charakterisierung von Ultrafiltern.

Lemma 7.19. *Ein Filter $F \subseteq R$ ist genau dann ein Ultrafilter, wenn für jedes $x \in R$ entweder $x \in F$ oder $1 + x \in F$ gilt.*

Beweis. Sei $x \in R$ und $F \subseteq R$ ein Filter. Der Filter F kann nicht sowohl x als auch $1 + x$ enthalten. Angenommen, F enthält weder x noch $1 + x$. Wir zeigen, dass F nicht maximal ist. Definiere hierfür die Menge $F' = \{z \in R \mid \exists y \in F: xy = xyz\}$. Es gilt $F \subseteq F'$ und $x \in F'$, da F nicht leer ist. Damit ist $F \neq F'$. Zu zeigen ist daher nur, dass F' ein Filter ist. Angenommen, es wäre $0 \in F'$. Dann ist $xy = xy0 = 0$ für ein $y \in F$ und damit $y(1 + x) = y$. Also ist $y \leq 1 + x \in F$. Dies war ausgeschlossen; und daher gilt $0 \notin F'$. Für $z \in F'$ und $z = zz'$ gilt für ein $y \in F$ die Gleichung $xy = xyz = xy(zz') = (xyz)z' = xyz'$, also $z' \in F'$. Damit ist F' ein Filter. $\square$

Die Menge der Ultrafilter $\mathcal{U}$ ist also genau die Menge der Filter, die für jedes $x \in R$ entweder x oder $1 + x$ enthalten. Im nächsten Schritt ordnen wir jedem $a \in R$ eine Menge $\rho(a) \subseteq \mathcal{U}$ von Ultrafiltern zu. Wir setzen

$$\rho(a) = \{U \in \mathcal{U} \mid a \in U\}$$

Dies liefert jetzt den allgemeinen Darstellungssatz.

Satz 7.20 (Stone). *Sei R ein boolescher Ring und $\mathcal{U}$ die Menge der Ultrafilter von R. Dann bettet die Zuordnung $a \mapsto \rho(a) = \{U \in \mathcal{U} \mid a \in U\}$ den Ring R als Unterring in das kartesische Produkt $2^{\mathcal{U}} = \mathbb{B}^{\mathcal{U}}$ ein. Insbesondere ist R ein Mengenring. Also ist jeder boolesche Verband isomorph zu einer Mengenalgebra.*

Beweis. Klar ist $\rho(0) = \varnothing$ und $\rho(1) = \mathcal{U}$. Wir zeigen als nächstes, dass ρ injektiv ist. Betrachte $a, b \in R$ mit $a \neq b$, dann können wir aus Symmetriegründen $a \neq ab$ annehmen, denn entweder ist $a \neq ab$ oder $b \neq ba = ab$. Hieraus folgt $a(1 + ab) = a + ab \neq 0$. Also gibt es einen Ultrafilter U_c, der den von $c = a(1 + ab)$ erzeugten Hauptfilter umfasst. Der Ultrafilter U_c enthält a und $1 + ab$. Er kann aber nicht b enthalten, denn sonst wäre $ab \in U_c$ im Widerspruch zu $1 + ab \in U_c$. Zu zeigen bleibt $\rho(ab) = \rho(a) \cap \rho(b)$ und $\rho(a + b) = \rho(a) \triangle \rho(b)$.

Die Ultrafilter, die ab enthalten sind genau diejenigen, die a und b enthalten. Die Gleichung $\rho(ab) = \rho(a) \cap \rho(b)$ ist daher erfüllt. Betrachte jetzt einen Ultrafilter U

mit $a + b \in U$. Angenommen, U enthielte weder a noch b. Dann gilt $1 + a, 1 + b \in U$, da ja ein Ultrafilter immer genau eines von zueinander komplementären Elementen enthält. Also gilt $(a + b)(1 + a)(1 + b) \in \mathcal{U}$. Dies ist nicht möglich, da $(a + b)(1 + a + b + ab) = a + a + ab + ab + b + ab + b + ab = 0$. Dies zeigt die Inklusion $\rho(a + b) \subseteq \rho(a) \cup \rho(b)$. Angenommen, U enthielte sowohl a als auch b. Dann gilt $ab \in U$. Dies ist unmöglich wegen $(a + b)ab = ab + ab = 0$. Also gehört U zur symmetrischen Differenz von $\rho(a)$ und $\rho(b)$ und wir schließen $\rho(a + b) \subseteq \rho(a) \triangle \rho(b)$. Für die umgekehrte Richtung starten wir mit einem Ultrafilter U mit $a, (1 + b) \in U$. Zu zeigen ist nur noch $a + b \in U$. Im anderen Fall wären $a, (1 + b), 1 + a + b \in U$. Dies ist jedoch unmöglich wegen $a(1 + b)(1 + a + b) = (a + ab)(1 + a + b) = a + a + ab + ab + ab + ab = 0$. Insgesamt erhalten wir $\rho(a + b) = \rho(a) \triangle \rho(b)$ und damit die Behauptung. □

Wir bemerken noch, dass Satz 7.13 für endliche boolesche Ringe (oder äquivalent für endliche boolesche Verbände) in der Tat ein Spezialfall von Satz 7.20 ist. Es reicht, sich zu überlegen, dass die Ultrafilter in endlichen booleschen Ringen genau von Atomen erzeugten Hauptfiltern entsprechen.

Der in Beispiel 7.18 hergestellte Bezug zur Topologie kann durch den Satz 7.20 weiter präzisiert werden. Für interessierte Leser, die mit den topologischen Grundbegriffen vertraut sind, können wir die wesentliche Idee erklären: Jeder Ultrafilter über R lässt sich als Abbildung in $\{0, 1\}^R$ auffassen. Das kartesische Produkt $\{0, 1\}^R$ wird mit der Produkttopologie über die diskreten Mengen $\{0, 1\}$ versehen. Die offenen Mengen sind dann beliebige Vereinigungen von Mengen der Form $N(f_S, S)$, wobei $S \subseteq R$ endlich und $f_S \in 2^S$ gilt. Wir setzen dann $N(f_S, S) = \{f \in 2^R \mid \forall s \in S: f(s) = f_S(s)\}$. Die Mengen $N(f_S, S)$ sind offen und abgeschlossen. Der Raum $2^R = \{0, 1\}^R$ bildet dann einen total unzusammenhängenden kompakten Hausdorff-Raum. Unter Ausnutzung der Kompaktheit erkennt man, dass die Mengen, die zugleich offen und abgeschlossen sind, sich als endliche Vereinigung von $N(f_S, S)$ schreiben lassen, wobei S fest gewählt werden kann. Wir wissen, dass die Menge der Ultrafilter $\mathcal{U}$ in 2^R enthalten ist. Die Eigenschaft nicht Ultrafilter zu sein, erkennen wir an maximal drei Komponenten, etwa durch $f(x) = f(y) = 1$, aber $f(xy) = 0$ oder $f(0) = 1$ und so fort. Daher ist das Komplement von $\mathcal{U}$ offen und $\mathcal{U}$ in 2^R abgeschlossen. Also ist $\mathcal{U}$ selbst ein total unzusammenhängender kompakter Hausdorff-Raum. Beispiel 7.18 und der Satz 7.20 sagen uns in dieser Interpretation, dass die booleschen Algebren genau den Mengenalgebren der offen-und-abgeschlossenen Mengen in total unzusammenhängenden kompakten Hausdorff-Räumen entsprechen. Die angesprochenen Begriffe und Sätze aus der Topologie findet man in Lehrbüchern wie etwa [32] oder [9].

Aufgaben

7.1. Die Kettenbedingung in dieser Übungsaufgabe kommt in der Algebra häufig vor. Sie ist benannt nach Marie Ennemond Camille Jordan (1838–1922) und Otto Ludwig Hölder (1859–1937). Sei $(M, \leq)$ eine Halbordnung mit kleinstem Element, in der jedes Element eine endliche Dimension habe. Dann sind folgende Aussagen äquivalent:

(a) Je zwei maximale Ketten mit den selben Endpunkten haben die gleiche Länge (Jordan-Hölder'sche Kettenbedingung).

(b) Für alle Elemente $a, b \in M$ gilt: Ist b oberer Nachbar von a, so gilt für die Dimensionen $\dim(b) = \dim(a) + 1$.

7.2. In den Bezeichnungen von Abschnitt 7.3 und Satz 7.4 sei $c \in \mathcal{F}$ eine partiell definierte Funktion von Σ nach Σ und $w = \textbf{while } b \textbf{ do } c \textbf{ od}$ eine while-Schleife. Wir wissen, dass die partiell definierte Funktion $w \in \mathcal{F}$ der kleinste Fixpunkt des Operators $\Gamma_{b,c}$ ist.

(a) Es sei $\bot$ die total undefinierte Funktion. Gilt $\Gamma_{b,c}(\bot) = \bot$?

(b) Wählen Sie b und c so, dass alle $f \in \mathcal{F}$ Fixpunkte des Operators $\Gamma_{b,c}$ sind. Wie sieht der Definitionsbereich von $\Gamma_{b,c}(\bot)$ in diesem Fall aus?

(c) Charakterisieren Sie sowohl $w \sqsubseteq \text{id}$ als auch $w \sqsubseteq \textbf{while } b' \textbf{ do } c \textbf{ od}$.

(d) Sei $c \in \mathcal{F}$ überall definiert. Zeigen Sie, dass $\Gamma_{b,c}$ genau dann mehrere Fixpunkte hat, wenn w nicht überall terminiert.

(e) Zeigen Sie, dass $\Gamma_{b,c}$ möglicherweise nur einen Fixpunkt hat, obwohl w nicht überall terminiert.

Hinweis: Nach Aufgabe 7.2. (d) ist $c \in \mathcal{F}$ nicht überall definiert.

7.3. Zeigen Sie die Existenz (unendlicher) boolescher Verbände, die zu keinem Potenzmengenverband isomorph sind.

7.4. Zeigen Sie die Existenz von unendlichen Verbänden (mit kleinstem Element), die keine irreduziblen Elemente haben.

7.5. Zeigen Sie die Existenz von unendlichen Mengenverbänden, die keine irreduziblen Elemente haben.

7.6. Sei M eine nichtleere Menge und 2^M die Potenzmenge von M. Zeigen Sie, dass wir keinen Ring erhalten, wenn wir Addition und Multiplikation definieren durch $A + B = A \cup B$ und $A \cdot B = A \cap B$.

7.7. Bestimmen Sie die Hasse-Diagramme aller Verbände mit 5 Elementen.

7.8. Zeigen Sie, dass der Verband M_5 nicht distributiv, aber modular und komplementär ist.

Verband M_5

Zusammenfassung

Begriffe

- Halbordnung, partielle Ordnung
- lineare/totale Ordnung
- wohlfundiert
- minimales, maximales Element
- Wohlordnung
- Nachbar
- Hasse-Diagramm
- Kette
- Dimension $\dim(x)$
- Verfeinerung
- topologische Sortierung
- gerichtete Teilmenge
- vollständige Halbordnung
- kleinstes Element $\bot$
- größtes Element $\top$
- monotone Abbildung
- stetige Abbildung

- Fixpunkt
- denotationale Semantik
- Verband
- Unterverband, Teilverband
- vollständiger Verband
- modularer Verband
- distributiver Verband
- Verbände M_5 und N_5
- irreduzible Elemente $\mathcal{J}(V)$
- Mengenverband
- komplementäre Elemente boolescher Verband
- Atom
- Mengenring
- boolescher Ring
- Mengenalgebra, Potenzmengenring
- Filter, Hauptfilter, Ultrafilter

Methoden und Resultate

- Jede abzählbare Halbordnung lässt sich injektiv und monoton in $\mathbb{Q}$ einbetten.
- Fixpunktsatz von Kleene: Jede stetige Abbildung f über einer vollständigen Halbordnung hat $\sup\{f^i(\bot) \mid i \geq 0\}$ als eindeutigen kleinsten Fixpunkt.
- Semantik von while-Schleifen als Fixpunkt des Operators $\Gamma_{b,c}$
- Jede monotone Abbildung über einer vollständigen Halbordnung hat einen eindeutigen kleinsten Fixpunkt.

- Die Rechenregeln (V2) bis (V4) charakterisieren Verbände.
- Fixpunktsatz von Knaster-Tarski: Sei f eine monotone Abbildung über einem vollständigen Verband. Dann sind die Fixpunkte von f ein vollständiger Teilverband.
- Dedekind: V modularer Verband $\Leftrightarrow$ V hat keinen Unterverband N_5
- Birkhoff: V distributiver Verband $\Leftrightarrow$ V hat keine Unterverbände M_5, N_5
- V ist endlicher, distributiver Verband $\Leftrightarrow$ V ist endlicher Mengenverband
- In jedem booleschen Verband gilt: irreduzibel $\Leftrightarrow$ Atom
- boolescher Ring = boolescher Verband
- Filter $F \subseteq R$ Ultrafilter $\Leftrightarrow$ $\forall x \in F$: entweder $x \in F$ oder $1 + x \in F$
- Satz von Stone: Jeder Boolesche Verband ist isomorph zu einer Mengenalgebra.

8 Boolesche Funktionen und Schaltkreise

Wir machen nun einen Ausflug in die Schaltkreistheorie. Zunächst beweisen wir ein fundamentales Resultat von Shannon aus dem Jahr 1949. Es besagt, dass n-stellige boolesche Funktionen durch Schaltkreise realisiert werden können, deren Größe mit der Ordnung $\Theta(2^n/n)$ wächst. Danach beweisen wir die schärfere Abschätzung von Oleg Borisovich Lupanov (1932–2006), nach der $2^n/n + o(2^n/n)$ Gatter genügen, um beliebige n-stellige boolesche Funktionen zu berechnen. Diese Schranke ist asymptotisch optimal.

Mit $\mathbb{B} = \{0, 1\}$ meinen wir den booleschen Verband mit $0 < 1$ und Operationen $\vee$ für das logische Oder, $\wedge$ für das logische Und und die Komplementbildung $\overline{x}$. Das komplementäre Element $\overline{x}$ (die Negation) von x kann als $1 - x$ geschrieben werden. Die Elemente in $\mathbb{B}$ interpretieren wir auch als *Wahrheitswerte*. Das kartesische Produkt $\mathbb{B}^n$ ist ein boolescher Verband, wobei die Operationen $\vee$ und $\wedge$ komponentenweise erklärt sind. Wir wissen bereits aus dem Darstellungssatz von Stone, dass die Verbände $\mathbb{B}^n$ mit $n \in \mathbb{N}$ alle endlichen booleschen Verbände (bis auf Isomorphie) eindeutig beschreiben. Gleichzeitig können wir ein $z \in \mathbb{B}^n$ als *Bit-Folge* $(z_1, \ldots, z_n)$ mit $z_i \in \{0, 1\}$ auffassen und diese wiederum als Binärdarstellung der Zahl $\sum_{i=1}^n z_i 2^{i-1} \in \{0, \ldots, 2^n - 1\}$ interpretieren. Die Hasse-Diagramme für $\mathbb{B}^1, \mathbb{B}^2$ und $\mathbb{B}^3$ stellen also jeweils die Zahlenbereiche $\{0, 1\}$, $\{0, 1, 2, 3\}$ und $\{0, \ldots, 7\}$ dar.

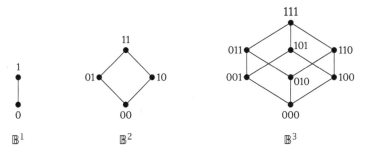

Unter einer *n-stelligen booleschen Funktion* wird eine Abbildung $f\colon \mathbb{B}^n \to \mathbb{B}$ verstanden. Wir können jede Funktion $f\colon \mathbb{B}^n \to \mathbb{B}$ auch als Eigenschaft der Zahlen zwischen 0 und $2^n - 1$ interpretieren. Für $n = 0$ sind diese Eigenschaften trivial und entsprechen den beiden Konstanten $\bot$ und $\top$. Im Folgenden nehmen wir $n \geq 1$ an. Schon für moderate n ist die Anzahl dieser Eigenschaften gewaltig; sie ist 2^{2^n}. Für $n = 8$ hat die Zahl 2^{256} schon 77 Dezimalstellen.

Schaltkreise nehmen eine Bit-Folge $(z_1, \ldots, z_n) \in \mathbb{B}^n$ als Eingabe und werten diese zu einem Wahrheitswert in $\mathbb{B}$ aus. Sie sind also geeignet, Eigenschaften von natürlichen Zahlen in dem Bereich von 0 bis $2^n - 1$ zu überprüfen. Die algorithmische Schwierigkeit einer Eigenschaft kann dann in der Größe der entsprechenden Schaltkreise gemessen werden. Wir werden sehen, dass sich alle n-stelligen booleschen

Funktionen durch Schaltkreise der Größe $\mathcal{O}(2^n/n)$ darstellen lassen und dass „fast alle" diese exponentielle Größe erfordern.

Diese sehr scharfe „generische" Schranke nützt leider für konkrete Funktionen nicht viel. Obwohl fast alle booleschen Funktionen extrem schwierig sind, ist keine einzige konkrete Familie von booleschen Funktionen f_n bekannt, deren Schaltkreisgröße mindestens quadratisch mit n wächst.

Im Folgenden bezeichnen wir allgemeiner alle Abbildungen von der Form $f\colon \mathbb{B}^n \to \mathbb{B}^m$ als boolesche Funktionen. Die formale Definition eines Schaltkreises über einer Menge von *booleschen Variablen* $\{b_1,\dots,b_n\}$ ist etwas technisch. Ein *Schaltkreis S* ist ein gerichteter Graph ohne Kreise, dessen Knoten *Gatter* genannt werden. Die Gatter haben entweder Eingangsgrad 0 oder 2, und jedes Gatter mit Eingangsgrad 2 besitzt einen *Typ*. Die Gatter mit Eingangsgrad 0 heißen *Eingangsgatter* und sind der Menge $\{g_1,\dots,g_n,\overline{g}_1,\dots,\overline{g}_n\}$ entnommen. Die Gatter mit Eingangsgrad 2 heißen *interne Gatter* und haben entweder den Typ $\vee$ (*logisches Oder*) oder $\wedge$ (*logisches Und*). Die internen Gatter mit Ausgangsgrad 0 heißen *Ausgabegatter* und sind durchnummeriert mit $o_1,\dots,o_m$. Die *Größe* eines Schaltkreises S ist die Anzahl der internen Gatter. Jeder Schaltkreis definiert eine boolesche Funktion $f\colon \mathbb{B}^n \to \mathbb{B}^m$, indem wir für alle $z \in \mathbb{B}^n$ und alle Gatter g einen Wert $g(z) \in \mathbb{B}$ definieren und dann $F(z)$ auf $(o_1(z),\dots,o_m(z))$ setzen. Die Definition von $g(z) \in \mathbb{B}$ erfolgt von den Eingangsgattern abwärts. Zunächst setzen wir für die Eingangsgatter $g_i(z) = z_i$ und $\overline{g}_i(z) = 1 - z_i$. Ist g ein internes Gatter mit gerichteten Kanten von den Gattern f und h, so können wir induktiv annehmen, dass $f(z)$ und $h(z)$ schon definiert sind. Ist g vom Typ $\vee$, so schreiben wir $g = f \vee h$ und setzen $g(z) = f(z) \vee h(z)$. Ist g vom Typ $\wedge$, so schreiben wir $g = f \wedge h$ und setzen $g(z) = f(z) \wedge h(z)$. Der Unterschied zwischen booleschen Formeln und Schaltkreisen ist, dass bei Schaltkreisen jedes Gatter das Eingangsgatter von mehreren anderen Gattern sein kann. Bei Formeln hingegen muss man identische Teilformeln bei Bedarf mehrfach hinschreiben, da man sie nicht wiederverwenden kann.

Ein Schaltkreis S *realisiert* eine boolesche Funktion $f\colon \mathbb{B}^n \to \mathbb{B}$, falls $o_1(z) = f(z)$ für alle $z \in \mathbb{B}^n$ gilt. Benutzt ein Schaltkreis nur Eingangsgatter aus der Menge $\{g_1,\dots,g_c,\overline{g}_1,\dots,\overline{g}_c\}$ und Ausgabegatter $o_1,\dots,o_d$, so stellt er boolesche Funktionen $f\colon \mathbb{B}^n \to \mathbb{B}^m$ für alle $n \geq c$ und $m \leq d$ dar.

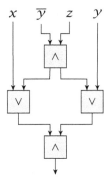

x	y	z	$f(x,y,z)$
0	0	0	0
0	0	1	1
0	1	0	0
0	1	1	0
1	0	0	0
1	0	1	1
1	1	0	1
1	1	1	1

Schaltkreis S von S realisierte Funktion f

8.1 Shannons obere Schranke für die Anzahl der Gatter

Sei $f\colon \mathbb{B}^n \to \mathbb{B}$ eine beliebige boolesche Funktion mit $n \geq 1$. Zum Aufwärmen wollen wir zwei einfache obere Schranken für die Anzahl der internen Gatter angeben, die man benötigt, um f zu realisieren. Zunächst zeigen wir, dass hierfür $2^{n+1} - 3$ Gatter genügen. Für $n = 1$ gibt es genau vier boolesche Funktionen, die jeweils durch ein internes Gatter dargestellt werden können. Also stimmt die Behauptung für $n = 1$. Ab jetzt sei $n \geq 2$ und die Behauptung richtig für $n - 1$. Die Funktion f definiert zwei Funktionen $f_0\colon \mathbb{B}^{n-1} \to \mathbb{B}$ und $f_1\colon \mathbb{B}^{n-1} \to \mathbb{B}$, indem wir $f_0(z_1, \ldots, z_{n-1}) = f(z_1, \ldots, z_{n-1}, 0)$ und $f_1(z_1, \ldots, z_{n-1}) = f(z_1, \ldots, z_{n-1}, 1)$ setzen. Induktiv können wir f_i durch einen Schaltkreis S_i realisieren, der weniger als $2^n - 3$ interne Gatter hat. Dann erhalten wir die *Shannon-Zerlegung* von f durch $(S_0 \wedge \overline{g}_n) \vee (S_1 \wedge g_n)$.

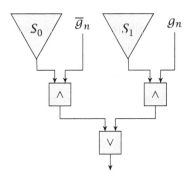

Schaltkreis für f

Dieser Schaltkreis benötigt höchstens $2(2^n - 3) + 3 = 2^{n+1} - 3$ interne Gatter. Damit ist die Behauptung für alle $n \geq 1$ gezeigt. Bei dieser Konstruktion hat jedes Gatter einen Ausgangsgrad von höchstens 1.

Durch Mehrfachverwendung von bereits berechneten Zwischenergebnissen (d. h., bei manchen Gattern ist der Ausgangsgrad größer als 1) erhalten wir bessere Schran-

ken. Ein einfacher Ansatz hierzu ist wie folgt. Zunächst bestimmen wir eine obere Schranke für die Anzahl der Gatter eines Schaltkreises, welcher *alle* booleschen Funktionen $\mathbb{B}^k \to \mathbb{B}$ berechnet; für jede dieser Funktionen hat der gesuchte Schaltkreis ein Ausgabegatter. Sei C_k die Größe des Schaltkreises. Wir behaupten:

$$C_k \leq 4 \cdot 2^{2^k}$$

Es gibt 4 boolesche Funktionen $\mathbb{B} \to \mathbb{B}$ und jede benötigt höchstens 1 internes Gatter. Damit gilt $C_1 \leq 4$ und die Behauptung gilt für $k = 1$. Sei nun $k \geq 2$. Wie zuvor können wir $f : \mathbb{B}^k \to \mathbb{B}$ in zwei Funktionen $f_0, f_1 : \mathbb{B}^{k-1} \to \mathbb{B}$ aufteilen und dann die Schaltkreise für f_1 und f_2 mit 3 Gattern zu einem Schaltkreis für f zusammensetzen. Wenn wir diese Technik für jede der 2^{2^k} Funktionen $\mathbb{B}^k \to \mathbb{B}$ anwenden, dann sehen wir:

$$C_k \leq C_{k-1} + 3 \cdot 2^{2^k}$$

Mit Induktion gilt nun $C_k \leq 4 \cdot 2^{2^{k-1}} + 3 \cdot 2^{2^k}$ und zusammen mit $4 \cdot 2^{2^{k-1}} \leq 2^{2^k}$ folgt daraus die Behauptung.

Wir verwenden nun die Schranke für C_k, um die Schranke bei einer einzelnen Funktion $f : \mathbb{B}^n \to \mathbb{B}$ etwas zu verfeinern. Nehmen wir an, wir hätten bereits einen einzelnen Schaltkreis für alle Funktionen $\mathbb{B}^k \to \mathbb{B}$. Wir wollen abschätzen, wie viele zusätzliche Gatter benötigt werden, um f zu realisieren, wenn $n \geq k$ gilt. Wir behaupten, dass $3 \cdot 2^{n-k} - 3$ zusätzliche Gatter ausreichen. Für $n = k$ benötigen wir keine weiteren Gatter, und mit $3 \cdot 2^0 - 3 = 0$ ist die Behauptung erfüllt. Sei nun $n > k$. Indem wir f wieder in Funktionen $f_0, f_1 : \mathbb{B}^{n-1} \to \mathbb{B}$ aufteilen, und die Schaltkreise für f_0 und f_1 mit 3 Gattern zu einem Schaltkreis für f kombinieren, sehen wir dass für f höchstens $2 \cdot (3 \cdot 2^{n-1-k} - 3) + 3 = 3 \cdot 2^{n-k} - 3$ zusätzliche Gatter benötigt werden. Zusammen mit den Gattern für die Funktionen $\mathbb{B}^k \to \mathbb{B}$ benötigen wir für f weniger als $3 \cdot 2^{n-k} + 4 \cdot 2^{2^k}$ Gatter. Mit $k = \lfloor \log_2 n \rfloor - 1$ sehen wir:

$$3 \cdot 2^{n-k} + 4 \cdot 2^{2^k} = 3 \cdot 2^{n - \lfloor \log_2 n \rfloor + 1} + 4 \cdot 2^{2^{\lfloor \log_2 n \rfloor - 1}}$$

$$\leq 3 \cdot 2^{n - \log_2 n + 2} + 4 \cdot 2^{2^{\log_2 n - 1}}$$

$$= 12 \cdot \frac{2^n}{n} + 4 \cdot 2^{n/2} \in \mathcal{O}(2^n / n)$$

8.2 Die untere Schranke von Shannon

Als Nächstes zeigen wir das Gegenstück zur oberen Schranke: Fast alle n-stelligen booleschen Funktionen benötigen Schaltkreise der Größe $2^n / n$, um sie zu realisieren. Der Grund ist eigentlich recht einfach; es gibt zu wenige Schaltkreise der Größe $2^n / n$, um 2^{2^n} Funktionen darzustellen. Wir dürfen sogar die Berechnungsfähigkeit der Gatter erweitern, ohne dass sich das Ergebnis ändert. Im Folgenden stellen wir Gatter für jede zweistellige boolesche Funktion zur Verfügung. Es gibt also nicht nur zwei verschiedene Typen, sondern $16 = 2^{2^2}$. Dafür können wir jetzt auf negierte Eingabegatter verzichten. Es gibt also nur noch n Eingabegatter $g_1, \ldots, g_n$. Jeweils ein

internes Gatter wird als Ausgabegatter spezifiziert, und damit berechnet jeder Schaltkreis mit diesen 16 möglichen Gattertypen eine n-stellige boolesche Funktion. Ein Gatter selbst kann als Tripel (T, ℓ, r) beschrieben werden, wobei T der Typ ist und ℓ und r das linke beziehungsweise rechte Eingangsgatter bezeichnet.

Das Ziel ist zu zeigen, dass fast alle Funktionen $2^n/n$ Gatter benötigen. Hierfür ist etwas Vorbereitung erforderlich. Wir nennen einen Schaltkreis *reduziert*, wenn je zwei verschiedene interne Gatter verschiedene boolesche Funktionen definieren. Man beachte, dass sich in reduzierten Schaltkreisen interne Gatter also durchaus wie Eingabegatter verhalten dürfen.

Lemma 8.1. *Sei S ein Schaltkreis mit s internen Gattern und n Eingabegattern. Dann existiert für jedes t mit $s \leq t \leq 2^{2^n}$ ein reduzierter Schaltkreis mit t internen Gattern, der dieselbe boolesche Funktion f berechnet wie S.*

Beweis. Wir ordnen die Gatter von S von links nach rechts, so dass für jedes Gatter die jeweiligen Eingangsgatter weiter links stehen. Die Eingabegatter stehen in dieser Ordnung also ganz links. (Wir haben die Gatter topologisch sortiert.) Danach erweitern wir die Ordnung auf der rechten Seite um weitere interne Gatter, bis am Ende jede der 2^{2^n} booleschen Funktionen durch ein internes Gatter repräsentiert wird. Die Ordnung, dass jeweilige Eingangsgatter weiter links stehen, behalten wir während der Erweiterung bei.

Diesen sehr großen neuen Schaltkreis reduzieren wir von links nach rechts. Wenn jetzt zwei interne Gatter dieselbe Funktion darstellen, so können wir auf das weiter rechts stehende Gatter verzichten, denn alle Leitungen, die von diesem Gatter ausgehen, können umdirigiert werden, ohne die Ordnung zu zerstören. Durch diesen Prozess wird der große Schaltkreis nach und nach reduziert. Wir stoppen den Prozess, wenn wir einen reduzierten Schaltkreis mit t internen Gattern erzeugt haben. Die noch nicht betrachteten Gatter am rechten Ende werden wieder gelöscht. Wegen $s \leq t$ befindet sich in dem reduzierten Schaltkreis ein internes Gatter, welches f berechnet. $\square$

Sei S ein Schaltkreis mit s internen Gattern. Wir identifizieren im nächsten Schritt die Gatter mit den Zahlen $1, \ldots, s, s+1, \ldots, s+n$, wobei wir jetzt die folgende Anordnung vornehmen. Wir benennen das Ausgabegatter mit s und mit $s+1, \ldots, s+n$ bezeichnen wir die Eingabegatter $g_1, \ldots, g_n$. Dies führt auf eine Beschreibung des Schaltkreises als Folge von Tripeln

$$S = ((T_1, \ell_1, r_1), \ldots, (T_s, \ell_s, r_s))$$

wobei T_i einer der möglichen 16 Gattertypen ist und ℓ_i und r_i die linken und rechten Vorgänger des internen Gatters i bezeichnen. Ist π eine Permutation von $\{1, \ldots, s+n\}$, welche die Zahlen $s, s+1, \ldots, s+n$ fest lässt, dann verändert π die Namen interner Gatter, aber nicht die Funktionsweise des Schaltkreises. Wir erhalten eine neue Schaltkreisbeschreibung $\pi(S)$. In $\pi(S)$ wertet sich jedes Gatter $\pi(i)$ ge-

nau wie i in S aus. Da π das Ausgabegatter fest lässt, berechnet das Gatter s weiterhin dieselbe boolesche Funktion. Die Schaltkreisbeschreibung nach Anwendung der Permutation π hat die Form $\pi(S) = ((T_1', \ell_1', r_1'), \ldots, (T_s', \ell_s', r_s'))$ und für $\pi(i) = j$ gilt $(T_j', \ell_j', r_j') = (T_i, \pi(\ell_i), \pi(r_i))$. Ist der Schaltkreis reduziert und $\pi \neq \mathrm{id}$, so gilt für die jeweiligen Beschreibungen $S \neq \pi(S)$. Dies sieht man wie folgt. Wegen $\pi \neq \mathrm{id}$ gilt $\pi(i) = j \neq i$ für zwei Gatter i und j. Angenommen, die Beschreibung wäre identisch, dann wäre insbesondere $(T_j, \ell_j, r_j) = (T_j', \ell_j', r_j') = (T_i, \pi(\ell_i), \pi(r_i))$. Also ist $T_i = T_j$, $\pi(\ell_i) = \ell_j$ und $\pi(r_i) = r_j$. Daher werten sich die ursprünglichen Gatter i und j genau gleich aus, was aber in reduzierten Schaltkreisen unmöglich ist.

Lemma 8.2. *Sei $s \geq n$. Dann ist die Anzahl der n-stelligen booleschen Funktionen, welche sich durch Schaltkreise mit höchstens s internen Gattern berechnen lassen, durch eine Funktion in $2^{s \log s + \mathcal{O}(s)}$ begrenzt.*

Beweis. Gehört $f \colon \mathbb{B}^n \to \mathbb{B}$ zu einer dieser Funktionen, so wird f nach Lemma 8.1 durch einen reduzierten Schaltkreis S mit genau s internen Gattern berechnet. Die Beschreibung von S ergibt eine Liste von Tripeln

$$((T_1, \ell_1, r_1), \ldots, (T_s, \ell_s, r_s))$$

und die Zahl dieser Folgen von Tripeln ist kleiner als $(16(s + n)(s + n))^s \leq k^s \cdot s^{2s}$ für eine Konstante $k \leq 64^2$. Zu jedem Schaltkreis S für die Funktion f gehören aber nun $(s - 1)!$ Beschreibungen $\pi(S)$, und diese sind alle paarweise verschieden. Bei der gewünschten Abschätzung dürfen wir also die Zahl $k^s \cdot s^{2s}$ durch $(s - 1)!$ teilen. Nach Gleichung (2.1) gilt $s! \geq (s/e)^s$. Damit erhalten wir:

$$k^s \cdot s^{2s}/(s - 1)! \leq s \cdot (ek)^s \cdot s^{2s}/s^s \in 2^{s \log s + \mathcal{O}(s)}$$

Also gilt die Behauptung. $\qquad\square$

Satz 8.3 (Shannon 1949). *Der Anteil der n-stelligen booleschen Funktionen, die sich durch Schaltkreise mit höchstens $2^n/n$ internen Gattern berechnen lassen, strebt gegen 0 für $n \to \infty$.*

Beweis. Nach Lemma 8.2 gibt es höchstens $2^{s \log s + \mathcal{O}(s)}$ Funktionen bis zur Größe s. Setzen wir $s = 2^n/n$ und bilden den Logarithmus, so erhalten wir $2^n - \frac{2^n \log n}{n} + \mathcal{O}(2^n/n)$. Um den Anteil der Funktionen der Größe s unter den 2^{2^n} Funktionen zu erkennen, müssen wir $\log(2^{2^n}) = 2^n$ subtrahieren und sehen, dass die Differenz gegen $-\infty$ strebt. Also geht der Anteil, wie behauptet, gegen 0. $\qquad\square$

Für eine boolesche Funktion $f \colon \mathbb{B}^n \to \mathbb{B}$ sei s_f die minimale Anzahl der Gatter, die ein Schaltkreis benötigt, der f realisiert. Ferner sei $s(n)$ das Maximum der s_f über alle $f \colon \mathbb{B}^n \to \mathbb{B}$. Dies bedeutet, $s(n)$ ist die minimale Anzahl an Gattern, die benötigt wird, um eine beliebige n-stellige boolesche Funktion zu realisieren. Für die Funktion $s(n)$ gilt $s(n) \in \Theta(2^n/n)$, denn nach Satz 8.3 gilt $s(n) \in \Omega(2^n/n)$, und die obere Schranke $s(n) \in \mathcal{O}(2^n/n)$ wurde bereits in Abschnitt 8.1 gezeigt.

8.3 Die obere Schranke von Lupanov

Die obere Schranke von $\mathcal{O}(2^n/n)$ aus Abschnitt 8.1 wurde von Lupanov weiter verbessert [27]. Er zeigte, dass sich jede n-stellige boolesche Funktion durch einen Schaltkreis mit nur $2^n/n + o(2^n/n)$ Gattern realisieren lässt. Wir beweisen als nächstes diese Schranke, welche aufgrund von Satz 8.3 asymptotisch optimal ist.

Als wichtiges Hilfsmittel für den Beweis betrachten wir boolesche Matrizen. Seien R und S endliche Indexmengen und sei F eine boolesche Matrix aus $\mathbb{B}^{R \times S}$. Wir fassen R und S als disjunkte Mengen boolescher Variablen auf. Die Elemente $\rho \in \mathbb{B}^R$ kann man als Teilmengen von Zeilen aus R auffassen. Damit ist $\rho(x)$ für $x \in R$ ein boolescher Wert, der sich genau dann zu 1 auswertet, wenn x zu der von ρ repräsentierten Menge gehört. Eine entsprechende Sichtweise ist auch für $\sigma \in \mathbb{B}^S$ möglich. Die Matrix F definiert nun eine boolesche Funktion $H\colon \mathbb{B}^R \times \mathbb{B}^S \to \mathbb{B}$ durch

$$H(\rho, \sigma) = \begin{cases} 1 & \text{falls } \exists x \in R\; \exists y \in S\colon 1 = \rho(x) = \sigma(y) = F(x, y) \\ 0 & \text{sonst} \end{cases}$$

Wenn wir $\rho(x) = 1$ lesen als „ρ aktiviert die Zeile x" und entsprechend $\sigma(y) = 1$ für Spalten interpretieren, bedeutet $H(\rho, \sigma) = 1$, dass $F(x, y) = 1$ für eine aktivierte Zeile x und eine aktivierte Spalte y ist. Wir sagen, ein Schaltkreis mit Eingabegattern h_x und h_y für $x \in R$ und $y \in S$ *realisiert* die Matrix F, wenn er die boolesche Funktion H realisiert. Die negierten Gatter $\overline{h}_x$ und $\overline{h}_y$ werden dabei nicht verwendet.

Lemma 8.4. *Seien R, S endliche Indexmengen und sei F eine boolesche Matrix aus $\mathbb{B}^{R \times S}$. Wenn in F höchstens p Zeilen nicht Null sind, dann lässt sich F mit einem Schaltkreis der Größe $3 \cdot 2^p + |S|$ realisieren.*

Beweis. Sei $P \subseteq R$ mit $|P| = p$ eine Teilmenge der Zeilen, welche alle Zeilen mit einem von Null verschiedenen Eintrag enthält. In F gibt es höchstens 2^p viele verschiedene Spalten. Wir definieren $H\colon \mathbb{B}^P \times \mathbb{B}^S \to \mathbb{B}$ durch

$$H(\rho, \sigma) = \begin{cases} 1 & \text{falls } \exists x \in P\; \exists y \in S\colon 1 = \rho(x) = \sigma(y) = F(x, y) \\ 0 & \text{sonst} \end{cases}$$

für $\rho \in \mathbb{B}^P$ und $\sigma \in \mathbb{B}^S$. Es genügt, H durch einen Schaltkreis mit Eingabegattern h_x und h_y für $x \in P$ und $y \in S$ zu realisieren. Die negierten Gatter $\overline{h}_x$ und $\overline{h}_y$ kommen in der Konstruktion nicht vor.

Für eine Spalte $s \in S$ sei $J(s) \subseteq S$ die Menge der Spalten von F, die mit s übereinstimmen. Formal ist

$$J(s) = \{ y \in S \mid \forall x \in P\colon F(x, y) = F(x, s) \}$$

Für $s \in S$ definieren wir außerdem zwei boolesche Funktionen:

$$E_s(\rho) = \bigvee \{\rho(x) \mid F(x, s) = 1\}$$
$$G_s(\sigma) = \bigvee \{\sigma(y) \mid y \in J(s)\}$$

Die Funktionen $E_s \colon \mathbb{B}^P \to \mathbb{B}$ und $G_s \colon \mathbb{B}^S \to \mathbb{B}$ können durch Schaltkreise von der Form $\bigvee \{ h_x \mid F(x,s) = 1 \}$ beziehungsweise $\bigvee \{ h_y \mid y \in J(s) \}$ realisiert werden. Durch ein weiteres Und-Gatter können wir folgende Funktion realisieren:

$$H_s(\rho, \sigma) = E_s(\rho) \wedge G_s(\sigma)$$

Damit ist $H_s(\rho, \sigma) = 1$ äquivalent dazu, dass es eine Zeile $x \in P$ und eine Spalte $y \in S$ mit den folgenden drei Eigenschaften gibt: (1) Es gilt $1 = \rho(x) = \sigma(y)$. (2) Die Spalten y und s sind identisch. (3) Es gilt $F(x,y) = 1$.

Im nächsten Schritt konstruieren wir einen Schaltkreis für H als Disjunktion von Schaltkreisen für H_s für eine geeignete Menge von Spalten s. Wir wählen dafür eine minimale Menge $Q \subseteq S$ mit $S = \bigcup \{ J(s) \mid s \in Q \}$. Dies bedeutet, jede Spalte in der Matrix F erscheint genau einmal als artgleiche Spalte in Q. Nun gilt:

$$H(\rho, \sigma) = \bigvee \{ H_s(\rho, \sigma) \mid s \in Q \}$$

Die Anzahl der Schaltkreise, die nötig ist, um alle Funktionen G_s zu realisieren, ist durch $|S|$ beschränkt, da die Mengen $J(s)$ eine Partition von S definieren. Für die Funktionen E_s genügen insgesamt 2^p Gatter. Mit $|Q| \le 2^p$ erhalten wir für die Größe des Schaltkreises für H die folgende obere Schranke:

$$\underbrace{2^p}_{E_s} + \underbrace{|S|}_{G_s} + \underbrace{2^p}_{\substack{\text{ein Und-Gatter} \\ \text{pro } H_s}} + \underbrace{2^p}_{\substack{\text{Oder-Gatter} \\ \text{bei } H}}$$

Dies beweist das Lemma. □

Satz 8.5 (Lupanov 1958). *Jede n-stellige boolesche Funktion wird durch einen Schaltkreis der Größe $2^n/n + o(2^n/n)$ realisiert.*

Beweis. Sei $f \colon \mathbb{B}^n \to \mathbb{B}$ eine n-stellige boolesche Funktion und $1 \le k < n$. Bei den Argumenten von f fassen wir die ersten k und die übrigen $n - k$ Parameter jeweils als eine Einheit auf. Sei $R = \mathbb{B}^k$ und $S = \mathbb{B}^{n-k}$, dann definieren wir eine Matrix $F \in \mathbb{B}^{R \times S}$ durch $F(x,y) = f(x,y)$ für einen Zeilenindex $x \in R$ und einen Spaltenindex $y \in S$. Weiter fassen wir in F jeweils p Zeilen zu einer neuen Matrix zusammen. Sei hierzu $R = A_1 \cup \cdots \cup A_\ell$ eine Partition von R mit $|A_i| \le p$ und $\ell = \lceil 2^k/p \rceil$. Wir definieren nun die Matrizen $F_i \in \mathbb{B}^{R \times S}$ durch

$$F_i(x,y) = \begin{cases} F(x,y) & \text{für } x \in A_i \\ 0 & \text{sonst} \end{cases}$$

Damit gilt $F = F_1 \vee \cdots \vee F_\ell$. Außerdem gibt es in F_i höchstens p Zeilen, die nicht Null sind. Mit Lemma 8.4 lässt sich jede Matrix F_i mit $3 \cdot 2^p + |S|$ Gattern realisieren. Damit existiert für F ein Schaltkreis der Größe:

$$\ell \cdot (3 \cdot 2^p + 2^{n-k}) + \ell$$

Um aus der Realisierung der Matrix F eine Realisierung der Funktion f zu erhalten, müssen wir noch die Eingangsgatter h_x und h_y für F zur Verfügung stellen. Seien $g_1, \ldots, g_n, \overline{g}_1, \ldots, \overline{g}_n$ die Eingangsgatter für den Schaltkreis zur Funktion f. Dann ergibt sich h_x als eine Konjunktion von k Gattern aus $\{g_1, \ldots, g_k, \overline{g}_1, \ldots, \overline{g}_k\}$ und h_y als eine Konjunktion von $n - k$ Gattern aus $\{g_{k+1}, \ldots, g_n, \overline{g}_{k+1}, \ldots, \overline{g}_n\}$. Damit benötigt die Berechnung aller h_x und h_y für $x \in R$ und $y \in S$ höchstens $k \cdot 2^k + (n - k) \cdot 2^{n-k}$ Gatter. Insgesamt ergibt sich damit für die Größe des Schaltkreises für f die folgende Schranke:

$$\left\lceil \frac{2^k}{p} \right\rceil (3 \cdot 2^p + 2^{n-k}) + \left\lceil \frac{2^k}{p} \right\rceil + k \cdot 2^k + (n - k) \cdot 2^{n-k}$$

Wir setzen jetzt $k = 3\lceil \log_2 n \rceil$ und $p = n - 5\lceil \log_2 n \rceil$. Für den hinteren Teil der obigen Schranke gilt nun

$$3 \cdot 2^p + 2^{n-k} + \left\lceil \frac{2^k}{p} \right\rceil + k \cdot 2^k + (n - k) \cdot 2^{n-k} \in o\left(\frac{2^n}{n}\right)$$

Für den restlichen Teil ist

$$\left(\left\lceil \frac{2^k}{p} \right\rceil - 1\right)(3 \cdot 2^p + 2^{n-k}) \leq \frac{2^k}{p}(3 \cdot 2^p + 2^{n-k})$$

$$= \frac{3 \cdot 2^{k+p}}{p} + \frac{2^n}{p} \in \frac{2^n}{n} + o\left(\frac{2^n}{n}\right)$$

Für die letzte Abschätzung beachte man, dass für jedes genügend große n die Ungleichung $1/(n - \log n) \leq 1/n + 1/(n \log n)$ gilt. Insgesamt benötigt der Schaltkreis für f höchstens $2^n/n + o(2^n/n)$ interne Gatter. $\square$

Korollar 8.6. *Sei $s(n)$ die kleinste natürliche Zahl, die ausreicht, damit jede n-stellige boolesche Funktion durch einen Schaltkreis mit $s(n)$ Gattern realisiert werden kann. Dann gilt die Asymptotik $s(n) \sim 2^n/n$.*

Beweis. Nach Satz 8.3 gilt $s(n) \geq 2^n/n$. Die Schranke von Lupanov besagt $s(n) \in 2^n/n + o(2^n/n)$. Zusammen ergibt dies die Aussage des Korollars. $\square$

Der Schaltkreis, den wir in Satz 8.5 konstruiert haben, verwendet vier logische Level; dies bedeutet es wird nur dreimal zwischen Und-Gattern und Oder-Gattern alterniert. Der oberste Level (ein Und-Level) ist die Konstruktion der h_x und h_y. Danach kommt durch die Konstruktion in Lemma 8.4 ein Oder-Level für E_s und G_s. Als dritter Level folgt wieder ein Und-Level zur Konstruktion der Schaltkreise für H_s. Als letztes folgt schließlich ein Oder-Level zur Konstruktion von H sowie zur Disjunktion der Matrizen F_i. Übliche Normalformen für boolesche Formeln, wie die disjunktive Normalform oder die konjunktive Normalform, verwenden nur zwei logische Level.

Für weiterführende Literatur zu Schaltkreisen verweisen wir auf die Bücher von Heribert Vollmer [31] und Ingo Wegener [33].

A Grundlagen

A.1 Mengen, Relationen und Abbildungen

Wir verwenden folgende Standardbezeichnungen: Mit $\mathbb{N}$, $\mathbb{Z}$, $\mathbb{Q}$, $\mathbb{R}$, $\mathbb{C}$ bezeichnen wir die *natürlichen, ganzen, rationalen, reellen und komplexen* Zahlen. In diesem Kapitel meint n stets eine natürliche und k eine ganze Zahl. Mit x, y und r meinen wir reelle oder komplexe Zahlen. Da viele Identitäten allgemeiner gelten, verzichten wir manchmal auf eine genaue Spezifikation. Wer mit komplexen Zahlen nicht vertraut ist, möge sich stets eine reelle Zahl vorstellen. Auf der anderen Seite gibt es keine guten Gründe, sie von vornherein auszuschließen. Wenn $x = a + bi$ mit $a, b \in \mathbb{R}$ und $i = \sqrt{-1}$ eine komplexe Zahl ist, so meint der *Betrag* $|x|$ wie üblich den Wert $\sqrt{a^2 + b^2}$. Insbesondere ist der Betrag $|r|$ einer reellen Zahl $r \in \mathbb{R}$ wie üblich definiert, also $|r| = r$ für $r \geq 0$ und $|r| = -r$ für $r < 0$.

Das *kartesische Produkt* $A \times B$ zweier Mengen A und B ist die Menge aller Paare (a, b) mit $a \in A$ und $b \in B$. Benannt ist diese Konstruktion nach dem französischen Mathematiker René Descartes (1596–1650). Sind A und B endlich mit n beziehungsweise m Elementen, so enthält $A \times B$ genau nm Elemente.

Eine *Relation* zwischen A und B ist eine Teilmenge von $A \times B$. Eine Relation ist also eine Menge von Paaren. Eine *Abbildung* (oder *Funktion*) $f\colon A \to B$ von einer Menge A nach B ist formal ein Tripel (A, B, R), wobei $R \subseteq A \times B$ eine Relation ist, für die gilt, dass es zu jedem $a \in A$ genau ein $b \in B$ mit $(a, b) \in R$ gibt. Wie üblich schreiben wir dann $f(a) = b$. Die *Hintereinanderausführung* von Abbildungen $f\colon A \to B$ und $g\colon B \to C$ ist eine Abbildung $g \circ f\colon A \to C$, welche durch die Vorschrift $(g \circ f)(a) = g(f(a))$ definiert ist.

Eine Abbildung $f\colon A \to B$ heißt *injektiv*, wenn für alle $b \in B$ höchstens ein $a \in A$ mit $f(a) = b$ existiert. Sie heißt *surjektiv*, wenn für alle $b \in B$ mindestens ein $a \in A$ mit $f(a) = b$ existiert. Sie heißt *bijektiv*, wenn sie sowohl injektiv als auch surjektiv ist. Wir sprechen auch von Injektionen, Surjektionen und Bijektionen. Eine Bijektion von einer endlichen Menge auf sich selbst wird auch *Permutation* genannt. Eine Menge A ist *abzählbar*, wenn eine Surjektion $f\colon \mathbb{N} \to A$ existiert.

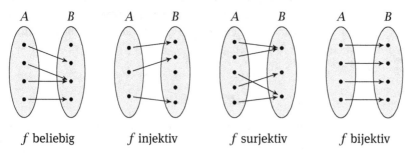

f beliebig f injektiv f surjektiv f bijektiv

A.2 Die $\mathcal{O}$-Notation

Häufig sind wir nicht an Funktionen selbst, sondern nur an ihrem Wachstumsverhalten interessiert. Hierfür haben sich die Landau-Symbole bewährt (Edmund Georg Hermann Landau, 1877–1938), welche Funktionsklassen beschreiben. Im Folgenden betrachten wir Funktionen $f, g, \ldots$ von $\mathbb{N}$ nach $\mathbb{R}$ oder nach $\mathbb{C}$. Die Funktionsklassen $\mathcal{O}(g)$, $\Omega(g)$ und $\Theta(g)$ (lies „Groß-Oh von g", „Groß-Omega von g" und „Theta von g") sind folgendermaßen definiert:

$$\mathcal{O}(g) = \{f : \mathbb{N} \to \mathbb{C} \mid \exists c > 0 \, \exists n_0 \geq 0 \, \forall n \geq n_0 : |f(n)| \leq c \cdot |g(n)|\}$$

$$\Omega(g) = \{f : \mathbb{N} \to \mathbb{C} \mid g \in \mathcal{O}(f)\}$$

$$\Theta(g) = \mathcal{O}(g) \cap \Omega(g)$$

Es gilt also $f \in \mathcal{O}(g)$, wenn f bis auf eine Konstante schließlich (ab einem n_0) nicht schneller wächst als g. Außerdem halten wir die Konvention fest, dass durch $\mathcal{O}$ definierte Klassen stets nach unten abgeschlossen sind. Damit vermeiden wir seltsame Effekte. Wenn man etwa eine Klasse $2^{\mathcal{O}(n)}$ definiert, so soll diese sicherlich auch alle Polynome enthalten. Analog gilt $f \in \Omega(g)$, wenn f bis auf eine Konstante schließlich nicht langsamer wächst als g und $f \in \Theta(g)$, wenn f bis auf Konstanten schließlich gleich schnell wächst wie g. Es gilt z. B. $\binom{2n}{3} \in \Theta(n^3) \subset \mathcal{O}(2^n)$.

Wir definieren $o(g)$ und $\omega(g)$ (lies „Klein-Oh" und „Klein-Omega") durch:

$$o(g) = \{f : \mathbb{N} \to \mathbb{C} \mid \forall c > 0 \, \exists n_0 > 0 \, \forall n \geq n_0 : |f(n)| \leq c \cdot |g(n)|\}$$

$$\omega(g) = \{f : \mathbb{N} \to \mathbb{C} \mid g \in o(f)\}$$

Damit enthält $o(g)$ die Funktionen, die (echt) langsamer wachsen als g und in $\omega(g)$ liegen die Funktionen, die schneller wachsen als g. Damit gilt z. B. $o(g) \subset \mathcal{O}(g)$ und $o(g) \cap \Theta(g) = \varnothing$.

Die Relationen, die sich durch $\mathcal{O}$, Ω und Θ ergeben, sind reflexiv: Es gilt $f \in \mathcal{O}(f)$, $f \in \Omega(f)$ und $f \in \Theta(f)$. Dies ist für o und ω falsch: $f \notin o(f)$ und $f \notin \omega(f)$. Sämtliche Relationen sind transitiv: aus $f \in \mathcal{O}(g)$ und $g \in \mathcal{O}(h)$ folgt $f \in \mathcal{O}(h)$, entsprechend für Ω, Θ, o, ω. Unter diesen Relationen ist nur Θ symmetrisch: aus $f \in \Theta(g)$ folgt $g \in \Theta(f)$. Insbesondere aufgrund dieser Symmetrieverletzung ist die in der Literatur übliche Bezeichnung $f = \mathcal{O}(g)$, also Gleichheitszeichen statt Element, mit Vorsicht zu gebrauchen.

Für alle Basen $a, b > 1$ gilt

$$\log_a(n) \in \Theta(\log_b(n))$$

Wir können also Klassen wie $\mathcal{O}(\log n)$ definieren, ohne die Basis zu spezifizieren. Weiterhin gilt noch die Beziehung:

$$\log^k(n) \in o(\log^{k+1}(n)) \subsetneq o(n)$$

Eine Funktion $f\colon \mathbb{N} \to \mathbb{C}$ heißt *polynomiell beschränkt*, wenn $f \in \mathcal{O}(n^k)$ für ein $k \geq 0$ gilt. Wir reden etwa von einem polynomiellen Algorithmus, wenn die Laufzeit polynomiell beschränkt ist. Häufig wird der Term *effizient* als Synonym hierfür benutzt. Diese Klasse von Algorithmen ist robust in dem Sinne, dass die Definition polynomieller Algorithmen weitgehend unabhängig vom Maschinenmodell (beziehungsweise von Implementationsdetails) ist.

Um *asymptotisch* gleiches Wachstum von Funktionen $f, g\colon \mathbb{N} \to \mathbb{C}$ auszudrücken, verwendet man die Bezeichnung $f \sim g$. Sie ist definiert durch die folgende Äquivalenz:

$$f \sim g \ \Leftrightarrow \ \lim_{n \to \infty} \frac{f(n)}{g(n)} = 1$$

Damit gilt zwar $\binom{n}{3} \in \Theta(n^3)$, aber $\binom{n}{3} \not\sim n^3$. Es gilt jedoch $\binom{n}{3} \sim \frac{n^3}{6}$. Allgemeiner können wir für $k \geq 0$ festhalten $\binom{n}{k} \sim \frac{n^k}{k!}$. Wir werden die Asymptotik nur auf Funktionen anwenden, die fast niemals den Wert Null annehmen, und vermeiden so hier die Diskussion, ob $0 \sim 0$ gilt.

B Lösungen der Aufgaben

Zu Kapitel 1

1.1. Es kommt gar nicht darauf an, dass p eine Primzahl ist. Wir benutzen nur, dass p keine Potenz von 10 ist, denn aus $\log_{10}(p) = r/s$ folgt $\log_{10}(p^s) = r$ und damit $p^s = 10^r$.

1.2. (a) Euklidischer Algorithmus:

$$56 = 2 \cdot 35 - 14, \qquad \text{d. h. } 14 = 2 \cdot 35 - 56$$
$$35 = 2 \cdot 14 + 7, \qquad \text{d. h. } 7 = 35 - 2 \cdot 14$$
$$14 = 2 \cdot 7, \qquad \text{d. h. } 7 = \mathrm{ggT}(35, 56)$$

Einsetzen der ersten Gleichung in die zweite liefert $7 = 35 - 2 \cdot (2 \cdot 35 - 56) = -3 \cdot 35 - (-2) \cdot 56$, d. h. $x = -3$, $y = -2$.

1.2. (b)

$$7 = -3 \cdot 35 + \frac{56}{7} \cdot 35 - \frac{35}{7} \cdot 56 + 2 \cdot 56$$
$$= \left(\frac{56}{7} - 3\right) \cdot 35 - \left(\frac{35}{7} - 2\right) \cdot 56$$
$$= 5 \cdot 35 - 3 \cdot 56, \quad \text{d. h. } x = 5, \ y = 3$$

1.3. Die Lösungen der beiden separaten Kongruenzen $3x \equiv 0 \mod 13$ und $-7y \equiv 11 \mod 13$ ergeben durch Addition auf alle Fälle Lösungen der Kongruenz $3x - 7y \equiv 11 \mod 13$. Sei $x_0 = 0$ eine spezielle Lösung von $3x \equiv 0 \mod 13$ und $y_0 = 4$ eine spezielle Lösung von $-7y \equiv 11 \mod 13$. Damit erhalten wir in $3x_0 - 7y_0$ eine zur Aufspaltung $11 = 0 + 11$ gehörige spezielle Lösung der ursprünglichen Kongruenz. Die zu x_0, y_0 gehörige allgemeine Lösung der ursprünglichen Kongruenz hat dann die Gestalt

$$3x_0 - 7y_0 + 13t = 3(x_0 + u) - 7(y_0 + v)$$

wobei $3u \equiv 7v \mod 13$ gelten muss. Für $u = 1$ ergibt sich $3 \equiv 7 \cdot 6 \mod 13$, also $3s \equiv 7 \cdot 6 \cdot s \mod 13$ mit beliebigem $s \in \mathbb{Z}$. Für die allgemeine Lösung erhalten wir also $x = s$, $y = 4 + 6s$, also $3s - 7(4 + 6s) \equiv 11 \mod 13$ für alle $s \in \mathbb{Z}$.

1.4. Jeder gemeinsame Teiler von $a + b$ und $a - b$ teilt auch $2a$ und $2b$. Die Behauptung folgt nun aus $\mathrm{ggT}(2a, 2b) = 2$.

1.5. (a) Sei $n = \sum_{k=0}^{\ell} a_k 10^k$. Dann ist die Quersumme $q(n) = \sum_{k=0}^{\ell} a_k$. Wegen $10 \equiv 1 \mod 3$ folgt $n \equiv q(n) \mod 3$. Dies beweist die *Dreierregel zur Division*.

1.5. (b) Sei wieder $n = \sum_{k=0}^{\ell} a_k 10^k$. Dann ist $10^k \equiv (-1)^k \mod 11$. Es folgt die *Elferregel zur Division*: $n \equiv \sum_{k=0}^{\ell} (-1)^k a_k \mod 11$. Ein Zahl ist also genau dann durch 11 teilbar, wenn ihre alternierende Quersumme durch 11 teilbar ist.

1.6. ⇒: Sei n zusammengesetzt, $n = pq$ mit $1 < p, q < n$. Dann ist p Teiler von $(n-1)!$, aber nicht von -1, da p kein Inverses in $(\mathbb{Z}/n\mathbb{Z})^*$ besitzt.

⇐: $(n-1)!$ ist das Produkt der Zahlen in $M = \{2, \ldots, n-1\}$. Ist n prim, so sind alle Elemente in M modulo n invertierbar und nur für $x \in M$ mit $x = n-1$ gilt $x \equiv x^{-1} \bmod n$. Fasse die Elemente aus $M \setminus \{n-1\}$ durch Umordnung paarweise mit ihren Inversen modulo n zusammen. Das Produkt über diese Zahlen ist damit 1 modulo n. Daher ist $(n-1)! \equiv n-1 \equiv -1 \bmod n$.

1.7. $n^4 + 4^n$ ist niemals 2, also können wir annehmen, dass $n = 2k+1$ ungerade ist mit $k \geq 1$. Setze $x = n$ und $y = 2^k$, dann ist $n^4 + 4^n = x^4 + 4y^4$, denn $4y^4 = 4 \cdot 2^{4k} = 4^{2k+1}$. Schließlich ergibt sich: $x^4 + 4y^4 = (x^2 + 2y^2)^2 - 4x^2y^2 = (x^2 + 2y^2 + 2xy)(x^2 + 2y^2 - 2xy)$. Sind jetzt $x, y \in \mathbb{N}$ mit $y > 1$, so gilt $x^2 + 2y^2 + 2xy \geq x^2 + 2y^2 - 2xy \geq (x-y)^2 + y^2 \geq 4$. Insbesondere ist ist $x^4 + 4y^4$ keine Primzahl.

1.8. (a) Angenommen $n = pq$ mit $p, q > 1$. Betrachte die Identität

$$x^q - y^q = (x-y) \sum_{i=0}^{q-1} x^i y^{q-1-i} \tag{B.1}$$

mit $x = 2^p$ und $y = 1$. Primzahlen der Form $2^n - 1$ bezeichnet man als Mersenne-Primzahlen (nach Marin Mersenne, 1588–1648).

1.8. (b) Angenommen $n = r2^m$ mit $r > 1$ ungerade. Dann gilt $2^{r2^m} + 1 = (2^{2^m})^r - (-1)^r$. Aus Gleichung (B.1) mit $x = 2^{2^m}$, $y = -1$ und $q = r$ ergibt sich wieder eine nichttriviale Faktorisierung. Primzahlen der Form $2^n + 1$ bezeichnet man als Fermat-Primzahlen.

1.8. (c) Sei ohne Einschränkung $1 \leq m < n$. Sei $d \in \mathbb{N}_1$ mit $d \mid f_m$ und $d \mid f_n$. Es gilt

$$\frac{f_n - 2}{f_m} = \frac{2^{2^n} - 1}{2^{2^m} + 1} = (2^{2^m})^{2^{n-m}-1} - (2^{2^m})^{2^{n-m}-2} + \cdots - 1$$

Damit ist $f_m \mid (f_n - 2)$ und folglich $d \mid (f_n - 2)$. Wegen $d \mid f_n$ folgt also $d \mid 2$. Aber es ist $d \neq 2$, da f_n und f_m beide ungerade sind. Da die Zahlen f_n paarweise teilerfremd sind und jedes f_n mindestens einen Primteiler hat, muss die Folge der Primzahlen unendlich sein.

1.9. Die Lösung lautet 111: Die beiden Forderungen $x \equiv 1 \bmod 2$ und $x \equiv 0 \bmod 3$ sind äquivalent mit $x \equiv 3 \bmod 6$. Nehmen wir die dritte Gleichung hinzu, so muss $3 + 6k \equiv 1 \bmod 5$ gelten. Daraus folgt $k \equiv 3 \bmod 5$ sowie $x \in 21 + 30\ell$ für ein $\ell \in \mathbb{N}$. Die letzte Gleichung verlangt $21 + 30\ell \equiv 6 \bmod 7$, also $2\ell \equiv 6 \bmod 7$. Dies ist äqivalent mit $\ell \equiv 3 \bmod 7$. Wir erhalten die eindeutige Lösung $x = 21 + 30 \cdot 3 = 111$. Sie ist die einzige positive Lösung im Bereich bis kgV$\{2, 3, 5, 7\} = 210$.

1.10. Sei x_0 eine Lösung des gegebenen Systems. Dann gilt $x_0 - a = rn$ und $x_0 - b = sm$ und folglich $a - b = (x_0 - b) - (x_0 - a) = sm - rn$. Wir setzen $d = \text{ggT}(n, m)$.

Die Linearkombination $sm - rn$ ist ein Vielfaches von d. Also gilt $d \mid (a - b)$. Sei umgekehrt $d \mid (a - b)$. Dann gilt $a - b = kd$. Ferner gibt es eine Darstellung $d = nx + my$. Folglich ist $a - knx = b + kmy = x_0$ eine Lösung des Systems. Sei $t = \text{kgV}(n, m)$. Dann gilt $dt = nm$. Ferner gibt es $v, w \in \mathbb{Z}$ mit $n = vd$, $m = wd$. Folglich ist $t = vwd$. Sei nun x_0 eine Lösung des gegebenen Systems. Dann gilt $x_0 \equiv a \mod vd$ und $x_0 \equiv b \mod wd$. Wegen $t = vwd$ sind damit die Kongruenzen $x_0 + kt \equiv a \mod vd$ und $x_0 + kt \equiv b \mod wd$ für beliebige $k \in \mathbb{Z}$ erfüllt; d. h., x_0 ist eindeutig modulo $t = \text{kgV}(n, m)$.

1.11. (a) Nach dem kleinen Satz von Fermat gilt $n^5 \equiv n \mod 2$, $n^5 \equiv n \mod 3$ und $n^5 \equiv n \mod 5$. Mit dem Chinesischen Restsatz folgt $n^5 \equiv n \mod 30$.

1.11. (b) Es gilt

$$3^{n^4+n^2+2n+4} \equiv 0 \equiv 21 \quad \mod 3 \tag{B.2}$$

$$3^{n^4+n^2+2n+4} \equiv 1 \equiv 21 \quad \mod 4 \tag{B.3}$$

$$3^{n^4+n^2+2n+4} \equiv 1 \equiv 21 \quad \mod 5 \tag{B.4}$$

Hierbei gilt (B.2), da die linke Seite für alle $n \in \mathbb{N}$ durch 3 teilbar ist. Die Gleichung (B.3) ist wahr, da für die Basis $3 \equiv -1 \mod 4$ gilt und der Exponent für alle $n \in \mathbb{N}$ gerade ist. Da $\text{ggT}(3, 5) = 1$ gilt, folgt (B.4) aus dem kleinen Satz von Fermat, denn für den Exponenten gilt $n^4 + n^2 + 2n + 4 \equiv 0 \mod 4$. Diese Kongruenz rechnet man leicht für alle $n \in \{-1, 0, 1, 2\}$ nach. Die Behauptung folgt nun aus dem Chinesischen Restsatz.

1.11. (c) Mit $64 \equiv 7 \mod 57$ ergibt sich $7^{n+2} + 8^{2n+1} = 49 \cdot 7^n + 8 \cdot (8^2)^n = 49 \cdot 7^n + 8 \cdot (64)^n \equiv 49 \cdot 7^n + 8 \cdot 7^n = 57 \cdot 7^n \equiv 0 \mod 57$.

1.12. Aus $\text{ggT}(a, p) = 1$ folgt mit dem kleinen Satz von Fermat, dass $a^{p-1} \equiv 1 \mod p$. Da $p - 1 = 2k$ gerade ist, folgt aus $\text{ggT}(a, 4) = 1$ mit dem Satz von Euler $a^{p-1} = a^{2k} = (a^{\varphi(4)})^k \equiv 1^k \equiv 1 \mod 4$. Wegen $\text{ggT}(4, p) = 1$ ergibt sich die Behauptung nun mit dem Chinesischen Restsatz.

1.13. (a) $|(\mathbb{Z}/51\mathbb{Z})^*| = \varphi(51) = \varphi(3)\varphi(17) = 2 \cdot 16 = 32$.

1.13. (b) Es gilt $\varphi(51) = 32 = 3 \cdot 11 - 1$. Damit gilt $3 \cdot 11 \equiv 1 \mod 32$ und es ergibt sich der geheime Schlüssel $s = 3$.

1.13. (c) $7^3 \equiv 49 \cdot 7 \equiv -2 \cdot 7 \equiv -14 \equiv 37 \mod 51$, d. h. $x = 37$.

1.13. (d) Nach dem Satz von Lagrange 1.22 teilt die Ordnung eines Elements die Gruppenordnung; da 10 kein Teiler von 32 ist, gibt es keine Elemente der Ordnung 10.

1.13. (e) Sowohl in $(\mathbb{Z}/3\mathbb{Z})^*$ als auch in $(\mathbb{Z}/17\mathbb{Z})^*$ haben alle Elemente a die Eigenschaft $a^{16} = 1$. Nach dem Chinesischen Restsatz haben auch in $(\mathbb{Z}/51\mathbb{Z})^*$ alle Elemente diese Eigenschaft. Insbesondere gibt es keine Elemente der Ordnung 32. Also ist $(\mathbb{Z}/51\mathbb{Z})^*$ nicht zyklisch.

1.14. Wider Erwarten hat der Haushaltsausschuss diesmal richtig gelegen. Es kommt nach dem chinesischen Restsatz und bis auf Symmetrie in p und q nur darauf an, dass immer $x^{es} \equiv x \mod p$ gilt. Dies ist schon erfüllt, wenn $es \equiv 1 \mod k$ für ein Vielfaches $k \in (p-1)\mathbb{Z}$ gilt. Nach Auflösung der Symmetrie reicht uns $k \in \mathrm{kgV}(p-1, q-1)\mathbb{Z}$. Für $p = 7$ und $q = 19$ mit $e = 5$ hätte der Haushaltsausschuss $s = 11$ statt $s = 65$ empfohlen.

1.15. (a) Es gilt $d(c(x)) = (x^e \mod n)^s \mod n \equiv x^{es} \equiv x^{1+k(p-1)} \equiv x \mod p$ für $k \in \mathbb{N}$. Analog zeigt man $d(c(x)) \equiv x \mod q$ und $d(c(x)) \equiv x \mod r$. Mit dem Chinesischen Restsatz folgt $d(c(x)) \equiv x \mod n$. Mit $x \in \{0, \ldots, n-1\}$ erhalten wir schließlich $d(c(x)) = x$.

1.15. (b) Es gilt $\varphi(66) = 20$ und $1 \equiv 21 = 3 \cdot 7 \equiv 3 \cdot 27 \mod 20$. Dies liefert den Enschlüsselungsexponenten $s = 3$. Es folgt $14^3 \equiv 0^3 \equiv 0 \mod 2$, $14^3 \equiv (-1)^3 \equiv -1 \equiv 2 \mod 3$ und $14^3 \equiv 3^3 \equiv 27 \equiv 5 \mod 11$. Mit dem chinesischen Restsatz erhalten wir $x = 38$.

1.16. Da Oskar die teilerfremden Zahlen e_1 und e_2 kennt, kann er mit Hilfe des euklidischen Algorithmus Zahlen $a, b \in \mathbb{Z}$ mit $ae_1 + be_2 = 1$ berechnen. Mit ihnen und den verschlüsselten Nachrichten $m^{e_1} \mod n$ und $m^{e_2} \mod n$ erhält er $(m^{e_1})^a \cdot (m^{e_2})^b = m^{ae_1+be_2} \equiv m \mod n$.

1.17. Wenn zwei der Zahlen n_1, n_2, n_3 nicht teilerfremd sind, kann Oskar deren größten gemeinsamen Teiler berechnen und erhält damit eine Faktorisierung. Nehmen wir also an, dass n_1, n_2 und n_3 teilerfremd sind. Dann kann aus den verschlüsselten Nachrichten $m^3 \mod n_i$ mit Hilfe des chinesischen Restsatzes eine Zahl $x \in \{1, \ldots, n_1 \cdot n_2 \cdot n_3\}$ mit $x \equiv m^3 \mod n_1 \cdot n_2 \cdot n_3$ bestimmt werden. Da $m < n_i$ ist, folgt $m^3 < n_1 \cdot n_2 \cdot n_3$ und somit $x = m^3$. Also kann die Nachricht $m = \sqrt[3]{x}$ durch Wurzelziehen bestimmt werden.

1.18. Aus Satz 1.19 folgt $a^{\varphi(b)} + b^{\varphi(a)} \equiv a^{\varphi(b)} \equiv 1 \mod b$. Analog gilt $a^{\varphi(b)} + b^{\varphi(a)} \equiv 1 \mod a$. Da a und b teilerfremd sind, impliziert der chinesische Restsatz $a^{\varphi(b)} + b^{\varphi(a)} \equiv 1 \mod ab$.

1.19. (a) Es ist $F_1 = F_3 - F_2, \ldots, F_{n-1} = F_{n+1} - F_n$. Aufsummieren ergibt $F_1 + \cdots + F_n = F_{n+2} - F_2 = F_{n+2} - 1$.

1.19. (b) Für $n = 0$ gilt $\sum_{k=0}^{0} F_k^2 = F_0^2 = 0 \cdot 0 = 0 = 0 \cdot 1 = F_0 F_1$. Für $n > 0$ ergibt sich $\sum_{k=0}^{n} F_k^2 = \sum_{k=0}^{n-1} F_k^2 + F_n^2 = F_{n-1} F_n + F_n^2 = F_n(F_{n-1} + F_n) = F_n F_{n+1}$.

1.19. (c) Für $k = 1$ ist dies trivial. Sei daher $k > 1$. Dann gilt:

$$
\begin{aligned}
F_k F_{n+1} + F_{k-1} F_n &= (F_{k-1} + F_{k-2}) F_{n+1} + F_{k-1} F_n \\
&= F_{k-1} F_{n+1} + F_{k-2} F_{n+1} + F_{k-1} F_n \\
&= F_{k-1}(F_{n+1} + F_n) + F_{k-2} F_{n+1} \\
&= F_{k-1} F_{n+2} + F_{k-2} F_{n+1} \\
&= F_{(n+1)+(k-1)} = F_{n+k}
\end{aligned}
$$

1.19. (d) 1. *Mit Induktion*: Es ist $F_2 F_0 - F_1^2 = -1$ und für $n > 1$ gilt:

$$
\begin{aligned}
F_{n+1} F_{n-1} - F_n^2 &= (F_n + F_{n-1}) F_{n-1} - F_n^2 \\
&= F_n F_{n-1} + F_{n-1}^2 - F_n^2 \\
&= F_n (F_{n-1} - F_n) + F_{n-1}^2 \\
&= -F_n F_{n-2} + F_{n-1}^2 \\
&\overset{\text{IV}}{=} -(-1)^{n-1} = (-1)^n
\end{aligned}
$$

2. *Matrixbeweis*: Nach Gleichung (1.4) gilt $\left(\begin{smallmatrix} F_{n-1} & F_n \\ F_n & F_{n+1} \end{smallmatrix} \right) = \left(\begin{smallmatrix} 0 & 1 \\ 1 & 1 \end{smallmatrix} \right)^n$. Die Determinante von $\left(\begin{smallmatrix} 0 & 1 \\ 1 & 1 \end{smallmatrix} \right)$ ist -1, also ist die Determinante der rechten Seite gerade $(-1)^n$. Die Determinante der linken Matrix ist $F_{n+1} F_{n-1} - F_n^2$. Dies zeigt die Behauptung.

1.20. (a) Da f^p die identische Abbildung ist, ist f eine Bijektion. Angenommen $f^i(m) = f^j(m)$ für ein $1 \le i < j \le p$. Dann gilt $f^q(m) = m$ für $q = j - i < p$. Es folgt $f^{\text{ggT}(p,q)}(m) = m$ und dann $f(m) = m$, denn $\text{ggT}(p, q) = 1$. Wegen $f(m) = m$ ist $f^k(m) = m$ für alle $k \in \mathbb{N}$.

1.20. (b) Die Relation $m \sim n$, falls $f^i(m) = f^j(n)$ für gewisse $i, j \in \mathbb{N}$, ist eine Äquivalenzrelation. Nach Aufgabe 1.20. (a) hat die Klasse von einem $m \in M \setminus F$ genau p Elemente. Die Anzahl der Nicht-Fixpunkte ist also durch p teilbar.

1.21. (a) Wir führen den Beweis induktiv nach n: Für $n \in \{1, 2\}$ ist die Gleichung jeweils erfüllt. Weiterhin gilt:

$$
\begin{aligned}
L_{n+2} = L_{n+1} + L_n &\overset{\text{IV}}{=} (F_{n+2} + F_n) + (F_{n+1} + F_{n-1}) \\
&= (F_{n+2} + F_{n+1}) + (F_n + F_{n-1}) = F_{n+3} + F_{n+1}
\end{aligned}
$$

1.21. (b) Es gilt $\mathcal{L}_1 = \{\varnothing\}$, denn 1 folgt nach 1 modulo 1. Die Menge $\mathcal{L}_2$ besteht aus den drei Teilmengen $\varnothing$, $\{1\}$ und $\{2\}$. Sei also $n \ge 3$. Rechnen wir nicht modulo n, so ist die Anzahl der entsprechenden Teilmengen von $\{1, \ldots, n\}$ gerade die Anzahl Wörter über den Buchstaben a, b, in denen keine zwei a's hintereinander stehen. Nach Beispiel 1.25 gibt es hiervon F_{n+2} Wörter. Angenommen, wir rechnen jetzt modulo n. Die Anzahl der Teilmengen $M \in \mathcal{L}_n$ mit $1 \notin M$ ist daher F_{n+1}, denn die Einschränkung, dass 1 der Nachfolger von n ist, kommt nicht zur Geltung. Betrachte jetzt die Teilmengen $M \in \mathcal{L}_n$ mit $1 \in M$. Dann können die Positionen 2 und n nicht besetzt werden. Ferner ist $2 < n$. Also ist die Anzahl solcher M (erneut nach Beispiel 1.25) gerade F_{n-1}. Die Behauptung folgt aus Aufgabe 1.21. (a).

1.21. (c) Sei $f : \mathcal{L}_p \to \mathcal{L}_p$ definiert durch $f(M) = \{i + 1 \bmod n \mid i \in M\}$. Dann gilt $f^p(M) = M$ für alle $M \in \mathcal{L}_p$. Der einzige Fixpunkt von f ist $M = \varnothing$. Nach Aufgabe 1.20. ist daher $|\mathcal{L}_p| \equiv 1 \bmod p$.

1.22. $F_0 = 0$ und $F_1 = 1$ zusammen mit Gleichung (1.2) definiert die Zahlen $F_n \in \mathbb{F}$ für alle $n \in \mathbb{Z}$. Wähle $q \in K$ mit $q^2 = 5$. Dann ist q eine Lösung der quadratischen

Gleichung $x^2 - 5$; und es gilt $x^2 - 5 = (x - q)(x + q)$ für alle $x \in \mathbb{F}$. Daher ist q bis auf das Vorzeichen eindeutig definiert. Wir schreiben $q = \sqrt{5}$ und setzen $\varphi = \frac{1+\sqrt{5}}{2}$ sowie $\hat{\varphi} = \frac{1-\sqrt{5}}{2}$. Dies ist möglich, denn 2 ist invertierbar. Ferner gilt $\varphi - \hat{\varphi} \neq 0$, da 5 invertierbar ist. Der Beweis von Gleichung (1.3) aus Abschnitt 1.11 kann nun wörtlich übernommen werden.

In $\mathbb{Z}/11\mathbb{Z}$ gilt $F_{10} = 0$ nach Gleichung (1.3) und dem kleinen Satz von Fermat. Außerdem gilt $4^2 = 16 \equiv 5 \bmod 11$ sowie $2^{-1} \equiv 6 \bmod 11$. Mit $\varphi = -3$ und $\hat{\varphi} = 4$ ist der goldene Schnitt -3 oder 4. Nach dem kleinen Satz von Fermat gilt nun $F_{12} = \frac{\varphi^2 - \hat{\varphi}^2}{\varphi - \hat{\varphi}} = \varphi + \hat{\varphi} = 1$.

1.23. (a) Zunächst sei $\sqrt{5}$ ein neues Symbol. Wir setzen $\mathbb{F} = \mathbb{F}_p \times \mathbb{F}_p$ und schreiben ein Paar $(a, b) \in \mathbb{F}$ als Summe $a + b\sqrt{5}$. Wir addieren komponentenweise und multiplizieren durch $(a + b\sqrt{5})(c + d\sqrt{5}) = ac + 5bd + (ad + bc)\sqrt{5}$. Assoziativ- und Distributivgesetze können direkt verifiziert werden. Wir können $\mathbb{Z}/p\mathbb{Z}$ in $\mathbb{F}$ vermöge $a \mapsto a + 0\sqrt{5}$ einbetten und sehen auch, dass $\sqrt{5}^2 = (0 + \sqrt{5})^2 = 5 + 0\sqrt{5} = 5$ gilt. In $\mathbb{F}$ ist 5 also ein Quadrat. Hierfür wird weder benötigt, dass p eine Primzahl ist noch dass 5 kein Quadrat in $\mathbb{Z}/p\mathbb{Z}$ ist. Wir benötigen diese Eigenschaften, um zu zeigen, dass $\mathbb{F}$ ein Körper ist. Zunächst sind alle Elemente $(a, 0)$ und $(0, a)$ invertierbar, sofern $a \neq 0$ gilt, da p eine Primzahl ungleich 5 ist. Es reicht daher, ein Inverses zu $1 + b\sqrt{5}$ zu finden. Nach der binomischen Formel gilt $(1 + b\sqrt{5})(1 - b\sqrt{5}) = 1 - 5b^2$. Da 5 kein Quadrat ist, ist dies ein von Null verschiedenes Element $c \in \mathbb{F}_p$. Das Inverse zu $1 + b\sqrt{5}$ erhalten wir nun durch $(1 - b\sqrt{5})c^{-1}$.

1.23. (b) Es gilt $\sqrt{5}^{2p} = 5^p = 5 \in \mathbb{F}$ nach dem kleinen Satz von Fermat. Also ist $\sqrt{5}^p \in \mathbb{F}$ ein Element, dessen Quadrat 5 ist. Hieraus folgt $\sqrt{5}^p = \pm\sqrt{5}$, denn $q^2 = 5$ ist in $\mathbb{F}$ äquivalent mit $(q - \sqrt{5})(q + \sqrt{5}) = 0$. Der Binomialsatz $(1 + y)^p = \sum_{k=0}^p \binom{p}{k} y^{p-k}$ kann sehr leicht mit Induktion im Vorgriff auf Satz 4.3 bewiesen werden. Alle Binomialkoeffizienten $\binom{p}{k}$ sind für $1 \leq k < p$ kongruent 0 modulo p, denn die Primzahl p kann bei $\binom{p}{k} = \frac{p!}{k!(p-k)!} = \frac{p(p-1)\cdots(p-k+1)}{k!}$ nicht gekürzt werden. Wir erhalten $(1 + \sqrt{5})^p = 1 + \sqrt{5}^p$ und $(1 - \sqrt{5})^p = 1 - \sqrt{5}^p$. Hieraus folgt die Behauptung, da nach dem kleinen Fermat $2^p = 2$ gilt.

1.24. Für $p = 2$ und $p = 5$ überprüfen wir die Behauptung direkt. Sei also $2 \neq p \neq 5$. Mit $\mathbb{F}_p$ bezeichnen wir den Körper $\mathbb{Z}/p\mathbb{Z}$ und setzen $\mathbb{F} = \mathbb{F}_p$, falls 5 ein Quadrat ist in $\mathbb{F}_p$. Ansonsten adjungieren wir $\sqrt{5}$ und betrachten $\mathbb{F} = \mathbb{F}_p(\sqrt{5})$ entsprechend Aufgabe 1.23.. Wir verwenden mehrfach den kleinen Satz von Fermat und setzen $\varphi = \frac{1+\sqrt{5}}{2}$.

1. *Herleitung*: Nach Aufgabe 1.22. können wir Gleichung (1.3) benutzen, und Aufgabe 1.23. (b) zeigt $\{\varphi^p, \hat{\varphi}^p\} = \{\varphi, \hat{\varphi}\}$. Wir unterscheiden zwei Fälle. Im ersten Fall gelte $\varphi^p = \varphi$, also auch $\hat{\varphi}^p = \hat{\varphi}$. (Dies ist insbesondere der Fall, wenn 5 ein Quadrat in $\mathbb{F}_p$ ist.) Es folgt $\varphi^{p-1} = \hat{\varphi}^{p-1} = 1$, also $F_{p-1} = 0$ und $F_p = 1$. Damit ist dann auch $F_{p+1} = 1$.

Im zweiten Fall gilt $\varphi^p = \hat{\varphi}$. (Insbesondere ist 5 kein Quadrat in $\mathbb{F}_p$.) Dann gilt auch $\hat{\varphi}^p = \varphi$. Es folgt $F_p = -1$. Wegen $\varphi \cdot \hat{\varphi} = -1$ ist ferner $F_{p+1} = 0$ und damit $F_{p-1} = 1$.

Sind wir also in einem Fall $F_{p-1} = 1$, wie etwa bei $p = 7$ mit $F_6 = 8$ oder $p = 13$ mit $F_{12} = 144$, so können wir schließen, dass 5 kein Quadrat in $\mathbb{F}_p$ ist. (Tatsächlich gilt $F_{p-1} \equiv 1 \bmod p$ genau dann, wenn 5 kein Quadrat in $\mathbb{F}_p$ ist; und $\sqrt{5}^p = -\sqrt{5}$ ist äquivalent mit $F_{p-1} = 1$.)

2. *Matrixbeweis*: Wir benutzen Kenntnisse der linearen Algebra. Die Spur der Matrix $\begin{pmatrix} F_{p-1} & F_p \\ F_p & F_{p+1} \end{pmatrix}$ ist die Summe der Diagonalelemente $F_{p+1} + F_{p-1}$. Nach Gleichung (1.4) reicht es, die Spur von $\begin{pmatrix} 0 & 1 \\ 1 & 1 \end{pmatrix}^p$ als $1 \in \mathbb{F}$ nachzuweisen. Hierfür diagonalisieren wir $\begin{pmatrix} 0 & 1 \\ 1 & 1 \end{pmatrix}$. Die Eigenwerte dieser Matrix berechnen sich aus der Lösung des linearen Gleichungssystems $\lambda x = y$ und $\lambda y = x + y$. Die Eigenwerte sind also gerade $\varphi = \frac{1+\sqrt{5}}{2}$ und $\hat{\varphi} = \frac{1-\sqrt{5}}{2}$. Es gilt $\varphi \neq \hat{\varphi}$ und die Matrix $\begin{pmatrix} 0 & 1 \\ 1 & 1 \end{pmatrix}$ kann über $\mathbb{F}$ durch $\begin{pmatrix} \varphi & 0 \\ 0 & \hat{\varphi} \end{pmatrix}$ diagonalisiert werden. Wir erhalten $\begin{pmatrix} \varphi & 0 \\ 0 & \hat{\varphi} \end{pmatrix}^p = \begin{pmatrix} \varphi^p & 0 \\ 0 & \hat{\varphi}^p \end{pmatrix}$. Die Exponentation mit p liefert nach Aufgabe 1.23. (b) nun $\{\varphi^p, \hat{\varphi}^p\} = \{\varphi, \hat{\varphi}\}$. Da die Spur einer Matrix nicht von der gewählten Basis abhängt, folgt die Behauptung wegen $\varphi + \hat{\varphi} = 1$.

1.25. Der euklidische Algorithmus berechnet auf eine Eingabe ℓ, k mit $\ell \geq k \geq 0$ wie üblich eine Folge $q_n, q_{n-1}, \dots, q_1, 0$ mit $\ell = q_n$, $k = q_{n-1}$ und $q_1 = \mathrm{ggT}(\ell, k)$. Setzen wir $g_i = |q_i|$, so gelten $g_0 = 0$, $g_1 \geq 1$ und $g_{m+1} \geq g_m + 2g_{m-1}$ für $1 \leq m < n$. Die quadratische Gleichung $x^2 = 1 + 2x$ hat die Lösungen $x = \sqrt{2} \pm 1$. Wie in Abschnitt 1.12 erhalten wir hieraus $g_n \leq ((\sqrt{2}+1)^n + (1-\sqrt{2})^n)/2$. Dies liefert die Behauptung.

Zu Kapitel 2

2.1. (a) Für $n = 0$ gilt $(1+x)^0 = 1 = 1 + 0x$. Sei jetzt $n > 0$. Mit Induktion gilt
$(1+x)^n = (1+x)(1+x)^{n-1} \geq (1+x)(1+(n-1)x)$
$= 1 + nx + (n-1)x^2 \geq 1 + nx$.

2.1. (b) Aus $e^x \geq 1 + x$ folgt $x = e^{\ln x} \geq 1 + \ln x$. Wir zeigen $e^x \geq 1 + x$.

1. *Reihendarstellung*: Es ist $e^x = \sum_{n \geq 0} x^n/n!$. Zu zeigen ist $\sum_{n \geq 2} x^n/n! \geq 0$. Nun ist $x^n/n! \geq -x^{n+1}/(n+1)!$ für gerade n äquivalent mit $n + 1 \geq -x$. Insbesondere gilt die Behauptung für $-1 \leq x$, indem wir immer zwei Summanden zusammenfassen. Also ist $e^{x/m} > 0$ für alle x, wenn nur m groß genug ist. Deshalb ist $e^x = (e^{x/m})^m > 0$ für alle x. Schließlich ist $1 + x < 0$ für $x < -1$. Damit gilt die Behauptung für alle $x \in \mathbb{R}$.

2. *Kurvendiskussion*: Die Funktion $f(x) = e^x - x - 1$ hat ein Minimum bei $x = 0$ (die Ableitung $e^x - 1$ wird nur dort Null). Ferner geht $f(x)$ gegen Unendlich für $x \to \pm\infty$. Da $x = 0$ eine Nullstelle von f ist, ist dies also die einzige Nullstelle, und sonst gilt $f(x) > 0$. Damit ist die Ungleichung gezeigt.

2.1. (c) Für $-x \le n \ne 0$ stehen auf beiden Seiten nicht negative Zahlen. Die Behauptung folgt, indem wir auf beiden Seiten der Ungleichung die n-te Wurzel ziehen und danach in Aufgabe 2.1. (b) einsetzen.

2.1. (d) Wir betrachten die Funktion $f(x) = \ln(x+1) - \frac{x}{x+1}$. Diese hat eine Nullstelle bei $x = 0$. Ferner nimmt f dort auch ihr Minimum an, denn die Ableitung $f'(x) = \frac{1}{x+1} - \frac{1}{(x+1)^2} = \frac{x}{(x+1)^2}$ ist positiv für $x > 0$ und negativ für $x < 0$.

2.2. Wir können π sortieren, indem wir nacheinander Situationen mit $b_{\pi(i)} > b_{\pi(i+1)}$ betrachten und dann $b_{\pi(i)}$ und $b_{\pi(i+1)}$ vertauschen. Es reicht zu zeigen, dass $S(\pi)$ bei einer solchen Vertauschung nicht abnimmt. Dies ist eine rein lokale Situation, daher dürfen wir $n = 2$ annehmen. Sei also $a_1 \le a_2$ und $b_1 \le b_2$. Zu vergleichen sind die Summen $S = a_1 b_1 + a_2 b_2$ und $S' = a_1 b_2 + a_2 b_1$. Die Differenz $S - S'$ ist nicht negativ wegen $a_1 b_1 + a_2 b_2 - a_1 b_2 - a_2 b_1 = a_1(b_1 - b_2) + a_2(b_2 - b_1) = (a_2 - a_1)(b_2 - b_1)$.

2.3. Es gilt

$$H = \frac{n}{\sum_i a_i^{-1}} \ge \frac{n}{\sum_{i=1}^n (\min_j a_j)^{-1}} = \min_j a_j$$

und

$$Q = \sqrt{n^{-1} \sum_{i=1}^n a_i^2} \le \sqrt{n^{-1} \sum_{i=1}^n (\max_j a_j)^2} = \max_j a_j$$

Als Nächstes zeigen wir $G \le A$. Der Beweis ist mit Induktion. Für $n = 1$ ist die Ungleichung erfüllt. Sei nun $n > 1$. Sind alle a_i gleich, so gilt auch $G = A$. Andernfalls können wir ohne Einschränkung annehmen, dass $a_1 > A$ und $a_2 < A$. Nun setzen wir $y = a_1 + a_2 - A$. Dann ist $(n-1)A = y + a_3 + \cdots + a_n$ und somit ist A auch das arithmetische Mittel von $y, a_3, \ldots a_n$. Ferner gilt $yA - a_1 a_2 = a_1 A + a_2 A - A^2 - a_1 a_2 = (a_1 - A)(A - a_2) > 0$. Also folgt mit Induktion $A^n = A \cdot A^{n-1} \ge A \cdot y \cdot a_3 \cdots a_n \ge a_1 \cdots a_n$. Zu $H \le G$: Es gilt $H = n / \sum_{i=1}^n a_i^{-1} = \prod_{j=1}^n a_j / (n^{-1} \sum_{i=1}^n \prod_{j \ne i} a_j)$. Im Nenner steht hier also ein arithmetisches Mittel. Nach dem eben gezeigten ist dies größer oder gleich dem geometrischen Mittel:

$$H \le \frac{\prod_{j=1}^n a_j}{\sqrt[n]{\prod_{i=1}^n \prod_{j \ne i} a_j}} = \frac{\prod_{j=1}^n a_j}{\sqrt[n]{\left(\prod_{j=1}^n a_j\right)^{n-1}}} = G$$

Zu $A \le Q$: Wir verwenden wieder $G \le A$ und erhalten

$$\sum_{i=1}^n \sqrt{n}^{-1} \cdot a_i \Big/ \sqrt{\sum_{j=1}^n a_j^2} = \sum_{i=1}^n \sqrt{\sqrt{n}^{-2} \cdot \left(a_i \Big/ \sqrt{\sum_{j=1}^n a_j^2}\right)^2}$$

$$\le \sum_{i=1}^n \left(\frac{1}{2n} + \frac{a_i^2}{2 \sum_{j=1}^n a_j^2}\right) = 1$$

Durch Multiplikation mit $\sqrt{\sum_{i=1}^n a_i^2} / \sqrt{n}$ ergibt sich die gewünschte Ungleichung.

2.4. Sei $s > 1$. Die Funktion $x \mapsto \frac{1}{x^s}$ ist monoton fallend für $x > 0$. Damit ist $\frac{1}{i^s} \leq \int_{i-1}^{i} \frac{1}{x^s}\,dx$ und es gilt also $\sum_{i \geq 1} \frac{1}{i^s} \leq 1 + \int_1^\infty \frac{1}{x^s}\,dx < \infty$. Sei nun $s = 1$ und $k \geq 1$, d.h., wir betrachten die harmonische Reihe. Dann ist $\sum_{i=2^{k-1}+1}^{2^k} i^{-1} \geq \sum_{i=2^{k-1}+1}^{2^k} 2^{-k} = 1/2$. Für alle $n \geq 1$ ist also $\sum_{i=1}^{2^n} \frac{1}{i} > n/2$.

2.5. Wir zeigen $n \ln n - n \leq n\,\bar t(n) \leq n \ln n + n$. Jede Zahl k wird in der Summe $\sum_{i=1}^{n} t(i)$ genau bei den Zahlen $k, 2k, \ldots, \lfloor \frac{n}{k} \rfloor k$ einmal als Teiler gezählt. Daraus folgt $n\,\bar t(n) = \sum_{k=1}^{n} \lfloor \frac{n}{k} \rfloor$. Hierfür erhalten wir schließlich die Abschätzungen

$$\sum_{k=1}^{n} \left\lfloor \frac{n}{k} \right\rfloor \leq \sum_{k=1}^{n} \frac{n}{k} \leq n \sum_{k=1}^{n} \frac{1}{k} \leq n\left(1 + \int_1^n \frac{1}{x}\,dx\right) \leq n + n \ln n$$

$$\sum_{k=1}^{n} \left\lfloor \frac{n}{k} \right\rfloor \geq \sum_{k=1}^{n} \left(\frac{n}{k} - 1\right) \geq -n + n \int_1^n \frac{1}{x}\,dx \geq -n + n \ln n$$

2.6. Nach Gleichung (2.9) gilt $\pi(n) \leq \frac{3n}{\log_2 n}$ für fast alle n. Damit muss sogar der mittlere Abstand zwischen Primzahlen wachsen. Eine elementare Lösung der Aufgabe kannte schon Euklid: Die $n-1$ Zahlen $n!+2, n!+3, \ldots, n!+n$ sind alle zusammengesetzt, da $n!+i$ für $1 \leq i \leq n$ durch i teilbar ist. Wenn p_i die größte Primzahl ist mit $p_i < n!+2$, dann folgt $p_{i+1} > n!+n$ und damit $p_{i+1} - p_i \geq n$.

2.7.(a) Aus $m/\log m \leq \pi(m)$ folgt mit $m = p_n$, dass $p_n \leq n \log p_n$ gilt (da $\pi(p_n) = n$). Für $p_n \leq 2n \log n$ ist diese Ungleichung erfüllt.

2.7.(b) Für jede genügend große Zahl m gilt $\pi(m) \leq 3m/\log m$ nach Gleichung (2.9). Mit $m = p_n$ ergibt sich $p_n \geq \frac{1}{3} n \log p_n$. Aus $p_n \geq n$ folgt daraus die Behauptung.

2.7.(c) *Elementare Lösung*: Angenommen, die Reihe konvergiert. Dann gilt $\sum_{i \geq k} \frac{1}{p_i} \leq \frac{1}{2}$ für einen genügend großen Index k. Für $n \in \mathbb{N}$ sei M_n die Menge der Zahlen aus $\{1, \ldots, n\}$, deren Primteiler alle kleiner als p_k sind. Jede Zahl $x \in M_n$ lässt sich eindeutig als Produkt $x = rs^2$ schreiben, wobei r quadratfrei ist. Für r gibt es nur konstant viele Möglichkeiten und für s gilt $s \leq \sqrt{n}$. Damit ist $|M_n| \in \mathcal{O}(\sqrt{n})$. Für jedes $i \geq 1$ ist die Anzahl der Zahlen aus $\{1, \ldots, n\}$, die durch p_i teilbar sind, kleiner oder gleich n/p_i, denn nur jede p_i-te Zahl ist durch p_i teilbar. Hieraus folgt

$$n \leq \sum_{i \geq k} \frac{n}{p_i} + \mathcal{O}(\sqrt{n}) \leq \frac{n}{2} + \mathcal{O}(\sqrt{n})$$

Dies ein Widerspruch.

Lösung mittels Primzahldichte: Mit p bezeichnen wir Primzahlen. Nach Satz 2.6 liegen zwischen n und $2n$ bereits $\Theta(n/\log n)$ Primzahlen. Daher ist $\sum_{2^k < p < 2^{k+1}} \frac{1}{p} \in \Theta(1/k)$ und damit

$$\sum_{p \leq 2^k} \frac{1}{p} \in \Theta(\log k)$$

Es folgt $\sum_{p \leq n} \frac{1}{p} \in \Theta(\log \log n)$.

Zu Kapitel 3

3.1. Es gibt $2^{10} = 1024$ Schussfolgen, die alle gleich wahrscheinlich sind. Hiervon ist eine Folge dabei, bei der er nie trifft. Bei zehn Folgen landet er genau einen Treffer und bei $\binom{10}{2} = 45$ Folgen sind es genau zwei. Also verbleiben $1024 - 56 = 968$ Schussfolgen mit mindestens drei Treffern. Die Wahrscheinlichkeit ergibt sich zu $\frac{968}{1024} = \frac{121}{128}$, dies sind nach kaufmännischer Rundung 95%.

3.2. (a) Es gibt insgesamt $2^4 = 16$ verschiedene Möglichkeiten welchen Geschlechts die vier Kinder sein können – von vier Jungen ($jjjj$) bis vier Mädchen ($mmmm$).
Damit gibt es vier verschiedene Möglichkeiten (das erstgeborene Kind ist ein Mädchen ($mjjj$), das zweitgeborene ist ein Mädchen ($jmjj$) usw.). Wir erhalten insgesamt:

$$\Pr[\text{genau ein Mädchen}] = 4 \cdot \frac{1}{16} = \frac{1}{4}$$

3.2. (b) Wenn die ersten beiden Geschlechter feststehen, gibt es für das dritte und vierte Kind genau vier Möglichkeiten, es ist deshalb

$$\Pr[\text{erstes und zweites Kind ein Junge}] = \frac{1}{4}$$

3.2. (c) Die Wahrscheinlichkeit, dass genau zwei Kinder männlich sind, ist $\frac{6}{16}$ (es ist $\binom{4}{2} = 6$), die Wahrscheinlichkeit, dass genau drei Kinder männlich sind, beträgt $\frac{4}{16}$ (es ist $\binom{4}{3} = 4$) und die Wahrscheinlichkeit, dass genau vier Kinder männlich sind, ist $\frac{1}{16}$. Insgesamt erhalten wir

$$\Pr[\text{mindestens zwei Kinder männlich}] = \frac{6}{16} + \frac{4}{16} + \frac{1}{16} = \frac{11}{16}$$

3.2. (d)

$$\Pr[\text{alle Kinder weiblich}] = \frac{1}{16}$$

3.3. Wir nehmen an, dass Alice a wählt und Bob b. Die Anzahl der Paare (a, b) mit $a = b$ ist m. Wir zählen jetzt zunächst die Paare mit $|a - b| \leq n$ und $a < b \leq n$. Deren Anzahl ist $\sum_{j=1}^{n-1} j = \frac{n(n-1)}{2}$. Die Anzahl der Paare mit $|a - b| \leq n$ und $a < b$ sowie $n + 1 \leq b \leq m$ ist nun $(m - n)n$. Für $a < b$ ergibt sich eine Mächtigkeit von $\frac{n(n-1)}{2} + (m-n)n$. Die Anzahl $|\{(a, b) \mid a, b \in M \text{ und } |a - b| \leq n\}|$ errechnet sich damit zu

$$m + 2\left(\frac{n(n-1)}{2} + n(m - n)\right) = m + 2mn - n^2 - n$$

Die Anzahl der Paare ist m^2. Also ergibt sich die gesuchte Wahrscheinlichkeit zu $\frac{1+2n}{m} - \frac{n^2-n}{m^2}$.

3.4. Es sei $Q(n)$ die mittlere Zahl der Vergleiche, wenn alle Positionen für das Pivotelement gleich wahrscheinlich sind. Die Lösung lautet $Q(n) = 2(n + 1)H_n - 4n$, wobei H_n die n-te harmonische Zahl ist.

Herleitung über Zufallsvariablen: Wir bezeichnen mit π eine Reihenfolge der Pivotelemente. Für $i < j$ sei X_{ij} die 0-1-wertige Zufallsvariable mit $X_{ij}(\pi) = $ „i wird mit j verglichen". Im Laufe von Quicksort werden i und j maximal einmal verglichen. Damit ist

$$Q(n) = \sum_{1 \le i < j \le n} E[X_{ij}]$$

Es gilt $X_{ij}(\pi) = 1$ genau dann, wenn eines der beiden Elemente i und j als ein frühestes Pivot-Element im Intervall $[i, j]$ gezogen wird. Hieraus folgt $E[X_{ij}] = \frac{2}{j-i+1}$; also

$$Q(n) = \sum_{1 \le i < j \le n} \frac{2}{j - i + 1} = \sum_{i=1}^{n-1} \sum_{d=1}^{n-i} \frac{2}{d + 1} = 2 \sum_{d=1}^{n-1} \frac{n - d}{d + 1}$$

$$= 2 \sum_{d=2}^{n} \frac{n + 1 - d}{d} = -2n + 2 \sum_{d=1}^{n} \frac{n + 1 - d}{d}$$

$$= -2n + 2(n + 1)H_n - 2n = 2(n + 1)H_n - 4n$$

Herleitung durch Rekursion: Wir benötigen $n - 1$ Vergleiche beim Pivotieren. Also gilt $Q(1) = 0$ und für $n \ge 2$:

$$Q(n) = (n - 1) + \frac{1}{n} \sum_{i=1}^{n} (Q(i - 1) + Q(n - i)) = (n - 1) + \frac{2}{n} \sum_{i=1}^{n} Q(i - 1)$$

Hieraus folgt $nQ(n) = n(n - 1) + 2 \sum_{i=1}^{n} Q(i - 1)$. Eine Subtraktion der jeweiligen Seiten für n und $n - 1$ liefert:

$$nQ(n) - (n - 1)Q(n - 1) = 2(n - 1) + 2Q(n - 1)$$

Wir erhalten:

$$nQ(n) = 2(n - 1) + 2Q(n - 1) + (n - 1)Q(n - 1)$$

$$= 2(n - 1) + (n + 1)Q(n - 1)$$

Eine weitere Umformung ergibt nun:

$$\frac{Q(n)}{n + 1} = \frac{2(n - 1)}{n(n + 1)} + \frac{Q(n - 1)}{n} = \frac{2(n - 1)}{n(n + 1)} + \frac{2(n - 2)}{(n - 1)n} + \frac{Q(n - 2)}{n - 1}$$

$$= \sum_{k=1}^{n} \frac{2(k - 1)}{k(k + 1)} = 2 \left(\sum_{k=1}^{n} \frac{2}{k + 1} - \sum_{k=1}^{n} \frac{1}{k} \right) = 2H_n + \frac{4}{n + 1} - 4$$

3.5. Es sei $Q(n)$ die durchschnittliche Anzahl an Vergleichen, um das k-te Element in π zu finden. Dabei halten wir k fest, und wir bezeichnen mit i, j Werte mit $1 \le i <$

$j \leq n$. Im Laufe von Quickselect werden i und j (genau wie bei Quicksort) maximal einmal verglichen. Damit ist

$$Q(n) = \sum_{1 \leq i < j \leq n} E[X_{ij}]$$

Falls i und j verglichen werden, ist i oder j aktuelles Pivotelement. Der Erwartungswert $E[X_{ij}]$ hängt von der relativen Position von i und j zu k ab. Wir unterscheiden drei Fälle.

1. Fall: Für $i < j \leq k$ gilt $X_{ij}(\pi) = 1$ genau dann, wenn eines der beiden Elemente i und j als ein frühestes Pivot-Element im Intervall $[i,k]$ gezogen wird. Hieraus folgt $E[X_{ij}] = \frac{2}{k-i+1}$; und damit ergibt sich:

$$\sum_{1 \leq i < k} \sum_{i < j \leq k} E[X_{ij}] = 2 \sum_{1 \leq i < k} \frac{k-i}{k-i+1} = 2 \sum_{1 \leq i < k} \left(1 - \frac{1}{k-i+1}\right)$$

$$= 2(k - H_k) < 2(k - \ln k)$$

2. Fall: Für $k \leq i < j$ folgt vollkommen analog

$$\sum_{k < j \leq n} \sum_{k \leq i < j} E[X_{ij}] = 2(n - k - H_{n-k}) < 2(n - k - \ln(n - k))$$

3. Fall: Für $i < k < j$ gilt jetzt $E[X_{ij}] = \frac{2}{j-i+1}$. Damit ergibt sich eine etwas kompliziertere Rechnung:

$$\sum_{1 \leq i < k} \sum_{k < j \leq n} E[X_{ij}] = 2 \sum_{1 \leq i < k} \sum_{k < j \leq n} \frac{1}{j - i + 1}$$

$$= 2 \sum_{1 \leq i < k} \left(\frac{1}{k-i+2} + \cdots + \frac{1}{n-i+1}\right)$$

$$< 2 \sum_{1 \leq i < k} (\ln(n-i+1) - \ln(k-i)) = 2 \ln\binom{n}{k-1}$$

Addieren wir nun die drei Fälle und benutzen $\binom{n}{k-1} < 2^n$, so ergibt sich die Behauptung

$$Q(n) < 2n + 2\ln\binom{n}{k-1} < 2(1 + \ln 2)n$$

Man kann aus den Rechnungen noch mehr herausholen. Ist k sehr nahe an $n/2$, so ist die Abschätzung bis auf log-Terme genau. Wir erwähnen ohne Beweis $2(1 + \ln 2)n \in Q(n) + \mathcal{O}(\log n)$.

3.6. Es gilt $E[X] = \sum_{k=1}^n k \Pr[X = k] = n/H_n \sim n/\ln n$. Für die Varianz erhalten wir

$$\mathrm{Var}[X] = \left(\sum_{k=1}^n k^2 \Pr[X = k]\right) - n^2/H_n^2 = \frac{H_n\binom{n+1}{2} - n^2}{H_n^2} \sim \frac{n^2}{2\ln n}$$

Damit strebt die Standardabweichung gegen $\frac{n}{\sqrt{2\ln n}}$.

Zu Kapitel 4

4.1. (a) Sei $f \in C^{(A \times B)}$. Für jedes $a \in A$ definieren wir die Funktion g_a mit $g_a(b) = f(a, b)$. Die zu f gehörige Funktion $\hat{f}$ in $(C^B)^A$ ist dann definiert durch $\hat{f}(a) = g_a$. Man sieht, dass die Zuordnung $f \mapsto \hat{f}$ injektiv ist, denn mit $f(a, b) = (\hat{f}(a))(b)$ kann man die Funktion f rekonstruieren. Umgekehrt sei $\hat{f} \in (C^B)^A$. Dann definiert man $f \in C^{(A \times B)}$ durch $f(a, b) = (\hat{f}(a))(b)$. Also ist die Zuordnung surjektiv.

4.1. (b) Für $f \in C^{A \cup B}$ sei $\hat{f} = (f|_A, f|_B)$ das Paar der beiden Einschränkungen von f auf A und B. Dann ist $f \mapsto \hat{f}$ eine Bijektion von $C^{A \cup B}$ nach $C^A \times C^B$. Dabei lässt sich aus einem Paar (f_1, f_2) die Funktion rekonstruieren, da $A \cap B = \varnothing$ gilt und man somit

$$f(x) = \begin{cases} f_1(x) & x \in A \\ f_2(x) & x \in B \end{cases}$$

als zu (f_1, f_2) gehörige Funktion finden kann.

4.1. (c) Wir wollen einen Widerspruch erzeugen und gehen davon aus, dass $f : A \to 2^A$ eine surjektive Abbildung ist. Dann betrachten wir die Menge $B = \{a \in A \mid a \notin f(a)\}$. Da f surjektiv ist, gibt es ein $b \in A$ mit $f(b) = B$. Wie man die Sache auch dreht und wendet, es ergibt sich ein Widerspruch:

$$b \in B \Leftrightarrow b \in f(b) \Leftrightarrow b \notin B$$

Die erste Äquivalenz ist die Definition von b, die zweite ergibt sich aus der Definition von f.

4.2. Verteile 9 Stellen für die Zahlen auf die 10 Ziffern, wobei die Ziffer 0 immer getroffen wird und keine Ziffer zweimal getroffen wird. Wir erhalten als gesuchte Zahl $\frac{10!}{(10-9)!} - 9! = 10! - 9! = 3265920$.

4.3. (a) Es müssen 4 der 15 Frauen und unabhängig davon 4 der 12 Männer ausgewählt werden. Dafür gibt es $\binom{15}{4}\binom{12}{4} = 1365 \cdot 495 = 675\,675$ Möglichkeiten.

4.3. (b) Es gibt $\binom{15}{8} + \binom{15}{7}\binom{12}{1} = 6435 + 6435 \cdot 12 = 83\,655$ Möglichkeiten, dass *maximal ein* Mann dabei ist. Daher gibt es $\binom{15+12}{8} - 83\,655 = 2\,136\,420$ Möglichkeiten, dass mindestens zwei Männer in der Kommission sind.

4.3. (c) Es gibt $\sum_{i=5}^{8} \binom{12}{i}\binom{15}{8-i} = 792 \cdot 455 + 924 \cdot 105 + 792 \cdot 15 + 495 \cdot 1 = 469\,755$ Möglichkeiten, dass mindestens 5 Männer enthalten sind.

4.4. Es gibt 81 Wörter der Länge 4 über dem Alphabet $\{b, s, w\}$. Jedem Wort ordnen wir ein Muster zu, indem wir die Seiten im Uhrzeigersinn lesen. Wörter bilden nur dann das gleiche Muster, wenn die Wörter zyklische Vertauschungen sind, aber nicht jede der vier zyklischen Vertauschungen liefert ein neues Wort.

Die Wörter, in denen nur jeweils ein Buchstabe vorkommt, entsprechen genau den drei Mustern, bei denen alle Seiten gleich sind. Es gibt ebenfalls drei Muster,

bei denen genau die gegenüberliegenden Seiten gleich sind. Jedem dieser Muster entsprechen zwei Wörter. Es verbleiben 72 Wörter, deren zyklische Vertauschung der Buchstaben jeweils ein neues Wort ergibt. Also gibt es insgesamt $24 = 72/4 + 6/2 + 3 = 18 + 3 + 3$ verschiedene Muster.

4.5.

$$\underbrace{\sum_{i \text{ gerade}} \binom{n}{i}}_{\substack{\text{Anzahl Teilmengen mit} \\ \text{gerade vielen Elementen}}} - \underbrace{\sum_{i \text{ ungerade}} \binom{n}{i}}_{\substack{\text{Anzahl Teilmengen mit} \\ \text{ungerade vielen Elementen}}} = \sum_i \binom{n}{i} (-1)^i = (1 - 1)^n = 0$$

4.6. (a) Wir wollen von n weißen Objekten einen Teil rot färben und m Objekte blau färben. Eine Vorgehensweise um eine solche Färbung zu erhalten, ist m Objekte blau zu färben. Von den verbleibenden $n - m$ Objekten färben wir einen Teil rot. Es gibt $\binom{n}{m} \cdot 2^{n-m}$ Möglichkeiten, auf diese Weise eine geeignete Färbung zu erzeugen. Eine andere Vorgehensweise ist, einen Teil der weißen Objekte rot zu färben und m der roten Objekte blau zu färben. Die Anzahl hierfür ist $\sum_k \binom{k}{m}\binom{n}{k}$. (Falls weniger als m Objekte rot gefärbt wurden, liefert dies 0 Möglichkeiten, um m davon blau zu färben.) Da beide Vorgehensweisen eindeutig eine gültige Färbung erzeugen, gilt die Behauptung.

4.6. (b) Zweimalige Anwendung des Binomialsatzes liefert

$$((x + 1) + 1)^n = \sum_k \binom{n}{k}(x + 1)^k = \sum_k \binom{n}{k}\sum_\ell \binom{k}{\ell} x^\ell = \sum_{k,\ell} \binom{n}{k}\binom{k}{\ell} x^\ell$$

Ableiten nach x und Einsetzen von $x = 1$ ergibt die Behauptung.

4.6. (c)

$$\sum_i \sum_j \binom{n}{i}\binom{n + i}{j} = \sum_i \binom{n}{i}\sum_j \binom{n + i}{j} = \sum_i \binom{n}{i} 2^{n+i}$$

$$= 2^n \sum_i \binom{n}{i} 2^i = 2^n (2 + 1)^n = 6^n$$

4.6. (d) Um $2n + 1$ Elemente aus der Menge $M = \{0, \ldots, 2m\}$ auszuwählen gibt es $\binom{2m+1}{2n+1}$ Möglichkeiten. Wir können diese Möglichkeiten auch auf eine alternative Weise zählen. Zuerst wählen wir das mittlere Element einer $2n + 1$-elementigen Teilmenge von M und nennen es $m - k$ mit $k \in \mathbb{Z}$. Dann sind links bzw. rechts von $m - k$ noch jeweils genau $m - k$ bzw. $m + k$ Elemente in $\{0, \ldots, 2m\}$ vorhanden. Also gibt es $\binom{m-k}{n}\binom{m+k}{n}$ Möglichkeiten jeweils n davon auszuwählen. Die Summe über alle k liefert nun die Gleichung.

4.6. (e)

$$\sum_{k=1}^{m} \binom{m+1}{k} \sum_{i=1}^{n} i^k = \sum_{i=1}^{n} \sum_{k=1}^{m} \binom{m+1}{k} i^k$$

$$= \sum_{i=1}^{n} \left(\sum_{k=0}^{m+1} \left(\binom{m+1}{k} i^k \right) - 1 - i^{m+1} \right)$$

$$= \sum_{i=1}^{n} \left((i+1)^{m+1} - 1 - i^{m+1} \right)$$

$$= (n+1)^{m+1} - (n+1)$$

Die letzte Gleichung ergibt sich aus einer *Teleskopsumme*, da sich Summanden in der vorletzten Zeile wechselseitig aufheben. Ein bijektiver Beweis der Aussage ist auch möglich; als kombinatorische Interpretation der beiden Seiten können dann die nicht-konstanten Abbildungen von $\{1, \ldots, m+1\}$ nach $\{1, \ldots, n+1\}$ verwendet werden.

4.7. (a) Wir zeigen die Identität für $n \geq -1$, denn für $n = -1$ ist sie trivial, und für $n = 0$ folgt sie wegen $F_1 = \binom{0}{0} = 1$. Sei jetzt $n \geq 1$.

$$\sum_{k \leq n} \binom{n-k}{k} = \sum_{k \leq n-1} \binom{n-k}{k}$$

$$= \sum_{k \leq n-1} \left[\binom{n-k-1}{k} + \binom{n-k-1}{k-1} \right]$$

$$= \sum_{k \leq n-1} \binom{n-k-1}{k} + \sum_{k \leq n-1} \binom{n-k-1}{k-1}$$

$$\overset{\text{Indexversch.}}{=} \sum_{k \leq n-1} \binom{n-k-1}{k} + \sum_{k \leq n-2} \binom{n-k-2}{k}$$

$$\overset{\text{Induktion}}{=} F_n + F_{n-1} = F_{n+1}$$

4.7. (b) Sei $M = \binom{0\ 1}{1\ 1}$ und E die (2×2)-Einheitsmatrix. Wir erinnern uns, dass für die n-te Potenz von M gilt $M^n = \binom{F_{n-1}\ \ F_n}{F_n\ \ F_{n+1}}$. Außerdem gilt $M^2 = \binom{1\ 1}{1\ 2}$, d.h. $M^2 = M + E$. Mit dem Binomialsatz 4.3 erhalten wir die n-te Potenz $M^{2n} = (M + E)^n = \sum_i \binom{n}{i} M^i$. Insbesondere sind die oberen rechten Einträge gleich und es folgt die Behauptung.

4.7. (c) Seien M und E die Matrizen aus Teilaufgabe (4.7. (b)). Nun gilt $M^3 = \binom{1\ 2}{2\ 3}$, d.h. $M^3 = 2M + E$. Damit erhalten wir $M^{3n} = (2M + E)^n = \sum_i \binom{n}{i} 2^i M^i$. Betrachten der Einträge der oberen rechten Ecken liefert die Behauptung.

4.7. (d) Seien M und E die Matrizen aus Teilaufgabe (4.7. (b)). Dort hatten wir uns bereits überzeugt, dass $M^2 = M + E$ gilt, also ist $E = M^2 - M$. Damit gilt $E = E^n = (M^2 - M)^n = \sum_j \binom{n}{j} (-M)^j (M^2)^{n-j} = \sum_j \binom{n}{j} (-1)^j M^{2n-j}$. Wir betrachten die

rechten oberen Einträge und erhalten $0 = \sum_j \binom{n}{j}(-1)^j F_{2n-j}$. Nach einer Indexverschiebung mit $j = n - i$ erhalten wir $0 = \sum_i \binom{n}{i}(-1)^{n-i} F_{n+i}$. Nach Kürzen mit $(-1)^{n-2i}$ ergibt sich die Behauptung.

4.8. Sei $A \subseteq \{1, \ldots, n\}$ eine Menge mit $|A| = 3$. Dann ist $\mathrm{sum}(A)$ genau dann gerade, wenn (i) A drei gerade Zahlen enthält oder (ii) A zwei ungerade Zahlen und eine gerade Zahl enthält. Die Menge $\{1, \ldots, n\}$ enthält genau $\lfloor n/2 \rfloor$ gerade Zahlen und $\lceil n/2 \rceil$ ungerade Zahlen. Also gilt

$$G^{(3)}(n) = \binom{\lfloor n/2 \rfloor}{3} + \lfloor n/2 \rfloor \binom{\lceil n/2 \rceil}{2}$$

4.9. (a) Wir betrachten eine Partition der Menge $\{1, \ldots, n+1\}$ in $m+1$ Klassen. Die Klasse, welche das Element $n+1$ enthält, sei ausgezeichnet. Die restlichen Klassen der Partition enthalten zusammen k Elemente. Es gibt $\binom{n}{k}$ Möglichkeiten, diese aus der Menge $\{1, \ldots, n\}$ zu wählen, und für jede dieser Möglichkeiten gibt es $\left\{{k \atop m}\right\}$ Möglichkeiten diese k Elemente auf m Klassen zu verteilen.

4.9. (b) Sei π eine Permutation von $\{1, \ldots, n\}$ mit k Zykeln. Wir zeichnen einen dieser Zykel speziell aus. Für diese Auswahl gibt es k Möglichkeiten. Nun kodieren wir die restlichen $k-1$ Zykel in einen einzigen Zykel. Dafür werden die $k-1$ nicht markierten Zykel von π so angeordnet, dass deren kleinstes Element vorne steht. Dann werden diese $k-1$ Zykel absteigend nach ihrem kleinsten Element sortiert. Diese Anordnung bestimmt nun den zweiten zur Verfügung stehenden Zykel. Dabei wird der Beginn dieser Sortierung vom Element $n+1$ bestimmt.
    Beispiel: Wir wählen $n = 7$ und $\pi = (12)(537)(64)$ und markieren den Zykel (12). Dann ergibt sich als Anordnung der Zykel (375) und (46). Da das kleinste Element des ersten Zykels mit 3 kleiner ist als das kleinste des Zykels (46) mit 4 ergibt sich der Zykel (846375). Der Beginn wird hier mit $n+1 = 8$ kodiert, das Ende des ersten Zykels (46) erkennt man, da die darauffolgende 3 kleiner ist als das kleinste vorherige Element, diese jedoch die jeweils kleinsten Elemente des Zykels sind.

4.9. (c) Diese Aufgabe lässt sich analog zu Teilaufgabe 4.9. (b) lösen. Anstatt nur einen der Zykel zu markieren, werden nun m von k Zykeln markiert. Es gibt $\binom{k}{m}$ Möglichkeiten für diese Markierung. Die restlichen Zykel werden wie eben zusammen mit dem Element $n+1$ in den letzten zur Verfügung stehenden Zykel kodiert.

4.10. (a) Die Schlümpfe kennen natürlich alle Ausweisnummern aller hundert Schlümpfe auswendig. Sie ordnen jeder Ausweisnummer zufällig eine eindeutige Zahl aus dem Bereich 1 bis 100 zu. Diese Zuordnung prägen sich die Schlümpfe rasch ein. Sie ändern ihren Namen auf diese Zahl und wenn sie irgendeinen Ausweis sehen, kennen sie sofort den zugehörigen (neuen) Namen. Sie sind nämlich wirklich ziemlich schlau.

Wenn der Schlumpf mit dem Namen i den Schubladenraum betritt, tut er das Folgende. Er beginnt die Suche nach seinem Ausweis in der Schublade i. Findet er dort seinen Ausweis, so ist er fertig. Ansonsten findet er dort einen Ausweis zu dem der Name j gehört. Als Nächstes schaut dieser Schlumpf dann in der Schublade j nach seinem Ausweis. Findet er dort nicht seinen Ausweis, so erkennt er dennoch einen Namen k, der zum Ausweis gehört. Also wendet er sich der Schublade k zu. Dieses Verfahren wird solange wiederholt bis in 50 Schubladen geschaut wurde oder der korrekte Ausweis gefunden wurde. Eine Zuordnung der Ausweise in die Schubladen wurde vom König festgelegt, der hat natürlich versucht, es den Schlümpfen schwer zu machen. Aber er hatte keine Chance, denn die Zuordnung der Ausweise zu Namen war zufällig. Es ist irrelevant, wie die Ausweise auf die Schubladen verteilt wurden. Wir können uns vorstellen, dass die Zuordnung der Schubladennummern zu Namen eine Zufallspermutation π der Menge $\{1, \ldots, 100\}$ ist.

Der Schlumpf i wird mit diesem Verfahren mit Sicherheit dann seinen Ausweis sehen, wenn i in einem Zykel von π mit Länge kleiner als 51 liegt. Damit sind die Schlümpfe genau dann erfolgreich, falls es in π keinen Zykel der Länge größer als 50 gibt.

Wir setzen $n = 100$ und berechnen die Wahrscheinlichkeit, dass ein Zykel der Länge größer als $n/2$ bei einer Zufallspermutation auftritt. Gibt es einen solchen, so ist er eindeutig bestimmt, denn es kann keine zwei verschiedenen Zykel dieser Länge geben. Wir betrachten zuerst die Anzahl der Permutationen, die einen Zykel der Länge k für $k > n/2$ haben. Es gibt $\binom{n}{k}$ Möglichkeiten für die Trägermenge dieses Zykels. Bei fester Trägermenge gibt es $(k-1)!$ verschiedene Zykel. Für jeden so gewählten Zykel Z gibt es damit genau $(n-k)!$ Permutationen, die den Zykel Z enthalten, denn die außerhalb von Z liegenden $n-k$ Elemente können beliebig permutiert sein. Insgesamt ergibt sich also die Wahrscheinlichkeit, einen Zykel mit mehr als $n/2$ Elementen zu finden, zu:

$$\frac{1}{n!} \sum_{k=\frac{n}{2}+1}^{n} \binom{n}{k}(k-1)!(n-k)! = \sum_{k=\frac{n}{2}+1}^{n} \frac{1}{k} < \int_{n/2}^{n} \frac{1}{t}\,dt = \ln 2 \approx 0{,}69$$

Tatsächlich liegt bei 100 Schlümpfen die Wahrscheinlichkeit, dass alle ihren Ausweis finden bei etwa 31,2%. Fortes fortuna adiuvat: den Tüchtigen hilft das Glück! So kamen die Schlümpfe wieder frei.

4.10. (b) Die hier vorgestellte Lösung ist von Eugene Curtin und Max Warshauer [10]. Die Folge der Ausweise in den geöffneten Schubladen legt den Ablauf des Spiels eindeutig fest. Sei $n_1, \ldots, n_{100}$ die Folge der Ausweise. Sei $n_{i_1} = 1$. Dann hat Schlumpf 1 die Schubladen $n_1, \ldots, n_{i_1}$ geöffnet. Sei $j_2 = \min(\{1, \ldots, 100\} \setminus \{n_1, \ldots, n_{i_1}\})$. Dann ist als Zweites der Schlumpf j_2 dran. Sei $n_{i_2} = j_2$. Der Schlumpf j_2 öffnet die Schubladen $n_{i_1+1}, \ldots, n_{i_2}$. Als Drittes ist $j_3 = \min(\{1, \ldots, 100\} \setminus \{n_1, \ldots, n_{i_2}\})$ dran, welcher die Schubladen $n_{i_2+1}, \ldots, n_{i_3}$ mit $n_{i_3} = j_3$ öffnet, und so fort. Dies

definiert die Zykeldarstellung

$$(n_1, \ldots, n_{i_1})(n_{i_1+1}, \ldots, n_{i_2}) \cdots (n_{i_k+1}, \ldots, n_{100})$$

einer Permutation. Wenn der König die Ausweise zufällig verteilt, dann ist hier jede Permutation (unabhängig von der Strategie der Schlümpfe) gleich wahrscheinlich. Die Schlümpfe kommen frei, wenn die Permutation keinen Zykel der Länge 51 oder mehr enthält. Die Wahrscheinlichkeit hierfür haben wir oben ausgerechnet. Sie ist genau $1 - H_{100} + H_{50} = 0{,}3118278 \cdots$. Die Schlümpfe können also beim modifizierten Spiel höchstens mit dieser Wahrscheinlichkeit entkommen; damit können sie auch im ursprünglichen Spiel nicht mit einer höheren Wahrscheinlichkeit frei kommen. Genauer zeigt dies, dass die in der Lösung von Aufgabe 4.10. (a) erläuterte Strategie optimal ist.

4.11. (a) Es gilt $R = \{i \in \{1, \ldots, n\} \mid R \cap \{i\} \neq \emptyset\}$, Bob stellt also für jedes $i \in \{1, \ldots, n\}$ die Frage „Ist $R \cap \{i\} = \emptyset$?"

4.11. (b) Wir können eine Strategie als binären Entscheidungsbaum auffassen. Dabei entsprechen innere Knoten einer Frage, Verzweigungen dem Ausgang der entsprechenden Frage und die Blätter entsprechen den Antworten Bobs. Der Baum einer Strategie, die mit t Fragen auskommt, kann höchstens 2^t Blätter besitzen.

Alice hat 2^n verschiedene Möglichkeiten R zu wählen. Angenommen, Bob hat eine Gewinnstrategie, die mit $t < n$ Fragen auskommt. Dann muss es Mengen $R_1 \neq R_2$ geben, die im Baum zum selben Blatt führen. Für mindestens eine der beiden Mengen ist Bobs Antwort dann falsch. Dies ist ein Widerspruch zur Korrektheit der Gewinnstrategie.

4.11. (c) Ja! Bob kann durch Fragen „Ist $R \cap \{i\} = \emptyset$?" für $1 \leq i < n$ die Menge R bis auf das Element n bestimmen. Hier muss er raten. Er gewinnt dann mit Wahrscheinlichkeit $1/2$. Diesen Erwartungswert kann er nicht verbessern, denn sein Strategiebaum aus der Lösung zu Aufgabe 4.11. (b) hat nach $n - 1$ Fragen höchstens 2^{n-1} Blätter. Auf diese verteilen sich 2^n Teilmengen, also im Mittel 2 pro Blatt. Fairness sagt, Alice und Bob können bei $r = n - 1$ genauso gut eine Münze werfen.

4.12. (a) Einsetzen der Definition von $C_n = \frac{1}{n+1}\binom{2n}{n}$ und Umordnen der Terme ergibt die gleichwertige Behauptung $\binom{2n}{n+1} = \frac{n}{n+1}\binom{2n}{n}$. Diese widerum folgt aus

$$\binom{2n}{n+1} = \frac{(2n)^{\underline{n+1}}}{(n+1)n!} = \frac{n}{n+1}\frac{(2n)^{\underline{n}}}{(n+1-1)!} = \frac{n}{n+1}\binom{2n}{n} \tag{B.5}$$

4.12. (b) Aus $\binom{n}{k} = \binom{n}{n-k}$ und der Vandermonde'schen Identität folgt

$$\sum_k \binom{n}{k}^2 = \sum_k \binom{n}{k}\binom{n}{n-k} = \binom{2n}{n} = (n+1)C_n$$

4.12. (c) Mit Gleichung (B.5) aus der vorletzten Teilaufgabe sehen wir

$$(n+2)C_{n+1} = \binom{2n+2}{n+1} \overset{(B.5)}{=} \frac{n+2}{n+1}\binom{2n+2}{n} = \frac{n+2}{n+1}\frac{(2n+2)^{\underline{n}}}{n!}$$

$$= \frac{n+2}{n+1} \cdot \frac{(2n+2)(2n+1)\cdot(2n)^{\underline{n}}}{(n+2)(n+1)\cdot n!} = \frac{2(2n+1)}{n+1}\binom{2n}{n}$$

und damit $C_{n+1} = \frac{2(2n+1)}{n+2}C_n$.

4.13. Sei T_n die Anzahl der Triangulierungen eines regelmäßigen n-Ecks mit der Knotenmenge $\{1,\dots,n\}$ und $n \geq 3$. Damit gilt $T_3 = C_1 = 1$. Sei jetzt $n \geq 4$. Die Kante $\{n,1\}$ ist in jeder Triangulierung an genau einem Dreieck beteiligt. Dieses Dreieck ist durch eine der $n-2$ Ecken k in $\{2,\dots,n-1\}$ spezifiziert. Durch Zerteilen des n-Ecks an dem Dreieck ergibt sich ein k-Eck und ein $(n-k+1)$-Eck die unabhängig voneinander trianguliert werden können. Setzen wir noch $T_2 = 1$, so erhalten wir $T_n = \sum_{k=2}^{n-1} T_k T_{n-k+1} = \sum_{\ell=0}^{n-3} C_\ell C_{n-3-\ell}$ Die zweite Gleichheit ergibt sich mit Induktion und einer Indexverschiebung. Mit Korollar 4.41 folgt $T_n = C_{n-2}$ und damit die Behauptung.

4.14. (a) Im Gegensatz zu einer Antikette ist eine *Kette* eine Folge $K_1,\dots,K_\ell$ mit $K_i \subsetneq K_{i+1}$, und eine maximale Kette entspricht genau einer Permutation $\pi = (\pi(1),\dots,\pi(n))$ mit $K_i = \{\pi(1),\dots,\pi(i)\}$. Ist nun $\mathcal{A}$ eine Antikette und π eine Permutation, so kommt maximal ein Mitglied $M \in \mathcal{A}$ als $K_i = \{\pi(1),\dots,\pi(i)\}$ vor. Umgekehrt, ist $M \in \mathcal{A}$ mit $|M| = k$, so gibt es genau $k!(n-k)!$ Permutationen π, bei denen M vorkommt. Wir erhalten

$$\sum_{M\in\mathcal{A}} |M|!(n-|M|)! \leq n! \tag{B.6}$$

Der Wert $|M|!(n-|M|)!$ wird minimal für $|M| = \lfloor\frac{n}{2}\rfloor$, da dann $\binom{n}{|M|}$ maximal ist. Also folgt $|\mathcal{A}| \cdot \lfloor\frac{n}{2}\rfloor!(n-\lfloor\frac{n}{2}\rfloor)! \leq n!$ und damit die Behauptung. (Kürzen der Gleichung (B.6) mit $n!$ ergibt die sogenannte LYM-Ungleichung $\sum_{M\in\mathcal{A}} \binom{n}{|M|}^{-1} \leq 1$ nach Lubell, Meshalkin und Yamamoto.)

4.14. (b) Die Menge $\binom{\{1,\dots,n\}}{\lfloor n/2\rfloor}$ ist eine Antikette mit $\binom{n}{\lfloor n/2\rfloor}$ Elementen.

Zu Kapitel 5

5.1. Die Reihe $f(z) = \sum_{n\geq 0} F_n z^n$ konvergiert absolut für $|z| < \Phi^{-1}$. Es ist $f(z) = \frac{z}{1-z-z^2}$. Setzen wir $z = 1/10$, so erhalten wir den Wert $10/89$.

5.2. (a) Wir gehen analog wie bei den Fibonacci-Zahlen F_n vor. Es ist

$$a(z) = \sum_{n \geq 0} a_n z^n = z + \sum_{n \geq 2} a_n z^n = z + \sum_{n \geq 2} c_1 a_{n-1} z^n + \sum_{n \geq 2} c_2 a_{n-2} z^n$$
$$= z + c_1 z a(z) + c_2 z^2 a(z),$$

also $a(z) = \frac{z}{1 - c_1 z - c_2 z^2}$.

5.2. (b) Die Nullstellen des Nenners sind $-\frac{c_1}{2c_2} \pm \frac{1}{c_2}\sqrt{(\frac{c_1}{2})^2 + c_2}$, also $-\frac{\lambda_1}{c_1}$ und $-\frac{\lambda_2}{c_2}$, und sie sind verschieden. Damit erhalten wir durch Partialbruchzerlegung und Koeffizientenvergleich die Form für a_n.

5.3. Wir definieren die erzeugende Funktion $a(z) = \sum_{n \geq 0} a_n z^n$. Es gilt

$$a(z) = 2 + 5z + \sum_{n \geq 2} a_n z^n = 2 + 5z + 5 \sum_{n \geq 2} a_{n-1} z^n - 6 \sum_{n \geq 2} a_{n-2} z^n$$
$$= 2 + 5z - 10z + 5z a(z) - 6z^2 a(z)$$

Auflösen nach $a(z)$ liefert $a(z) = (2 - 5z)/(1 - 5z + 6z^2)$. Mit Hilfe von Partialbruchzerlegung erhält man dann $a(z) = 1/(1 - 2z) + 1/(1 - 3z)$. Ein Koeffizientenvergleich mit den zugehörigen geometrischen Reihen liefert schließlich die Formel $a_n = 2^n + 3^n$.

5.4. Wir definieren die erzeugende Funktion $a(z) = \sum_{n \geq 0} a_n z^n$. Es gilt

$$a(z) = z + \sum_{n \geq 2} a_n z^n = z + \sum_{n \geq 2} (3a_{n-1} - 2a_{n-2} + 2^{n-1}) z^n$$
$$= z + 3z a(z) - 2z^2 a(z) + \frac{2z^2}{1 - 2z}$$

Auflösen nach $a(z)$ liefert $a(z) = z/((1 - 3z + 2z^2)(1 - 2z))$. Mit Hilfe einer Partialbruchzerlegung erhält man daraus $a(z) = 1/(1 - z) + 1/(1 - 2z)^2 - 2/(1 - 2z)$. Wir benutzen die Formel $\sum_{n \geq 0} (n + 1) z^n = 1/(1 - z)^2$ für $|z| < 1$, die sich durch Ableiten der geometrischen Reihe ergibt. Zusammen mit der Summenformel für die geometrische Reihe erhalten wir

$$a(z) = \sum_{n \geq 0} z^n + \sum_{n \geq 0} (n + 1) 2^n z^n - 2 \sum_{n \geq 0} 2^n z^n$$

Ein Koeffizientenvergleich liefert wie gewünscht $a_n = 1 + (n - 1)2^n$.

5.5. Es gilt:

$$h(z) = \sum_{n \geq 0} \sum_{k=1}^{n} z^n / k = \left(\sum_{n \geq 1} \sum_{k=1}^{n-1} z^n / k \right) + \sum_{n \geq 1} z^n / n = z h(z) - \ln(1 - z)$$

Hieraus folgt $h(z) = \frac{\ln(1-z)}{z-1}$.

5.6. 1. *Standardlösung*: Setze $G_n = F_{2n}$, damit $G_0 = 0$ und damit $G_1 = F_2 = 1$. Für $n \geq 2$ erhalten wir $G_n = 3G_{n-1} - G_{n-2}$, denn $F_{2n+2} = 2F_{2n} + F_{2n-1}$ und $F_{2n} = F_{2n-1} + F_{2n-2}$, also ist $F_{2n-1} = F_{2n} - F_{2n-2}$ und damit $F_{2n+2} = 3F_{2n} - F_{2n-2}$. Sei $g(z)$ die erzeugende Funktion der Fibonacci-Zahlen G_n mit geradem Index. Dann gilt

$$g(z) = z + \sum_{n \geq 2} G_n z^n = z + 3 \left(\sum_{n \geq 2} G_{n-1} z^n \right) - \left(\sum_{n \geq 2} G_{n-2} z^n \right)$$

$$= z + 3zg(z) - z^2 g(z)$$

Hieraus folgt $g(z) = z/(z^2 - 3z + 1)$.

2. *Alternativlösung mit Magie*: Sei $f(z) = z/(1 - z - z^2)$ die erzeugende Funktion der Fibonacci-Zahlen. Betrachte die Funktion $h(z) = f(z) + f(-z)$. Es gilt $h(z) = \sum_{n \geq 0} F_n(z^n + (-1)^n z^n) = 2 \sum_{n \geq 0} F_{2n}(z^2)^n$ Damit ist $g(z) = \frac{h(\sqrt{z})}{2}$ also die erzeugende Funktion der Fibonacci-Zahlen mit geradem Index. Es ist

$$h(\sqrt{z}) = \sqrt{z} \left(\frac{1}{1 - \sqrt{z} - z} - \frac{1}{1 + \sqrt{z} - z} \right)$$

$$= \sqrt{z} \left(\frac{2\sqrt{z}}{(1-z)^2 - (\sqrt{z})^2} \right) = \frac{2z}{z^2 - 3z + 1}$$

Damit ergibt sich erneut $g(z) = \frac{h(\sqrt{z})}{2} = z/(z^2 - 3z + 1)$.

5.7. Wir definieren die erzeugende Funktion $a(z) = \sum_{n \geq 0} a_n z^n$. Zunächst beobachten wir, dass $\sum_{i=0}^{n} (n - i)a_i$ für $n > 0$ genau a_n entspricht und für $n = 0$ den Wert 0 ergibt. Wir stellen $a(z)$ als Faltung mit sich selbst dar:

$$a(z) = \sum_{n \geq 0} a_n z^n = 1 + \sum_{n \geq 0} \sum_{i=0}^{n} (n - i)a_i z^n$$

$$= 1 + \left(\sum_{n \geq 0} a_n z^n \right) \left(\sum_{n \geq 0} n z^n \right) = 1 + a(z) \cdot \frac{z}{(1-z)^2}$$

Es ergibt sich somit $a(z) = \frac{(1-z)^2}{(1-z)^2 - z} = \frac{(1-z)^2}{z^2 - 3z + 1}$.

5.8. Sei $r(z)$ die exponentielle erzeugende Funktion der Rencontres-Zahlen. Nach Satz 4.18 ist bekannt, dass $R_n = n! \sum_{k=0}^{n} \frac{(-1)^k}{k!}$ gilt. Damit erhalten wir

$$r(z) = \sum_{n \geq 0} \frac{R_n}{n!} z^n = \sum_{n \geq 0} \sum_{k=0}^{n} \frac{(-1)^k}{k!} z^n = \left(\sum_{n \geq 0} \frac{(-1)^n}{n!} z^n \right) \left(\sum_{n \geq 0} z^n \right) = \frac{e^{-z}}{1 - z}$$

5.9. (a) Da $\delta(q_0, \varepsilon) = q_0$ gilt, folgt

$$L_{q_0} = \{\varepsilon\} \cup \{wa \mid w \in \Sigma^*, a \in \Sigma, \delta(q_0, wa) = q_0\}$$

$$= \{\varepsilon\} \cup \{wa \mid w \in \Sigma^*, a \in \Sigma, \delta(\delta(q_0, w), a) = q_0\}$$

$$= \{\varepsilon\} \cup \bigcup_p \{wa \mid w \in \Sigma^*, a \in \Sigma, \delta(q_0, w) = p, \delta(p, a) = q_0\}$$

$$= \{\varepsilon\} \cup \bigcup_{\delta(p,a)=q_0} L_p \cdot a$$

Für $q \neq q_0$ ist $\delta(q_0, \varepsilon) \neq q$ und damit insbesondere $\varepsilon \notin L_q$. Die Rechnung ist nun ganz analog zu oben, nur ohne das leere Wort.

5.9. (b) Da jedes Wort einen eindeutigen Pfad durch den Automaten definiert, sind die Vereinigungen in 5.9. (a) disjunkt. Es gilt also $a_0^{q_0} = 1$ und $a_0^q = 0$ für $q \neq q_0$ sowie $a_n^q = \sum_{\delta(p,a)=q} a_{n-1}^p$ für alle $q \in Q$ und $n > 0$.

5.9. (c) Die Anzahl der Wörter der Länge n in $L(\mathcal{A})$ ist gerade die Summe $\sum_{q \in F} a_n^q$, denn $L(\mathcal{A}) = \bigcup_{q \in F} L_q$, wobei die Vereinigung disjunkt ist.

5.9. (d) Mit Hilfe der Formeln aus Teil 5.9. (b) erhalten wir folgendes Gleichungssystem:

$$a^{q_0}(z) = 1 + a^{q_0}(z) \cdot z + a^{q_1}(z) \cdot z$$
$$a^{q_1}(z) = a^{q_0}(z) \cdot z$$

Wir müssen $a^{q_2}(z)$ nicht betrachten, da es von q_2 keinen Weg in einen Endzustand gibt. Durch Lösen des Gleichungssystems ergibt sich:

$$a^{q_0}(z) = \frac{1}{1 - z - z^2} \quad \text{und} \quad a^{q_1}(z) = \frac{z}{1 - z - z^2}$$

Sei $f(z)$ die erzeugende Funktion der Fibonacci-Zahlen. Dann gilt $f(z) = z \cdot a^{q_0}(z)$ $= a^{q_1}(z)$. Nach Teil 5.9. (c) ist $b(z) = a^{q_0}(z) + a^{q_1}(z) = \sum_{n \geq 0}(F_{n+1} + F_n)z^n = \sum_{n \geq 0} F_{n+2}z^n$ die gesucht erzeugende Funktion. Der Automat akzeptiert also genau F_{n+2} Wörter der Länge n. Dies haben wir bereits in Beispiel 1.25 auf einem anderen Weg berechnet.

Zu Kapitel 6

6.1. Die Menge der Kanten ist eine Teilmenge von $\binom{V}{2}$. Da $|V| = n$ gibt es $2^{\binom{n}{2}}$ solche Teilmengen und somit $2^{\binom{n}{2}}$ Graphen. Man beachte, dass hierbei isomorphe Graphen mehrfach gezählt werden.

6.2. (a)

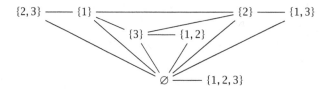

6.2. (b) Die Anzahl der Knoten ist $|V_n| = 2^n$. Enthält eine Menge i Elemente, so gibt es zu allen Teilmengen der restlichen $n-i$ Elemente Kanten. Dies liefert $\sum_{i=1}^{n} \binom{n}{i} 2^{n-i}$ $= 3^n$ Paare. Hierin ist allerdings noch die Schlinge $\emptyset - \emptyset$ enthalten. Außerdem werden die Kanten doppelt gezählt. Die Anzahl an Kanten ergibt sich damit zu $\frac{3^n - 1}{2}$.

6.3. (a) Sei d_x der Grad von Knoten $x \in V$. Dann gilt $4|V| \leq \sum_{x \in V} d_x = 2|E|$. Daraus folgt $2|V| \leq |E|$.

6.3. (b) Für $n = 5$ erfüllt der vollständige Graph mit 5 Knoten die Behauptung. Sei nun (V, E) ein Graph mit mindestens 5 Knoten, bei dem alle Knoten den Grad 4 haben. Seien $x_1, x_2, y_1, y_2 \in V$ vier verschiedene Knoten mit $(x_1, x_2), (y_1, y_2) \in E$. Wir entfernen die Kanten (x_1, x_2) und (y_1, y_2). Hiernach haben x_1, x_2, y_1, y_2 den Grad 3. Alle anderen Knoten haben den Grad 4. Wir nehmen einen neuen Knoten $z \notin V$ hinzu. Außerdem fügen wir die 4 Kanten (z, x_1), (z, x_2), (z, y_1) und (z, y_2) ein. In dem entstandenen Graphen (V', E') mit $V' = V \cup \{z\}$ und $E' = (E \setminus \{(x_1, x_2), (y_1, y_2)\}) \cup \{(z, x_1), (z, x_2), (z, y_1), (z, y_2)\}$ haben alle Knoten den Grad 4. Dieses induktive Vorgehen zeigt die Behauptung.

6.4. Da G nicht vollständig ist, existieren Knoten $a, b \in V$ mit $ab \notin E$. Da G zusammenhängend ist, gibt es in G einen kürzesten Weg $a = x_1, x_2, \ldots, x_n = b$ mit $n \geq 3$. Zwischen x_1 und x_3 gibt es keine Kante, sonst ließe sich dieser Weg verkürzen. Also gilt die Behauptung mit $(x_1, x_2, x_3) = (u, v, w)$.

6.5. Angenommen G ist nicht zusammenhängend, dann lässt sich die Knotenmenge V in zwei disjunkte, nichtleere Mengen A, B zerlegen, so dass keine Kanten zwischen A und B erlaufen. Wir zeigen jetzt, dass $\overline{G}$ zusammenhängend ist. Seien $u, v \in A$. Es existiert ein Knoten $x \in B$ mit $ux, vx \in \overline{E}$. Die Knoten u und v sind also zusammenhängend. Analog gilt dies für $u, v \in B$. Für $u \in A, v \in B$ ergibt sich direkt $uv \in \overline{E}$.

6.6. Angenommen, es gäbe zwei längste disjunkte einfache Wege $u_0 \cdots u_{\ell(G)}$ und $v_0 \cdots v_{\ell(G)}$ in G. Da G zusammenhängend ist, gibt es zwei Knoten u_i und v_j, so dass ein doppelpunktfreier Weg $u_i = w_0, w_1, \ldots, w_{k-1}, w_k = v_j$ existiert mit $\{w_1, \ldots, w_{k-1}\} \cap \{u_0, \ldots, u_{\ell(G)}, v_0, \ldots, v_{\ell(G)}\} = \emptyset$. Ohne Einschränkung sei $i \geq \ell(G)/2$ und $j \geq \ell(G)/2$, andernfalls nummerieren wir den jeweiligen Weg in der entgegengesetzten Richtung. Dann ist aber die Länge des Weges $u_0, \ldots, u_i, w_1, \ldots, w_{k-1}, v_j, \ldots, v_0$ mindestens $\ell(G)/2 + 1 + \ell(G)/2 > \ell(G)$, ein Widerspruch!

6.7. Für den Grad eines Knotens $x \in V$ gilt $0 \leq d_x \leq |V| - 1$. Es sind also $|V|$ verschiedene Werte für d_x möglich. Haben nun alle Knoten einen unterschiedlichen Grad, so muss es insbesondere einen Knoten vom Grad 0 und einen Knoten vom Grad $|V| - 1$ geben. Für $|V| \geq 2$ ist dies nicht möglich, denn der Knoten mit Grad $|V| - 1$ ist mit allen anderen Knoten verbunden.

6.8. Sei G ein zusammenhängender Graph, in dem alle Knoten geraden Grad haben. Der Graph G enthält einen Eulerkreis. Nach Entfernen einer beliebigen Kante existiert noch ein Eulerweg. Insbesondere ist der entstandene Graph zusammenhängend. Der Graph G enthält also keine Brücke.

6.9. Sei $G = (V, E)$ ein gerichteter Graph mit $V = \Sigma^{k-1}$ und beschrifteten Kanten $x_1 \cdots x_{k-1} \xrightarrow{a} x_2 \cdots x_{k-1} a$ für jedes $a \in \Sigma$ (Schleifen sind möglich). Bei jedem

Knoten ist der Ein- und Ausgangsgrad $|\Sigma|$. Der Graph G ist zusammenhängend, denn es gilt $x_1 \cdots x_{k-1} \xrightarrow{y_1} \cdots \xrightarrow{y_{k-1}} y_1 \cdots y_{k-1}$. Es existiert deshalb ein gerichteter Eulerkreis $z_1 \cdots z_\ell z_1$ der jede Kante genau einmal besucht. Auf diesem Kreis kommt bei der Sequenz der Beschriftungen jedes Wort genau einmal vor. Aus $|E| = |\Sigma|^k$ folgt $\ell = |\Sigma|^k$. Sei $y_1 \ldots y_\ell$ die Sequenz der Kantenbeschriftungen, dann ist $w = z_1 y_1 \ldots y_\ell$ das gesuchte Wort. Man bezeichnet w oft auch als De Bruijn-Folge der Ordnung k (nach Nicolaas Govert de Bruijn, 1918–2012).

6.10. Wir beschreiben den Algorithmus von Carl Hierholzer (1840–1871), welcher posthum erschien [25]. In diesem Verfahren starten wir bei einem beliebigen Knoten v und konstruieren einen Kreis K, indem wir zu dem anfangs leeren Pfad immer weitere Kanten hinzufügen und diese aus dem ursprünglichen Graphen entfernen. Durch die Gradbedingung ist garantiert, dass wir irgendwann wieder zu v zurückkommen. Falls wir keine Kanten mehr hinzufügen können, haben wir also einen geschlossenen Pfad von v nach v gefunden. Sind alle Kanten aus G entfernt, so benutzt K alle Kanten und wir sind fertig. Andernfalls gehen wir den gefundenen Pfad zurück, bis wir auf den ersten Knoten u treffen, der noch ausgehende Kanten hat. Da G zusammenhängend ist, muss u existieren. Von u ausgehend führen wir den Algorithmus rekursiv aus und fügen den rekursiv berechneten Kreis anstelle von u in K ein. Dann laufen wir den entstandenen Kreis weiter zurück (insbesondere laufen wir erst durch den neu eingefügten Teil zurück) und bearbeiten auf die gleiche Weise alle Knoten mit noch ausgehenden Kanten. Am Ende dieses Vorgehens haben wir alle Kanten des ursprünglichen Graphen in den Kreis übernommen, der damit ein Eulerkreis ist. Die Laufzeit ergibt sich dadurch, dass wir in jedem Schritt eine Kante besuchen. Aus dem ursprünglichen Graphen wird nach einem Besuch einer Kante diese aus G gelöscht. Im Kreis K besuchen wir durch ausschließliches Rückwärtslaufen jede Kante maximal einmal. Damit wird jede Kante insgesamt höchstens zweimal betrachtet.

6.11. Für eine beliebige bijektive Funktion $\varphi : E \to \{1, \ldots, 12\}$ gilt

$$\sum_{v \in V} \sum_{vw \in E} \varphi(vw) = 2 \cdot \sum_{i=1}^{12} i = 156$$

Bei gleicher Gewichtssumme an allen 8 Ecken, müsste dieses Eckengewicht $156/8 = 19{,}5$ sein, ein Widerspruch!

6.12. Für $n = 0$ und $n = 1$ ist die Äquivalenz trivial erfüllt, da Bäume nicht leer sind und Bäume mit nur einem Knoten keine Kanten haben. Sei also $n \geq 2$.

$\Rightarrow$: Ein Baum mit n Knoten hat $n - 1$ Kanten. Da in der Summe jede Kante genau zweimal gezählt wird, folgt $\sum_{i=1}^{n} d_i = 2n - 2$.

$\Leftarrow$: Für $n = 2$ gilt die Behauptung. Sei $n \geq 3$. Dann existieren $i, j \in \{1, \ldots, n\}$ mit $d_i = 1$ und $d_j > 1$. Ohne Einschränkung sei $i = n$ und $j = n - 1$. Durch Induktion existiert ein Baum $(\{1, \ldots, n-1\}, E)$, so dass für $1 \leq i \leq n-2$ der Knoten i den Grad d_i hat, und Knoten $n - 1$ hat Grad $d_{n-1} - 1$. Nun ist $(\{1, \ldots, n\}, E \cup \{(n-1, n)\})$ ein Baum, bei dem Knoten i den Grade d_i hat.

6.13. Für $|V| = 1$ und $|V| = 2$ ist die Aussage wahr. Sei nun $|V| > 2$. Die Abbildung φ permutiert die Blätter $B = \{x \in V \mid d_x = 1\}$. Damit ist die Einschränkung $\varphi|_{V \setminus B}$ ein Automorphismus auf dem von $V \setminus B$ induzierten Untergraphen, welcher selbst wieder ein Baum ist. Mit Induktion lässt $\varphi|_{V \setminus B}$ einen Knoten y oder eine Kante e fix. Also lässt auch φ einen Knoten y oder eine Kante e fix.

6.14. Als erstes beobachten wir, dass zwei verschiedene Mengen A_i und A_j durch Entfernen von x genau dann gleich werden, wenn sie sich nur durch x unterscheiden (d. h., die symmetrische Differenz von A_i und A_j ist $\{x\}$). Ohne Einschränkung sei $\ell = n$. Wir konstruieren einen kantengefärbten Graph mit Knotenmenge $\{A_1, \ldots, A_n\}$. Eine Kante zwischen A_i und A_j existiert genau dann, wenn sich A_i und A_j nur um ein Element $k \in M$ unterscheiden. Diese Kante wird mit k gefärbt. Sei A_i ein Knoten auf einem Kreis und k die Farbe einer von A_i ausgehenden Kante e auf diesem Kreis. Dann muss auf diesem Kreis eine weitere Kante mit dieser Farbe existieren (wenn z. B. der Nachbar von A_i auf e das Element k nicht enthält, muss es irgendwann auf diesem Kreis wieder „dazukommen") . Diese Kante entfernen wir aus dem Graph. Indem wir so sukzessiv Kanten aus dem Graph entfernen, erhalten wir einen Wald (keine Kreise). In diesem Wald kommen alle Farben noch vor, die im ursprünglichen Graph aufgetreten sind. Da ein Wald mit n Knoten höchstens $n - 1$ Kanten besitzt, gibt es eine Farbe x, die in dem ursprünglichen Graph nicht vorgekommen ist. Wenn wir diese Farbe x aus den A_i, $1 \leq i \leq n$ entfernen, bleiben diese nach obiger Beobachtung weiterhin alle verschieden.

6.15. Ist $P_i = Q_j$, so können wir diese Menge entfernen, daher sind ohne Einschränkung P_i und Q_j paarweise verschieden. Wir bilden einen bipartiten Graphen mit der Knotenmenge $\{P_1, \ldots, P_m\} \cup \{Q_1, \ldots, Q_m\}$ und Kantenmenge $E = \{(P_i, Q_j) \mid P_i \cap Q_j \neq \emptyset\}$. Ein gemeinsames Vertretersystem definiert ein perfektes Matching und umgekehrt. Wir müssen also die Existenz eines perfekten Matchings nachweisen. Hierfür verwenden wir den Heiratssatz 6.11. Zu zeigen ist nur, dass für $S \subseteq \{P_1, \ldots P_m\}$ die Heiratsbedingung $|N(S)| \geq |S|$ erfüllt ist. Für jedes $x \in P_i \in S$ gibt es ein Q_j mit $x \in Q_j \in N(S)$. Also gilt

$$\bigcup_{P_i \in S} P_i \subseteq \bigcup_{Q_j \in N(S)} Q_j$$

Links und rechts stehen disjunkte Vereinigungen von Klassen mit jeweils genau k Elementen, daher ist schließlich $|S| \leq |N(S)|$.

6.16. Angenommen, es gäbe eine stabile Heirat M mit $(a, b'), (a', b) \in M$, in der die Präferenz von a für b' noch niedriger als für b ist. Wir wissen aus der Bemerkung 6.13, dass b im Gale-Shapley-Verfahren seine optimale Partnerin a gefunden hat, folglich muss die Präferenz von b für a' niedriger als für a sein. Es würde also beim Zusammentreffen von den Paaren (a, b') und (a', b) zu einer Scheidung und anschließender neuer Bindung (a, b) kommen. Dies ist ein Widerspruch zur Stabilität von M.

6.17. (a) Aus der Abschätzung $e \le 3n - 6$ und der Eulerformel $n - e + f = 2$ folgt

$$f = 2 - n + e \le 2 - n + 3n - 6 = 2n - 4$$

6.17. (b) Die Summe der Knotengrade ist $dn/2 + 2dn/2 = 3dn/2$. Dies liefert $3dn/2 = 2e$ und damit $3dn = 4e$. Mit $e \le 3n - 6$ folgt $3dn \le 12n - 24$ und $24 \le (12 - 3d)n$. Für $d \ge 4$ ist die Ungleichung nicht erfüllt.

6.17. (c)

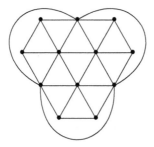

6.18. Zunächst entfernen wir alle Kanten. Dann ist $m = 0$. Für $m = 0$ ist nun $f = 1$ und $n = z$. Fügen wir nacheinander die Kanten wieder ein, so wird jeweils entweder z um 1 verringert und f bleibt gleich oder eine Facette wird zerteilt. Dann bleibt z gleich und f wird um 1 vergrößert.

6.19. Wir betrachten zunächst den Graph K_7. Wir stellen die Torusoberfläche als Rechteck dar, welches wir jeweils entlang der gegenüberliegenden Seiten verkleben. Insbesondere stimmt der obere Punkt x mit dem unteren Punkt x überein, und der Knoten ganz links ist mit dem Knoten ganz rechts identisch.

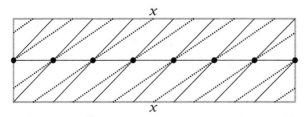

Die horizontalen Kanten verbinden jeweils nebeneinander liegende Knoten, die durchgezogenen schräg verlaufenden Kanten überspringen einen Knoten und die gestrichelten Kanten überspringen jeweils zwei Knoten. Es gibt keine Kreuzungen und, wie man gut an einem mittleren Knoten erkennt, ist der Grad aller Knoten 6. Damit sind alle Kanten des K_7 vorhanden.

Um den Graph $K_{4,4}$ zu zeichnen, gehen wir ganz analog vor. Die einen vier Knoten $\{1, 2, 3, 4\}$ (rund) zeichnen wir auf den Rand des Rechtecks; insbesondere entsprechen die vier Punkte in den Ecken nur einem einzigen Knoten. Die anderen vier Knoten (eckig) sind in der Mitte des Rechtecks.

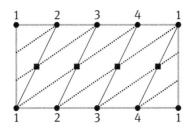

Der besseren Übersichtlichkeit wegen haben wir einen Teil der Kanten gestrichelt gezeichnet.

6.20. Nach der Eulerformel gilt in planaren Graphen mit mindestens drei Knoten, dass $|E| \leq 3|V| - 6$. Für den Komplementärgraphen gilt somit

$$\frac{|V|(|V| - 1)}{2} - 3|V| + 6 \leq \left| \binom{V}{2} \setminus E \right| \leq 3|V| - 6$$

Für $|V| \geq 11$ ist die Ungleichung nicht erfüllt. Also ist jeder Graph mit mehr als 10 Knoten nicht planar oder aber sein Komplementärgraph $\overline{G}$ ist nicht planar. Da es (bis auf Isomorphie) nur endlich viele Graphen mit 10 Knoten oder weniger gibt, folgt die Behauptung.

6.21. Durch Hinzufügen von weiteren Kanten können wir annehmen, dass alle Facetten (auch die äußere) nur von drei Kanten begrenzt werden; dies erhöht den Grad nur. Insbesondere gibt es $2|E|/3$ Facetten. Jeder Knoten x erhält zunächst das Startgewicht $6 - d_x$. Für das Gesamtgewicht gilt $\sum_{x \in V}(6 - d_x) = 6|V| - \sum_{x \in V} d_x = 6|V| - 2|E| = 12$, wobei die letzte Gleichung aus der Eulerformel folgt.

Nun verteilen wir die Gewichte um. Jeder Knoten mit Grad 5 gibt jedem seiner Nachbarn $\frac{1}{5}$ von seinem Gewicht. Da das Gesamtgewicht positiv ist, existiert nach diesem Umverteilen ein Knoten y mit positivem Gewicht. Es gilt $d_y \leq 7$, da zum Startgewicht $6 - d_y$ von y maximal $\frac{d_y}{5}$ hinzu kommen, so dass für das aktuelle Gewicht y von y die Abschätzung $0 < y \leq 6 - 4d_y/5$ gilt.

Falls $d_y = 7$ gilt, dann hat y mindestens 6 Nachbarn mit Grad 5 (andernfalls hätte y zu wenig abbekommen, um nun positives Gewicht zu haben). Da G trianguliert ist, sind zwei dieser Nachbarn von y durch eine Kante verbunden, wodurch die Aussage in diesem Fall bewiesen ist. Falls $d_y \leq 6$ gilt, dann hat y mindestens einen Nachbarn x mit Grad 5, so dass xy die Forderung der Aufgabe erfüllt.

6.22. (a) Für den leeren Graph gilt die Eigenschaft. Sei $V \neq \emptyset$ und $u \in V$. Wir definieren zwei Teilgraphen durch

$$V_1 = \{v \in V \mid (v, u) \in E\}, \quad E_1 = E \cap (V_1 \times V_1)$$
$$V_2 = \{v \in V \mid (u, v) \in E\}, \quad E_2 = E \cap (V_2 \times V_2)$$

Induktiv seien $a_1 \cdots a_k$ und $b_1 \cdots b_\ell$ einfache Wege in (V_1, E_1) bzw. (V_2, E_2), die jeden Knoten einmal besuchen. Dann ist $a_1 \cdots a_k u b_1 \cdots b_\ell$ ein einfacher Weg in G, der jeden Knoten einmal besucht.

6.22. (b) Wir zeigen die Behauptung mit Induktion. Für $|V| = 2$ ist die Aussage trivial. Sei also $|V| > 2$. Entnehme $x \in V$ und setze $G' = G - x$. Dies bedeutet, $G' = (V', E')$ ist der durch $V' = V \setminus \{x\}$ induzierte Untergraph. Nach Induktion enthält V' einen Knoten y, von dem aus jeder andere Knoten $u \in V'$ in höchstens 2 Schritten erreichbar ist. Ist auch x von y in höchstens 2 Schritten erreichbar, so ist y der gesuchte Knoten. Andernfalls betrachte $u \in V'$ mit $u = y$ oder $(y, u) \in E$. Dann muss $(x, u) \in E$ gelten, ansonsten hätten wir x von y aus in maximal zwei Schritten erreicht. Für alle anderen $v \in V'$, die nicht direkt von y aus erreichbar sind, gibt es ein $u \in V'$ und $(y, u), (u, v) \in E$. Also gilt auch $(x, u), (u, v) \in E$ und v ist von x in zwei Schritten erreichbar. Damit ist x der gesuchte Knoten.

6.23. (a) Wir betrachten die n Schubfächer $\{2i - 1, 2i\}$ für $1 \leq i \leq n$. Dadurch wird $\{1, \ldots, 2n\}$ in n Zweiermengen partitioniert. Ein Element x wird in Schubfach S gelegt, falls $x \in S$ gilt. Da wir $n + 1$ Elemente aber nur n Schubfächer haben, liegen am Ende in einem Schubfach zwei Elemente. Diese sind benachbart.

6.23. (b) Wir betrachten die Schubfächer $\{i, 2n + 1 - i\}$ für $1 \leq i \leq n$. Auch hier gehören zwei Elemente aus M zum selben Schubfach.

6.23. (c) Wir betrachten die n Schubfächer $\{u \, 2^s \mid s \geq 0\}$ für alle ungeraden Zahlen $u \in \{1, \ldots, 2n\}$. Seien $k < \ell$ mit $k, \ell \in \{u \, 2^s \mid s \geq 0\}$. Dann unterscheiden sich k und ℓ nur um eine Zweierpotenz, und es gilt $k \mid \ell$.

6.24. (a) Im folgenden Graphen gibt es keine Dreiecke, da der mittlere Punkt in der Zeichnung nicht zum Graphen gehört. Maximale unabhängige Mengen enthalten drei Knoten.

6.24. (b) Es gibt mindestens einen Knoten x vom Grad $d_x \neq 3$, denn sonst wäre $\sum_{x \in V} d_x = 9 \cdot 3$ ungerade, was dem Handschlaglemma widerspricht. Wir unterscheiden nun zwei Fälle. Ist $d_x \geq 4$, dann bilden die 4 Nachbarn von x entweder eine

unabhängige Menge der Größe 4 oder zwei der Knoten bilden mit x zusammen eine Clique der Größe 3. Ist $d_x \leq 2$, dann verbleiben 6 andere Knoten im Graph. Diese 6 Knoten enthalten entweder eine Clique oder eine unabhängige Menge der Größe 3. Enthalten sie eine unabhängige Menge, dann bilden diese 3 Knoten zusammen mit x eine unabhängige Menge der Größe 4.

6.25. Wir definieren eine Färbung b von $\binom{\mathbb{N}}{2}$ durch $b(\{i,j\}) = c(|j - i|)$. Nach dem Satz von Ramsey existiert eine unendliche Teilmenge $X \subseteq \mathbb{N}$, so dass $\binom{X}{2}$ bezüglich b monochromatisch gefärbt ist. Wähle $i, j, k \in X$ mit $i < j < k$ und setze $x = j - i$, $y = k - j$ sowie $z = k - i$. Es ist $z = x + y$ und aus $b(i,j) = b(j,k) = b(i,k)$ folgt $c(x) = c(y) = c(z)$.

6.26. Wir zeigen, dass mit beliebig hoher Wahrscheinlichkeit alle Knoten paarweise zueinander einen Abstand kleiner oder gleich zwei haben, wenn n genügend groß ist. Betrachte zunächst zwei feste Knoten x und y. Wenn n wächst, nimmt die Wahrscheinlichkeit exponentiell ab, dass kein dritter Knoten z vorhanden ist, der sowohl eine Kante zu x als auch zu y hat. Auf der anderen Seite existieren nur quadratisch viele Paare von Knoten. Die Gesamtwahrscheinlichkeit, dass nicht alle Knoten einen Höchstabstand von zwei haben, nimmt also für $n \to \infty$ exponentiell ab.

Zu Kapitel 7

7.1. Sei $\bot$ das kleinste Element. Betrachte zunächst $a < b$ und $\dim(a) = d$. Dann gibt es eine maximale Kette K von $\bot$ nach a der Länge d und $K \cup \{b\}$ ist eine maximale Kette von $\bot$ nach b. Haben nun je zwei maximale Ketten mit den selben Endpunkten die gleiche Länge, so folgt nach Definition der Dimension $\dim(b) = \dim(a) + 1$.

Für die Rückrichtung betrachte zwei maximale Ketten K_1 und K_2 von c nach b. Die Längen seien jeweils ℓ_i für $i = 1,2$ und wir nehmen $\ell_1 \leq \ell_2$ an. Es sei $d = \dim(c)$ die Dimension von c. Wir zeigen $\dim(b) = d + \ell_1$. Hieraus folgt dann sofort $\ell_1 = \ell_2$, da in jedem Fall $\dim(b) \geq d + \ell_2$ ist.

Für $\ell_1 = 0$ ist die Behauptung klar, da dann $c = b$ sein muss. Sei jetzt $\ell_1 \geq 1$. Die Kette K_1 läuft durch einen vorletzten Punkt a mit $a < b$, sonst wäre K_1 nicht maximal. Nach Induktion gilt $\dim(a) = d + \ell_1 - 1$. Folgt nun aus $a < b$ schon $\dim(b) = \dim(a) + 1$, so ist $\dim(b) = d + \ell_1$, wie behauptet.

7.2. (a) Es ist $\Gamma_{b,c}(\bot)(\sigma) = \sigma$ für $b(\sigma) = 0$ und undefiniert für $b(\sigma) = 1$. Also ist $\Gamma_{b,c}(\bot) = \bot$ genau dann, wenn w nirgends terminiert.

7.2. (b) Wähle b und c mit $b(\sigma) = 1$ und $c(\sigma) = \sigma$ für alle $\sigma \in \Sigma$, dann ist $w =$ **while** true **do** id_Σ **od** und $\Gamma_{b,c}(f) = f$. Also sind alle $f \in \mathcal{F}$ Fixpunkte. Insbesondere gilt $\Gamma_{b,c}(\bot) = \bot$ und der Definitionsbereich von $\Gamma_{b,c}(\bot)$ ist leer.

7.2. (c) Aus $w(\sigma) = \tau$ folgt $b(\tau) = 0$. Also ist $w(\sigma) = \sigma$ gleichbedeutend mit $b(\sigma) = 0$. Dies bedeutet $w \sqsubseteq \mathrm{id} \iff \mathrm{dom}(w) = \{\sigma \in \Sigma \mid b(\sigma) = 0\}$. Sei jetzt

$w' = $ **while** b' **do** c **od** und $w \sqsubseteq w'$. Dann ist $w(\sigma) = w'(\sigma)$ für alle $\sigma \in \mathrm{dom}(w)$. Ist also $b(\sigma) = 0$ oder $b'(\sigma) = 0$, so gilt dies für beide. Dies impliziert $b(\sigma) = b'(\sigma)$ für alle $\sigma \in \mathrm{dom}(w)$. Umgekehrt, gilt $b(\sigma) = b'(\sigma)$ für alle $\sigma \in \mathrm{dom}(w)$, so folgt $w \sqsubseteq w'$. Damit gilt $w \sqsubseteq w'$ genau dann, wenn b und b' auf $\mathrm{dom}(w)$ übereinstimmen.

7.2. (d) Terminiert w überall, so ist w überall definiert und aus $w \sqsubseteq f$ folgt $w = f$. Der kleinste Fixpunkt w ist also der einzige Fixpunkt von $\Gamma_{b,c}$. Umgekehrt, sei jetzt $w(\sigma)$ undefiniert. Da c überall definiert ist, gilt dies auch für c^i. Setze $f(c^i(\sigma)) = \sigma$ für alle $i \in \mathbb{N}$ und lasse f undefiniert sonst. Dann ist zunächst $b(c^i(\sigma)) = 1$ für alle $i \in \mathbb{N}$, denn $w(\sigma)$ ist undefiniert. Hieraus folgt $\Gamma_{b,c}(f)(c^i(\sigma)) = f(c^{i+1}(\sigma)) = \sigma = f(c^i(\sigma))$. Es ergibt sich $f \sqsubseteq \Gamma_{b,c}(f)$. Die Kette $f \sqsubseteq \Gamma_{b,c}(f) \sqsubseteq \Gamma_{b,c}^2(f) \sqsubseteq \cdots$ liefert einen Fixpunkt von $\Gamma_{b,c}$, der echt oberhalb von w liegt.

7.2. (e) Sei $c \in \mathcal{F}$ überall undefiniert. Wir können c beispielsweise darstellen durch $c = $ **while** true **do** id_Σ **od**. Wir zeigen, dass $\Gamma_{b,c}$ genau einen Fixpunkt hat. Sei hierfür $\Gamma_{b,c}(f) = f$. Betrachte ein σ mit $b(\sigma) = 0$, dann gilt $\Gamma_{b,c}(f)(\sigma) = \sigma = f(\sigma)$. Für $b(\sigma) = 1$ ist $\Gamma_{b,c}(f)(\sigma) = f(c(\sigma))$ undefiniert, also auch $f(\sigma)$. Damit ist f durch die Bedingung b eindeutig festgelegt.

7.3. Es gibt abzählbar unendliche Verbände, etwa $(\mathbb{N} \times \mathbb{N}, \leq)$, $(\mathbb{Z} \times \mathbb{Z}, \leq)$ oder der Verband (aus der vorigen Aufgabe) aller Teilmengen von $\mathbb{N}$, die entweder endlich sind oder ein endliches Komplement haben. Diese Verbände sind unendlich, aber abzählbar. Potenzmengenverbände haben die Form 2^A. Sie sind also endlich oder überabzählbar. Insbesondere ist jeder abzählbar unendliche Verband ein Beispiel für einen booleschen Verband, der nicht isomorph ist zu einem Potenzmengenverband 2^A für irgendeine Menge A ist.

7.4. In $(\mathbb{Z} \times \mathbb{Z}, \leq)$ ist kein Paar (m, n) irreduzibel, da $(m, n) = (m - 1, n) \vee (m, n - 1)$ gilt. Auch die Hinzunahme eines kleinsten Elements $\perp$ ergibt keine irreduziblen Elemente.

7.5. Sei X unendlich und M der Verband aller Teilmengen von X, die ein endliches Komplement haben. Dann ist M ein Mengenverband, denn es gilt:
1. $A, B \in M \Rightarrow A \cup B \in M$,
2. $A, B \in M \Rightarrow A \cap B \in M$,
Da $A \in M$ unendlich ist, gibt es $a, b \in A$ mit $a \neq b$. Also können wir $A = (A \setminus \{a\}) \cup (A \setminus \{b\})$ schreiben; und A ist nicht irreduzibel.

7.6. Wäre $(2^M, \cup, \cap, 0, 1)$ ein Ring, so wäre $\emptyset = 0$ und $A(A + B) = AA + AB$. Andererseits ist $A(A + B) = A \cap (A \cup B) = A = A \cap A = AA$, also $A \cap B = \emptyset$ für alle $A, B \in 2^M$. Dies widerspricht $M \neq \emptyset$.

7.7.

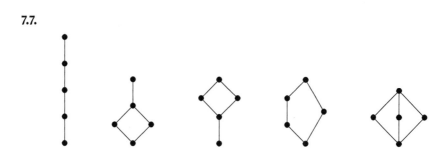

7.8. M_5 ist nicht distributiv, denn $(a \vee b) \wedge c = c$, aber $(a \wedge c) \vee (b \wedge c) = \bot$. M_5 ist modular: Sei z. B. $a < \top \Rightarrow a \vee (b \wedge \top) = a \vee b = \top = \top \wedge \top = (a \vee b) \wedge \top$. M_5 ist komplementär: Denn z. B. $b \vee c = \top$ und $b \wedge c = \bot$.

Literaturverzeichnis

[1]   M. Agrawal, N. Kayal und N. Saxena: *PRIMES is in P*. Ann. of Math., 160:781–793, 2004.

[2]   M. Aigner und G. M. Ziegler: *Das Buch der Beweise*. Springer, Berlin, 2009.

[3]   N. Alon, P. D. Seymour und R. Thomas: *Planar Separators*. SIAM J. Discrete Math., 7(2):184–193, 1994.

[4]   K. I. Appel und W. Haken: *Every planar map is four colorable*. Bull. Amer. Math. Soc., 82(5):711–712, 1976.

[5]   K. I. Appel und W. Haken: *Every planar map is four colorable*, Band 98 von *Contemporary mathematics*. American Mathematical Society, 1989.

[6]   G. D. Birkhoff: *Aesthetic measure*. Harvard University Press, 1933.

[7]   D. Boneh: *Twenty years of attacks on the RSA cryptosystem*. Notices Amer. Math. Soc., 46:203–213, 1999.

[8]   D. Boneh und G. Durfee: *Cryptanalysis of RSA with private key d less than $N^{0.292}$*. IEEE Transactions on Information Theory, 46:1339–1349, 2000.

[9]   T. Camps, S. Kühling und G. Rosenberger: *Einführung in die mengentheoretische und die algebraische Topologie*. Berliner Studienreihe zur Mathematik 15. Heldermann, 2011.

[10]   E. Curtin und M. Warshauer: *The locker puzzle*. Math. Intell., 28(1):28–31, 2006.

[11]   C.-J. de la Vallée Poussin: *Recherches analytiques sur la théorie des nombres premiers*. Ann. Soc. Sci. Bruxelles, 20:183–256, 1896.

[12]   V. Diekert, M. Kufleitner und G. Rosenberger: *Diskrete algebraische Methoden*. Walter de Gruyter, 2013.

[13]   R. Diestel: *Graphentheorie*. Springer-Lehrbuch. Springer, 4. Auflage, 2010.

[14]   Y. Dinitz: *Algorithm for solution of a problem of maximum flow in a network with power estimation*. Soviet Math. Doklady (Doklady), 11:1277–1280, 1970.

[15]   Y. Dinitz: *Dinitz' Algorithm: The Original Version and Even's Version*. In: O. Goldreich, A. L. Rosenberg und A. L. Selman (Herausgeber), *Essays in Memory of Shimon Even*, Band 3895 von *Lecture Notes in Computer Science*, S. 218–240. Springer, 2006.

[16]   J. Edmonds und R. M. Karp: *Theoretical improvements in algorithmic efficiency for network flow problems*. Journal of the Association for Computing Machinery, 19:248–264, 1972.

[17]   P. Elias, A. Feinstein und C. E. Shannon: *A note on the maximum flow through a network*. IRE Transactions on Information Theory, 2(4):117–119, 1956.

[18]   P. Erdős: *Beweis eines Satzes von Tschebyschef*. Acta Litt. Sci. Szeged, 5:194–198, 1932.

[19]   L. R. Ford, Jr. und D. R. Fulkerson: *Maximal flow through a network*. Research Memorandum RM-1400, The RAND Corporation, 1954.

[20]   L. R. Ford, Jr. und D. R. Fulkerson: *Maximal flow through a network*. Canadian Journal of Mathematics, 8:399–404, 1956.

[21]   F. Göring: *Short proof of Menger's Theorem*. Discrete Mathematics, 219(1–3):295–296, 2000.

[22]   R. L. Graham, D. E. Knuth und O. Patashnik: *Concrete Mathematics: A Foundation for Computer Science*. Addison-Wesley, 1994.

[23]   J. Hadamard: *Sur la distribution des zéros de la fonction $\zeta(s)$ et ses conséquences arithmétiques*. Bull. Soc. Math. France, 24:199–220, 1896.

[24]   J. Håstad: *Solving Simultaneous Modular Equations of Low Degree*. SIAM J. Comput., 17:336–341, 1988.

[25]   C. Hierholzer: *Über die Möglichkeit, einen Linienzug ohne Wiederholung und ohne Unterbrechung zu umfahren*. Math. Ann., VI:30–32, 1873.

[26]   R. J. Lipton und R. E. Tarjan: *A separator theorem for planar graphs*. SIAM Journal on Applied Mathematics, 36(2):177–189, 1979.

[27]   O. B. Lupanov: *A method of circuit synthesis*. Izvesitya VUZ, Radiofizika, 1:120–140, 1958.

[28]   J. Matoušek und J. Nešetřil: *Diskrete Mathematik – Eine Entdeckungsreise*. Springer-Verlag, 2002.

[29]   M. Nair: *On Chebyshev-type inequalities for primes*. The American Mathematical Monthly, 89(2):126–129, 1982.

[30]   P. L. Tschebyschev: *Sur la fonction qui détermine la totalité des nombres premiers inférieurs à une limite donnée*. Mémoires présentés à l'Académie Impériale des Sciences de St.-Pétersbourg par divers Savants et lus dans ses Assemblées, Bd. 6, S. 141–157, 1851.

[31]   H. Vollmer: *Introduction to Circuit Complexity*. Springer, Berlin, 1999.

[32]   B. von Querenburg: *Mengentheoretische Topologie*. Hochschultexte. Springer-Verlag, 1973.

[33]   I. Wegener: *The complexity of Boolean functions*. Wiley-Teubner, 1987.

Symbolverzeichnis

Mengen

$\varnothing$	leere Menge
$\lvert A \rvert$	Mächtigkeit der Menge A, S. 54
$A \cup B$	Vereinigung der Mengen A und B
$A \cap B$	Durchschnitt der Mengen A und B
$A \setminus B$	Elemente aus A, welche nicht in B vorkommen
$A \times B$	kartesisches Produkt, S. 199
B^A	Menge der Abbildungen $f : A \to B$, S. 54
2^A	Potenzmenge von A, S. 55
$\binom{A}{k}$	Menge der k-elementigen Teilmengen von A, S. 56
$[a, b]$	abgeschlossenes Intervall
$(\Sigma \to_p \Sigma)$	partielle Abbildungen von Σ nach Σ, S. 166
$\mathbb{C}$	komplexe Zahlen, S. 2
$\mathbb{F}_p$	Körper mit p Elementen, S. 29
$\mathbb{N}$	natürliche Zahlen, inklusive 0, S. 1
$\mathbb{Q}$	rationale Zahlen, S. 2
$\mathbb{R}$	reelle Zahlen, S. 2
$\mathbb{Z}$	ganze Zahlen, S. 2
$\mathbb{Z}/n\mathbb{Z}$	Restklassen modulo n, S. 6
$(\mathbb{Z}/n\mathbb{Z})^*$	Einheiten in $\mathbb{Z}/n\mathbb{Z}$, S. 7

Relationen

$f \in \mathcal{O}(g)$	f wächst höchstens so schnell wie g, S. 200
$f \in o(g)$	f wächst echt langsamer als g, S. 200
$f \in \Omega(g)$	f wächst mindestens so schnell wie g, S. 200
$f \in \omega(g)$	f wächst echt schneller als g, S. 200

$f \in \Theta(g)$	f wächst genauso schnell wie g, S. 200
$f \sim g$	asymptotisch gleiches Wachstum, $\lim\limits_{n \to \infty} \dfrac{f(n)}{g(n)} = 1$, S. 201
$f \sqsubseteq g$	Ordnung auf partiellen Abbildungen, S. 167
$k \equiv \ell \bmod n$	k und ℓ sind kongruent modulo n, S. 6
$k \mid \ell$	k teilt ℓ, S. 3
$x \lessdot y$	y ist oberer Nachbar von x, S. 162

Zahlen, Abbildungen und Operationen

$\bot$	kleinstes Element, S. 165
$\top$	größtes Element, S. 165
$|x|$	Betrag von x, S. 199
$\lfloor x \rfloor$	x abgerundet
$\lceil x \rceil$	x aufgerundet
$n!$	Fakultät $n(n-1)\cdots 1$, S. 55
$x^{\underline{k}}$	fallende Faktorielle $x(x-1)\cdots(x-k+1)$, S. 57
$x^{\overline{k}}$	steigende Faktorielle $x(x+1)\cdots(x+k-1)$, S. 80
$\dbinom{n}{k}$	Binomialkoeffizienten, n über k, S. 57
$\dbinom{n}{k_1,\ldots,k_d}$	Multinomialkoeffizienten, S. 67
$\begin{bmatrix} n \\ k \end{bmatrix}$	Stirling-Zahlen der 1. Art, n in k Zykel, S. 78
$\begin{Bmatrix} n \\ k \end{Bmatrix}$	Stirling-Zahlen der 2. Art, n in k Klassen, S. 74
$x \vee y$	kleinste obere Schranke von x und y, S. 171
$x \wedge y$	größte untere Schranke von x und y, S. 171
$\overline{x}$	zu x komplementäres Element, S. 180
B_n	Bell-Zahlen, S. 82
C_n	Catalan-Zahlen, S. 86

$\dim(x)$	Dimension von x, S. 163
D_n	Dyck-Wörter der Länge $2n$, S. 87
$\mathrm{dom}(f)$	Definitionsbereich von f, S. 167
d_x	Grad von Knoten x, S. 122
e	Euler'sche Zahl, S. 55
$E[X]$	Erwartungswert der Zufallsvariable X, S. 45
f'	Ableitung von f
F_n	Fibonacci-Zahlen, S. 21
$\varphi(n)$	Euler'sche φ-Funktion, S. 17
$\mathrm{ggT}(k, \ell)$	größter gemeinsamer Teiler von k und ℓ, S. 3
$g \circ f$	Hintereinanderausführung von Funktionen, S. 199
$\Gamma_{b,c}(g)$	Operator für while-Schleifen, S. 168
H_n	harmonische Zahlen, S. 81
i	imaginäre Zahl, S. 2
$\inf(D)$	größte untere Schranke, S. 165
$\mathcal{J}(V)$	irreduzible Elemente von V, S. 178
$\mathrm{kgV}(n)$	kleinstes gemeinsames Vielfaches von $\{1, \ldots, n\}$, S. 32
$\lim$	Grenzwert
$\log n$	Logarithmus zur Basis 2
$\ln n$	Logarithmus zur Basis e
$\max(A)$	größtes Element der Menge A
$\min(A)$	kleinstes Element der Menge A
π	Kreiszahl, S. 55
$\pi(x)$	Anzahl der Primzahlen $\leq x$, S. 32
$p(n, k)$	untere Partitionszahlen, S. 85
$P(n)$	summatorische Partitionszahlen, S. 84
$P(n, k)$	arithmetische Partitionszahlen, S. 84
$\mathrm{Pr}[A]$	Wahrscheinlichkeit des Ereignisses A, S. 45
R_n	Rencontres-Zahlen, S. 72
$R_{k,c}(n)$	Ramsey-Zahlen, S. 153

σ_X	Standardabweichung der Zufallsvariable X, S. 48
$\sup(D)$	kleinste obere Schranke, S. 165
$\mathrm{Var}[X]$	Varianz der Zufallsvariable X, S. 48
$Z_M(n)$	n in Summanden aus M, S. 103

Graphen

C_n	einfacher Kreis mit n Knoten, S. 121
$\overline{G}$	komplementärer Graph von G, S. 120
K_n	vollständiger Graph mit n Knoten, S. 121
$K_{m,n}$	vollständig bipartiter Graph, S. 122
P_n	einfacher Weg mit n Knoten, S. 121

Verbände

$\mathbb{B}$	boolescher Verband $\{0,1\}$, S. 181
M_5	modularer, nicht distributiver Verband, S. 175
N_5	nicht modularer Verband, S. 175

Index

Weitere empfehlenswerte Titel

Diskrete algebraische Methoden
Arithmetik, Kryptographie, Automaten und Gruppen
Volker Diekert, Manfred Kufleitner, Gerhard Rosenberger, 2013
ISBN 978-3-11-031260-7, e-ISBN 978-3-11-031261-4

Differenzengleichungen und diskrete dynamische Systeme
Eine Einführung in Theorie und Anwendungen
Ulrich Krause, Tim Nesemann, 2. Auflage, 2012
ISBN 978-3-11-025038-1, e-ISBN 978-3-11-025039-8

Mathematische Optimierungsverfahren des Operations Research
Matthias Gerdts, Frank Lempio, 2011
ISBN 978-3-11-024994-1, e-ISBN 978-3-11-024998-9

Approximative Algorithmen und Nichtapproximierbarkeit
Klaus Jansen, Marian Margraf, 2008
ISBN 978-3-11-020316-5, e-ISBN 978-3-11-020317-2

Erfolgreich recherchieren
Mathematik
Astrid Teichert, 2013
ISBN 978-3-11-029896-3, e-ISBN 978-3-11-029896-3

www.ingramcontent.com/pod-product-compliance
Lightning Source LLC
LaVergne TN
LVHW062312060326
832902LV00013B/2168

* 9 7 8 3 1 1 0 2 7 7 6 7 8 *